AF767005

Mathematics and Computer Science Volume 1

Scrivener Publishing
100 Cummings Center, Suite 541J
Beverly, MA 01915-6106

Advances in Data Engineering and Machine Learning

Series Editors: Niranjanamurthy M, PhD, Juanying XIE, PhD, and Ramiz Aliguliyev, PhD

Scope: Data engineering is the aspect of data science that focuses on practical applications of data collection and analysis. For all the work that data scientists do to answer questions using large sets of information, there have to be mechanisms for collecting and validating that information. Data engineers are responsible for finding trends in data sets and developing algorithms to help make raw data more useful to the enterprise.

It is important to have business goals in line when working with data, especially for companies that handle large and complex datasets and databases. Data Engineering Contains DevOps, Data Science, and Machine Learning Engineering. DevOps (development and operations) is an enterprise software development phrase used to mean a type of agile relationship between development and IT operations. The goal of DevOps is to change and improve the relationship by advocating better communication and collaboration between these two business units. Data science is the study of data. It involves developing methods of recording, storing, and analyzing data to effectively extract useful information. The goal of data science is to gain insights and knowledge from any type of data — both structured and unstructured.

Machine learning engineers are sophisticated programmers who develop machines and systems that can learn and apply knowledge without specific direction. Machine learning engineering is the process of using software engineering principles, and analytical and data science knowledge, and combining both of those in order to take an ML model that's created and making it available for use by the product or the consumers. "Advances in Data Engineering and Machine Learning Engineering" will reach a wide audience including data scientists, engineers, industry, researchers and students working in the field of Data Engineering and Machine Learning Engineering.

Publishers at Scrivener
Martin Scrivener (martin@scrivenerpublishing.com)
Phillip Carmical (pcarmical@scrivenerpublishing.com)

Mathematics and Computer Science Volume 1

Edited by

Sharmistha Ghosh

M. Niranjanamurthy

Krishanu Deyasi

Biswadip Basu Mallik

and

Santanu Das

Scrivener Publishing

WILEY

This edition first published 2023 by John Wiley & Sons, Inc., 111 River Street, Hoboken, NJ 07030, USA and Scrivener Publishing LLC, 100 Cummings Center, Suite 541J, Beverly, MA 01915, USA
© 2023 Scrivener Publishing LLC
For more information about Scrivener publications please visit www.scrivenerpublishing.com.

All rights reserved. No part of this publication may be reproduced, stored in a retrieval system, or transmitted, in any form or by any means, electronic, mechanical, photocopying, recording, or otherwise, except as permitted by law. Advice on how to obtain permission to reuse material from this title is available at http://www.wiley.com/go/permissions.

Wiley Global Headquarters
111 River Street, Hoboken, NJ 07030, USA

For details of our global editorial offices, customer services, and more information about Wiley products visit us at www.wiley.com.

Limit of Liability/Disclaimer of Warranty
While the publisher and authors have used their best efforts in preparing this work, they make no representations or warranties with respect to the accuracy or completeness of the contents of this work and specifically disclaim all warranties, including without limitation any implied warranties of merchantability or fitness for a particular purpose. No warranty may be created or extended by sales representatives, written sales materials, or promotional statements for this work. The fact that an organization, website, or product is referred to in this work as a citation and/or potential source of further information does not mean that the publisher and authors endorse the information or services the organization, website, or product may provide or recommendations it may make. This work is sold with the understanding that the publisher is not engaged in rendering professional services. The advice and strategies contained herein may not be suitable for your situation. You should consult with a specialist where appropriate. Neither the publisher nor authors shall be liable for any loss of profit or any other commercial damages, including but not limited to special, incidental, consequential, or other damages. Further, readers should be aware that websites listed in this work may have changed or disappeared between when this work was written and when it is read.

Library of Congress Cataloging-in-Publication Data

ISBN 9781119879671

Front cover images supplied by Wikimedia Commons
Cover design by Russell Richardson

Set in size of 11pt and Minion Pro by Manila Typesetting Company, Makati, Philippines

Printed in the USA

10 9 8 7 6 5 4 3 2 1

Contents

Preface

The mathematical sciences are part of nearly all aspects of everyday life. The discipline has underpinned such beneficial modern capabilities as internet searching, medical imaging, computer animation, weather prediction, and all types of digital communications. Mathematics is an essential component of computer science. Without it, you would find it challenging to make sense of abstract language, algorithms, data structures, or differential equations, all of which are necessary to fully appreciate how computers work. In a sense, computer science is just another field of mathematics. It does incorporate various other fields of mathematics, but then focuses those other fields on their use in computer science. Mathematics matters for computer science because it teaches readers how to use abstract language, work with algorithms, self-analyze their computational thinking, and accurately model real-world solutions. Algebra is used in computer programming to develop algorithms and software for working with math functions. It is also involved in design programs for numerical programs. Statistics is a field of math that deploys quantified models, representations, and synopses to conclude from data sets.

This book focuses on mathematics, computer science, and where the two intersect, including heir concepts and applications. It also represents how to apply mathematical models in various areas with case studies. The contents include 29 peer-reviewed papers, selected by the editorial team.

1

Error Estimation of the Function by $\left(\mathcal{Z}_r^u, r \geq 1\right)$ Using Product Means $(E,s)\left(\overline{\mathcal{N}}, p_n, q_n\right)$ of the Conjugate Fourier Series

Aradhana Dutt Jauhari* **and Pankaj Tiwari**

Division of Mathematics, Department of Basic Sciences, Galgotias University, Greater Noida, G. B. Nagar, U.P., India

Abstract

The purpose of the current chapter is to attain the best result on a new way and the best approximation of different classes of the work, which has been discussed by different mathematician under different summability means. Here, we are presenting the theorems established under $(E,s)\left(\overline{\mathcal{N}}, p_n, q_n\right)$ means of the CFS of a signal, belongs to $\left(\mathcal{Z}_r^\mu, r \geq 1\right)$. Some known and unknown results have been proven by many mathematicians. But this is a new and unique way of proving a new result.

Keywords: Generalized Zygmund class $\left(\mathcal{Z}_r^\mu, r \geq 1\right)$, product means, Degree of Approximation (DoA), Conjugate Fourier Series (CFS)

1.1 Introduction

Let, us take $\sum a_n$ – an infinite series and s_n – the sequence of partial sums. Also, $\{p_n\}$ and $\{q_n\}$ are the sequences of positive real numbers s.t.- $P_n = \sum_{k=0}^{n} p_k$ and $Q_k = \sum_{k=0}^{n} q_k$ and let $R_n = p_o q_n + p_1 q_{n-1} + \cdots + p_n q_o \neq 0, p_{-1} = q_{-1} = R_{-1}$

**Corresponding author*: draradhana27@gmail.com

Sharmistha Ghosh, M. Niranjanamurthy, Krishanu Deyasi, Biswadip Basu Mallik, and Santanu Das (eds.) Mathematics and Computer Science Volume 1, (1–20) © 2023 Scrivener Publishing LLC

1.1.1 Definition 1

Sequence-to-sequence of the transformation:

$$E_n^s = \frac{1}{(1+s)^n}\sum_{v=0}^{n}\binom{n}{v}s^{n-v}s_v \tag{1.1}$$

defines, $E_n^s,(E,s)$ mean of $\{s_n\}$.

If $E_n^s \to s \; as \; n \to \infty$ the Σa_n is summable to s by (E, s) method regular ([12] Hardy, 1949).

1.1.2 Definition 2

Sequence-to-sequence transformation:

$$t_n^{\mathcal{N}} = \frac{1}{R_n}\sum_{k=0}^{n} p_k q_k s_k \tag{1.2}$$

where,

$R_n = p_o q_n + p_1 q_{n-1} + \cdots + p_n q_o \,(\neq), p_{-1} = q_{-1} = R_{-1}$ then the series Σa_n is $(\overline{\mathcal{N}}, p_n, q_n)$ summable to s. Regularity conditions [12]-

1. $\frac{p_k q_k}{R_n} \to 0, \forall k \geq 0, as\; n \to \infty$
2. $\left|\sum_{k=0}^{n} p_k q_k s_k\right| < M\,|\,R_n\,|$, where M > 0 and independent of n.

1.1.3 Definition 3

The transform (E, s) over $(\overline{\mathcal{N}}, p_n, q_n)$ summability and is given by [7]-

$$\mathrm{T}_n^{E\mathcal{N}} = \frac{1}{\mathrm{R}_n}\sum_{\mathrm{k}=0}^{\mathrm{n}} p_k q_k \left\{\frac{1}{(1+s)^{\mathrm{k}}}\sum_{\vartheta=0}^{\mathrm{k}}\binom{k}{v}s^{\mathrm{k}-v}s_v\right\} \tag{1.3}$$

$$\mathrm{T}_n^{E\mathcal{N}} \to s, n \to \infty$$

and Σa_n is said to summable to s by the product means $(E,s)(\overline{\mathcal{N}}, p_n, q_n)$.

Remark. If we put $q_n = 1$ in equation (1.3) then $(\overline{\mathcal{N}}, p_n, q_n)$ summability mean reduces to $(\overline{\mathcal{N}}, p_n)$ mean and for $p_n = 1$ it becomes to $(\overline{\mathcal{N}}, q_n)$ mean.

Now we take Fourier and its CFS (Conjugate Fourier series) as-

$$g = \frac{a_0}{2} + \sum_{n=1}^{k} (a_n cosnx + b_n sinnx)$$
$$\tilde{g} = \sum_{n=1}^{k} (a_n sinnx - b_n cosnx) \quad (1.4)$$

The partial sum is given by,

$$\tilde{g} = -\frac{1}{2\pi}\int_0^{\pi} \phi_x(t)\cot\left(\frac{t}{2}\right)dt$$

Let $L^z[\mathbf{0},\ \mathbf{2\pi}] = \left\{\tilde{g}:[\mathbf{0},\ \mathbf{2\pi}] \to \mathcal{R} : \int_0^{2\pi} |g(x)|^{\mathcal{Z}}\, dx < \infty, \mathcal{Z} \geq 1\right\}$ known as space function which is 2π - periodic and also integrable. Norm $||.||_r$ is defined as,

$$||\tilde{g}||_z = \begin{cases} \left\{\frac{1}{2\pi}\int_0^{2\pi} ||g(x)|^r| dx\right\} for\ 1 \leq z < \infty \\ ess\ sup_{x\in(0,2\pi)} |g(x)|\ for\ z = \infty \end{cases}$$

also,

$$w(\tilde{g}, h) = sup_{0\leq h, x\in\mathcal{R}} |g(x+t) + g(x-t)|.$$

again, $C_{2\pi}$- The Banach space of 2π- periodic functions, defined on $[\mathbf{0},\ \mathbf{2\pi}]$ under the sup. norm.

Where, $0 < \alpha \leq 1$,

$\mathcal{Z}_{(\alpha)} = \left\{\tilde{g}\epsilon C_{2\pi} : |g(x+t) + g(x-t)| = O\left(|(t)|^{\alpha}\right)\right\}$

the function space [11].

$$||\tilde{g}|| = sup_{0\leq x\leq 2\pi} |g(x)| + sup_{x,t\neq 0} \frac{g(x+t) + g(x-t)}{|t|^{\alpha}}$$

For $g \in L^r\ [\mathbf{0},\ \mathbf{2\pi}]$, $r \geq 1$,

$$w_r(\tilde{g},h) = sup_{0\le x\le 2\pi}\left\{\frac{1}{2\pi}\int_0^{2\pi}|g(x+t)+g(x-t)|^r\,dx\right\}^{\frac{1}{r}}.$$

For $g \in C_{2\pi}$ and $r = \infty$

$$w_\infty(\tilde{g},h) = g_{0\le t\le h} max|g(x+t)+g(x-t)|$$

also, $w_\infty(\tilde{g},h)\to 0$ as $r\to 0$

Now,

$$\mathcal{Z}_{(\alpha),r} = \left\{\tilde{g}\in L^r[\mathbf{0},\ \mathbf{2\pi}]\text{:}\left(\int_0^{2\pi}|g(x+t)+g(x-t)|^r\,dx\right)^{\frac{1}{r}} = O\left(|x|^\alpha\right)\right\}$$

also, Banach space with the norm $||.||_{(\alpha),\,r}$ can be defined as the space $Z_{(\alpha)}$, $r \ge 1, 0 < \alpha \le 1$

$$||\tilde{g}||_{(\alpha),r} = ||g||_r + sup_{t\ne 0}\frac{||g(.+t)+g(.-t)||_r}{|t|^\alpha}$$

The class function $\mathcal{Z}^{(w)} = \left\{\tilde{g}\epsilon C_{2\pi}\text{: } |g(x+t)+g(x-t)| = O(w,(t))\right\}$,
Where, $w > 0$, continuous function having sub-linear property, that is

(*i*) $w(0) = 0$,
(*ii*) $w(t_1 + t_2) \le w(t_1) + w(t_2)$

Let w: $[\mathbf{0}, \mathbf{2\pi}] \to \boldsymbol{R}$ s.t. $w(t) > 0$ for $0 \le t < 2\pi$.

$$\lim_{t\to 0^+} w(t) = 0$$

$$\mathcal{Z}_r^{(w)} = \left\{\tilde{g}\in L_r\text{:}L\le r\le\infty, sup_{t\ne 0}\frac{||g(.+t)+g(.-t)||_r}{w(t)} < \infty\right\}$$

$$\tilde{g}_r^{(w)} = \tilde{g}_r + sup_{t\ne 0}\frac{||g(.+t)+g(.-t)||_r}{w(t)}, r\ge 1.$$

Clearly $||\cdot||_r^{(w)}$ is a norm on $\mathcal{Z}_r^{(w)}$.

As we know L_r ($r \ge 1$) is complete, the space $\mathcal{Z}_r^{(w)}$ is also complete Banach space under the norm $||.||_r^{(w)}$.

Used notations in this paper:

$$\phi_x(t) = g(x+t) + g(x-t),$$

$$\tilde{Q}_n^{\mathcal{E}\mathcal{N}}(t) = \frac{1}{2\pi R_n} \sum_{k=0}^{n} p_k q_k \left\{ \frac{1}{(1+s)^k} \sum_{k=0}^{n} \binom{k}{v} s^{k-v} \left(\frac{\cos\frac{t}{2} - \cos\left(v+\frac{1}{2}\right)t}{\sin\frac{t}{2}} \right) \right\}$$

1.2 Theorems

Many mathematicians and the researcher studies and worked on Error Estimation of FS and its CFS in different Lip class, Weightage class as well as Zygmund class using various product means. Amongst them some are Nigam [1–3], Singh [4], Değer [8], Lal [5], Pradhan [6, 7] and Mishra [9, 10] etc.

We are going to prove a new result for conjugate Fourier series belonging to $\left(\mathcal{Z}_r^{\mu}, r \geq 1\right)$ with $(E,s)\left(\overline{\mathcal{N}}, p_n, q_n\right)$ means.

1.2.1 Theorem 1

Let $\tilde{g}$ is 2π-periodic function and Lebesgue integrable in [0, 2π] and belonging to $\mathcal{Z}_r^{(w)}(r \geq 1)$, then using the product mean $(E,s)\left(\overline{\mathcal{N}}, p_n, q_n\right)$ and the DoA of the function $\tilde{g}$ is given by-

$$E_n(\tilde{u}) = inf \,||\, M_r(.) \,||_r^v = O\left(\int_{\frac{1}{n+1}}^{\pi} \frac{w(t)}{tv(t)} dt \right)$$

Where $w(t)$ and $v(t)$ denoted the Zygmund modulus of continuity s.t. $\frac{w(t)}{v(t)}$ positive.

1.2.2 Theorem 2

Let $\tilde{g}$ is 2π – periodic function and Lebesgue integrable in **[0, 2π]** and belonging to $\mathcal{Z}_r^{(w)}(r \geq 1)$, then using the product mean $(E,s)\left(\overline{\mathcal{N}}, p_n, q_n\right)$ the DoA of $\tilde{g}$ is given by-

$$E_n(\tilde{u}) = infM_r(.)_r^v = O\left(\frac{w\left(\frac{1}{n+1}\right)}{(n+1)^2 v\left(\frac{1}{n+1}\right)}(\pi(n+1)-1)\right)$$

1.3 Lemmas

1.3.1 *Lemma 1*

$|\tilde{Q}_n(t)| = O(n)$ for $0 \le t \le \frac{1}{(n+1)}$

Proof.

$$|\tilde{Q}_n(t)| = \left|\frac{1}{2\pi R_n}\sum_{R=0}^{n} p_q q_k \left\{\frac{1}{(1+s)^k}\sum_{v=0}^{k}\binom{k}{v}s^{k-v}\left(\frac{cos\frac{t}{2} - \cos\left(v+\frac{1}{2}\right)t}{sin\frac{t}{2}}\right)\right\}\right|$$

$$= \frac{1}{2\pi R_n}\left|\sum\nolimits_{R=0}^{n} p_q q_k \left\{\frac{1}{(1+s)^k}\sum\nolimits_{v=0}^{k}\binom{k}{v}s^{k-v}\left(\frac{\cos\left(v+\frac{1}{2}\right)t}{sin\frac{t}{2}}\right)\right\}\right|$$

$$\le \frac{1}{2\pi R_n}\left(\sum_{k=0}^{n} p_q q_k \frac{1}{(1+s)^k}\right)\left|\sum_{v=0}^{k}\binom{k}{v}s^{k-v}\right| O(n)$$

$$\le O\left\{\frac{1}{2\pi R_n}\sum\nolimits_{k=0}^{n} p_k q_k \frac{1}{(1+s)^k}(1+s)^k\right\}$$

$$= O(n)$$

1.3.2 *Lemma 2*

$|\tilde{Q}_n(t)| = O\left(\frac{1}{t}\right)$ for $\frac{1}{(n+1)} \le t \le \pi$

Proof. Since, $\frac{1}{(n+1)} \le t \le \pi$, using Jordan lemma $sin\frac{t}{2} \ge \frac{t}{\pi}$ and $sinnt \le 1$

$$
\begin{aligned}
\left|\tilde{Q}_n(t)\right| &= \frac{1}{2\pi R_n}\left|\sum_{R=0}^{n} p_q q_k \left\{\frac{1}{(1+s)^k}\sum_{v=0}^{k}\binom{k}{v} s^{k-v}\left(\frac{cos\frac{t}{2}-\cos\left(v+\frac{1}{2}\right)t}{sin\frac{t}{2}}\right)\right\}\right| \\
&\le \frac{1}{2\pi R_n}\sum\nolimits_{k=0}^{n} p_q q_k \left\{\frac{1}{(1+s)^k}\sum\nolimits_{v=0}^{k}\binom{k}{v} s^{k-v}\left(\frac{cos\frac{t}{2}-\cos vt cos\frac{t}{2}+sin vt \sin\frac{t}{2}}{sin\frac{t}{2}}\right)\right\} \\
&\le \frac{1}{2\pi R_n}\sum_{k=0}^{n} p_q q_k \left\{\frac{1}{(1+s)^k}\sum_{v=0}^{k}\binom{k}{v} s^{k-v}\frac{\cos\left(v+\frac{1}{2}\right)t}{sin\frac{t}{2}}\right\} \\
&\le \frac{1}{2\pi R_n}\sum_{k=0}^{n} p_q q_k \left\{\frac{1}{(1+s)^k}\sum_{v=0}^{k}\binom{k}{v} s^{k-v}\frac{\cos\left(v+\frac{1}{2}\right)t}{t}\right\} \\
&\le \frac{1}{2tR_n}\sum_{k=0}^{n} p_q q_k \left\{\frac{1}{(1+s)^k}\sum_{v=0}^{k}\binom{k}{v} s^{k-v}\cos\left(v+\frac{1}{2}\right)t\right\} \\
&\le \frac{1}{2tR_n}\sum_{k=0}^{n} p_q q_k \left\{\frac{1}{(1+s)^k}\sum_{v=0}^{k}\binom{k}{v} s^{k-v}\left[Re.e^{i\left(v+\frac{1}{2}\right)t}\right]\right\} \\
&\le \frac{1}{2tR_n}\sum_{k=0}^{n} p_q q_k \left\{\frac{1}{(1+s)^k}\sum_{v=0}^{k}\binom{k}{v} s^{k-v}\left[Re.\left(e^{ivt}e^{i\frac{t}{2}}\right)\right]\right\} \\
&\le \frac{1}{2tR_n}\sum_{k=0}^{n} p_q q_k \left\{\frac{1}{(1+s)^k}\sum_{v=0}^{k}\binom{k}{v} s^{k-v}[Re.(e^{ivt})]\right\} \\
&\le \frac{1}{2tR_n}\sum_{k=0}^{n} p_q q_k \left\{\frac{1}{(1+s)^k}Re\left[\sum_{v=0}^{k}\binom{k}{v} s^{k-v}e^{ivt}\right]\right\} \\
&\le \frac{1}{2t}\left|\sum\nolimits_{k=0}^{\tau-1}\left[\binom{k}{v}\frac{1}{(1+s)^k}\sum\nolimits_{v=0}^{k} e^{ivt}s^{k-v}\right]\right| \\
&+\frac{1}{2t}\left|\sum\nolimits_{k=\tau}^{n}\left[\binom{k}{v}\frac{1}{(1+s)^k}Re\left\{\sum\nolimits_{v=0}^{k} e^{ivt}s^{k-v}\right\}\right]\right| \qquad (1.5)
\end{aligned}
$$

Taking, first term of (1.5)-

$$\leq \frac{1}{2t}\left|\sum_{k=0}^{\tau-1}\left[\binom{n}{k}\frac{1}{(1+s)^k}Re\left\{\sum_{v=0}^{k}e^{ivt}\right\}\right]\right|$$

$$\leq \frac{1}{2t}\left|\sum_{k=0}^{\tau-1}\left[\binom{n}{k}\frac{1}{(1+s)^k}\left\{\sum_{v=0}^{k}1\right\}\right]|e^{ivt}|\right|$$

$$\leq \frac{1}{2t}\left|\sum\nolimits_{k=0}^{\tau-1}\left[\binom{n}{k}\right]\right| \tag{1.6}$$

Taking, next term of (1.5), by Abel's lemma-

$$\leq \frac{1}{2t}\left|\sum_{k=\tau}^{n}\left[\binom{n}{k}\frac{1}{(1+s)^k}Re\left\{\sum_{v=0}^{k}e^{ivt}\right\}\right]\right|$$

$$\leq \frac{1}{2t}\sum_{k=\tau}^{n}\binom{n}{k}\frac{1}{(1+s)^k}\max_{0\leq m\leq k}\left|\sum_{v=0}^{k}e^{ivt}\right|$$

$$\leq \frac{1}{2t}\sum_{k=\tau}^{n}\binom{n}{k}\frac{1}{(1+s)^k}(1+s)^k$$

$$\leq \frac{1}{2t}\sum\nolimits_{k=\tau}^{n}\binom{n}{k} \tag{1.7}$$

Combining (1.5), (1.6), and (1.7) we get

$$\left|\tilde{Q}_n(t)\right| \leq \frac{1}{2t}\sum_{k=0}^{\tau-1}\binom{n}{k}+\frac{1}{2t}\sum_{k=\tau}^{n}\binom{n}{k}$$

$$\left|\tilde{Q}_n(t)\right| = O\left(\frac{1}{t}\right)$$

1.3.3 *Lemma 3*

Let, $\tilde{g} \in \mathcal{Z}_r^{(w)}$, where $0 < t \leq \pi$,

i. $\| \phi_x(.,t) \|_r = O(w(t))$

ii. $\| \phi_x(.+y,t) + \phi_x(.-y,t) \|_r = \begin{cases} O(w(t)) \\ O(v(t)) \end{cases}$.

1.4 Proof of the Theorems

1.4.1 Proof of the Theorem 1

Using Riemann-Lebesgue theorem,

$$\tilde{S}_k(\tilde{g},x) = \frac{1}{2\pi} \int_0^{\pi} \phi_x(x,t) \frac{\cos\frac{t}{2} - \cos\left(v + \frac{1}{2}\right)t}{\sin\frac{t}{2}} dt$$

Further,

$$M_r(x) = T_r^{\bar{N}E^s}(x) - \tilde{g}(x)$$

$$M_r(x) = \frac{1}{2\pi R_n} \int_0^{\pi} \sum_{k=0}^{n} p_q q_k \left\{ \frac{1}{(1+s)^k} \sum_{v=0}^{k} \binom{k}{v} s^{k-v} \frac{\cos\left(v + \frac{1}{2}\right)t}{\sin\frac{t}{2}} \right\} dt$$

$$M_r(x) = T_r^{\bar{N}E^s}(x) - \tilde{g}(x) = \int_0^{\pi} \phi_x(x,t) \tilde{Q}_n^{E^s\bar{N}}(t) dt$$

$$M_r(x+y) + M_r(x-y) = \int_0^{\pi} [\phi_x(x+y,t) + \phi_x(x-y,t)] \tilde{Q}_n^{E^s\bar{N}}(t) dt$$

Using GMI, we have

$$\| M_r(x+y)+M_r(x-y)\|_p=\left\{\frac{1}{2\pi}\int_0^{2\pi}|M_r(x+y)+M_r(x-y)|^p\,dx\right\}^{\frac{1}{p}}$$

$$=\left\{\frac{1}{2\pi}\int_0^{2\pi}\left|\int_0^{\pi}[\phi_x(x+y,t)+\phi_x(x-y,t)]\tilde{Q}_n^{E^s\bar{N}}(t)dt\right|^p dx\right\}^{\frac{1}{p}}$$

$$\leq\left\{\frac{1}{2\pi}\int_0^{2\pi}\int_0^{\pi}\left|[\phi_x(x+y,t)+\phi_x(x-y,t)]\tilde{Q}_n^{E^s\bar{N}}(t)dt\right|^p\right\}^{\frac{1}{p}} dt$$

$$=\int_0^{\pi}\left(\tilde{Q}_n^{E^s\bar{N}}(t)\right)^{\frac{1}{p}}\left\{\frac{1}{2\pi}\int_0^{2\pi}\left|[\phi_x(x+y,t)+\phi_x(x-y,t)]\tilde{Q}_n^{E^s\bar{N}}(t)dx\right|^{\bar{p}}\right\}^{\frac{1}{p}} dt$$

$$=\int_0^{\pi}\|\phi_x(.+y,t)+\phi_x(.-y,t)\|_p\left|\tilde{Q}_n^{E^s\bar{N}}(t)\right|dt$$

$$=\int_0^{\frac{1}{n+1}}\|\phi_x(.+y,t)+\phi_x(.-y,t)\|_p\left|]\tilde{Q}_n^{E^s\bar{N}}(t)\right|$$

$$+\int_{\frac{1}{n+1}}^{\pi}\|\phi_x(.+y,t)+\phi_x(.-y,t)\|_p\left|]\tilde{Q}_n^{E^s\bar{N}}(t)\right|dt$$

$$=I_1+I_2(say). \tag{1.8}$$

Now, using Lemma 1 and Lemma 3 with monotonicity of $\frac{w(t)}{v(t)}$ with the respect to t,

$$I_1 = \int_0^{\frac{1}{n+1}} \|\phi_x(.+y,t)+\phi_x(.-y,t)\|_p \left|\tilde{Q}_n^{E^s\bar{N}}(t)\right| dt$$

$$= \int_0^{\frac{1}{n+1}} O\left(v(y)\frac{w(t)}{v(t)}\right) O(n)dt$$

$$\leq O\left(nv(y)\int_0^{\frac{1}{n+1}} \frac{w(t)}{v(t)}dt\right)$$

Using, 2nd MVT of integral, we have

$$I_1 \leq O\left(nv(y)\frac{w\left(\frac{1}{n+1}\right)}{v\left(\frac{1}{n+1}\right)}\int_0^{\frac{1}{n+1}} dt\right)$$

$$= O\left(\frac{1}{n+1}v(y)\frac{w\left(\frac{1}{n+1}\right)}{v\left(\frac{1}{n+1}\right)}\right)$$

$$= O\left(v(y)\frac{w\left(\frac{1}{n+1}\right)}{v\left(\frac{1}{n+1}\right)}\right)\left(\frac{n}{n+1}\right)$$

$$= O\left(v(y)\frac{w\left(\frac{1}{n+1}\right)}{v\left(\frac{1}{n+1}\right)}\right) \tag{1.9}$$

Using, Lemma 2 and Lemma 3-

$$I_2 = \int_{\frac{1}{n+1}}^{\pi} \phi_x(.+y,t)+\phi_x(.-y,t)_p \left|\tilde{Q}_n^{E^s\bar{N}}(t)\right| dt$$

$$= \int_{\frac{1}{n+1}}^{\pi} O\left(v(y)\frac{w(t)}{v(t)}\right)\frac{1}{t}dt$$

$$= O\left(v(y)\int_{\frac{1}{n+1}}^{\pi} \frac{w(t)}{v(t)}\frac{1}{t}dt\right) \tag{1.10}$$

From (1.8), (1.9) and (1.10) we have

$$sup_{y\neq 0}\frac{\|M_r(.+y)+M_r(.-y)\|_r}{v(y)} = O\left(v(y)\frac{w\left(\frac{1}{n+1}\right)}{v\left(\frac{1}{n+1}\right)}\right)+O\left(v(y)\int_{\frac{1}{n+1}}^{\pi}\frac{w(t)}{v(t)}\frac{1}{t}dt\right) \tag{1.11}$$

Clearly,

$$\phi x\,(x, t) = |g\,(x + t) + g(x - t)|.$$

Applying Minikowski's inequality,

$$\begin{aligned}\|\phi x\,(., t)\|_r &= \|g\,(x + t) + g(x - t)\|_r.\\ &= O\,(w, (t)).\end{aligned}$$

Using Lemma 2 and Lemma 3, we get

$$\| M_r(.)\|_r \leq \left(\int_0^{\frac{1}{n+1}} + \int_{\frac{1}{n+1}}^{\pi} \phi_x(.,t)_r \left| \tilde{Q}_n^{E^s \bar{N}}(t) \right| dt \right)$$

$$= O\left(n \int_0^{\frac{1}{n+1}} w(t)dt \right) + O\left(\int_{\frac{1}{n+1}}^{\pi} \frac{w(t)}{t} dt \right)$$

$$= O\left(nw\left(\frac{1}{n+1}\right) \int_0^{\frac{1}{n+1}} w(t)dt \right) + O\left(\int_{\frac{1}{n+1}}^{\pi} \frac{w(t)}{t} dt \right)$$

$$= O\left(\frac{n}{n+1} w\left(\frac{1}{n+1}\right)\right) + O\left(\int_{\frac{1}{n+1}}^{\pi} \frac{w(t)}{t} dt \right)$$

$$= O\left(w\left(\frac{1}{n+1}\right)\right) + O\left(\int_{\frac{1}{n+1}}^{\pi} \frac{w(t)}{t} dt \right) \tag{1.12}$$

From (1.11) and (1.12), we have

$$\| M_r(.)\|_r^v = \| M_r(.)\|_r = sup_{y \neq 0} \frac{\| M_r(.+y) + M_r(.-y)\|_r}{v(y)}$$

$$= O\left(w\left(\frac{1}{n+1}\right)\right) + O\left(\int_{\frac{1}{n+1}}^{\pi} \frac{w(t)}{t} dt \right)$$

$$= O\left(\frac{w\left(\frac{1}{n+1}\right)}{v\left(\frac{1}{n+1}\right)} \right) + O\left(\int_{\frac{1}{n+1}}^{\pi} \frac{w(t)}{tv(t)} dt \right)$$

$$= \sum_{i=1}^{4} H_i \tag{1.13}$$

We write H_1 in terms of H_3 and H_2, H_3 in terms of H_4.
In show of monotonicity of v (t) for $0 < t \leq \pi$, we have

$$w(t) = \frac{w(t)}{v(t)}.v(t) \leq v(\pi)\frac{w(t)}{v(t)}.v(t) = O\left(\frac{w(t)}{v(t)}\right) \tag{1.14}$$

Therefore, we can write (1.14)

$$H_1 = O\,(H_3)$$

Again, using monotonicity of v (t)

$$\begin{aligned} H_2 &= \int_{\frac{1}{n+1}}^{\pi} \frac{w(t)}{t} dt = \int_{\frac{1}{n+1}}^{\pi} \frac{w(t)}{v(t)t} v(t) dt \\ &\leq \int_{\frac{1}{n+1}}^{\pi} v(\pi) \frac{w(t)}{v(t)t} v(t) dt \\ &= O(H_4). \end{aligned} \tag{1.15}$$

By using the fact that $\frac{w(t)}{v(t)} > 0$, we have

$$H_2 = \int_{\frac{1}{n+1}}^{\pi} \frac{w(t)}{tv(t)} dt = \frac{w\left(\frac{1}{n+1}\right)}{v\left(\frac{1}{n+1}\right)} \int_{\frac{1}{n+1}}^{\pi} \frac{dt}{t} \geq \frac{w\left(\frac{1}{n+1}\right)}{v\left(\frac{1}{n+1}\right)} \tag{1.16}$$

Therefore, we write

$$H_3 = O\,(H_4) \tag{1.17}$$

Now combining, (1.15) and (1.17) we have-

$$\| M_r(.) \|_r^v = O(H_4) = O\left(\int_{\frac{1}{n+1}}^{\pi} \frac{w(t)}{tv(t)} dt \right) \tag{1.18}$$

Hence,

$$E_n(\tilde{u}) = inf \| M_r(.) \|_r^v = O\left(\int_{\frac{1}{n+1}}^{\pi} \frac{w(t)}{tv(t)} dt \right) \tag{1.19}$$

Proof of the theorem 1 is completed.

1.4.2 Proof of the Theorem 2

We have,

$$E_n(\tilde{u}) = inf \| M_r(.) \|_r^v = O\left(\int_{\frac{1}{n+1}}^{\pi} \frac{w(t)}{tv(t)} dt \right) \tag{1.20}$$

we have assumed that the $\frac{w(t)}{tv(t)}$ is positive and decreasing in t. Hence,

$$E_n(\tilde{u}) = inf \| M_r(.) \|_r^v = O\left(\frac{w\left(\frac{1}{n+1}\right)}{(n+1)v\left(\frac{1}{n+1}\right)} \int_{\frac{1}{n+1}}^{\pi} dt \right)$$

$$= O\left(\frac{w\left(\frac{1}{n+1}\right)}{(n+1)v\left(\frac{1}{n+1}\right)} [t]_{\frac{1}{n+1}}^{\pi} \right)$$

$$= O\left(\frac{w\left(\frac{1}{n+1}\right)}{(n+1)^2 v\left(\frac{1}{n+1}\right)} (\pi(n+1)-1) \right) \tag{1.21}$$

Proof of the theorem 2 is completed.

1.5 Corollaries

1.5.1 Corollary 1

If we put (E, 1), (C, 1) mean in place of $(E,s)(\bar{\mathcal{N}},p_n,q_n)$ mean, in the theorem-1, the DoA of the function $\tilde{g} \in \mathcal{Z}_r^{(w)}$ by (E, 1), (C, 1) mean-

$$\mathrm{T}_n^{EC} = \frac{1}{2^n}\sum_{\mathrm{k}=0}^{n}\left\{\binom{k}{v}\frac{1}{\mathrm{k}+1}\sum_{\vartheta=0}^{\mathrm{k}} s_v\right\}$$

Of the CFS is,

$$E_n(\tilde{u}) = O\left(\int_{\frac{1}{n+1}}^{\pi}\frac{w(t)}{tv(t)}dt\right)$$

Proof: [8]

1.5.2 Corollary 2

If we put $(E,1)(\bar{\mathcal{N}},p_n)$ mean in place of $(E,s)(\bar{\mathcal{N}},p_n,q_n)$ mean in the Theorem 1, the DoA of the function $\tilde{g} \in \mathcal{Z}_r^{(w)}$ by (E, 1), (C, 1) mean

$$\mathrm{T}_n^{E\mathcal{N}} = \frac{1}{2^n}\sum_{\mathrm{k}=0}^{n}\binom{k}{v}\left\{\frac{1}{\mathcal{P}_k}\sum_{v=0}^{\mathrm{k}} p_v s_v\right\}$$

of CFS is,

$$E_n(\tilde{u}) = O\left(\int_{\frac{1}{n+1}}^{\pi}\frac{w(t)}{tv(t)}dt\right)$$

Proof: [8]

1.6 Example

In this example, we are showing (*E*, *s*) and $(\bar{\mathcal{N}},p_n,q_n)(f;x)$ (Nörlund (N_p)) summability of the partial sums of a FS is better behaved than the sequence of partial sum $s_n(x)$.

Let

$$f(x)=\begin{cases}-1\, if\, -\pi\leq x\leq -\frac{\pi}{2}\\ 0\, if\, -\frac{\pi}{2}<x\leq\frac{\pi}{2}\\ 1\, if\, \frac{\pi}{2}<x\leq\pi\end{cases}$$

with $f(x+2\pi)=f(x), \forall\, x$.

Fourier series of

$$f(x)-\frac{2}{\pi}\sum_{n=1}^{\infty}\left(cos\frac{n\pi}{2}-cosn\pi\right)sinnx, \quad -\pi\leq x<\pi$$

The *nth* partial sum $s_n(x)$ of FS is -

$$s_n(x)=\frac{2}{\pi}\left[\left(cos\frac{x}{2}-cosx\right)sinx+(cosx-cos2x)\frac{1}{2}sin2x+\ldots+\left(cos\frac{nx}{2}-cosnx\right)\frac{1}{n}sinx\right]$$

Figure 1.1 [10]: The graph of $f(x)$ function and the Fourier series expansion for $n = 10$

$$E_n^s=\frac{1}{(1+s)^n}\sum_{v=0}^{n}\binom{n}{v}s^{n-v}s_v\rightarrow s, \quad as\, n\rightarrow\infty$$

If Nörland method N_p has increasing weights $\{p_n\}$, its means N_p is given by

$$t_n^{\mathcal{N}}=\frac{2}{(n+1)(n+2)}\sum_{k=0}^{n}(n-k+1)s_k$$

We observed that E_n^s and $t_n^{\mathcal{N}}$ converge to $f(x)$ faster than $s_n(x)$ in $\left[-\frac{\pi}{2},\frac{\pi}{2}\right]$.

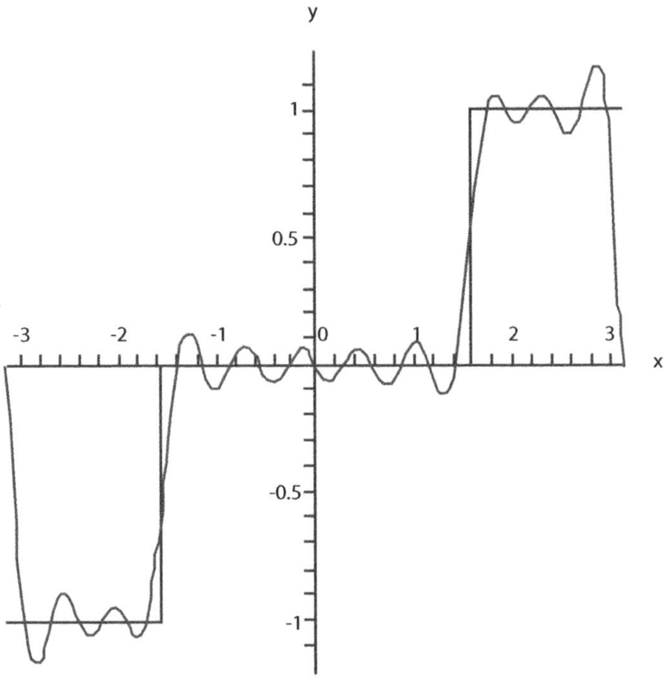

Figure 1.1 Radial basis neural network architecture. Reprint with copyright permission from [11].

1.7 Conclusion

The proved results in this paper are an attempt to study the approximations of the functions in $\left(Z_r^{\mu}, r \geq 1\right)$ by using product means $(E,s)\left(\bar{N}, p_n q_n\right)$ of conjugate Fourier series, which generalize several results. Further, the result can be extended for other functions belonging to $\left(Z_r^{\mu}, r \geq 1\right)$.

References

1. Nigam, H.K., Best approximation conjugate of a function in generalized Zygmund class. *Tamkang J. Math.*, 50, 4, 417–428, Dec. 2019. doi:10.5556/j.tkjm.50.2019.3006.
2. Nigam, H.K. and Sharma, K., On (e, 1) (c, 1) summability of fourier series and its conjugate series. *Int. J. Pure Appl. Math.*, 82, 3, 365–375, ISSN: 1311-8080, 2013.

3. Nigam, H.K., On approximation in generalized Zygmund class. De Gruyter. *Demonstratio Mathematica*, 52, 370–387, 2019. http://doi.org/10.1515/dem-2019-0022.
4. Singh, M.V., Mittal, M.L., Rhoades, B.E., Approximation of function in the generalized Zygmund class using hausdroff means. *J. Inequal. Appl.*, 2017, 101, 2017. DOI 10.1186/s13660-017-1361-8.
5. Lal, S. and Shireen, Best approximations of functions of generalized Zygmund class by Matrix- Euler Summability means of fourier series. *Bulletin of Mathematical Analysis & Applications*, ISSN: 1821-1291, 5, 4, 1–13. 13p, 2013. URL: http//www.bmathaa.org.
6. Pradhan, T., Paikray, S., Das, A., Dutta, H., Approximation of signals in generalised Zygmund class via (e, 1) (N&xxx0305;, pn)-summability mean of conjugate fourier series. *Proyecciones*, 38, 5, 981–998, Dec. 2019. doi:10.22199/iss.0717-6279-2019-05-0063.
7. Pradhan, T., Paikray, S.K., Das, A., Mishra, U., Approximation of signals belonging to generalised Lipschitz class via (n, p_n, q_n) (e, s) - summability mean of fourier series. *Cogent Math.*, 3, 1250343, 2016. http://dx.doiorg/10.1080/23311835.2016.1250343.
8. Değer, U., On approximation to functions in the W $(L^p, \xi(t))$ class by a new matrix mean. *Novi Sad J. Math.*, 6, 1, 1–14, 2016.
9. Mishra, V.N. and Sonavane, V., Approximation of functions of lipschitz class by (n, pn) (e, 1) summability means of conjugate series of fourier series. *J. Class. Anal.*, 6, 137–151, 2015.
10. V.N. Mishra and K. Khatri: *Degree of Approximation of Functions Class by the Means in the Hölder Metric*, International Journal of Mathematics and Mathematical Sciences Volume 2014, Article ID 837408, 9 pages. https://doi.org/10.1155/2014/837408.
11. A .Zygmund: *Trigonometric Series*, second ed., Cambridge University Press, Cambridge, 1959.
12. G. H. Hardy: *Divergent Series*, Oxford University Press 1st Ed. 1949.

2

Blow Up and Decay of Solutions for a Klein-Gordon Equation With Delay and Variable Exponents

Hazal Yüksekkaya* **and Erhan Pişkin**

Department of Mathematics, Dicle University, Diyarbakır, Turkey

Abstract

In this article, we deal with a Klein-Gordon equation with delay and variable exponents. Under appropriate conditions, we establish the blow up of solutions in a finite time. Also, we get the decay results utilizing the Komornik integral inequality.

Keywords: Blow up, decay, delay term, Klein-Gordon equation, variable exponents

2.1 Introduction

In this article, we deal with the Klein-Gordon equation with delay and variable-exponents as follows:

$$\begin{cases} u_{tt}-\Delta u-\Delta u_t+m^2u+\mu_1 u_t\,|u_t|^{r(x)-2} \\ +\mu_2 u_t(x,t-\tau)\,|u_t|^{r(x)-2}(x,t-\tau) \\ =bu\,|u|^{p(x)-2} \qquad in\,\Omega\times(0,\infty), \\ u(x,t)=0 \qquad in\,\partial\Omega\times[0,\infty), \\ u(x,0)=u_0(x),\quad u_t(x,0)=u_1(x) \qquad in\,\Omega, \\ u_t(x,t-\tau)=f_0(x,t-\tau) \qquad in\,\Omega\times(0,\tau), \end{cases} \tag{2.1}$$

**Corresponding author*: hazally.kaya@gmail.com

Sharmistha Ghosh, M. Niranjanamurthy, Krishanu Deyasi, Biswadip Basu Mallik, and Santanu Das (eds.) Mathematics and Computer Science Volume 1, (21–44) © 2023 Scrivener Publishing LLC

where Ω is bounded domain in R^n, $n \geq 1$, with $\partial\Omega$. $\tau > 0$ is time delay term, $b \geq 0$ is a constant, $m \neq 0$ is a real constant, $\mu_1 > 0$ is a constant and μ_2 is a real number. The functions u_0, u_1, f_0 are the initial data to be specified later.

r (.) and p (.) are variable exponents which given as measurable functions on $\bar{\Omega}$ satisfy:

$$2 \leq r^- \leq r(x) \leq r^+ \leq p^- \leq p(x) \leq p^+ \leq p^* \tag{2.2}$$

Where

$$r^- = essinf_{x\in\Omega} r(x), \qquad r^+ = esssup_{x\in\Omega} r(x),$$
$$p^- = essinf_{x\in\Omega} p(x), \qquad p^+ = esssup_{x\in\Omega} p(x),$$

and

$$p^* = \begin{cases} \infty, & if\ n=1,2, \\ \dfrac{2(n-1)}{n-2} & if\ n \geq 3. \end{cases}$$

The Klein-Gordon equation appears in many scientific applications such as quantum field theory, nonlinear optics and solid state physics. The problems with variable-exponents seem in, generally, image processing, electrorheological fluids and nonlinear elasticity theory [1, 2]. Time delay often appears in physical, chemical, thermal, biological and economic phenomena [3].

Our goal is to study the Klein-Gordon equation with delay ($\mu_2 u_t(x, t - \tau)$) and variable exponents which make the problem more interesting than from those concerned in the literature.

Kafini and Messaoudi [4] looked into the equation with variable exponents and time delay as follows:

$$u_{tt} - \Delta u + \mu_1 u_t |u_t|^{m(x)-2} + \mu_2 u_t(x, t-\tau) |u_t|^{m(x)-2}(x, t-\tau) = bu|u|^{p(x)-2} \tag{2.3}$$

They established the global nonexistence and decay results for the equation (2.3).

Without delay term ($\mu_2 u_t (x, t - \tau) |u_t|^{r(x)-2} (x, t - \tau)$), Pişkin [5] considered the Klein-Gordon equation with variable exponents:

$$u_{tt} - \Delta u - \Delta u_t + m^2 u + u_t |u_t|^{p(x)-2} = u|u|^{q(x)-2}. \tag{2.4}$$

The author proved the nonexistence of solutions for the equation (2.4).

Antontsev *et al.* [6] concerned with the Petrovsky equation with variable exponents:

$$u_{tt} + \Delta^2 u - \Delta u_t + u_t |u_t|^{p(x)-2} = u|u|^{q(x)-2}. \tag{2.5}$$

The authors proved the existence and the blow up of solutions of the equation (2.5).

Pişkin and Yüksekkaya [7], studied the Petrovsky equation with time delay and variable exponents:

$$u_{tt} + \Delta^2 u - \Delta u_t + \mu_1 u_t |u_t|^{m(x)-2} + \mu_2 u_t (x, t - \tau) |u_t|^{m(x)-2} (x, t - \tau) = 0. \tag{2.6}$$

They obtained the decay results by utilizing Komornik integral inequality of (2.6). Recently, some other authors investigated some related wave equations (see [8–13]).

The plan of the paper is: Firstly, in Sect. 2.2, the definition of variable exponent Sobolev and Lebesque spaces are introduced. Then, in Sect. 2.3, we get the blow-up of solutions. Finally, in Sect. 2.4, we establish the decay results.

2.2 Preliminaries

In this part, we denote the materials of Lebesgue $L^{p(.)}(\Omega)$ and Sobolev $W^{1p(.)}(\Omega)$ spaces with variable exponents (see [2, 14–16]).

Suppose that $p: \Omega \rightarrow [1, \infty)$ is measurable function. We define variable exponent Lebesgue space with $p(.)$

$$L^{p(.)}(\Omega) = \left\{ u: \Omega \rightarrow R; \, measurable \; in \; \Omega: \int_{\Omega} |u|^{p(.)} \, dx < \infty \right\},$$

With

$$u_{p(.)} = inf\left\{\lambda > 0: \int_{\Omega}\left|\frac{u}{\lambda}\right|^{p(x)} dx \leq 1\right\},$$

(Luxemburg-type norm) with this norm, $L^{p}(.)$ (Ω) is Banach space. (see [2]).

Next, we define the variable-exponent Sobolev space $W^{1,p(.)}$ (Ω) as follows:

$$W^{1,p(.)}(\Omega) = \{ u \in L^{p}(.)(\Omega): \nabla u \textit{ exists and } |\nabla u| \in L^{p}(.)(\Omega)\}$$

Variable exponent Sobolev space with the norm

$$||u||_{1,\,p(.)} = ||u||_{p(.)} + ||\nabla u||_{p(.)},$$

is a Banach space. $W_0^{1,p(.)}(\Omega)$ is the space which is defined as the closure of $C_0^{\infty}(\Omega)$ in $W^{1,\,p(.)}(\Omega)$.

For $W_0^{-1,p'(.)}(\Omega)$, we can define an equivalent norm

$$||u||_{1,\,p(.)} = ||\nabla u||_{p(.)}.$$

The dual of $W_0^{1,p(.)}(\Omega)$ is defined as $W_0^{-1,p'(.)}(\Omega)$, similar to Sobolev spaces, where

$$\frac{1}{p(.)} + \frac{1}{p'(.)} = 1$$

We also assume that:

$$|p(x) - p(y)| \leq -\frac{A}{log|x-y|} \tag{2.7}$$

$A > 0$ and $0 < \delta < 1$ with $|x - y| < \delta$. (log-Hölder condition)

Lemma 1 [15] (Poincare inequality) Assume that $p(.)$ satisfies (2.7). Then,

$$||u||_{p(.)} \leq c||\nabla u||_{p(.)} \text{ for all } u \in W_0^{1,p(.)}(\Omega),$$

Where $c = c\,(p^-, p^+, |\Omega|) > 0$.

Lemma 2 [15] If $p\colon \bar{\Omega} \to [1,\infty)$ is continuous,

$$2 \leq p^- \leq p(x) \leq p^+ \leq \frac{2n}{n-2}, n \geq 3, \tag{2.8}$$

satisfies, then the embedding $H_0^1(\Omega) \hookrightarrow L^{p(.)}(\Omega)$ is continuous.

Lemma 3 [14] If $p^+ < \infty$ and $p\colon \Omega \to [1, \infty)$ is measurable function, thus, $C_0^\infty(\Omega)$ is dense in $L^p(.)\,(\Omega)$.

Lemma 4 [14] (Holder' inequality) Suppose that measurable functions $p, q, s \geq 1$ and

$$\frac{1}{s(y)} = \frac{1}{p(y)} + \frac{1}{q(y)}, \; y \in \Omega,$$

holds. In the case, $f \in L^{p(.)}\,(\Omega)$ and $g \in L^{q(.)}\,(\Omega)$, hence, $fg \in L^{s(.)}\,(\Omega)$,

$$||fg||_{s(.)} \leq 2||f||_{p(.)}||g||_{q(.)}$$

Lemma 5 [14] (Unit ball property) Suppose that $p \geq 1$ is measurable function on Ω. Thus,

$$||fg||_{p(.)} \leq \textit{if and only if } \boldsymbol{\rho}_p(.)\;(f) \leq 1,$$

where

$$\rho_{p(.)}(f) = \int_\Omega |f(x)|^{p(x)}\,dx$$

Lemma 6 [15] Assume that $p \geq 1$ is measurable function on Ω, thus

$$min\left\{||u||_{p(.)}^{p^-}, ||u||_{p(.)}^{p^+}\right\} \leq \rho_{p(.)}(u) \leq max\left\{||u||_{p(.)}^{p^-}, ||u||_{p(.)}^{p^+}\right\},$$

for any $u \in L^{p(.)}\,(\Omega)$ and for a.e. $x \in \Omega$.

Lemma 7 [2] Let $q(.)\in C(\bar{\Omega})$ and $p\colon \Omega \to [1, \infty)$ be measurable function and satisfies

$$essinf_{x\in\bar{\Omega}}(q^*(x)-p(x))>0.$$

Then, the Sobolev embedding $W_0^{1,q(x)}(\Omega) \hookrightarrow L^{p(x)}(\Omega)$ is compact and continuous, where

$$q^*(x)=\begin{cases} \dfrac{nq^-}{n-q^-}, & if\ q^- < n, \\ \infty, & if\ q^- \geq n. \end{cases}$$

2.3 Blow Up of Solutions

In this part, for the case $b > 0$, we establish the blow-up result for our main problem (2.1). Now we introduce, as in the work of [17], a function

$$z(x, \rho, t) = u_t(x, t - \tau\rho),\ x\in\Omega,\ \rho\in(0, 1),\ t > 0,$$

hence, we get

$$\tau z_t(x, \rho, t) + z_\rho(x, \rho, t) = 0,\ x\in\Omega,\ \rho\in(0, 1),\ t > 0.$$

Consequently, the problem (2.1) can be transformed as follows:

$$\begin{cases} u_{tt}-\Delta u-\Delta u_t+m^2u+\mu_1u_t\,|u_t|^{r(x)-2} \\ +\mu_2z(x,1,t)\,|z(x,1,t)|^{r(x)-2} \\ =bu\,|u|^{p(x)-2} & in\,\Omega\times(0,\infty), \\ \tau z_t(x,\rho,t)+z_\rho(x,\rho,t)=0\ in\,\Omega\times(0,1)\times(0,\infty), \\ z(x,\rho,0)=f_0(x,-\rho\tau) & in\ \Omega\times(0,1), \\ u(x,t)=0 & on\,\partial\Omega\times[0,\infty), \\ u(x,0)=u_0(x),\ u_t(x,0)=u_1(x) & in\,\Omega. \end{cases} \tag{2.9}$$

We define the energy functional of (2.9) as

$$E=\frac{1}{2}\|u_t\|^2+\frac{1}{2}\|\nabla u\|^2+\frac{m^2}{2}\|u\|^2$$
$$+\int_0^1\int_\Omega \frac{\xi(x)|z(x,\rho,t)|^{r(x)}}{r(x)}dxd\rho-b\int_\Omega \frac{|u|^{p(x)}}{p(x)}dx \qquad (2.10)$$

For $t \geq 0$, where ξ is a continuous function satisfies

$$\tau|\mu_2|\,(r\,(x)-1)<\xi\,(x)<\tau\,(\mu_1\,r\,(x)-|\mu_2|). \qquad (2.11)$$

The following lemma gives that, under the condition $\mu_1 > |\mu_2|$, $E\,(t)$, is non-increasing.

Lemma 8 Suppose that (u, z) is a solution of (2.9). Then, for $C_0 > 0$

$$E'(t)\leq -C_0\int_\Omega \left(|u_t|^{r(x)}+|z(x,1,t)|^{r(x)}\right)dx\leq 0 \qquad (2.12)$$

Proof. We multiply the first equ. in (2.9) by u_t, integrate over Ω, then multiply the second eq. of (2.9) by $\frac{1}{\tau}\,\xi(x)|z|^{r(x)-2}\,z$ and integrate over $\Omega\times(0, 1)$, summing up, we get

$$\frac{d}{dt}\left[\frac{1}{2}\|u_t\|^2+\frac{1}{2}\|\nabla u\|^2+\frac{m^2}{2}\|u\|^2\right.$$
$$\left.+\int_0^1\int_\Omega \frac{\xi(x)|z(x,\rho,t)|^{r(x)}}{r(x)}dxd\rho-b\int_\Omega \frac{|u|^{p(x)}}{p(x)}dx\right]$$
$$=-\mu_1\int_\Omega |u_t|^{r(x)}\,dx$$
$$-\frac{1}{\tau}\int_\Omega\int_0^1 \xi(x)|z(x,\rho,t)|^{r(x)-2}\,zz_\rho(x,\rho,t)d\rho dx+\|\nabla u_t\|^2$$
$$-\mu_2\int_\Omega u_t z(x,1,t)|z(x,1,t)|^{r(x)-2}\,dx. \qquad (2.13)$$

Next, the terms in the right-hand side of (2.13) are estimated as follows:

$$-\frac{1}{\tau}\int_\Omega\int_0^1 \xi(x)\,|z(x,\rho,t)|^{r(x)-2}\, zz_\rho(x,\rho,t)d\rho dx$$
$$=-\frac{1}{\tau}\int_\Omega\int_0^1 \frac{\partial}{\partial\rho}\left(\frac{\xi(x)\,|z(x,\rho,t)|^{r(x)}}{r(x)}\right)d\rho dx$$
$$=\frac{1}{\tau}\int_\Omega \frac{\xi(x)}{r(x)}\left(|z(x,0,t)|^{r(x)}-|z(x,1,t)|^{r(x)}\right)dx$$
$$=\int_\Omega \frac{\xi(x)}{\tau r(x)}|u_t|^{r(x)}\,dx-\int_\Omega \frac{\xi(x)}{\tau r(x)}|z(x,1,t)|^{r(x)}.$$

Using the Young's inequality, $q=\dfrac{r(x)}{r(x)-1}$ and $q'=r(x)$ for the last term to obtain

$$|u_t||z(x,1,t)|^{r(x)-1}\le\frac{1}{r(x)}|u_t|^{r(x)}+\frac{r(x)-1}{r(x)}|z(x,1,t)|^{r(x)}.$$

Consequently, we deduce that

$$-\mu_2\int_\Omega u_t z\,|z(x,1,t)|^{r(x)-2}\,dx\le$$
$$|\mu_2|\left(\int_\Omega\frac{1}{r(x)}|u_t(t)|^{r(x)}\,dx+\int_\Omega\frac{r(x)-1}{r(x)}|z(x,1,t)|^{r(x)-2}\,dx\right).$$

So,

$$\frac{dE(t)}{dt}\le-\int_\Omega\left[\mu_1-\left(\frac{\xi(x)}{\tau r(x)}+\frac{|\mu_2|}{r(x)}\right)\right]|u_t(t)|^{r(x)}\,dx$$
$$-\frac{1}{2}\|\nabla u_t\|^2-\int_\Omega\left(\frac{\xi(x)}{\tau r(x)}+\frac{|\mu_2|(r(x)-1)}{r(x)}\right)|z(x,1,t)|^{r(x)}\,dx.$$

As a result, for all $x\in\bar{\Omega}$, the relation (2.11) satisfies,

$$f_1(x) = \mu_1 - \left(\frac{\xi(x)}{\tau r(x)} + \frac{|\mu_2|}{r(x)}\right) > 0,$$

$$f_2(x) = \frac{\xi(x)}{\tau r(x)} - \frac{|\mu_2|(r(x)-1)}{r(x)} > 0.$$

Since $r(x)$, and hence $\xi(x)$, is bounded, we infer that $f_1(x)$ and $f_2(x)$ are also bounded. So, if we define

$$C_0(x) = min\{f_1(x), f_2(x)\} > 0, \qquad x \in \bar{\Omega},$$

and take $C_0(x) = inf_{\bar{\Omega}} C_0(x)$, so $C_0(x) \geq C_0 > 0$. Hence,

$$E'(t) \leq -C_0 \left[\int_\Omega |u_t(t)|^{r(x)}\, dx + \int_\Omega z(x,1,t)^{r(x)} dx\right] \leq 0$$

To get blow-up result, let $E(0) < 0$ in addition to (2.2).
Set

$$H(t) = -E(t), \tag{2.14}$$

hence,

$$H'(t) = -E'(t) \geq 0,$$

$$0 < H(0) \leq H(t) \leq b \int_\Omega \frac{|u|^{p(x)}}{p(x)} dx \leq \frac{b}{p^-} \rho(u), \tag{2.15}$$

where

$$\rho(u) = \rho_{p(.)}(u) = \int_\Omega |u|^{p(x)}\, dx.$$

Lemma 9 [4] Suppose that the condition (2.2) satisfies. Then, depending on Ω only, for $C > 1$

$$\rho^{s/p^-}(u) \leq C(\|\nabla u\|^2 + \rho(u)). \tag{2.16}$$

Then, we have following inequalities:

i)

$$\|u\|_{p^-}^s \leq C\left(\|\nabla u\|^2 + \|u(t)\|_{p^-}^{p^-}\right), \tag{2.17}$$

ii) $$\rho^{s/p^-}(u) \leq C\left(|H(t)| + \|u_t\|^2 + \rho(u) + \int_0^1\int_\Omega \frac{\xi(x)|z(x,\rho,t)|^{r(x)}}{r(x)} dxd\rho\right), \tag{2.18}$$

iii) $$\|u\|_{p^-}^s \leq C\left(|H(t)| + \|u_t\|^2 + \|u\|_{p^-}^{p^-} + \int_0^1\int_\Omega \frac{\xi(x)|z(x,\rho,t)|^{r(x)}}{r(x)} dxd\rho\right), \tag{2.19}$$

for any $u \in H_0^1(\Omega)$ and $2 \leq s \leq p^-$. Suppose that (u, z) is a solution of (2.9), thus

iv)

$$\rho(u) \geq \|u\|_{p^-}^{p^-}, \tag{2.20}$$

v)

$$\int_\Omega |u|^{r(x)} dx \leq C\left(\rho^{r^-/p^-}(u) + \rho^{r^+/p^-}(u)\right) \tag{2.21}$$

Theorem 10 Suppose that the conditions (2.2) and (2.7) are provided and assume that $E(0) < 0$.

Thus, the solution (2.9) blows up in finite time

$$T^* \leq \frac{1-\alpha}{\Psi\alpha[L(0)]^{\alpha/(1-\alpha)}}$$

Where $L(t)$ and $\boldsymbol{\alpha}$ are given in (2.22) and (2.23), respectively.

Proof. Define

$$L(t) = H^{1-\alpha}(t) + \varepsilon\int_{\Omega} uu_t dx + \frac{\varepsilon}{2}\|\nabla u\|^2, \tag{2.22}$$

where ε small enough and

$$0 < \alpha \le min\left\{\frac{p^- - 2}{2p^-}, \frac{p^- - r^+}{p^-(r^+ - 1)}\right\} \tag{2.23}$$

Differentiation $L(t)$ and use the first equation in (2.9), we get

$$\begin{aligned}L' &= (1-\alpha)H^{-\alpha}H' + \varepsilon\|u\|_t^2 - \varepsilon\|\nabla u\|^2 - \varepsilon m^2\|u\|^2 \\ &+ \varepsilon b\int_{\Omega}|u|^{p(x)}dx - \varepsilon\mu_1\int_{\Omega} uu_t|u_t|^{r(x)-2}dx \\ &- \varepsilon\mu_2\int_{\Omega} uz(x,1,t)|z(x,1,t)|^{r(x)-2}dx.\end{aligned}$$

By using the definition of the $H(t)$ and for $0 < a < 1$, such that

$$\begin{aligned}&L'(t) \ge C_0(1-\alpha)H^{-\alpha}(t)\left[\int_{\Omega}|u_t|^{r(x)}dx + \int_{\Omega}|z(x,1,t)|^{r(x)}dx\right] \\ &+\varepsilon(1-a)p^-H(t) + \varepsilon\frac{(1-a)p^- + 2}{2}\|u_t\|^2 + \varepsilon\frac{(1-a)p^- - 2}{2}\|\nabla u\|^2 \\ &+\varepsilon m^2\frac{(1-a)p^- - 2}{2}u^2 + \varepsilon(1-a)p^-\int_0^1\int_{\Omega}\frac{\xi(x)|z(x,\rho,t)|^{r(x)}}{r(x)}dxd\rho + \varepsilon abp(u) \\ &-\varepsilon\mu_1\int_{\Omega}uu_t|u_t|^{r(x)-2}dx - \varepsilon\mu_2\int_{\Omega}uz(x,1,t)\left|z(x,1,t)\right|^{r(x)-2}dx\end{aligned} \tag{2.24}$$

Utilizing Young's inequality, we get

$$\int_{\Omega}|u_t|^{r(x)-1}|u|dx \le \frac{1}{r^-}\int_{\Omega}\delta^{r(x)}|u|^{r(x)}dx + \frac{r^+ - 1}{r^+}\int_{\Omega}\delta^{-\frac{r(x)}{r(x)-1}}|u_t|^{r(x)}dx \tag{2.25}$$

and

$$\int_\Omega |z(x,1,t)|^{r(x)-1}|u|\,dx \le \frac{1}{r^-}\int_\Omega \delta^{r(x)}|u|^{r(x)}\,dx$$
$$+\frac{r^+-1}{r^+}\int_\Omega \delta^{-\frac{r(x)}{r(x)-1}}|z(x,1,t)|^{r(x)}\,dx. \tag{2.26}$$

As in [18], the estimates (2.25) and (2.26) remain valid if **δ** is time-dependent. Let us choose **δ** so that

$$\delta^{-\frac{r(x)}{r(x)-1}} = kH^{-\alpha}(t),$$

where $k \ge 1$ is specified later, we get

$$\int_\Omega \delta^{-\frac{r(x)}{r(x)-1}}|u_t|^{r(x)}\,dx = kH^{-\alpha}(u_t)\int_\Omega |u_t|^{r(x)}\,dx, \tag{2.27}$$

$$\int_\Omega \delta^{-\frac{r(x)}{r(x)-1}}|z(x,1,t)|^{r(x)}\,dx = kH^{-\alpha}(t)\int_\Omega |z(x,1,t)|^{r(x)}\,dx \tag{2.28}$$

and

$$\int_\Omega \delta^{r(x)}|u|^{r(x)}\,dx = \int_\Omega k^{1-r(x)}H^{\alpha(r(x)-1)}(t)|u|^{r(x)}\,dx$$
$$\le \int_\Omega k^{1-r^-}H^{\alpha(r^+-1)}(t)\int_\Omega |u|^{r(x)}\,dx. \tag{2.29}$$

By using (2.20) and (2.21), we get

$$H^{\alpha(r^+-1)}(t)\int_\Omega (u)^{r(x)}dx \le C\left[(\rho(u))^{r^-/p^-+\alpha(r^+-1)} + (\rho(u))^{r^+/p^-+\alpha(r^+-1)}\right] \tag{2.30}$$

From (2.23), we deduce that

$$s = r^- + \alpha p^-(r^+-1) \le p^- \text{ and } s = r^+ + \alpha p^-(r^+-1) \le p^-.$$

Then, by using lemma 9, satisfies

$$H^{\alpha(r^{+}-1)}(t)\int_{\Omega}|u|^{r(x)}\,dx\leq C\left[\|\nabla u\|^{2}+\rho(u)\right] \tag{2.31}$$

Combining (2.25)-(2.31), we get

$$\begin{aligned}
L'(t)\geq&(1-\alpha)H^{-\alpha}(t)\left[C_{0}-\varepsilon\left(\frac{r^{+}-1}{r^{+}}\right)ck\right]\int_{\Omega}|u_{t}|^{r(x)}\,dx\\
&+(1-\alpha)H^{-\alpha}(1-\alpha)\left[C_{0}-\varepsilon\left(\frac{r^{+}-1}{r^{+}}\right)ck\right]\int_{\Omega}|z(x,1,t)|^{r(x)}\,dx\\
&+\varepsilon\left(\frac{(1-a)p^{-}-2}{2}-\frac{C}{r^{-}k^{1-r^{-}}}\right)\|\nabla u\|^{2}+\varepsilon(1-a)p^{-}H(t)+\varepsilon\frac{(1-a)p^{-}+2}{2}u_{t}^{2}\\
&+\varepsilon m^{2}\frac{(1-a)p^{-}-2}{2}u^{2}+\varepsilon\left(ab-\frac{C}{r^{-}k^{1-r^{-}}}\right)\rho(u)\\
&+\varepsilon(1-a)p^{-}\int_{0}^{1}\int_{\Omega}\frac{\xi(x)|z(x,\rho,t)|^{r(x)}}{r(x)}\,dxd\rho.
\end{aligned} \tag{2.32}$$

Let us choose a small enough such that

$$\frac{(1-a)p^{-}+2}{2}>0$$

and k enough large,

$$\frac{(1-a)p^{-}-2}{2}-\frac{C}{r^{-}k^{1-r^{-}}}>0 \text{ and } ab-\frac{C}{r^{-}k^{1-r^{-}}}>0.$$

Once k and a are fixed, pick ε small enough such that

$$C_{0}-\varepsilon\left(\frac{r^{+}-1}{r^{+}}\right)ck>0,\quad C_{0}-\varepsilon\left(\frac{r^{+}-1}{r^{+}}\right)ck>0$$

and

$$L(0)=H^{1-\alpha}(0)+\varepsilon\int_{\Omega}u_{0}u_{1}\,dx>0.$$

Consequently, (2.32) yields

$$L'(t) \geq \varepsilon\eta \Bigg[H(t) + ||u_t||^2 + ||\nabla u||^2 + m^2||u||^2 + \rho(u) + \int_0^1 \int_\Omega \frac{\xi(x)|z(x,\rho,t)|^{r(x)}}{r(x)} dx d\rho \Bigg] \tag{2.33}$$

for a constant $\eta > \mathbf{0}$. Thus, we get

$$L(t) \geq L(0) > 0, \forall t \geq 0.$$

Now, for some constants $\omega, \Gamma > 0$ we denote

$$L'(t) \geq \Gamma L\omega(t).$$

Utilizing Hölder inequality, we get

$$\left| \int_\Omega u u_t dx \right|^{1/(1-\alpha)} \leq C||u||_{p^-}^{1/(1-\alpha)} ||u_t||_2^{1/(1-\alpha)},$$

and by using Young's inequality satisfies

$$\left| \int_\Omega u u_t dx \right|^{1/(1-\alpha)} \leq C\left[||u||_{p^-}^{ru/(1-\alpha)} + ||u_t||_2^{\theta/(1-\alpha)} \right],$$

where $\frac{1}{\mu} + \frac{1}{\theta} = 1$. From (2.23), the choice of $\theta = 2(1-\alpha)$ will make

$$\frac{\mu}{(1-\alpha)} = \frac{2}{(1-2\alpha)} \leq p^-.$$

Hence,

$$\left| \int_\Omega u u_t dx \right|^{1/(1-\alpha)} \leq C\left[||u||_{p^-}^{s} + ||u_t||^2 \right],$$

where $s = \mu / (1-\alpha)$. From (2.19), we have

$$\left|\int_\Omega uu_t dx\right|^{1/(1-\alpha)}$$
$$\leq C\left[H(t)+\|u_t\|^2+\rho(u)+\int_0^1\int_\Omega \frac{\xi(x)|z(x,\rho,t)|^{r(x)}}{r(x)}dxd\rho\right]. \quad (2.34)$$

Hence, we get

$$L^{\frac{1}{(1-\alpha)}}(t)=\left[H^{(1-\alpha)}(t)+\varepsilon\int_\Omega uu_t dx+\frac{\varepsilon}{2}\|\nabla u\|^2\right]^{\frac{1}{(1-\alpha)}}$$
$$\leq 2^{\frac{\alpha}{1-\alpha}}\left[H(t)+\varepsilon^{\frac{1}{(1-\alpha)}}\left|\int_\Omega uu_t dx\right|^{\frac{1}{(1-\alpha)}}+\left(\frac{\varepsilon}{2}\|\nabla u\|^2\right)^{\frac{1}{(1-\alpha)}}\right]$$
$$\leq C\Big[H(t)+\|u_t\|^2+\|\nabla u\|^2+m^2\|u\|^2+\rho(u)$$
$$+\int_0^1\int_\Omega \frac{\xi(x)|z(x,\rho,t)|^{r(x)}}{r(x)}dxd\rho\Big]. \quad (2.35)$$

So, for some $\Psi > 0$, from (2.33) we arrive

$$L'(t)\geq \Psi L^{1/(1-\alpha)}(t). \quad (2.36)$$

An integration of (2.36) over $(0, t)$ satisfies

$$L^{\alpha/(1-\alpha)}\geq\frac{1}{L^{-\alpha/(1-\alpha)}(0)-\Psi\alpha t/(1-\alpha)},$$

with

$$T^*\leq\frac{1-\alpha}{\Psi\alpha[L(0)]^{\alpha/(1-\alpha)}},$$

thus, the solution blows up. As a result, we completed the proof.

2.4 Decay of Solutions

In this part, for the case $b = 0$, we obtain the decay results for the problem (2.9). When $b = 0$, the energy functional of (2.9) is:

$$E(t) = \frac{1}{2}\|u_t\|^2 + \frac{1}{2}\|\nabla u\|^2 + \frac{m^2}{2}\|u\|^2 + \int_0^1\int_\Omega \frac{\xi(x)|z(x,\rho,t)|^{r(x)}}{r(x)}dxd\rho \quad (2.37)$$

here ξ is the continuous function given in (2.11).

Similar to Lemma 8, we establish, for $\mu_1 > |\mu_2|$ and $C_0 > 0$, for

$$E'(t) \le -C_0\int_\Omega \left(|u_t|^{r(x)} + |z(x,1,t)|^{r(x)}\right)dx \le 0. \quad (2.38)$$

We need the following technical lemmas before we get our main decay results.

Lemma 11 [19] Suppose that $E: R^+ \to R^+$ is nonincreasing function and suppose $\sigma, \omega > 0$,

$$\int_s^\infty E^{1+\sigma}(t)dt \le \frac{1}{\Omega}E^\sigma(0)E(s) = cE(s) \quad \forall s > 0.$$

Therefore,

$$\begin{cases} E(t) \le \dfrac{cE(0)}{(1+t)^{\frac{1}{\sigma}}} & if\ \sigma > 0, \\ E(t) \le cE(0)e^{-\omega t} & if\ \sigma = 0, \end{cases}$$

where $t \ge 0$.

Lemma 12 [4]. The functional

$$F(t) = \tau\int_0^1\int_\Omega e^{-\rho\tau}\xi(x)|z(x,\rho,t)|^{r(x)}\,dxd\rho$$

satisfies,

$$F'(t)\leq \int_{\Omega}\xi(x)|u_t|^{r(x)}dx-\tau e^{-\tau}\int_0^1\int_{\Omega}\xi(x)\left|z(x,\rho,t)\right|^{r(x)}dxd\rho,$$

along the solution of (2.9).

Theorem 13 Assume that the conditions (2.2) and (2.7) are satisfied. Then, for c, $\alpha > 0$ constants, that any global solution of (2.9) holds,

$$\begin{cases} E(t)\leq ce^{-\alpha t} & \text{if } r(.)=2, \\ E(t)\leq \dfrac{cE(0)}{(1+t)^{\frac{2}{(r^+-2)}}} & \text{if } r^+>2. \end{cases}$$

Proof. Multiply the first equation of (2.9) by $uE^q(t)$, for $q > 0$, and integrating over $\Omega \times (s, T)$, $s < T$, to have

$$\int_s^T E^q\int_{\Omega}\Big[uu_{tt}-u\Delta u-u\Delta u_t+m^2u^2+\mu_1uu_t|u_t|^{r(x)-2}$$
$$+\mu_2uz(x,1,t)|z(x,1,t)|^{r(x)-2}\Big]dxdt=0$$

which implies that

$$\int_s^T E^q\int_{\Omega}\Big(\frac{d}{dt}(uu_t)-u_t^2+|\nabla u|^2+\nabla u\nabla u_t+m^2u^2+\mu_1uu_t|u_t|^{r(x)-2}$$
$$+\mu_2uz(x,1,t)|z(x,1,t)|^{r(x)-2}\Big)dxdt=0. \tag{2.39}$$

Recalling the definition of $E(t)$, given in (2.37) adding and subtracting some terms and using the relation

$$\frac{d}{dt}\left(E^q\int_{\Omega}uu_tdx\right)=qE^{q-1}(t)E'(t)\int_{\Omega}uu_tdx+E^q(t)\frac{d}{dt}\int_{\Omega}uu_tdx,$$

the equation (2.39) satisfies

$$2\int_s^T E^{q+1}dt = -\int_s^T \frac{d}{dt}\left(E^q \int_\Omega uu_t dx\right)dt + q\int_s^T E^{q-1}E'\int_\Omega uu_t dxdt$$
$$+2\int_s^T E^q \int_\Omega u_t^2 dxdt - \frac{1}{2}\int_s^T \frac{d}{dt}\left(E^q \int_\Omega |\nabla u|^2\, dx\right)dt$$
$$+\frac{q}{2}\int_s^T E^{q-1}E'\int_\Omega |\nabla u|^2\, dxdt - \mu_1\int_s^T E^q \int_\Omega uu_t\,|u_t|^{r(x)-2}\, dxdt$$
$$-\mu_2\int_s^T E^q \int_\Omega uz(x,1,t)\,|z(x,1,t)|^{r(x)-2}\, dxdt$$
$$+2\int_s^T E^q \int_0^1\int_\Omega \frac{\xi(x)\,|z(x,1,t)|^{r(x)}}{r(x)}dxd\rho dt. \tag{2.40}$$

The first term is estimated as following:

$$\left|-\int_s^T \frac{d}{dt}\left(E^q(t)\int_\Omega uu_t dx\right)dt\right| = \left|E^q(s)\int_\Omega uu_t(x,s)dx - E^q(T)\int_\Omega uu_t(x,T)dx\right|$$
$$\le \frac{1}{2}E^q(s)\left[\int_\Omega u^2(x,s)dx + \int_\Omega u_t^2(x,s)dx\right]$$
$$+\frac{1}{2}E^q(T)\left[\int_\Omega u^2(x,T)dx + \int_\Omega u_t^2(x,T)dx\right]$$
$$\le \frac{1}{2}E^q(s)\left[C_p\,||\nabla u(s)||_2^2 + 2E(s)\right] + \frac{1}{2}E^q(T)\left[C_p\,||\nabla u(T)||_2^2 + 2E(T)\right]$$
$$\le E^q(s)[C_pE(s)+E(s)] + E^q(T)[C_pE(T)+E(T)],$$

where C_p is the Poincare's constant. Because of $E\,(t)$ is nonincreasing, we infer that

$$\left|-\int_s^T \frac{d}{dt}\left(E^q \int_\Omega uu_t dx\right)dt\right| \le cE^{q+1}(s) \le cE^q(0)E(s) \le cE(s) \tag{2.41}$$

In similar way, we handle the term

$$\left|q\int_s^T E^{q-1}E'\int_\Omega uu_t dxdt\right| \le -q\int_s^T E^{q-1}E'[C_pE(t)+E(t)]dt$$
$$\le -c\int_s^T E^q E' \le cE^{q+1}(s) \le cE(s). \tag{2.42}$$

To treat the other term, we set

$$\Omega_{+} = \{x{\in}\Omega, |ut\ (x, t)| \geq 1\} \text{ and } \Omega_{-} = \{x{\in}\Omega, |ut\ (x, t)| < 1\}.$$

Then, utilizing the Young's and Hölder's inequalities, to get

$$\left|\int_s^T E^q \int_\Omega u_t^2 dxdt\right| = \left|\int_s^T E^q \left[\int_{\Omega_+} u_t^2 dx + \int_{\Omega_-} u_t^2 dx\right] dt\right|$$

$$\leq c\int_s^T E^q \left[\left(\int_{\Omega_+} |u_t|^{r(x)}\, dx\right)^{\frac{2}{r^-}} + \left(\int_{\Omega_+} |u_t|^{r(x)}\, dx\right)^{\frac{2}{r^+}}\right] dt$$

$$\leq c\int_s^T E^q \left[(-E')^{\frac{2}{r^-}} + (-E')^{\frac{2}{r^+}}\right] dt$$

$$\leq c\varepsilon\int_s^T [E]^{\frac{qr^-}{(r^- -2)}} dt + c(\varepsilon)\int_s^T (-E')dt + c\varepsilon\int_s^T E^{q+1} dt$$

$$+\, c(\varepsilon)\int_s^T (-E')^{2(q+1)/r^+}\, dt.$$

For $r^- > 2$ and the choice of $q = \frac{r^+}{2} - 1$ will give $\frac{qr^-}{r^- - 2} = q + 1 + \frac{r^+ - r^-}{r^- - 2}$. Therefore,

$$\left|\int_s^T E^q \int_\Omega u_t^2 dxdt\right| \leq c\varepsilon\int_s^T E^{q+1} dt + c\varepsilon[E(0)]^{\frac{r^+ - r^-}{r^- - 2}} \int_s^T E^{q+1} dt$$

$$+\, c(\varepsilon)E(s) \leq c\int_s^T E^{q+1} dt + c(\varepsilon)E(s) \tag{2.43}$$

For the case $r^- = 2$ and the choice of $q = \frac{r^+}{2} - 1$ will give the similar result.

The other term is estimated:

$$\begin{aligned}
&\left|-\frac{1}{2}\int_s^T \frac{d}{dt}\left(E^q(t)\int_\Omega |\nabla u|^2\,dx\right)dt\right| \\
&\le \frac{1}{2}E^q(s)\int_\Omega |\nabla u(s)|^2\,dx + \frac{1}{2}E^q(T)\int_\Omega |\nabla u(T)|^2\,dx \\
&\le cE^{q+\frac{2}{r^+}}(s) \\
&\le c\left(E^{q-1+\frac{2}{r^+}}(0)\right)E(s) \\
&\le \lambda E(s),
\end{aligned} \tag{2.44}$$

Where $c, \lambda > 0$.

Similarly,

$$\begin{aligned}
\int_s^T E^{q-1}(t)E'(t)\int_\Omega |\nabla u|^2\,dxdt &\le cE^{q+\frac{2}{r^+}}(s) \\
&\le c\left(E^{q-1+\frac{2}{r^+}}(0)\right)E(s) \le \lambda_1 E(s),
\end{aligned} \tag{2.45}$$

where c and λ_1 are positive constants.

For the other term, from Young's inequality, we conclude

$$\begin{aligned}
&\left|-\mu_1\int_s^T E^q\int_\Omega u|u_t|^{r(x)-1}\,dxdt\right| \\
&\le \varepsilon\int_s^T E^q\int_\Omega |u_t|^{r(x)}\,dxdt + c\int_s^T E^q(t)\int_{\Omega_-} c_\varepsilon |u|^{r^+}\,dxdt \\
&\le \varepsilon\int_s^T E^q\left[\int_{\Omega_+} |u|^{r^-}\,dx + \int_{\Omega_-} |u|^{r^+}\,dx\right]dt + c\int_s^T E^q\int_\Omega c_\varepsilon |u_t|^{r(x)}\,dxdt,
\end{aligned}$$

here we used Young's inequality and

$$p(x) = \frac{r(x)}{r(x)-1},\ p'(x) = r(x),$$

therefore,

$$c_\varepsilon(x) = (r(x)-1)\,r(x)^{r(x)/(1-r(x))}\,\varepsilon^{1/(1-r(x))}.$$

That is why, by using the embeddings $H_0^1(\Omega)\hookrightarrow L^{r^-}(\Omega)$ and $H_0^1(\Omega)\hookrightarrow L^{r^+}(\Omega)$ we obtain

$$\begin{aligned}
&\left|-\mu_1\int_s^T E^q\int_\Omega u\,|u_t|^{r(x)-1}\,dxdt\right| \\
&\leq \varepsilon\int_s^T E^q\left[c\,\|\nabla u(s)\|_2^{r^-}+c\,\|\nabla u(s)\|_2^{r^+}\right]dt \\
&+c\int_s^T E^q\int_\Omega c_\varepsilon(x)\,|u_t|^{r(x)}\,dxdt \leq c\varepsilon\int_s^T E^{q+1}dt \\
&+\int_s^T E^q\int_\Omega c_\varepsilon(x)\,|u_t|^{r(x)}\,dxdt.
\end{aligned} \tag{2.46}$$

The next term of (2.40) can be estimated in a similar attitude to get

$$\begin{aligned}
&\left|-\mu_2\int_s^T E^q\int_\Omega u|z(x,1,t)|^{r(x)-1}dxdt\right| \leq \varepsilon\int_s^T E^q\left[c\|\nabla u(s)\|_2^{r^-}+c\|\nabla u(s)\|_2^{r^+}\right]dt \\
&+c\int_s^T E^q\int_\Omega c_\varepsilon|z(x,1,t)|^{r(x)}dxdt \leq c\varepsilon\int_s^T E^{q+1}dt+\int_s^T E^q\int_\Omega c_\varepsilon|z(x,1,t)|^{r(x)}dxdt.
\end{aligned} \tag{2.47}$$

For the last term of (2.40), from Lemma 12, we get

$$\begin{aligned}
&2\int_s^T E^q\int_0^1\int_\Omega \frac{\xi(x)\,|z(x,\rho,t)|^{r(x)}}{r(x)}dxd\rho dt \\
&\leq \frac{2}{r^-}\int_s^T E^q\int_0^1\int_\Omega \xi(x)\,|z(x,\rho,t)|^{r(x)}\,dxd\rho dt \\
&\leq -\frac{2\tau}{r^-}\int_s^T E^q\frac{d}{dt}\left(\int_0^1\int_\Omega e^{-\rho\tau}\xi(x)\,|z|^{r(x)}\,dxd\rho\right)dt \\
&+\frac{2}{r^-}\int_s^T E^q\int_\Omega \xi(x)\,|u_t|^{r(x)}\,dxd\rho \leq -\frac{2\tau}{r^-}\left[E^q\int_0^1\int_\Omega e^{-\rho\tau}\xi(x)\,|z|^{r(x)}\,dxd\rho\right]_{t=s}^{t=T} \\
&+\frac{2}{r^-}\int_s^T E^q\int_\Omega \xi(x)\,|u_t|^{r(x)}\,dxdt.
\end{aligned}$$

As $\xi(x)$ is bounded, by (2.37) we have

$$2\int_s^T E^q \int_0^1 \int_\Omega \frac{\xi(x)\,|z(x,\rho,t)|^{r(x)}}{r(x)}\,dx\,d\rho\,dt \le \frac{2\tau e^{-\tau}}{r^-} E^q(s)E(s)$$
$$+\frac{2c}{r^-}E^{q+1}(T) \le \frac{2\tau e^{-\tau}}{r^-}E^q(0)E(s) + \frac{2c}{r^-}E^q(T)E(s) \le cE(s), \tag{2.48}$$

for some $c > 0$.

By combining (2.40)-(2.48), we infer that

$$\int_s^T E^{q+1}dt \le \varepsilon \int_s^T E^{q+1}dt + cE(s) + c\int_s^T E^q \int_\Omega c_\varepsilon(x)\,|z(x,1,t)|^{r(x)}\,dxdt. \tag{2.49}$$

Choosing ε so small

$$\int_s^T E^{q+1}dt \le cE(s) + c\int_s^T E^q \int_\Omega c_\varepsilon(x)\,|z(x,1,t)|^{r(x)}\,dxdt.$$

Once ε is fixed, $c_\varepsilon(x) \le M$, since $r(x)$ is bounded. Therefore, we infer

$$\int_s^T E^{q+1}dt \le cE(s) + cM\int_s^T E^q \int_\Omega |z(x,1,t)|^{r(x)}\,dxdt \le cE(s)$$
$$-C_0M\int_s^T E^q E'dt \le cE(s) + \frac{C_0M}{q+1}[E^{q+1}(s) - E^{q+1}(T)] \le cE(s). \tag{2.50}$$

By taking $T \to \infty$, we obtain

$$\int_s^\infty E^{q+1}(t)dt \le cE(s)$$

Thus, Komornik's Lemma $\left(\text{with } \sigma = q = \frac{r^+}{2} - 1\right)$ implies the desired result.

Acknowledgment

The authors are grateful to DUBAP (ZGEF.20.009) for research funds.

References

1. Chen, Y., Levine, S., Rao, M., Variable exponent, linear growth functionals in image restoration. *SIAM J. Appl. Math.*, 66, 1383–1406, 2006.
2. Diening, L., Hästo, P., Harjulehto, P., Ruzicka, M.M., *Lebesque and Sobolev Spaces with Variable Exponents*, Springer-Verlag, Berlin, Heidelberg, 2011.
3. Kafini, M. and Messaoudi, S.A., A blow-up result in a nonlinear wave equation with delay. *Mediterr. J. Math.*, 13, 237–247, 2016.
4. Kafini, M. and Messaoudi, S.A., On the decay and global nonexistence of solutions to a damped wave equation with variable-exponent nonlinearity and delay. *Ann. Pol. Math.*, 122, 1, 49–70, 2019.
5. Pişkin, E., Finite time blow up of solutions for a strongly damped nonlinear Klein-Grdon equation with variable exponents. *Honam Math. J.*, 40, 4, 771–783, 2018.
6. Antontsev, S., Ferreira, J., Pişkin, E., Existence and blow up of solutions for a strongly damped petrovsky equation with variable-exponent nonlinearities. *Electron J. Differ. Eq.*, 6, 1–18, 2021.
7. Pişkin, E. and Yüksekkaya, H., Decay of solutions for a nonlinear Petrovsky equation with delay term and variable exponents. *Aligarh Bull. Math.*, 39, 2, 63–78, 2020c.
8. Kang, J.R., Existence and blow-up of solutions for von Karman equations with time delay and variable exponents. *Appl. Math. Comput.*, 371, 124917, 2020.
9. Pişkin, E. and Yüksekkaya, H., Nonexistence of solutions of a delayed wave equation with variable-exponents. *C-POST*, 3, 1, 97–101, 2020a.
10. Pişkin, E. and Yüksekkaya, H., Decay and blow up of solutions for a delayed wave equation with variable-exponents. *C-POST*, 3, 1, 91–96, 2020b.
11. Pişkin, E. and Yüksekkaya, H., Local existence and blow up of solutions for a logarithmic nonlinear viscoelastic wave equation with delay. *Comput. Methods Differ. Equ.*, 9, 2, 623–636, 2021a.
12. Pişkin, E. and Yüksekkaya, H., Nonexistence of global solutions of a delayed wave equation with variable-exponents. *Miskolc Math. Notes*, 22, 2, 841–859, 2021b. (Accepted).
13. Pişkin, E. and Yüksekkaya, H., Blow-up of solutions for a logarithmic quasilinear hyperbolic equation with delay term. *J. Math. Anal.*, 12, 1, 56–64, 2021c.
14. Antontsev, S., Wave equation with $p(x, t)$–laplacian and damping term: Blow up of solutions. *C.R. Mecanique*, 339, 751–755, 2011a.

15. Antontsev, S., Wave equation with p (x, t)–laplacian and damping term: Existence and blow-up. *Differ. Equ. Appl.*, 3, 503–525, 2011b.
16. Pişkin, E., *Sobolev Spaces*, Seçkin Publishing, Ankara, 2017.
17. Nicaise, S. and Pignotti, C., Stability and instability results of the wave equation with a delay term in the boundary or internal feedbacks. *SIAM J. Control Optim.*, 45, 1561–1585, 2006.
18. Messaoudi, S.A., Talahmeh, A.A., Al-Smail, J.H., Nonlinear damped wave equation: Existence and blow-up. *Comput. Math. Appl.*, 74, 3024–3041, 2017.
19. Komornik, V., *Exact Controllability and Stabilization. The Multiplier Method*, Masson and Wiley, Paris, 1994.

3

Some New Inequalities Via Extended Generalized Fractional Integral Operator for Chebyshev Functional

Bhagwat R. Yewale* **and Deepak B. Pachpatte**

Department of Mathematics, Dr. B. A. M. University, Aurangabad, Maharashtra, India

Abstract

In present article, we prove some integral inequalities for Chebyshev functional using extended generalized fractional operator. The result obtained in the case of differentiable as well as Lipschitz functions.

Keywords: Chebyshev functional, Integral inequalities, Extended generalized fractional integral operator

3.1 Introduction

In 1882, Chebyshev introduced the following inequality [13]:

Let $\zeta, \eta : [\bar{c}, \tau] \to \mathbb{R}^+$ be differentiable functions such that $\zeta', \eta' \in \mathcal{L}_\infty[\bar{c}, \tau]$ and

$$S(\zeta,\eta) = \frac{1}{\tau - \bar{c}} \int_{\bar{c}}^{\tau} \zeta(\theta)\eta(\theta)d\theta - \left(\frac{1}{\tau - \bar{c}} \int_{\bar{c}}^{\tau} \zeta(\theta)d\theta\right)\left(\frac{1}{\tau - \bar{c}} \int_{\bar{c}}^{\tau} \eta(\theta)d\theta\right), \tag{3.1}$$

then

$$|S(\zeta,\eta)| \le \frac{1}{12} \|\zeta'\|_\infty \|\eta'\|_\infty (\tau - \bar{c})^2. \tag{3.2}$$

**Corresponding author*: yewale.bhagwat@gmail.com

Sharmistha Ghosh, M. Niranjanamurthy, Krishanu Deyasi, Biswadip Basu Mallik, and Santanu Das (eds.) Mathematics and Computer Science Volume 1, (45–66) © 2023 Scrivener Publishing LLC

The constant $\frac{1}{12}$ is the best possible.

The functional (3.1) have large number of applications in the field of statistics and probability. Many researcher's provided lot of integral inequalities related to this functional in their literature (see [9, 12, 14, 15]).

Integrals and derivatives of any positive order are allowed in fractional calculus. Due to its applications in a variety of domains, many authors have contributed to the development of fractional calculus. For more details, one may refer [11, 16, 20].

Integral inequalities in the sense of fractional operators (fractional integrals and fractional derivatives) have proved to be one of the most important and powerful tool for studying various problems in different branches of Mathematics. Dahmani [5–7] established certain Chebyshev type integral inequalities using the Riemann-Liouville fractional integral operator. In [21], Sarikaya *et al.* studied some integral inequalities by using (k, s)-Riemann Liouville fractional integral operator. Purohit and Raina used the Saigo fractional integral operator to study Chebyshev-like inequalities [17]. Using generalized Katugampola operator, Aljaaidi and Pachpatte established Gruss-type inequalities in [1]. Sousa *et al.* [25] derived Gruss-type inequalities by means of generalized fractional integrals. Since then many researchers have established large number of inequalities by employing various fractional integral operators, see [2, 4, 8, 10, 19, 22–24] and the references therein.

Inspired by aforementioned work, in this article, we obtain some new integral inequalities by employing extended generalized fractional integral operator. The remaining paper is organized as follows: In section 3.2, we give some preliminaries which will be useful in the sequel. In section 3.3, we establish some new inequalities involving extended generalized fractional integral operator related to the functional (3.1). Integral inequalities associated with the extended version of the functional (3.1) are derived in section 3.4 and in section 3.5.

3.2 Preliminaries

In this section, we mention some preliminary facts and definitions that are used to establish our main results:

Here, $\mathcal{L}_1([\bar{c},\tau])$ denotes the space of all Lebesgue measurable functions such that $\| g \|_1 = \int_{\bar{c}}^{\tau} |g(y)|\, dy < \infty$ and $\mathcal{L}_\infty([0,\infty))$ denotes the space of all bounded functions on $[0,\infty))$, with the norm defined by $\| g(y) \|_\infty = \sup_{y\in[0,\infty)} |g(y)|$.

Definition 3.2.1. [3] Let $\vartheta, \bar{\theta}_1, \iota, \gamma, \varsigma \in \mathbb{C}, \mathcal{R}(\vartheta), \mathcal{R}(\bar{\theta}_1), \mathcal{R}(\iota) > 0, \mathcal{R}(\varsigma) > \mathcal{R}(\gamma) > 0$ with $\mathcal{P} \geq 0, \lambda > 0$ and $0 < \kappa \leq \lambda + R(\vartheta)$. Then the extended generalized Mittag-Leffler function is denoted by $E^{\gamma,\lambda,\kappa,\varsigma}_{\vartheta,\bar{\theta}_1,\iota}$ and is defined as

$$E^{\gamma,\lambda,\kappa,\varsigma}_{\vartheta,\bar{\theta}_1,\iota}(\varkappa;\mathcal{P}) = \sum_{n=0}^{\infty} \frac{\beta_{\mathcal{P}}(\gamma+nk,\varsigma-\gamma)}{\beta(\gamma,\varsigma-\gamma)} \frac{(\varsigma)_{n\kappa}}{\Gamma(\vartheta n+\bar{\theta}_1)} \frac{\varkappa^n}{(\iota)_{n\lambda}}, \tag{3.3}$$

where $\beta_{\mathcal{P}}$ is an extension of the beta function

$$\beta_{\mathcal{P}}(r,s) = \int_0^1 \varkappa^{r-1}(1-\varkappa)^{s-1} e^{-\frac{\mathcal{P}}{\varkappa(1-\varkappa)}} d\varkappa, \quad (\mathcal{R}(r), \mathcal{R}(s), \mathcal{R}(\mathcal{P}) > 0).$$

Definition 3.2.2 [3] Let $\omega, \vartheta, \bar{\theta}_1, \iota, \gamma, \varsigma \in \mathbb{C}, \mathcal{R}(\vartheta), \mathcal{R}(\bar{\theta}_1), \mathcal{R}(\iota) > 0, \mathcal{R}(\varsigma) > \mathcal{R}(\gamma) > 0$ with $\mathcal{P} \geq 0, \lambda > 0$ and $0 < \kappa \leq \lambda + \mathcal{R}(\vartheta)$. Let $h \in \mathcal{L}_1[\bar{c},\tau]$ and $z \in [\bar{c},\tau]$. Then the extended generalized fractional integral operator is denoted by $I^{\omega,\gamma,\lambda,\kappa,\varsigma}_{\bar{c},\vartheta,\bar{\theta}_1,\iota}$ and is defined as

$$\left(I^{\omega,\gamma,\lambda,\kappa,\varsigma}_{\bar{c},\vartheta,\bar{\theta}_1,\iota} h\right)(z;\mathcal{P}) = \int_{\bar{c}}^{z} (z-\varkappa)^{\bar{\theta}_1-1} E^{\gamma,\lambda,\kappa,\varsigma}_{\vartheta,\bar{\theta}_1,\iota}(\omega(z-\varkappa)^{\vartheta};\mathcal{P}) h(\varkappa) d\varkappa. \tag{3.4}$$

For convenience, we use the following notation:

$$\begin{aligned}
\boldsymbol{E}_{\bar{\theta}_1}(\varkappa;\mathcal{P}) &:= E^{\gamma,\lambda,\kappa,\varsigma}_{\vartheta,\bar{\theta}_1,\iota}(\varkappa;\mathcal{P}) \\
&= \sum_{n=0}^{\infty} \frac{\beta_{\mathcal{P}}(\gamma+nk,\varsigma-\gamma)}{\beta(\gamma,\varsigma-\gamma)} \frac{(\varsigma)_{n\kappa}}{\Gamma(\vartheta n+\bar{\theta}_1)} \frac{\varkappa^n}{(\iota)_{n\lambda}}, \\
(\boldsymbol{I}_{\bar{\theta}_1} h)(z;\mathcal{P}) &:= (I^{\omega,\gamma,\lambda,\kappa,\varsigma}_{\bar{c},\vartheta,\bar{\theta}_1,\iota} h)(z;\mathcal{P}) \\
&= \int_{\bar{c}}^{z} (z-\varkappa)^{\bar{\theta}_1-1} E^{\gamma,\lambda,\kappa,\varsigma}_{\vartheta,\bar{\theta}_1,\iota}(\omega(z-\varkappa);\mathcal{P}) h(\varkappa) d\varkappa.
\end{aligned}$$

3.3 Fractional Inequalities for the Chebyshev Functional

This section devoted some Chebyshev type inequalities for differentiable function related to the functional (3.1) by using extended generalized integral operator.

Theorem 3.3.1 Let the functions ζ and η be differentiable on $[0, \infty)$ with $\zeta', \eta' \in \mathcal{L}_\infty([0, \infty))$. Then for all $\varkappa > \bar{c} \geq 0$, $\bar{\theta}_1 > 0$, following inequalities hold:

$$\begin{aligned} &2|(\boldsymbol{I}_{\bar{\theta}_1}1)(\varkappa;\mathcal{P})(\boldsymbol{I}_{\bar{\theta}_1}\zeta\eta)(\varkappa;\mathcal{P}) \\ &-(\boldsymbol{I}_{\bar{\theta}_1}\zeta)(\varkappa;\mathcal{P})(\boldsymbol{I}_{\bar{\theta}_1}\eta)(\varkappa;\mathcal{P})| \\ \leq{}& \|\zeta'\|_\infty \|\eta'\|_\infty [(\boldsymbol{I}_{\bar{\theta}_1}1)(\varkappa;\mathcal{P})(\boldsymbol{I}_{\bar{\theta}_1}\rho^2)(\varkappa;\mathcal{P}) \\ &-2(\boldsymbol{I}_{\bar{\theta}_1}\rho)(\varkappa;\mathcal{P})(\boldsymbol{I}_{\bar{\theta}_1}\sigma)(\varkappa;\mathcal{P}) \\ &+(\boldsymbol{I}_{\bar{\theta}_1}1)(\varkappa;\mathcal{P})(\boldsymbol{I}_{\bar{\theta}_1}\sigma^2)(\varkappa;\mathcal{P})]. \end{aligned} \tag{3.5}$$

Proof. Define

$$B(\rho, \sigma) := (\zeta(\rho) - \zeta(\sigma))(\eta(\rho) - \eta(\sigma)), \tag{3.6}$$

for all $\varkappa > \bar{c} \geq 0$, $\rho, \sigma \in (\bar{c}, \varkappa)$.

Therefore

$$B(\rho, \sigma) = \zeta(\rho)\eta(\rho) - \zeta(\sigma)\eta(\rho) - \zeta(\rho)\eta(\sigma) + \zeta(\sigma)\eta(\sigma). \tag{3.7}$$

Multiplyingby$(\varkappa - \rho)^{\bar{\theta}_1-1}\boldsymbol{E}_{\bar{\theta}_1}(\omega(\varkappa - \rho)^{\vartheta};\mathcal{P})(\varkappa - \sigma)^{\bar{\theta}_1-1}\boldsymbol{E}_{\bar{\theta}_1}(\omega(\varkappa - \sigma)^{\vartheta};\mathcal{P})$, $\rho, \sigma \in (\bar{c}, \varkappa)$ to (3.10) and then integrating w. r. t. ρ and σ over $(\bar{c}, \varkappa)^2$ respectively, we get

$$\begin{aligned} &\int_{\bar{c}}^{\varkappa} (x-\rho)^{\bar{\theta}_1-1}\boldsymbol{E}_{\bar{\theta}_1}(\omega(x-\rho)^{\vartheta};\mathcal{P})B(\rho,\sigma)d\rho \\ &\quad = (\boldsymbol{I}_{\bar{\theta}_1}\zeta\eta)(\varkappa;\mathcal{P}) - \zeta(\sigma)(\boldsymbol{I}_{\bar{\theta}_1}\eta) \\ &\quad -\eta(\sigma)(\boldsymbol{I}_{\bar{\theta}_1}\zeta) + \zeta(\sigma)\eta(\sigma)(\boldsymbol{I}_{\bar{\theta}_1}1). \end{aligned} \tag{3.8}$$

Multiplying by $(\varkappa - \sigma)^{\bar{\theta}_1-1}\boldsymbol{E}_{\bar{\theta}_1}(\omega(\varkappa - \sigma)^{\vartheta};\mathcal{P})$, $\sigma \in (\bar{c}, \varkappa)$ to (3.8) and then integrating w. r. t. σ from $\bar{c}$ to $\varkappa$, we get

$$\begin{aligned} &\int_{\bar{c}}^{\varkappa}\int_{\bar{c}}^{\varkappa} (\varkappa - \rho)^{\bar{\theta}_1-1}(\varkappa - \sigma)^{\bar{\theta}_1-1}\boldsymbol{E}_{\bar{\theta}_1}(\omega(\varkappa - \rho)^{\vartheta};\mathcal{P}) \\ &\quad \boldsymbol{E}_{\bar{\theta}_1}(\omega(\varkappa - \sigma)^{\vartheta};\mathcal{P})B(\rho,\sigma)d\rho d\sigma \\ &\quad = (\boldsymbol{I}_{\bar{\theta}_1}\zeta\eta)(\varkappa;\mathcal{P})(\boldsymbol{I}_{\bar{\theta}_1}1)(\varkappa;\mathcal{P}) \\ &\quad - (\boldsymbol{I}_{\bar{\theta}_1}\eta)(\varkappa;\mathcal{P})(\boldsymbol{I}_{\bar{\theta}_1}\zeta)(\varkappa;\mathcal{P}) \\ &\quad - (\boldsymbol{I}_{\bar{\theta}_1}\zeta)(\varkappa;\mathcal{P})(\boldsymbol{I}_{\bar{\theta}_1}\eta)(\varkappa;\mathcal{P}) \\ &\quad + (\boldsymbol{I}_{\bar{\theta}_1}1)(\varkappa;\mathcal{P})(\boldsymbol{I}_{\bar{\theta}_1}\zeta\eta)(\varkappa;\mathcal{P}). \end{aligned}$$

Consequently

$$\int_{\bar{c}}^{\varkappa}\int_{\bar{c}}^{\varkappa}(\varkappa-\rho)^{\bar{\theta}_1-1}(\varkappa-\sigma)^{\bar{\theta}_1-1}\boldsymbol{E}_{\bar{\theta}_1}(\omega(\varkappa-\rho)^{\vartheta};\mathcal{P})$$
$$\boldsymbol{E}_{\bar{\theta}_1}(\omega(\varkappa-\sigma)^{\vartheta};\mathcal{P})B(\rho,\sigma)d\rho d\sigma$$
$$=2[(\boldsymbol{I}_{\bar{\theta}_1}1)(\bar{\varkappa};p)(\boldsymbol{I}_{\bar{\theta}_1}\zeta\eta)(\varkappa;\mathcal{P})$$
$$-(\boldsymbol{I}_{\bar{\theta}_1}\zeta)(\varkappa;\mathcal{P})(\boldsymbol{I}_{\bar{\theta}_1}\eta)(\varkappa;\mathcal{P})]. \tag{3.9}$$

Now from (3.6), we obtain

$$B(\rho,\sigma)=\int_{\rho}^{\sigma}\int_{\rho}^{\sigma}\zeta'(m)\eta'(n)dmdn.$$

Since $\zeta', \eta' \in \mathcal{L}_{\infty}([0,\infty))$, it follows

$$|B(\rho,\sigma)|\le\left|\int_{\rho}^{\sigma}\zeta'(\rho)d\rho\right|.\left|\int_{\rho}^{\sigma}\eta'(\sigma)d\sigma\right|$$
$$\le\|\zeta'\|_{\infty}\|\eta'\|_{\infty}(\rho-\sigma)^2. \tag{3.10}$$

Multiplying by $(\varkappa-\rho)^{\bar{\theta}_1-1}\boldsymbol{E}_{\bar{\theta}_1}(\omega(\bar{\varkappa}-\rho)^{\vartheta};\mathcal{P})(\varkappa-\sigma)^{\bar{\theta}_1-1}\boldsymbol{E}_{\bar{\theta}_1}(\omega(\varkappa-\sigma)^{\vartheta};\mathcal{P})$, $\rho,\sigma\in(\bar{c},\varkappa)$ to (3.10) and then integrating w. r. t. ρ and σ over $(\bar{c},\varkappa)^2$ respectively, we get

$$\int_{\bar{c}}^{\varkappa}\int_{\bar{c}}^{\varkappa}(\varkappa-\rho)^{\bar{\theta}_1-1}\boldsymbol{E}_{\bar{\theta}_1}(\omega(\varkappa-\rho)^{\vartheta};\mathcal{P})(\varkappa-\sigma)^{\bar{\theta}_1-1}$$
$$\boldsymbol{E}_{\bar{\theta}_1}(\omega(\varkappa-\sigma)^{\vartheta};\mathcal{P})|B(\rho,\sigma)|d\rho d\sigma$$
$$\le\|\zeta'\|_{\infty}\|\eta'\|_{\infty}\int_{\bar{c}}^{\varkappa}\int_{\bar{c}}^{\varkappa}(\varkappa-\rho)^{\bar{\theta}_1-1}\boldsymbol{E}_{\bar{\theta}_1}(\omega(\varkappa-\rho)^{\vartheta};\mathcal{P})$$
$$(\varkappa-\sigma)^{\bar{\theta}_1-1}\boldsymbol{E}_{\bar{\theta}_1}(\omega(\varkappa-\sigma)^{\vartheta};\mathcal{P})(\rho^2-2\rho\sigma+\sigma^2)d\rho d\sigma. \tag{3.11}$$

From above we get

$$\int_{\bar{c}}^{\varkappa}\int_{\bar{c}}^{\varkappa}(\varkappa-\rho)^{\bar{\theta}_1-1}\boldsymbol{E}_{\bar{\theta}_1}(\omega(\varkappa-\rho)^{\vartheta};\mathcal{P})(\varkappa-\sigma)^{\bar{\theta}_1-1}$$
$$\boldsymbol{E}_{\bar{\theta}_1}(\omega(\varkappa-\sigma)^{\vartheta};\mathcal{P})\,|B(\rho,\sigma)|\,d\rho d\sigma$$
$$\leq \|\zeta'\|_{\infty}\|\eta'\|_{\infty}[(\boldsymbol{I}_{\bar{\theta}_1}1)(\varkappa;\mathcal{P})(\boldsymbol{I}_{\bar{\theta}_1}\rho^2)(\varkappa;p)$$
$$-2(\boldsymbol{I}_{\bar{\theta}_1}\rho)(\varkappa;\mathcal{P})(\boldsymbol{I}_{\bar{\theta}_1}\sigma)(\varkappa;\mathcal{P})+(\boldsymbol{I}_{\bar{\theta}_1}1)(\varkappa;\mathcal{P})$$
$$(\boldsymbol{I}_{\bar{\theta}_1}\sigma^2)(\varkappa;\mathcal{P})].$$

By using (3.9) and (3.11) we get required inequality (3.5). □

Theorem 3.3.2 Let the functions ζ and η be differentiable on $[0,\infty)$ with ζ', $\eta' \in \mathcal{L}_{\infty}([0,\infty))$. Then for $\bar{\theta}_1>0$, $\bar{\theta}_2>0$ we have

$$\begin{aligned}&|(\boldsymbol{I}_{\bar{\theta}_2}1)(\varkappa;\mathcal{P})(\boldsymbol{I}_{\bar{\theta}_1}\zeta\eta)(\varkappa;\mathcal{P})\\&-(\boldsymbol{I}_{\bar{\theta}_1}\eta)(\varkappa;\mathcal{P})(\boldsymbol{I}_{\bar{\theta}_2}\zeta)(\varkappa;\mathcal{P})\\&-(\boldsymbol{I}_{\bar{\theta}_1}\zeta)(\varkappa;\mathcal{P})(\boldsymbol{I}_{\bar{\theta}_2}\eta)(\varkappa;\mathcal{P})\\&+(\boldsymbol{I}_{\bar{\theta}_1}1)(\varkappa;\mathcal{P})(\boldsymbol{I}_{\bar{\theta}_2}\zeta\eta)(\varkappa;\mathcal{P})|\\ \leq\ &\|\zeta'\|_{\infty}\|\eta'\|_{\infty}[(\boldsymbol{I}_{\bar{\theta}_2}1)(\varkappa;\mathcal{P})(\boldsymbol{I}_{\bar{\theta}_1}\rho^2)(\varkappa;\mathcal{P})\\&-2(\boldsymbol{I}_{\bar{\theta}_1}\rho)(\varkappa;\mathcal{P})(\boldsymbol{I}_{\bar{\theta}_2}\sigma)(\varkappa;\mathcal{P})\\&+(\boldsymbol{I}_{\bar{\theta}_1}1)(\varkappa;\mathcal{P})(\boldsymbol{I}_{\bar{\theta}_2}\sigma^2)(\varkappa;\mathcal{P})].\end{aligned} \tag{3.12}$$

Proof. Multiplying by $(\varkappa-\sigma)^{\bar{\theta}_2-1}\boldsymbol{E}_{\bar{\theta}_2}(\omega(\varkappa-\sigma)^{\vartheta};\mathcal{P})$, $\sigma\in(\bar{c},\varkappa)$ to (3.8) and then integrating w. r. t. σ from $\bar{c}$ to $\varkappa$, we get

$$\begin{aligned}&\int_{\bar{c}}^{\varkappa}\int_{\bar{c}}^{\varkappa}(\varkappa-\rho)^{\bar{\theta}_1-1}(\varkappa-\rho)^{\bar{\theta}_2-1}\boldsymbol{E}_{\bar{\theta}_1}(\omega(\varkappa-\rho)^{\vartheta};\mathcal{P})\\&\boldsymbol{E}_{\bar{\theta}_2}(\omega(\varkappa-\sigma)^{\vartheta};\mathcal{P})B(\rho,\sigma)d\rho d\sigma\\&=(\boldsymbol{I}_{\bar{\theta}_2}1)(\varkappa;\mathcal{P})(\boldsymbol{I}_{\bar{\theta}_1}\zeta\eta)(\varkappa;\mathcal{P})\\&-(\boldsymbol{I}_{\bar{\theta}_1}\eta)(\varkappa;\mathcal{P})(\boldsymbol{I}_{\bar{\theta}_2}\zeta)(\varkappa;\mathcal{P})\\&-(\boldsymbol{I}_{\bar{\theta}_1}\zeta)(\varkappa;\mathcal{P})(\boldsymbol{I}_{\bar{\theta}_2}\eta)(\varkappa;\mathcal{P})\\&+(\boldsymbol{I}_{\bar{\theta}_1}1)(\varkappa;\mathcal{P})(\boldsymbol{I}_{\bar{\theta}_2}\zeta\eta)(\varkappa;\mathcal{P})\end{aligned} \tag{3.13}$$

Now multiplying by $(\varkappa-\rho)^{\bar{\theta}_1-1}\mathbf{E}_{\bar{\theta}_1}(\omega(\varkappa-\rho)^{\vartheta};\mathcal{P})(\varkappa-\sigma)^{\bar{\theta}_2-1}\mathbf{E}_{\bar{\theta}_2}(\omega(\varkappa-\sigma)^{\vartheta};\mathcal{P})$ $\rho,\sigma\in(\bar{c},\varkappa)$ to (3.10) and then integrating w. r. t. ρ and σ over $(\bar{c},\varkappa)^2$ respectively, we get

$$\begin{aligned}
&\int_{\bar{c}}^{\varkappa}\int_{\bar{c}}^{\varkappa}(\varkappa-\rho)^{\bar{\theta}_1-1}\mathbf{E}_{\bar{\theta}_1}(\omega(\varkappa-\rho)^{\vartheta};\mathcal{P})(\varkappa-\sigma)^{\bar{\theta}_2-1}\\
&\quad\mathbf{E}_{\bar{\theta}_2}(\omega(\varkappa-\sigma)^{\vartheta};\mathcal{P})\,|\,B(\rho,\sigma\,|\,d\rho d\sigma\\
&\quad\le\|\zeta'\|_{\infty}\|\eta'\|_{\infty}[(\mathbf{I}_{\bar{\theta}_2}1)(\varkappa;\mathcal{P})(\mathbf{I}_{\bar{\theta}_1}\rho^2)(\varkappa;\mathcal{P}) \\
&\quad-2(\mathbf{I}_{\bar{\theta}_1}\rho)(\varkappa;\mathcal{P})(\mathbf{I}_{\bar{\theta}_2}\sigma)(\varkappa;\mathcal{P})\\
&\quad+(\mathbf{I}_{\bar{\theta}_1}1)(\varkappa;\mathcal{P})(\mathbf{I}_{\bar{\theta}_2}\sigma^2)(\varkappa;\mathcal{P})].
\end{aligned}\tag{3.14}$$

From (3.13) and (3.14) we get (3.12). □

Corollary 3.3.2.1 If we put $\bar{\theta}_1=\bar{\theta}_2$, then the inequality (3.12) reduces to the inequality (3.5).

Theorem 3.3.3 Let the functions ζ and η be differentiable on $[0,\infty)$ with $\left|\frac{\zeta'(\varkappa)}{\eta'(\varkappa)}\right|\le L$, for some $L>0$ and $\eta'(\varkappa)\neq 0,\varkappa\in[0,\infty)$. Then, $\bar{\theta}_1>0,\bar{\theta}_2>0$, we have.

$$\begin{aligned}
&|(\mathbf{I}_{\bar{\theta}_1}1)(\varkappa;\mathcal{P})(\mathbf{I}_{\bar{\theta}_2}\zeta\eta)(\varkappa;\mathcal{P})\\
&+(\mathbf{I}_{\bar{\theta}_2}1)(\varkappa;\mathcal{P})(\mathbf{I}_{\bar{\theta}_1}\zeta\eta)(\varkappa;\mathcal{P})\\
&-(\mathbf{I}_{\bar{\theta}_1}\zeta)(\varkappa;\mathcal{P})(\mathbf{I}_{\bar{\theta}_2}\eta)(\varkappa;\mathcal{P})\\
&-(\mathbf{I}_{\bar{\theta}_2}\zeta)(\varkappa;\mathcal{P})(\mathbf{I}_{\bar{\theta}_1}\eta)(\varkappa;\mathcal{P})|\\
&\le L[(\mathbf{I}_{\bar{\theta}_2}1)(\bar{\varkappa};\mathcal{P})(\mathbf{I}_{\bar{\theta}_1}\eta^2)(\varkappa;\mathcal{P})\\
&+(\mathbf{I}_{\bar{\theta}_1}1)(\varkappa;\mathcal{P})(\mathbf{I}_{\bar{\theta}_2}\eta^2)(\varkappa;\mathcal{P})\\
&-2(\mathbf{I}_{\bar{\theta}_1}\eta)(\varkappa;\mathcal{P})(\mathbf{I}_{\bar{\theta}_2}\eta)(\varkappa;\mathcal{P})].
\end{aligned}\tag{3.15}$$

Proof. By Cauchy mean value theorem for every $\rho,\sigma\in[\bar{c},\varkappa];\rho=\sigma,\ \bar{\varkappa}>0$ there exists $r\in(\rho,\sigma)$ such that

$$\frac{\zeta(\rho)-\zeta(\sigma)}{\eta(\rho)-\eta(\sigma)}=\frac{\zeta'(r)}{\eta'(r)}.$$

Therefore

$$|\zeta(\rho)-\zeta(\sigma)| \leq L|\eta(\rho)-\eta(\sigma)|, \quad \text{for all} \quad \rho,\sigma \in [\bar{c},\varkappa], \quad \varkappa > 0.$$

From above,

$$|(\zeta(\rho)-\zeta(\sigma))(\eta(\rho)-\eta(\sigma))| \\ \leq L(\eta(\rho)-\eta(\sigma))^2, \quad \text{for all} \quad \rho,\sigma \in [\bar{c},\varkappa], \quad \varkappa > 0.$$

That is

$$|B(\rho,\sigma)| \leq L(\eta^2(\rho) - 2\eta(\rho)\eta(\sigma) + \eta^2(\sigma)). \tag{3.16}$$

Multiplying by $(\varkappa-\rho)^{\bar{\theta}_1-1}\boldsymbol{E}_{\bar{\theta}_1}(\omega(\varkappa-\rho)^{\vartheta};\mathcal{P})(\varkappa-\sigma)^{\bar{\theta}_2-1}\boldsymbol{E}_{\bar{\theta}_2}(\omega(\varkappa-\sigma)^{\vartheta};\mathcal{P})$, $\rho,\sigma \in [\bar{c},\varkappa]$ to (3.16) and then integrating w. r. t. ρ and σ from $\bar{c}$ to $\varkappa$ respectively, we obtain

$$\int_{\bar{c}}^{\varkappa}\int_{\bar{c}}^{\varkappa}(\varkappa-\rho)^{\bar{\theta}_1-1}(\varkappa-\sigma)^{\bar{\theta}_2-1}\boldsymbol{E}_{\bar{\theta}_1}(\omega(\varkappa-\rho)^{\vartheta};\mathcal{P}) \\ \boldsymbol{E}_{\bar{\theta}_2}(\omega(\varkappa-\sigma)^{\vartheta};\mathcal{P})\,|B(\rho,\sigma)|\,d\rho d\sigma \\ \leq L\int_{\bar{c}}^{\varkappa}\int_{\bar{c}}^{\varkappa}(\varkappa-\rho)^{\bar{\theta}_1-1}(\varkappa-\sigma)^{\bar{\theta}_2-1}\boldsymbol{E}_{\bar{\theta}_1}(\omega(\varkappa-\rho)^{\vartheta};\mathcal{P}) \\ \boldsymbol{E}_{\bar{\theta}_2}(\omega(\varkappa-\sigma)^{\vartheta};\mathcal{P})(\eta^2(\rho)-2\eta(\rho)\eta(\sigma)+\eta^2(n))d\rho d\sigma.$$

Hence

$$\int_{\bar{c}}^{\varkappa}\int_{\bar{c}}^{\varkappa}(\varkappa-\rho)^{\bar{\theta}_1-1}\boldsymbol{E}_{\bar{\theta}_1}(\omega(\varkappa-\rho)^{\vartheta};\mathcal{P})(\varkappa-\sigma)^{\bar{\theta}_2-1} \\ \boldsymbol{E}_{\bar{\theta}_2}(\omega(\varkappa-\sigma)^{\vartheta};\mathcal{P})\,|B(\rho,\sigma)|\,d\rho d\sigma \\ \leq L[(\boldsymbol{I}_{\bar{\theta}_2}1)(\varkappa;\mathcal{P})(\boldsymbol{I}_{\bar{\theta}_1}\eta^2)(\varkappa;\mathcal{P})+(\boldsymbol{I}_{\bar{\theta}_1}1)(\varkappa;\mathcal{P})(\boldsymbol{I}_{\bar{\theta}_2}\eta^2)(\varkappa;\mathcal{P}) \\ -2(\boldsymbol{I}_{\bar{\theta}_1}\eta)(\varkappa;\mathcal{P})(\boldsymbol{I}_{\bar{\theta}_2}\eta)(\varkappa;\mathcal{P})]. \tag{3.17}$$

From (3.17) we get (3.15). □

Corollary 3.3.3.1 Let the functions ζ and η be differentiable on $[0, \infty)$ with $\left|\frac{\zeta'(\varkappa)}{\eta'(\varkappa)}\right| \leq L$, for some $L > 0$ and $\eta'(\varkappa) \neq 0, \varkappa \in [0,\infty)$. Then for all $\varkappa > \bar{c} \geq 0$, $\bar{\theta}_1 > 0$, we have

$$\begin{aligned}&|(I_{\bar{\theta}_1}1)(\varkappa;p)(I_{\bar{\theta}_1}\zeta\eta)(\varkappa;p)\\&-(I_{\bar{\theta}_1}\zeta)(\varkappa;p)(I_{\bar{\theta}_1}\eta)(\varkappa;p)|\\&\leq L[(I_{\bar{\theta}_1}1)(\varkappa;p)(I_{\bar{\theta}_1}\eta^2)(\varkappa;p)\\&-((I_{\bar{\theta}_1}\eta)(\varkappa;\mathcal{P}))^2].\end{aligned} \tag{3.18}$$

Proof. If we put $\bar{\theta}_1 = \bar{\theta}_2$, then (3.15) reduces to (3.18).

3.4 Fractional Inequalities in the Case of Extended Chebyshev Functional

In this section, we use extended generalized fractional integral operator to prove some Chebyshev type inequalities in the case of differentiable function.

Theorem 3.4.1 Let $\bar{\theta}_1 > 0$ and the functions ζ, η be differentiable on $[0, \infty)$ having the same sense of variation with $\zeta', \eta' \in \mathcal{L}_\infty([0, \infty))$. Let the function u be a positive on $[0, \infty)$, then for all $\varkappa > \bar{c} \geq 0$, following inequalities hold:

$$\begin{aligned}0 &\leq 2[(I_{\bar{\theta}_1}u)(\varkappa;\mathcal{P})(I_{\bar{\theta}_1}u\zeta\eta)(\varkappa;\mathcal{P})\\&-(I_{\bar{\theta}_1}u\zeta)(\varkappa;\mathcal{P})(I_{\bar{\theta}_1}u\eta)(\varkappa;\mathcal{P})]\\&\leq \|\zeta'\|_\infty \|\eta'\|_\infty [(I_{\bar{\theta}_1}u)(\varkappa;\mathcal{P})(I_{\bar{\theta}_1}\rho^2u)(\varkappa;\mathcal{P})\\&-2(I_{\bar{\theta}_1}\rho u)(\varkappa;\mathcal{P})(I_{\bar{\theta}_1}\sigma u)(\varkappa;\mathcal{P})\\&+(I_{\bar{\theta}_1}u)(\varkappa;\mathcal{P})(I_{\bar{\theta}_1}\sigma^2u)(\varkappa;\mathcal{P})].\end{aligned} \tag{3.19}$$

Proof. Since ζ and η have same sense of variation on $[0, \infty)$, we have

$$(\zeta(\rho)-\zeta(\sigma))(\eta(\rho)-\eta(\sigma)\geq 0, \quad \forall\ \rho,\sigma\in(\bar{c},\varkappa).$$

Therefore

$$B(\rho,\sigma)\geq 0. \tag{3.20}$$

Multiplying by $(\varkappa-\rho)^{\bar{\theta}_1-1}E_{\bar{\theta}_1}(\omega(\varkappa-\rho)^{\vartheta};p)\ u(\rho)$, $\rho\in(\bar{c},\varkappa)$ to (3.20) and then integrating w. r. t. ρ from $\bar{c}$ to $\varkappa$, we get

$$\int_{\bar{c}}^{\varkappa}(\varkappa-\rho)^{\bar{\theta}_1-1}E_{\bar{\theta}_1}(\omega(\varkappa-\rho)^{\vartheta};\mathcal{P})u(\rho)B(\rho,\sigma)d\rho$$
$$=(I_{\bar{\theta}_1}u\zeta\eta)(\varkappa;p)-\eta(\sigma)(I_{\bar{\theta}_1}u\zeta)(\varkappa;p) \tag{3.21}$$
$$-\zeta(\sigma)(I_{\bar{\theta}_1}u\eta)(\varkappa;p)+\zeta(\sigma)\eta(\sigma)(I_{\bar{\theta}_1}u)(\varkappa;p)\geq 0.$$

Multiplying by $(\varkappa-\sigma)^{\bar{\theta}_1-1}E_{\bar{\theta}_1}(\omega(\varkappa-\sigma)^{\vartheta};\mathcal{P})\,u(\sigma),\sigma\in(\bar{c},\varkappa)$ to (3.21) and then integrating w. r. t. σ from $\bar{c}$ to $\varkappa$, we obtain

$$\int_{\bar{c}}^{\varkappa}\int_{\bar{c}}^{\varkappa}(\varkappa-\rho)^{\bar{\theta}_1-1}(\varkappa-\sigma)^{\bar{\theta}_1-1}E_{\bar{\theta}_1}(\omega(\varkappa-\rho)^{\vartheta};\mathcal{P})$$
$$E_{\bar{\theta}_1}(\omega(\varkappa-\sigma)^{\vartheta};\mathcal{P})u(\rho)u(\sigma)B(\rho,\sigma)d\rho d\sigma$$
$$=(I_{\bar{\theta}_1}u\zeta\eta)(\varkappa;\mathcal{P})(I_{\bar{\theta}_1}u)(\varkappa;\mathcal{P})$$
$$-(I_{\bar{\theta}_1}u\zeta)(\varkappa;\mathcal{P})(I_{\bar{\theta}_1}u\eta)(\varkappa;\mathcal{P})$$
$$-(I_{\bar{\theta}_1}u\eta)(\varkappa;\mathcal{P})(I_{\bar{\theta}_1}u\zeta)(\varkappa;\mathcal{P})$$
$$+(I_{\bar{\theta}_1}u)(\varkappa;\mathcal{P})(I_{\bar{\theta}_1}u\zeta\eta)(\varkappa;\mathcal{P})\geq 0.$$

This gives

$$\int_{\bar{c}}^{\varkappa}\int_{\bar{c}}^{\varkappa}(\varkappa-\rho)^{\bar{\theta}_1-1}(\varkappa-\sigma)^{\bar{\theta}_1-1}E_{\bar{\theta}_1}(\omega(\varkappa-\rho)^{\vartheta};\mathcal{P})$$
$$E_{\bar{\theta}_1}(\omega(\varkappa-\sigma)^{\vartheta};\mathcal{P})u(\rho)u(\sigma)B(\rho,\sigma))d\rho d\sigma \tag{3.22}$$
$$=2[(I_{\bar{\theta}_1}u\zeta\eta)(\varkappa;P)(I_{\bar{\theta}_1}u)(\varkappa;\mathcal{P})$$
$$-(I_{\bar{\theta}_1}u\zeta)(\varkappa;\mathcal{P})(I_{\bar{\theta}_1}u\eta)(\varkappa;\mathcal{P})]\geq 0.$$

From (3.20), we have

$$B(\rho,\sigma)=\int_{\rho}^{\sigma}\int_{\rho}^{\sigma}\zeta'(m)\eta'(n)dmdn. \tag{3.23}$$

Also $\zeta', \eta' \in \mathcal{L}_{\infty}([0,\infty))$, then we have

$$|B(\rho,\sigma)|\leq\left|\int_{\rho}^{\sigma}\zeta'(m)dm\right|.\left|\int_{\rho}^{\sigma}\eta'(n)dn\right|$$
$$\leq\|\zeta'\|_{\infty}\|\eta'\|_{\infty}(\rho-\sigma)^2. \tag{3.24}$$

Therefore

$$\int_{\bar{c}}^{\varkappa}\int_{\bar{c}}^{\varkappa}(\varkappa-\rho)^{\bar{\theta}_1-1}(\varkappa-\sigma)^{\bar{\theta}_1-1}\boldsymbol{E}_{\bar{\theta}_1}(\omega(\varkappa-\rho)^{\vartheta};\mathcal{P})$$
$$\boldsymbol{E}_{\bar{\theta}_1}(\omega(\varkappa-\sigma)^{\vartheta};\mathcal{P})u(\rho)u(\sigma)B(\rho,\sigma)d\rho d\sigma$$
$$\leq\|\zeta'\|_{\infty}\|\eta'\|_{\infty}\int_{\bar{c}}^{\varkappa}\int_{\bar{c}}^{\varkappa}(\varkappa-m)^{\bar{\theta}_1-1}(\varkappa-n)^{\bar{\theta}_1-1}\boldsymbol{E}_{\bar{\theta}_1}(\omega(\varkappa-\rho)^{\vartheta};\mathcal{P})$$
$$\boldsymbol{E}_{\bar{\theta}_1}(\omega(\varkappa-\sigma)^{\vartheta};\mathcal{P})(\rho^2-2\rho\sigma+\sigma^2)u(\rho)u(\sigma)d\rho d\sigma.$$

Consequently

$$\int_{\bar{c}}^{\varkappa}\int_{\bar{c}}^{\varkappa}(\varkappa-\rho)^{\bar{\theta}_1-1}(\varkappa-\sigma)^{\bar{\theta}_1-1}\boldsymbol{E}_{\bar{\theta}_1}(\omega(\varkappa-\rho)^{\vartheta};\mathcal{P})$$
$$\boldsymbol{E}_{\bar{\theta}_1}(\omega(\varkappa-\sigma)^{\vartheta};\mathcal{P})u(\rho)u(\sigma)B(\rho,\sigma)d\rho d\sigma$$
$$\leq\|\zeta'\|_{\infty}\|\eta'\|_{\infty}[(\boldsymbol{I}_{\bar{\theta}_1}u)(\varkappa;\mathcal{P})(\boldsymbol{I}_{\bar{\theta}_1}\rho^2u)(\varkappa;\mathcal{P}) \tag{3.25}$$
$$-2(\boldsymbol{I}_{\bar{\theta}_1}\rho u)(\varkappa;\mathcal{P})(\boldsymbol{I}_{\bar{\theta}_1}\sigma u)(\varkappa;\mathcal{P})$$
$$+(\boldsymbol{I}_{\bar{\theta}_1}u)(\varkappa;\mathcal{P})(\boldsymbol{I}_{\bar{\theta}_1}\sigma^2u)(\varkappa;\mathcal{P})].$$

From (3.24) and (3.25) we get the required inequality (3.19). □

In our next result, we use extended generalized fractional integral operator to prove chebyshev inequalities for differentiable functions in the case of two positive parameters.

Theorem 3.4.2 Let $\bar{\theta}_1$, $\bar{\theta}_2>0$ and the functions ζ, η be differentiable on $[0,\infty)$ having the same sense of variation with $\zeta',\eta'\in\mathcal{L}_{\infty}([0,\infty))$. Let u be a positive function on $[0,\infty)$, then

$$0\leq(\boldsymbol{I}_{\bar{\theta}_1}u)(\varkappa;p)(\boldsymbol{I}_{\bar{\theta}_2}u\zeta\eta)(\varkappa;p)$$
$$+(\boldsymbol{I}_{\bar{\theta}_2}u)(\varkappa;p)(\boldsymbol{I}_{\bar{\theta}_1}u\zeta\eta)(\varkappa;p)$$
$$-(\boldsymbol{I}_{\bar{\theta}_1}u\zeta)(\varkappa;p)(\boldsymbol{I}_{\bar{\theta}_2}u\eta)(\varkappa;p)$$
$$-(\boldsymbol{I}_{\bar{\theta}_2}u\zeta)(\varkappa;p)(\boldsymbol{I}_{\bar{\theta}_1}u\eta)(\varkappa;p)$$
$$\leq\|\zeta'\|_{\infty}\|\eta'\|_{\infty}[(\boldsymbol{I}_{\bar{\theta}_2}u)(\varkappa;p)(\boldsymbol{I}_{\bar{\theta}_1}u)(\varkappa;p) \tag{3.26}$$
$$-2(\boldsymbol{I}_{\bar{\theta}_1}\rho u)(\varkappa;p)(\boldsymbol{I}_{\bar{\theta}_2}\sigma u)(\varkappa;p)$$
$$+(\boldsymbol{I}_{\bar{\theta}_1}u)(\varkappa;p)(\boldsymbol{I}_{\bar{\theta}_2}\sigma^2u)(\varkappa;p)].$$

Proof. By the same fashion of Theorem (3.19), we have

$$\int_{\bar{c}}^{\varkappa}(\varkappa-\rho)^{\bar{\theta}_1-1}\boldsymbol{E}_{\bar{\theta}_1}(\omega(\varkappa-\rho)^{\vartheta};\mathcal{P})u(\rho)B(\rho,\sigma)d\rho$$
$$=(\boldsymbol{I}_{\bar{\theta}_1}u\zeta\eta)(\varkappa;\mathcal{P})-\eta(n)(\boldsymbol{I}_{\bar{\theta}_1}u\zeta)(\varkappa;\mathcal{P}) \tag{3.27}$$
$$-\zeta(n)(\boldsymbol{I}_{\bar{\theta}_1}u\eta)(\varkappa;\mathcal{P})+\zeta(n)\eta(n)(\boldsymbol{I}_{\bar{\theta}_1}u)(\varkappa;\mathcal{P})\geq 0.$$

Multiplying by $(\varkappa-\sigma)^{\bar{\theta}_2-1}\boldsymbol{E}_{\bar{\theta}_2}(\omega(\varkappa-\sigma)^{\vartheta};\mathcal{P})\,u(\sigma),\sigma\in(\bar{c},\varkappa)$ to (3.27) and then integrating w.r.t. σ from $\bar{c}$ to $\varkappa$, we obtain

$$\int_{\bar{c}}^{\varkappa}\int_{\bar{c}}^{\varkappa}(\varkappa-\rho)^{\bar{\theta}_1-1}(\varkappa-\sigma)^{\bar{\theta}_2-1}\boldsymbol{E}_{\bar{\theta}_1}(\omega(\varkappa-\rho)^{\vartheta};\mathcal{P})$$
$$\boldsymbol{E}_{\bar{\theta}_2}(\omega(\varkappa-\sigma)^{\vartheta};\mathcal{P})u(\rho)u(\sigma)B(\rho,\sigma)d\rho d\sigma$$
$$=(\boldsymbol{I}_{\bar{\theta}_1}u\zeta\eta)(\varkappa;\mathcal{P})(\boldsymbol{I}_{\bar{\theta}_2}u)(\varkappa;\mathcal{P}) \tag{3.28}$$
$$-(\boldsymbol{I}_{\bar{\theta}_1}u\zeta)(\varkappa;\mathcal{P})(\boldsymbol{I}_{\bar{\theta}_2}\eta)(\varkappa;\mathcal{P})$$
$$-(\boldsymbol{I}_{\bar{\theta}_1}u\eta)(\varkappa;\mathcal{P})(\boldsymbol{I}_{\bar{\theta}_2}\zeta)(\varkappa;\mathcal{P})$$
$$+(\boldsymbol{I}_{\bar{\theta}_1}u)(\varkappa;\mathcal{P})(\boldsymbol{I}_{\bar{\theta}_2}\zeta\eta)(\varkappa;\mathcal{P})\geq 0.$$

Now from (3.23) and (3.24), we can write

$$\int_{\bar{c}}^{\varkappa}\int_{\bar{c}}^{\varkappa}(\varkappa-\rho)^{\bar{\theta}_1-1}(\varkappa-\sigma)^{\bar{\theta}_2-1}\boldsymbol{E}_{\bar{\theta}_1}(\omega(\varkappa-\rho)^{\vartheta};\mathcal{P})$$
$$\boldsymbol{E}_{\bar{\theta}_2}(\omega(\varkappa-\sigma)^{\vartheta};\mathcal{P})u(\rho)u(\sigma)B(\rho,\sigma)d\rho d\sigma$$
$$\leq\|\zeta'\|_{\infty}\|\eta'\|_{\infty}\int_{\bar{c}}^{\varkappa}\int_{\bar{c}}^{\varkappa}(\varkappa-\rho)^{\bar{\theta}_1-1}(\varkappa-\sigma)^{\bar{\theta}_2-1}$$
$$\boldsymbol{E}_{\bar{\theta}_1}(\omega(\varkappa-\rho)^{\vartheta};\mathcal{P})\boldsymbol{E}_{\bar{\theta}_2}(\omega(\varkappa-\sigma)^{\vartheta};\mathcal{P})$$
$$(\rho^2-2\rho\sigma+\sigma^2)u(\rho)u(\sigma)d\rho d\sigma.$$

Then

$$\int_{c}^{\varkappa}\int_{c}^{\varkappa}(\varkappa-\rho)^{\bar{\theta}_1-1}(\varkappa-\sigma)^{\bar{\theta}_2-1}\boldsymbol{E}_{\bar{\theta}_1}(\omega(\varkappa-\rho)^{\vartheta};\mathcal{P})$$
$$\boldsymbol{E}_{\bar{\theta}_2}(\omega(\varkappa-\sigma)^{\vartheta};\mathcal{P})u(\rho)u(\sigma)B(\rho,\sigma)d\rho d\sigma$$
$$\leq\|\zeta'\|_{\infty}\|\eta'\|_{\infty}[(\boldsymbol{I}_{\bar{\theta}_2}u)(\varkappa;\mathcal{P})(\boldsymbol{I}_{\bar{\theta}_1}\rho^2u)(\varkappa;\mathcal{P})$$
$$-2(\boldsymbol{I}_{\bar{\theta}_1}\rho u)(\varkappa;\mathcal{P})(\boldsymbol{I}_{\bar{\theta}_2}\sigma u)(\varkappa;\mathcal{P})$$
$$+(\boldsymbol{I}_{\bar{\theta}_1}u)(\varkappa;\mathcal{P})(\boldsymbol{I}_{\bar{\theta}_2}\sigma^2u)(\varkappa;\mathcal{P})]. \tag{3.29}$$

From inequality (3.28) and (3.29) we get (3.26). □

Corollary 3.4.2.1 If we put $\bar{\theta}_1=\bar{\theta}_2$, then inequality (3.26) reduces to (3.19).

3.5 Some Other Fracional Inequalities Related to the Extended Chebyshev Functional

In this section, we use extended generalized fractional integral operator to prove some new inequalities in the case of integrable functions having some specific condition and lipschitzian function.

Theorem 3.5.1 Let $\bar{\theta}_1>0$. Let the functions ζ and η be integrable on $[0,\infty)$ satisfying:

$$\mathcal{W}_1\leq\zeta(m)\leq\mathcal{W}_2,\quad \mathcal{Q}_1\leq\eta(m)\leq\mathcal{Q}_2,$$

for all $\mathcal{W}_1,\mathcal{W}_2,\mathcal{Q}_1,\mathcal{Q}_2\in\mathbb{R}$, $m\in[0,\infty)$. Let u > 0, v > 0 be any two functions defined on $[0,\infty)$. Then

$$|(\boldsymbol{I}_{\bar{\theta}_1}1)(\varkappa;p)(\boldsymbol{I}_{\bar{\theta}_1}u\zeta\eta)(\varkappa;p)$$
$$+(\boldsymbol{I}_{\bar{\theta}_1}u)(\varkappa;p)(\boldsymbol{I}_{\bar{\theta}_1}v\zeta\eta)(\varkappa;p)$$
$$-(\boldsymbol{I}_{\bar{\theta}_1}u\zeta)(\varkappa;p)(\boldsymbol{I}_{\bar{\theta}_1}v\eta)(\varkappa;p) \tag{3.30}$$
$$-(\boldsymbol{I}_{\bar{\theta}_1}v\zeta)(\varkappa;p)(\boldsymbol{I}_{\bar{\theta}_1}u\eta)(\varkappa;p)|$$
$$\leq(\boldsymbol{I}_{\bar{\theta}_1}u)(\varkappa;p)(\boldsymbol{I}_{\bar{\theta}_1}v)(\varkappa;p)(\mathcal{W}_2-\mathcal{W}_1)(\mathcal{Q}_2-\mathcal{Q}_1).$$

Proof. From given condition we can state following

$$\left|\zeta(\rho)-\zeta(\sigma)\right| \leq \mathcal{W}_2-\mathcal{W}_1,$$
$$\left|\eta(\rho)-\eta(\sigma)\right| \leq \mathcal{Q}_2-\mathcal{Q}_1, \quad \forall \rho,\sigma \in (\bar{c},\varkappa).$$

Therefore

$$\left|(\zeta(\rho)-\zeta(\sigma))(\eta(\rho)-\eta(\sigma))\right| \leq (\mathcal{W}_2-\mathcal{W}_1)(\mathcal{Q}_2-\mathcal{Q}_1),$$
$$\left|\zeta(\rho)\eta(\rho)+\zeta(\sigma)\eta(\sigma)-\zeta(\rho)\eta(\sigma)-\zeta(\sigma)\eta(\rho)\right| \leq (\mathcal{W}_2-\mathcal{W}_1)(\mathcal{Q}_2-\mathcal{Q}_1).$$

Define

$$B(\rho,\sigma) := \zeta(\rho)\eta(\rho) + \zeta(\sigma)\eta(\sigma) - \zeta(\rho)\eta(\sigma) - \zeta(\sigma)\eta(\rho), \tag{3.31}$$

for all $\rho,\sigma \in (\bar{c},\varkappa)$.

Multiplying by $(\varkappa-\rho)^{\bar{\theta}_1-1}\boldsymbol{E}_{\bar{\theta}_1}(\omega(\varkappa-\rho)^{\vartheta};\mathcal{P})\, u(\rho)$, $\rho \in (\bar{c},\varkappa)$ to (3.31) and then integrating w. r. t. ρ from $\bar{c}$ to $\varkappa$, we get

$$\begin{aligned}
&\int_{\bar{c}}^{\varkappa} (\varkappa-\rho)^{\bar{\theta}_1-1}\boldsymbol{E}_{\bar{\theta}_1}(\omega(\varkappa-\rho)^{\vartheta};\mathcal{P})u(\rho)B(\rho,\sigma)d\rho \\
&= \int_{\bar{c}}^{\varkappa} (\varkappa-\rho)^{\bar{\theta}_1-1}\boldsymbol{E}_{\bar{\theta}_1}(\omega(\varkappa-\rho)^{\vartheta};\mathcal{P})u(\rho)\zeta(\rho)\eta(\rho)d\rho \\
&+\zeta(\sigma)\eta(\sigma)\int_{\bar{c}}^{\varkappa} (\varkappa-\rho)^{\bar{\theta}_1-1}\boldsymbol{E}_{\bar{\theta}_1}(\omega(\varkappa-\rho)^{\vartheta};\mathcal{P})u(\rho)d\rho \\
&-\eta(\sigma)\int_{\bar{c}}^{\varkappa} (\varkappa-\rho)^{\bar{\theta}_1-1}\boldsymbol{E}_{\bar{\theta}_1}(\omega(\varkappa-\rho)^{\vartheta};\mathcal{P})u(\rho)\zeta(\rho)d\rho \\
&-\zeta(\sigma)\int_{\bar{c}}^{\varkappa} (\varkappa-\rho)^{\bar{\theta}_1-1}\boldsymbol{E}_{\bar{\theta}_1}(\omega(\varkappa-\rho)^{\vartheta};\mathcal{P})u(\rho)\eta(\rho)d\rho,
\end{aligned}$$

which can be written as

$$\begin{aligned}
&\int_{\bar{c}}^{\varkappa} (\varkappa-\rho)^{\bar{\theta}_1-1}\boldsymbol{E}_{\bar{\theta}_1}(\omega(\varkappa-\rho)^{\vartheta};\mathcal{P})u(\rho)B(\rho,\sigma)d\rho \\
&\quad = (\boldsymbol{I}_{\bar{\theta}_1}u\zeta\eta)(\varkappa;\mathcal{P}) + \zeta(\sigma)\eta(\sigma)(\boldsymbol{I}_{\bar{\theta}_1}u)(\varkappa;\mathcal{P}) \\
&\quad -\eta(\sigma)(\boldsymbol{I}_{\bar{\theta}_1}u\zeta)(\varkappa;\mathcal{P}) - \zeta(\sigma)(\boldsymbol{I}_{\bar{\theta}_1}u\eta)(\varkappa;\mathcal{P}).
\end{aligned} \tag{3.32}$$

Multiplying by $(\varkappa-\sigma)^{\bar{\theta}_1-1}\boldsymbol{E}_{\bar{\theta}_1}(\omega(\varkappa-\sigma)^{\vartheta};\mathcal{P})\,v(\sigma),\ \sigma\in(\bar{c},\varkappa)$ to (3.32) and then integrating w. r. t. σ from $\bar{c}$ to $\varkappa$, we get

$$\int_{\bar{c}}^{\varkappa}\int_{\bar{c}}^{\varkappa}(\varkappa-\rho)^{\bar{\theta}_1-1}(\varkappa-\sigma)^{\bar{\theta}_1-1}\boldsymbol{E}_{\bar{\theta}_1}(\omega(\varkappa-\rho)^{\vartheta};\mathcal{P})$$
$$\boldsymbol{E}_{\bar{\theta}_1}(\omega(\varkappa-\rho)^{\vartheta};\mathcal{P})u(\rho)v(\sigma)B(\rho,\sigma)d\rho d\sigma$$
$$=(\boldsymbol{I}_{\bar{\theta}_1}u\zeta\eta)(\varkappa;\mathcal{P})\int_{\bar{c}}^{\varkappa}(\varkappa-\rho)^{\bar{\theta}_1-1}\boldsymbol{E}_{\bar{\theta}_1}(\omega(\varkappa-\rho)^{\vartheta};\mathcal{P})v(\sigma)d\sigma$$
$$+(\boldsymbol{I}_{\bar{\theta}_1}u)(\varkappa;\mathcal{P})\int_{\bar{c}}^{\varkappa}(\varkappa-\rho)^{\bar{\theta}_1-1}\boldsymbol{E}_{\bar{\theta}_1}(\omega(\varkappa-\rho)^{\vartheta};\mathcal{P})v(\sigma)\zeta(\sigma)\eta(\sigma)d\sigma$$
$$-(\boldsymbol{I}_{\bar{\theta}_1}u\zeta)(\varkappa;\mathcal{P})\int_{\bar{c}}^{\varkappa}(\varkappa-\rho)^{\bar{\theta}_1-1}\boldsymbol{E}_{\bar{\theta}_1}(\omega(\varkappa-\rho)^{\vartheta};\mathcal{P})v(\sigma)\eta(\sigma)d\sigma$$
$$-(\boldsymbol{I}_{\bar{\theta}_1}u\eta)(\varkappa;\mathcal{P})\int_{\bar{c}}^{\varkappa}(\varkappa-\rho)^{\bar{\theta}_1-1}\boldsymbol{E}_{\bar{\theta}_1}(\omega(\varkappa-\rho)^{\vartheta};\mathcal{P})v(\sigma)\zeta(\sigma)d\sigma.$$

From above

$$\begin{aligned}&\int_{\bar{c}}^{\varkappa}\int_{\bar{c}}^{\varkappa}(\varkappa-\rho)^{\bar{\theta}_1-1}(\varkappa-\sigma)^{\bar{\theta}_1-1}\boldsymbol{E}_{\bar{\theta}_1}(\omega(\varkappa-\rho)^{\vartheta};\mathcal{P})\\&\boldsymbol{E}_{\bar{\theta}_1}(\omega(\varkappa-\sigma)^{\vartheta};\mathcal{P})u(\rho)v(\sigma)B(\rho,\sigma)d\rho d\sigma\\&=(\boldsymbol{I}_{\bar{\theta}_1}u\zeta\eta)(\varkappa;\mathcal{P})(\boldsymbol{I}_{\bar{\theta}_1}v)(\varkappa;\mathcal{P})\\&+(\boldsymbol{I}_{\bar{\theta}_1}u)(\varkappa;\mathcal{P})(\boldsymbol{I}_{\bar{\theta}_1}v\zeta\eta)(\varkappa;\mathcal{P})\\&-(\boldsymbol{I}_{\bar{\theta}_1}u\zeta)(\varkappa;\mathcal{P})(\boldsymbol{I}_{\bar{\theta}_1}v\eta)(\varkappa;\mathcal{P})\\&-(\boldsymbol{I}_{\bar{\theta}_1}u\eta)(\varkappa;\mathcal{P})(\boldsymbol{I}_{\bar{\theta}_1}v\zeta)(\varkappa;\mathcal{P}).\end{aligned}\tag{3.33}$$

By using (3.31), we can estimate (3.33) as follows

$$\left|\int_{\bar{c}}^{\varkappa}\int_{\bar{c}}^{\varkappa}(\varkappa-\rho)^{\bar{\theta}_1-1}(\varkappa-\sigma)^{\bar{\theta}_1-1}\boldsymbol{E}_{\bar{\theta}_1}(\omega(\varkappa-\rho)^{\vartheta};\mathcal{P})\right.$$
$$\left.\boldsymbol{E}_{\bar{\theta}_1}(\omega(\varkappa-\sigma)^{\vartheta};\mathcal{P})u(\rho)v(\sigma)B(\rho,\sigma)d\rho d\sigma\right|$$
$$\leq(\mathcal{W}_2-\mathcal{W}_1)(\mathcal{Q}_2-\mathcal{Q}_1)\int_{\bar{c}}^{\varkappa}\int_{\bar{c}}^{\varkappa}(\varkappa-\rho)^{\bar{\theta}_1-1}(\varkappa-\sigma)^{\bar{\theta}_1-1}$$
$$\boldsymbol{E}_{\bar{\theta}_1}(\omega(\varkappa-\rho)^{\vartheta};\mathcal{P})\boldsymbol{E}_{\bar{\theta}_1}(\omega(\varkappa-\sigma)^{\vartheta};\mathcal{P})u(\rho)v(\sigma)d\rho d\sigma.$$

Consequently

$$|\int_{\bar{c}}^{\varkappa}\int_{\bar{c}}^{\varkappa}(\varkappa-\rho)^{\bar{\theta}_1-1}(\varkappa-\sigma)^{\bar{\theta}_1-1}\boldsymbol{E}_{\bar{\theta}_1}(\omega(\varkappa-\rho)^{\vartheta};\mathcal{P})$$
$$\boldsymbol{E}_{\bar{\theta}_1}(\omega(\varkappa-\sigma)^{\vartheta};\mathcal{P})u(\rho)v(\sigma)B(\rho,\sigma)d\rho d\sigma|$$
$$\leq(\boldsymbol{I}_{\bar{\theta}_1}u)(\varkappa;\mathcal{P})(\boldsymbol{I}_{\bar{\theta}_1}v)(\varkappa;\mathcal{P})((\mathcal{W}_2-\mathcal{W}_1)(\mathcal{Q}_2-\mathcal{Q}_1)).$$

From above we get required inequality (3.30). □

Theorem 3.5.2 Let $\bar{\theta}_1>0$ and the functions ζ and η be lipschitzian with the constants $K>0$, $L>0$ on $[0,\infty)$ and suppose that u and v defined as in Theorem 3.5.1. Then

$$\begin{aligned}
&|(\boldsymbol{I}_{\bar{\theta}_1}v)(\varkappa;p)(\boldsymbol{I}_{\bar{\theta}_1}u\zeta\eta)(\varkappa;p)\\
&+(\boldsymbol{I}_{\bar{\theta}_1}u)(\varkappa;p)(\boldsymbol{I}_{\bar{\theta}_1}v\zeta\eta)(\varkappa;p)\\
&-(\boldsymbol{I}_{\bar{\theta}_1}u\zeta)(\varkappa;p)(\boldsymbol{I}_{\bar{\theta}_1}v\eta)(\varkappa;p)\\
&-(\boldsymbol{I}_{\bar{\theta}_1}v\zeta)(\varkappa;p)(\boldsymbol{I}_{\bar{\theta}_1}u\eta)(\varkappa;p)|\\
&\leq KL[(\boldsymbol{I}_{\bar{\theta}_1}\rho^2u)(\varkappa;p)(\boldsymbol{I}_{\bar{\theta}_1}v)(\varkappa;p)\\
&-2(\boldsymbol{I}_{\bar{\theta}_1}\rho u)(\varkappa;p)(\boldsymbol{I}_{\bar{\theta}_1}\sigma v)(\varkappa;p)\\
&+(\boldsymbol{I}_{\bar{\theta}_1}u)(\varkappa;p)(\boldsymbol{I}_{\bar{\theta}_1}\sigma^2v)(\varkappa;p)],\quad\forall\,\varkappa>\bar{c}\geq 0.
\end{aligned}\tag{3.34}$$

Proof. Since ζ and η are lipschitzian functions, we have

$$|\zeta(\rho)-\zeta(\sigma)|\leq K|\rho-\sigma|,\ |\zeta(\rho)-\eta(\sigma)|\leq L|\rho-\sigma|.\tag{3.35}$$

Therefore

$$|(\zeta(\rho)-\zeta(\sigma))(\eta(\rho)-\eta(\sigma))|\leq KL(\rho-\sigma)^2.$$

From (3.31), we can write

$$|B(\rho,\sigma)|\leq KL(\rho-\sigma)^2\tag{3.36}$$

Setting

$$R(\rho,\sigma):=KL(\rho-\sigma)^2.\tag{3.37}$$

Multiplying by $(\varkappa-\rho)^{\bar{\theta}_1-1}\boldsymbol{E}_{\bar{\theta}_1}(\omega(\varkappa-\rho)^{\vartheta};\mathcal{P})\,u(\rho),\quad \rho\in(\bar{c},\varkappa)$ to (3.37) and then integrating w. r. t. ρ from $\bar{c}$ to $\varkappa$, we obtain

$$\int_{\bar{c}}^{\varkappa}(\varkappa-\rho)^{\bar{\theta}_1-1}\boldsymbol{E}_{\bar{\theta}_1}(\omega(\varkappa-\sigma)^{\vartheta};\mathcal{P})u(\rho)R(\rho,\sigma)d\rho$$
$$=KL[(\boldsymbol{I}_{\bar{\theta}_1}\rho^2u)(\varkappa;\mathcal{P})-2\sigma(\boldsymbol{I}_{\bar{\theta}_1}\rho u)(\varkappa;\mathcal{P})+\sigma^2(\boldsymbol{I}_{\bar{\theta}_1}u)(\varkappa;\mathcal{P})]. \tag{3.38}$$

Multiplying by $(\varkappa-\sigma)^{\bar{\theta}_1-1}\boldsymbol{E}_{\bar{\theta}_1}(\omega(\varkappa-\sigma)^{\vartheta};\mathcal{P})\,v(\sigma),\quad \sigma\in(\bar{c},\varkappa)$ to (3.38) and then integrating w. r. t. σ from $\bar{c}$ to $\varkappa$, we get

$$\begin{aligned}\int_{\bar{c}}^{\varkappa}\int_{\bar{c}}^{\varkappa}&(\varkappa-\rho)^{\bar{\theta}_1-1}(\varkappa-\sigma)^{\bar{\theta}_1-1}\boldsymbol{E}_{\bar{\theta}_1}(\omega(\varkappa-\rho)^{\vartheta};\mathcal{P})\\&\boldsymbol{E}_{\bar{\theta}_1}(\omega(\varkappa-\sigma)^{\vartheta};\mathcal{P})u(\rho)v(\sigma)R(\rho,\sigma)d\rho\\&=KL[(\boldsymbol{I}_{\bar{\theta}_1}\rho^2u)(\varkappa;\mathcal{P})(\boldsymbol{I}_{\bar{\theta}_1}v)(\varkappa;\mathcal{P})\\&-2(\boldsymbol{I}_{\bar{\theta}_1}\rho u)(\varkappa;\mathcal{P})(\boldsymbol{I}_{\bar{\theta}_1}\sigma v)(\varkappa;\mathcal{P})\\&+(\boldsymbol{I}_{\bar{\theta}_1}u)(\varkappa;\mathcal{P})(\boldsymbol{I}_{\bar{\theta}_1}\sigma^2v)(\varkappa;\mathcal{P})].\end{aligned} \tag{3.39}$$

From (3.36) and (3.39) we get (3.34). □

Theorem 3.5.3 Let $\bar{\theta}_1>0,\ \bar{\theta}_2>0$ and the functions ζ and η be an integrable such that:

$$\mathcal{W}_1\le\zeta(m)\le\mathcal{W}_2,\quad \mathcal{Q}_1\le\eta(m)\le\mathcal{Q}_2,$$

for all $\mathcal{W}_2,\mathcal{W}_1,\mathcal{Q}_2,\mathcal{Q}_1\in\mathbb{R}$, $m\in[0,\infty)$ and suppose that u and v defined as in Theorem (3.5.1). Then

$$\begin{aligned}&|(\boldsymbol{I}_{\bar{\theta}_2}v)(\varkappa;p)(\boldsymbol{I}_{\bar{\theta}_1}u\zeta\eta)(\varkappa;p)\\&+(\boldsymbol{I}_{\bar{\theta}_1}u)(\varkappa;p)(\boldsymbol{I}_{\bar{\theta}_2}v\zeta\eta)(\varkappa;p)\\&-(\boldsymbol{I}_{\bar{\theta}_1}u\zeta)(\varkappa;p)(\boldsymbol{I}_{\bar{\theta}_2}v\eta)(\varkappa;p)\\&-(\boldsymbol{I}_{\bar{\theta}_2}v\zeta)(\varkappa;p)(\boldsymbol{I}_{\bar{\theta}_1}u\eta)(\varkappa;p)|\\&\le(\boldsymbol{I}_{\bar{\theta}_1}u)(\varkappa;p)(\boldsymbol{I}_{\bar{\theta}_2}v)(\varkappa;p)(\mathcal{W}_2-\mathcal{W}_1)(\mathcal{Q}_2-\mathcal{Q}_1).\end{aligned} \tag{3.40}$$

Proof. From the given condition we can state the following

$$|\zeta(\rho)-\zeta(\sigma)|\leq \mathcal{W}_2-\mathcal{W}_1,\quad |\eta(\rho)-\eta(\sigma)|\leq \mathcal{Q}_2-\mathcal{Q}_1.$$

Therefore

$$|B(\rho,\sigma)|\leq(\mathcal{W}_2-\mathcal{W}_1)\,(\mathcal{Q}_2-\mathcal{Q}_1). \tag{3.41}$$

Multiplying by $(\varkappa-\sigma)^{\bar{\theta}_2-1}\boldsymbol{E}_{\bar{\theta}_2}(\omega(\varkappa-\sigma)^{\vartheta};\mathcal{P})\,v(\sigma),\quad \sigma\in(\bar{c},\varkappa)$ to (3.41) and then integrating w. r. t. σ from $\bar{c}$ to $\varkappa$, we obtain

$$\begin{aligned}\int_{\bar{c}}^{\varkappa}\int_{\bar{c}}^{\varkappa}&(\varkappa-\rho)^{\bar{\theta}_1-1}(\varkappa-\sigma)^{\bar{\theta}_2-1}\boldsymbol{E}_{\bar{\theta}_1}(\omega(\varkappa-\rho)^{\vartheta};\mathcal{P})\\ &\boldsymbol{E}_{\bar{\theta}_2}(\omega(\varkappa-\sigma)^{\vartheta};\mathcal{P})u(\rho)v(\sigma)B(\rho,\sigma)d\rho d\sigma\\ &=(\boldsymbol{I}_{\bar{\theta}_1}u\zeta\eta)(\varkappa;\mathcal{P})(\boldsymbol{I}_{\bar{\theta}_2}v)(\varkappa;\mathcal{P})\\ &+(\boldsymbol{I}_{\bar{\theta}_1}u)(\varkappa;\mathcal{P})(\boldsymbol{I}_{\bar{\theta}_2}v\zeta\eta)(\varkappa;\mathcal{P})\\ &-(\boldsymbol{I}_{\bar{\theta}_1}u\zeta)(\varkappa;\mathcal{P})(\boldsymbol{I}_{\bar{\theta}_2}v\eta)(\varkappa;\mathcal{P})\\ &-(\boldsymbol{I}_{\bar{\theta}_1}u\eta)(\varkappa;\mathcal{P})(\boldsymbol{I}_{\bar{\theta}_2}v\zeta)(\varkappa;\mathcal{P}).\end{aligned} \tag{3.42}$$

Also,

$$\begin{aligned}(\mathcal{W}_2-\mathcal{W}_1)(\mathcal{Q}_2-\mathcal{Q}_1)&\int_{\bar{c}}^{\varkappa}\int_{\bar{c}}^{\varkappa}(\varkappa-\rho)^{\bar{\theta}_1-1}(\varkappa-\sigma)^{\bar{\theta}_2-1}\\ &\boldsymbol{E}_{\bar{\theta}_1}(\omega(\varkappa-\sigma)^{\vartheta};\mathcal{P})\boldsymbol{E}_{\bar{\theta}_2}(\omega(\varkappa-\sigma)^{\vartheta};\mathcal{P})\\ &u(\rho)v(\sigma)B(\rho,\sigma)d\rho d\sigma\\ =(\boldsymbol{I}_{\bar{\theta}_1}u)(\varkappa;\mathcal{P})&(\boldsymbol{I}_{\bar{\theta}_2}v)(\varkappa;\mathcal{P})(\mathcal{W}_2-\mathcal{W}_1)(\mathcal{Q}_2-\mathcal{Q}_1).\end{aligned}$$

Finally, from (3.41), (3.42) and (3.43) we get (3.40). □

Corollary 3.5.3.1 If we put $\bar{\theta}_1=\bar{\theta}_2$, then the inequality (3.40) reduces to the inequality (3.30).

Theorem 3.5.4 Let $\bar{\theta}_1>0,\ \bar{\theta}_2>0$ and the functions ζ and η be lipschitzian the with constants $K>0$, $L>0$ on $[0,\infty)$ and suppose that u and v defined as in Theorem (3.5.1). Then

$$\begin{aligned}&|(\boldsymbol{I}_{\bar{\theta}_2}v)(\varkappa;p)(\boldsymbol{I}_{\bar{\theta}_1}u\zeta\eta)(\varkappa;p)\\&+(\boldsymbol{I}_{\bar{\theta}_1}u)(\varkappa;p)(\boldsymbol{I}_{\bar{\theta}_2}v\zeta\eta)(\varkappa;p)\\&-(\boldsymbol{I}_{\bar{\theta}_1}u\zeta)(\varkappa;p)(\boldsymbol{I}_{\bar{\theta}_2}v\eta)(\varkappa;p)\\&-(\boldsymbol{I}_{\bar{\theta}_2}v\zeta)(\varkappa;p)(\boldsymbol{I}_{\bar{\theta}_1}u\eta)(\varkappa;p)|\\&\le KL[(\boldsymbol{I}_{\bar{\theta}_1}\rho^2u)(\varkappa;p)(\boldsymbol{I}_{\bar{\theta}_2}\,v)(\varkappa;p)\\&-2(\boldsymbol{I}_{\bar{\theta}_1}\rho u)(\varkappa;p)(\boldsymbol{I}_{\bar{\theta}_2}\sigma v)(\varkappa;p)\\&+(\boldsymbol{I}_{\bar{\theta}_1}u)(\varkappa;p)(\boldsymbol{I}_{\bar{\theta}_2}\sigma^2v)(\varkappa;p)].\end{aligned}\tag{3.44}$$

Proof. Multiplying by $(\varkappa-\rho)^{\bar{\theta}_1-1}(\varkappa-\sigma)^{\bar{\theta}_2-1}\boldsymbol{E}_{\bar{\theta}_1}(\omega(\varkappa-\rho)^{\vartheta};\mathcal{P})$ $\boldsymbol{E}_{\bar{\theta}_2}(\omega(\varkappa-\sigma)^{\vartheta};\mathcal{P})u(\rho)v(\sigma),\rho,\sigma\in(\bar{c},\varkappa)$ to (3.36) and then integrating w. r. t. ρ and σ respectively over $(\bar{c},\varkappa)^2$, we get

$$\begin{aligned}&\int_{\bar{c}}^{\varkappa}\int_{\bar{c}}^{\varkappa}(\varkappa-\rho)^{\bar{\theta}_1-1}(\varkappa-\sigma)^{\bar{\theta}_2-1}\boldsymbol{E}_{\bar{\theta}_1}(\omega(\varkappa-\rho)^{\vartheta};\mathcal{P})\\&\quad\boldsymbol{E}_{\bar{\theta}_2}(\omega(\varkappa-\sigma)^{\vartheta};\mathcal{P})u(\rho)v(\sigma)\,|B(\rho,\sigma)|\,d\rho d\sigma\\&\le KL\int_{\bar{c}}^{\varkappa}\int_{\bar{c}}^{\varkappa}(\varkappa-\rho)^{\bar{\theta}_1-1}(\varkappa-\sigma)^{\bar{\theta}_2-1}\boldsymbol{E}_{\bar{\theta}_1}(\omega(\varkappa-\rho)^{\vartheta};\mathcal{P})\\&\quad\boldsymbol{E}_{\bar{\theta}_2}(\omega(\varkappa-\sigma)^{\vartheta};\mathcal{P})u(\rho)v(\sigma)(\rho-\sigma)^2d\rho d\sigma.\end{aligned}\tag{3.45}$$

From (3.45) we get the required inequality (3.44). □

Corollary 3.5.4.1 If we put $\bar{\theta}_1=\bar{\theta}_2$, then the inequality (3.44) reduces to (3.34).

3.6 Concluding Remark

By choosing different values of the parameters, the extended generalized fractional integral operator used in the present article reduces to the existing fractional integral operators and consequently, we obtain Chebyshev inequalities for these operators. For example, if we put (i) $\mathcal{P}=\omega=0$, we get inequalities for the Riemann-Liouville fractional integral [11], |(ii) $\mathcal{P}=0$ and $\imath=\kappa=\lambda=1$, we get inequalities for fractional integral established by Prabhakar in [18], (iii) $\mathcal{P}=0$ and $\imath=\kappa=1$, we get inequalities for Srivasatva-Tomovski fractional integral [26]. There are some articles that carried out a brief study of Chebyshev type inequalities via various fractional integral operators and one of the objective of this paper is contribute so that it can have a greater extent of studies within the theory of Mathematical inequalities.

References

1. Aljaaidi, T.A. and Pachpatte, D.B., Some Gruss-type inequalities using generalized Katugampola fractional integral. *AIMS Math.*, 5, 2, 1011–1024, 2020, doi: 10.3934/math.2020070.
2. Agarwal, R.P., Luo, M.J., Raina, R.K., On Ostrowski type inequalities. *Fasc. Math.*, 56, 1, 5–27, 2016, https://doi.org/10.1016/j.mcm.2007.12.004.
3. Andric, M., Farid, G., Pecaric, J., A further extension of Mittag-Leffler function. *Fract. Calc. Appl. Anal.*, 21, 5, 1377–1395, 2018, https://doi.org/10.1515/fca-2018-0072.
4. Belarbi, S. and Dahmani, Z., On some new fractional integral inequalities. *J. Inequal. Pure Appl. Math.*, 10, 3, 1–9, Article 86, 2009.
5. Dahmani, Z., New inequalities for a class of differentiable functions. *Int. J. Nonlinear Anal. Appl.*, 2, 2, 19–23, 2010, doi.10.22075/ijnaa.2011.89.
6. Dahmani, Z., The Riemann-Liouville operator to generate some new inequalities. *Int. J. Nonlinear Sci.*, 12, 4, 452–455, 2011.
7. Dahmani, Z., Tabharit, L., Taf, S., Some fractional integral inequalities. *Nonlinear Sci. Lett. A*, 1, 2, 155–160, 2010.
8. Dahmani, Z., Mechouar, O., Brahami, S., Certain inequalities related to the Chebyshev's functional involving a Riemann-Liouville operator. *Bull. Math. Anal. Appl.*, 3, 4, 38–44, 2011.
9. Dragomir, S.S., Some integral inequalities of Gruss type. *Indian J. Pure Appl. Math.*, 31, 4, 397–415, 2000.
10. Farid, G., Rehman, U., Mishra, V.N., Mehmood, S., Fractional integral inequalities of Gruss type via generalized Mittag-Leffler function. *Int. J. Anal. Appl.*, 17, 4, 548–558, 2019.
11. Kilbas, A.A., Srivastava, H.M., Trujillo, J.J., *Theory and applications of fractional differential equations*, p. 204, Elsevier, Amsterdam, 2016.
12. Mitrinovic, D.S., *Analytic inequalities*, Springer Verlag, Berlin, 1970.
13. Mitrinovic, D.S., Pecaric, J.E., Fink, A.M., *Classical and new inequalities in analysis*, p. 297, Kluwer Academic Publishers, Dordrecht, 1993.
14. Pachpatte, B.G., A note on Chebyshev-Gruss type inequalities for differential functions. *Tamsui Oxf. J. Math. Sci.*, 22, 1, 29–36, 2006.
15. Pachpatte, B.G., New Chebyshev type inequalities via Trapezoidal like rules. *J. Inequal. Pure Appl. Math.*, 7, 1, 1–15, Article 31, 2006.
16. Podlubny, I., *Fractional differential equations*, p. 198, Academic Press, San Diego, 1999.

17. Purohit, S.D. and Raina, R.K., Chebyshev type inequalities for the Saigo fractional integrals and their q-analogues. *J. Math. Inequal.*, 7, 2, 239–249, 2013, dx.doi.org/10.7153/jmi-07-22.
18. Prabhakar, T.R., A singular integral equation with a generalized Mittag-Leffler function in the kernel. *Yokohama Math. J.*, 19, 7–15, 1971.
19. Rahman, G., Khan, A., Abdeljawad, T., Nisar, K.S., The Minkowski inequalities via generalized proportional fractional integral operators. *Adv. Differ. Equ.*, 287, 2019, 1–14, 2019, https://doi.org/10.1186/s13662-019-2229-7.
20. Samko, S.G., Kilbas, A.A., Marichev, O.I., *Fractional integrals and derivatives: Theory and applications*, Gordon, and Breach, Yverdon, 1987.
21. Sarikaya, M.Z., Dahmani, Z., Kiris, M.E., Ahmad, F., (k, s)-Riemann-Liouville fractional integral and applications. *Hacettepe J. Math. Stat.*, 45, 1, 77–89, 2016, DOI: 10.15672/HJMS.20164512484.
22. Sarikaya, M.Z., Set, E., Yaldiz, H., Basak, N., Hermite-Hadamard's inequalities for fractional integrals and related fractional inequalities. *Math. Comput. Model.*, 57, 9–10, 2403–2407, 2013, https://doi.org/10.1016/j.mcm.2011.12.048.
23. Set, E., Choi, J., Demirbas, S., Some new Chebyshev type inequalities for fractional integral operators containing a further extension of Mittag-Leffler function in the kernel. *Afr. Mat.*, 33, 1–9, Article 42, 2022, https://doi.org/10.1007/s13370-022-00972-3.
24. Set, E., Ozdemir, M.E., Demirbas, S., Chebyshev type inequalities involving extended generalized fractional integral operators. *AIMS Math.*, 5, 4, 3573–3583, 2020, doi: 10.3934/math.2020232.
25. Sousa, J.V.C., Oliveira, D.S., de Oliveira, E.C., Gruss-type inequalities by means of generalized fractional integrals. *Bull. Braz. Math. Soc.*, 50, 2, 1029–1047, 2019, https://doi.org/10.1007/s00574-019-00138-z.
26. Srivastava, H.M. and Tomovski, Z., Fractional calculus with an integral operator containing a generalized Mittag-Leffler function in the kernel. *Appl. Math. Comput.*, 211, 1, 198–210, 2009.

4

Blow Up of the Higher-Order Kirchhoff-Type System With Logarithmic Nonlinearities

Nazlı Irkil* and Erhan Pişkin

Department of Mathematics, Dicle University, Diyarbakır, Turkey

Abstract
This contribution investigates the solution of high-order Kirchhoff-type system with logarithmic nonlinearities. Under the appropriate assumptions, we establish the global nonexistence of the solution at low initial energy level $\boldsymbol{E(0) < d}$.

***Keywords*:** Blow up, system of higher-order, Kirchhoff type equation

4.1 Introduction

This article deals with the following nonlinear higher-order Kirchhoff equations with logarithmic nonlinearities and with nonlinear damping term

$$\begin{cases} u_{tt}+M(\|D^m u\|^2+\|D^m v\|^2)(-\Delta)^m u+(-\Delta)^m u_t=|u|^{p-2}\,u ln|u|, & x\in\Omega,\, t>0, \\ v_{tt}+M(\|D^m u\|^2+\|D^m v\|^2)(-\Delta)^m v+(-\Delta)^m v_t=|v|^{p-2}\,v ln|v|, & x\in\Omega,\, t>0, \end{cases} \tag{4.1}$$

with initial data for $x \in \Omega$

$$u\,(x, 0) = u_0\,(x),\, u_t\,(x, 0) = u_1\,(x),$$

**Corresponding author*: nazliirkil@gmail.com

Sharmistha Ghosh, M. Niranjanamurthy, Krishanu Deyasi, Biswadip Basu Mallik, and Santanu Das (eds.) Mathematics and Computer Science Volume 1, (67–88) © 2023 Scrivener Publishing LLC

$$v(x,0) = v^0(x),\ v_t(x,0) = v^1(x),$$

and boundary value for $x \in \partial\Omega$, $t \geq 0$ *and* $k = 0, 1, 2, \ldots m-1$,

$$\frac{\partial^k}{\partial r^k}u(x,t)=0,\quad \frac{\partial^k}{\partial r^k}v(x,t)=0,$$

where $p \geq 2\gamma + 2$ are real numbers and $m \geq 1$ are nonnegative integers. The Kirchhoff term $M(s) = \beta_1 + \beta_2 s^\gamma$, $\gamma > 0$, $\beta_1 \geq 1$, $\beta_2 \geq 0$. We will take $\beta_1 = \beta_2 = 1$ for simplicity. $\Omega \subset R^n$ is a regular and bounded domain with smooth boundary $\partial\Omega$. r indicates the unit outward normal vector on $\partial\Omega$, and $\frac{\partial^k}{\partial r^k}$ shows the kth order normal derivation. D is the gradient operator, which was defined $Du=\nabla u=\frac{\partial u}{\partial x_1},\frac{\partial u}{\partial x_2},\ldots\frac{\partial u}{\partial x_n}$.

The single wave equation form of the problem (4.1) is denoted such that

$$u_{tt} + M(\|D^m u\|^2 + \|D^m v\|^2)(-\Delta)^m u + (-\Delta)^m u_t = |u|^{p-2}\, u ln\,|u|. \tag{4.2}$$

In the case $M(s) = 1$, $m = 1$ and for a logarithmic source term and damping term (4.2) was considered

$$u_{tt} - \Delta u + g(u_t) = u ln\,|u| \tag{4.3}$$

This kind of problems are related with many applications in many branches of physics [1–3, 5, 10], the authors proved the uniqueness and existence of a solution in problem (4.3) for $g(u_t) = 0$. Ma and Fang [16], established the solutions of the decay results and infinite blow-up for problem (4.3) with a strong damping term. A lot of paper have been dedicated to the work of logarithmic source tem, during the past decades (see [7–9, 24–27]). Related with blow up of solutions, we refer to the studies [4, 6, 11, 12, 15, 22, 23, 28].

The system of wave equation of high order Kirchhoff type without logarithmic source term was expressed as such that

$$\begin{cases} u_{tt} + \Phi\left(\| D^{m_1}u\|^2 + \| D^{m_2}v\|^2\right)(-\Delta)^{m_1}u + \alpha|u_t|^{q-2}u_t = f_1(u,v), & x\in\Omega, t>0,\\ v_{tt} + \Phi\left(\| D^{m_1}u\|^2 + \| D^{m_2}v\|^2\right)(-\Delta)^{m_2}v + \alpha|v_t|^{q-2}v_t = f_2(u,v), & x\in\Omega, t>0. \end{cases} \tag{4.4}$$

Ye [29] concerned the existence and uniqueness of solution for (4.4). There have been some papers related with the higher-order Kirchoff system [13, 14, 17–21].

Inserting a logarithmic source term makes the problem (4.1) different from the one (4.4). Because of this, unsatisfactory results are, at the present time, known for the high-order Kirchhoff type system with logarithmic source term and many problems remain unsolved. As far as we know, only Wang *et al.* [30] studied the problem (4.1) for $m = 1$ and damping term

$$\begin{cases} u_{tt} + M\left(\|\nabla u\|^2 + \|\nabla v\|^2\right)\Delta u + u_t = |u|^{k-2} u ln|u|, \\ v_{tt} + M\left(\|\nabla u\|^2 + \|\nabla v\|^2\right)\Delta v + v_t = |v|^{k-2} v ln|v|, \end{cases}$$

with $M(s) = \alpha + \beta s^{\gamma}$, $\gamma > 0$, $\alpha \geq 1$, $\beta \geq 0$ *and* $k \geq 2\gamma + 2$. They proved global existence and finite time blow up under the different conditions by employing potential well technique and concavity technique.

Motivated on the above, our aim is consider the global nonexistence solution at low energy level in section 4.3. In section 4.2, we give some lemmas which will be useful.

4.2 Preliminaries

Now, we define the potential energy of problem (4.1)

$$J(u,v) = \frac{1}{2}\left(\|D^m u\|^2 + \|D^m v\|^2\right) + \frac{1}{2\gamma+2}\left(\|D^m u\|^2 + \|D^m v\|^2\right)^{\gamma+1} - \frac{1}{p}\left(\int_\Omega |u|^p ln|u|dx + \int_\Omega |v|^p ln|v|dx\right) + \frac{1}{p^2}\left(\|u\|_p^p + \|v\|_p^p\right) \tag{4.5}$$

and

$$I(u,v) = \left(\|D^m u\|^2 + \|D^m v\|^2\right) + \left(\|D^m u\|^2 + \|D^m v\|^2\right)^{\gamma+1} - \frac{1}{p}\left(\int_\Omega |u|^p ln|u|dx + \int_\Omega |v|^p ln|v|dx\right). \tag{4.6}$$

By Eqs. (4.5) and (4.6), we obtain

$$J(u,v)=\frac{I(u,v)}{p}+\frac{p-2}{2p}\left(\|D^m u\|^2+\|D^m v\|^2\right)+\frac{p-2\gamma-2}{2\gamma+2}$$
$$\left(\|D^m u\|^2+\|D^m v\|^2\right)^{\gamma+1}+\frac{1}{p^2}\left(\|u\|_p^p+\|v\|_p^p\right). \tag{4.7}$$

Then we can produce the stable set

$$W=\left\{(u,v)\in H_0^m(\Omega)\times H_0^m(\Omega): J(u,v)\langle d, I(u,v)\rangle 0\right\}\cup\{0\},$$

the instable set of the potential well

$$V=\left\{(u,v)\in H_0^m(\Omega)\times H_0^m(\Omega): J(u,v)\langle d, I(u,v)\langle 0\right\}.$$

We produce the total energy

$$E(u,v)=E(t)=\frac{1}{2}\left(\|u_t\|^2+\|v_t\|^2\right)+\frac{1}{2}\left(\|D^m u\|^2+\|D^m v\|^2\right)$$
$$+\frac{1}{2\gamma+2}\left(\|D^m u\|^2+\|D^m v\|^2\right)^{\gamma+1}+\frac{1}{p^2}\left(\|u\|_p^p+\|v\|_p^p\right)$$
$$-\frac{1}{p}\left(\int_\Omega |u|^p \ln|u|\,dx+\int_\Omega |v|^p \ln|v|\,dx\right), \tag{4.8}$$

and for $(u,v)\in H_0^m(\Omega)\times H_0^m(\Omega), t\geq 0$. And we can obtain the initial total energy such that

$$E(0)=\frac{1}{2}\left(\|u_1\|^2+\|v_1\|^2\right)+\frac{1}{2}\left(\|D^m u_0\|^2+\|D^m v_0\|^2\right)$$
$$+\frac{1}{2\gamma+2}\left(\|D^m u_0\|^2+\|D^m v_0\|^2\right)^{\gamma+1}+\frac{1}{p^2}\left(\|u_0\|_p^p+\|v_0\|_p^p\right)$$
$$-\frac{1}{p}\left(\int_\Omega |u_0|^p \ln|u_0|\,dx+\int_\Omega |v_0|^p \ln|v_0|\,dx\right). \tag{4.9}$$

We introduce by (4.8) and (4.7)

$$E(t)=E(u,v)=\frac{1}{2}\left(\|u_t\|^2+\|v_t\|^2\right)+J(u,v). \tag{4.10}$$

Lemma 4.1. Assume that r is a number which satisfies $2\leq r<\infty$ *if* $n\leq 2k$ *and* $2\leq r\leq \frac{2n}{n-2r}$ *if* $n>2k$. Then, there exists a fixed according as such that

$$\|u\|_r\leq C\|D^m u\|, \forall (u,v)\in H_0^m(\Omega)\times H_0^m(\Omega).$$

Lemma 4.2. If (u, v) is a solution of the (4.1), the $E(t)$ is a nonincreasing for $t \geq 0$ and

$$E'(t) = -\left(\|D^m u_t\|^2 + \|D^m v_t\|^2\right) \leq 0. \tag{4.11}$$

Proof. To establish the proof of the Lemma 4.2, we multiply the first equation (4.1) by u_t and the second equation (4.1) by v_t. Later, we integrate on $\boldsymbol{\Omega}$. So that, it can be written in the following form

$$\frac{1}{2}\frac{d}{dt}\|u_t\|^2+\frac{d}{dt}\left(\frac{1}{p^2}\|u\|_p^p-\frac{1}{p}\int_\Omega |u|^p \ln|u|\,dx\right)\frac{1}{2}$$
$$\left(1+\left(\|D^m u\|^2+\|D^m v\|^2\right)^\gamma\right)\frac{d}{dt}\|D^m u\|^2=-\int_\Omega \|D^m u_t\|^2\,dx, \tag{4.12}$$

and

$$\frac{1}{2}\frac{d}{dt}\|v_t\|^2+\frac{d}{dt}\left(\frac{1}{p^2}\|v\|_p^p-\frac{1}{p}\int_\Omega |v|^p \ln|v|\,dx\right)\frac{1}{2}$$
$$\left(1+\left(\|D^m u\|^2+\|D^m v\|^2\right)^\gamma\right)\frac{d}{dt}\|D^m v\|^2=-\int_\Omega \|D^m v_t\|^2\,dx. \tag{4.13}$$

A summarization of (4.12) and (4.13) hence gives

$$\frac{1}{2}\frac{d}{dt}\left(\|u_t\|^2+\|v_t\|^2\right)+\frac{d}{dt}\left(\frac{1}{p^2}\left(\|u\|_p^p\|+v\|_p^p\right)\right)$$

$$+\frac{d}{dt}\left(-\frac{1}{p}\left(\int_\Omega |u|^p\, ln|u|\,dx+\int_\Omega |v^p ln|v|\,dx\right)\right) \tag{4.14}$$

$$\frac{1}{2}\left(1+\left(\|D^m u\|^2+\|D^m v\|^2\right)^\gamma\right)\frac{d}{dt}\left(\|D^m u\|^2+\|D^m v\|^2\right)$$

$$=-\left(\int_\Omega \|D^m u_t\|^2\,dx+\int_\Omega \|D^m v_t\|^2\,dx\right).$$

Integrating (4.14) with respect to t on $[0, t]$ we arrive at

$$\frac{1}{2}\left(\|u_t\|^2+\|v_t\|^2\right)+\frac{1}{p^2}\left(\|u\|_p^p\|+v\|_p^p\right)$$

$$-\frac{1}{p}\left(\int_\Omega |u|^p\, ln|u|\,dx+\int_\Omega |v^p ln|v|\,dx\right)$$

$$\frac{1}{2}\int_0^t\left(1+\left(\|D^m u\|^2+\|D^m v\|^2\right)^\gamma\right)d\left(\|D^m u\|^2+\|D^m v\|^2\right)$$

$$=\frac{1}{2}\left(\|u_1\|^2+\|v_1\|^2\right)+\frac{1}{p^2}\left(\|u\|_p^p\|+v\|_p^p\right) \tag{4.15}$$

$$-\frac{1}{p}\left(\int_\Omega |u_0|^p\, ln|u_0|\,dx+\int_\Omega |v_0{}^p ln|v_0|\,dx\right)$$

$$-\left(\int_0^t \|D^m u_\tau\|^2\,d\tau+\int_0^t \|D^m v_\tau\|^2\,d\tau\right).$$

We compute the second term in the left hand

$$\frac{1}{2}\int_0^t 1+\left(\|D^m u\|^2+\|D^m v\|^2\right)^\gamma d\left(\|D^m u\|^2+\|D^m v\|^2\right)$$

$$=\frac{1}{2}\left(\|D^m u\|^2+\|D^m v\|^2\right)$$

$$-\frac{1}{2}\left(\|D^m u_0\|^2+\|D^m v_0\|^2\right)+\frac{1}{2\gamma+2}\left(\|D^m u\|^2+\|D^m v\|^2\right)^{\gamma+1}$$

$$-\frac{1}{2\gamma+2}\left(\|D^m u_0\|^2+\|D^m v_0\|^2\right)^{\gamma+1},$$

which makes Eq. (4.15) become

$$
\begin{aligned}
&\frac{1}{2}\left(\|u_t\|^2+\|v_t\|^2\right)+\frac{1}{2}\left(\|D^m u\|^2+\|D^m v\|^2\right)\\
&+\frac{1}{2\gamma+2}\left(\|D^m u\|^2+\|D^m v\|^2\right)^{\gamma+1}+\left(\int_0^t\|D^m u_\tau\|^2\,d\tau+\int_0^t\|D^m v_\tau\|^2\,d\tau\right)\\
&-\frac{1}{p}\left(\int_\Omega|u|^p\,ln|u|\,dx+\int_\Omega|v|^p\,ln|v|\,dx\right)+\frac{1}{p^2}\left(\|u\|_p^p+\|v\|_p^p\right)\\
&+\frac{1}{2\gamma+2}=\frac{1}{2}\left(\|u_1\|^2+\|v_1\|^2\right)+\frac{1}{2}\left(\|D^m u_0\|^2+\|D^m v_0\|^2\right)\\
&\left(\|D^m u_0\|^2+\|D^m v_0\|^2\right)^{\gamma+1}+\frac{1}{p^2}\left(\|u_0\|_p^p+\|v_0\|_p^p\right)\\
&-\frac{1}{p}\left(\int_\Omega|u_0|^p\,ln|u_0|\,dx+\int_\Omega|v_0|^p\,ln|v_0|\,dx\right),
\end{aligned}
\tag{4.16}
$$

that is

$$
E(t)+\int_0^t\|D^m u_\tau\|^2\,d\tau+\int_0^t\|D^m v_\tau\|^2\,d\tau=E(0). \tag{4.17}
$$

Later, we demonstrate some features of the $J(u, v)$ and $I(u, v)$ respectively.

Lemma 4.3. For any $(u,v)\in H_0^m(\Omega)\times H_0^m(\Omega)$, $\|D^m u\|\neq 0,\|D^m v\|\neq 0.$ Moreover, we have

i) $\lim\limits_{\lambda\to 0} J(\lambda u,\lambda v)=0,\ \lim\limits_{\lambda\to\infty,} J(\lambda u,\lambda v)=-\infty,$

ii) There is a unique λ_1 such that $\dfrac{dJ(\lambda u,\lambda v)}{d\lambda}=0,$

iii) Then we have

$$
I(\lambda u,\lambda v)=\lambda\frac{dJ(\lambda u,\lambda v)}{d\lambda}\begin{cases}>0, & 0\leq\lambda<\lambda_1\\ =0, & \lambda=\lambda_1,\\ <0, & \lambda_1<\lambda.\end{cases}
$$

Proof. From the definition of $J(u, v)$ we obtain

$$J(\lambda u,\lambda v)=\frac{1}{2}\left(\|D^m\lambda u\|^2+\|D^m\lambda v\|^2\right)$$
$$+\frac{1}{2\gamma+2}\left(\|D^m\lambda u\|^2+\|D^m\lambda v\|^2\right)^{\gamma+1}$$
$$-\frac{1}{p}\left(\int_\Omega|\lambda u|^p\, ln|\lambda u|dx+\int_\Omega|\lambda v|^p\, ln|\lambda v|dx\right)$$
$$+\frac{1}{p^2}\left(\|\lambda u\|_p^p+\|\lambda v\|_p^p\right)=\frac{\lambda^2}{2}\left(\|D^m u\|^2+\|D^m v\|^2\right) \tag{4.18}$$
$$+\frac{\lambda^{2\gamma+2}}{2\gamma+2}\left(\|D^m u\|^2+\|D^m v\|^2\right)^{\gamma+1}$$
$$-\frac{\lambda^p}{p}\left(\int_\Omega|u|^p\, ln|u|dx+\int_\Omega|v|^p\, ln|v|dx\right)$$
$$+\frac{\lambda^p}{p^2}\left(\|u\|_p^p+\|v\|_p^p\right)-\frac{\lambda^p}{p}ln|\lambda|\left(\|u\|_p^p+\|v\|_p^p\right).$$

Since $\|D^m u\|^2 \neq 0$, and $+\|D^m v\|^2\neq 0, \lim_{\lambda\to 0} J(\lambda u,\lambda v)=0, \lim_{\lambda\to\infty}, J(\lambda u,\lambda v) =-\infty$, Now, differentiating $J(\boldsymbol{\lambda} u, \boldsymbol{\lambda} v)$ with respect to $\boldsymbol{\lambda}$, we have

$$\frac{dJ(\lambda u,\lambda v)}{d\lambda}=\lambda\left(\|D^m u\|^2+\|D^m v\|^2\right)$$
$$+\lambda^{2\gamma+1}\left(\|D^m u\|^2+\|D^m v\|^2\right)^{\gamma+1}$$
$$-\lambda^{p-1}\left(\int_\Omega|u|^p\, ln|u|dx+\int_\Omega|v|^p\, ln|v|dx\right) \tag{4.19}$$
$$-\lambda^{p-1}ln|\lambda|\left(\|u\|_p^p+\|v\|_p^p\right)=\lambda\Big[\left(\|D^m u\|^2+\|D^m v\|^2\right)$$
$$+\lambda^{2\gamma}\left(\|D^m u\|^2+\|D^m v\|^2\right)^{\gamma+1}-\lambda^{p-2}\left(\int_\Omega|u|^p\, ln|u|dx\right.$$
$$\left.+\int_\Omega|v|^p\, ln|v|dx\right)-ln|\lambda|\left(\|u\|_p^p+\|v\|_p^p\right)\Big].$$

Let

$$\psi(\lambda)=\lambda^{2\gamma}\left(\|D^m u\|^2+\|D^m v\|^2\right)^{\gamma+1}-\lambda^{p-2}ln|\lambda|\left(\|u\|_p^p+\|v\|_p^p\right)$$

$$-\lambda^{p-2}\left(\int_\Omega |u|^p\, ln|u|\,dx+\int_\Omega |v|^p\, ln|v|\,dx\right).$$

Then from $2\gamma \le p-2$ we can deduce that $\lim_{\lambda\to\infty}\psi(\lambda)=-\infty, \psi(\lambda)$ is monotone decreasing when $\lambda > \lambda^*$ and there exists only a λ^* such that $\psi(\lambda^*) = 0$. Then we obtain there is a $\lambda_1 > \lambda^*$ such that $\lambda\,[(\|D^m u\|^2 + \|D^m v\|^2) + \psi(\lambda)] = 0$, which means $\frac{dJ(\lambda u,\lambda v)}{d\lambda}=0$.

The last feature (iii), is only a basic result of the fact that

$$\lambda\frac{dJ(\lambda u,\lambda v)}{d\lambda}=I(\lambda u,\lambda v). \tag{4.20}$$

Lemma 4.4. *i)* We put the definition the depth of potential well

$$d=\inf_{(u,v)\in N} J(u,v) \tag{4.21}$$

where

$$N=\left\{(u,v):(u,v)\in H_0^m(\Omega)\times H_0^m(\Omega)\setminus\{0\}: I(u,v)=0\right\}$$

is to equal

$$d=\inf\left\{\sup_{\lambda\ge 0} J(\lambda u,\lambda v):(u,v)\in H_0^m(\Omega)\times H_0^m(\Omega), \|D^m u\|^2\neq 0, \|D^m v\|^2\neq 0\right\}. \tag{4.22}$$

ii) The d which is defined in (4.21) satisfies

$$d=\frac{(p-2)}{2p}\left(\frac{1}{C_1^{p+1}}\right)^{\frac{2}{p-1}}$$

Where C_1 is the optimal constant of Lemma 4.1 $\left(H_0^m(\Omega) \hookrightarrow L^{p+1}(\Omega)\right)$ and

$$\begin{cases} 2\gamma+2 \leq p \leq \infty, & n \leq 2m, \\ 2\gamma+2 \leq p \leq \dfrac{n+2m}{n-2m}, & n > 2m. \end{cases} \tag{4.23}$$

Proof. *i)* Because of from *(iii)* of Lemma 4.3, for any $(u,v) \in H_0^m(\Omega) \times H_0^m(\Omega)$ it supposes that, there is a λ_1 such that $I(\lambda_1 u, \lambda_1 v) = 0$, that is $(\lambda_1 u, \lambda_1 v) \in N$. From the definition of d we have for any $(u,v) \in H_0^m(\Omega) \times H_0^m(\Omega) \,/\, \{0\}$

$$J(\lambda_1 u, \lambda_1 v) \geq d \tag{4.24}$$

and because of Lemma 4.3

$$\sup_{\lambda \geq 0} J(\lambda u, \lambda v) = J(\lambda_1 u, \lambda_1 v)$$

which by virtue of Eq. (4.24) means

$$\inf_{(u,v) \in H_0^m(\Omega) \times H_0^m(\Omega)} \sup_{\lambda \geq 0} J(\lambda u, \lambda v) = \geq d. \tag{4.25}$$

As $(u,v) \in H_0^m(\Omega) \times H_0^m(\Omega) \,/\, \{0\}$, we find d is not to equal to 0, which is given in Eq. (4.22). Otherwise, from the (4.22) it implies that there is a λ^* such that

$$\sup_{\lambda \geq 0} J(\lambda u, \lambda v) = J(\lambda^* u, \lambda^* v).$$

After that, from Lemma 4.3, we can conclude $\lambda^* = \lambda_1$. And it clearly shows that

$$I(\lambda^* u, \lambda^* v) = I(\lambda_1 u, \lambda_1 v) = 0,$$

which means $(\lambda^* u, \lambda^* v) \in N$. So that, by (4.22), we obtain

$$d = \inf_{(\lambda^* u, \lambda^* v) \in N} J(\lambda^* u, \lambda^* v),$$

that is

$$d = \inf_{(u,v) \in N} J(u,v). \tag{4.26}$$

So that the proof is completed for *(i)*.

ii) Due to $I(u, v) = 0$ and definition of $I(u, v)$ and Sobolev embedding theorems we acquire

$$\left(\|D^m u\|^2 + \|D^m v\|^2\right) + \left(\|D^m u\|^2 + \|D^m v\|^2\right)^{\gamma+1}$$

$$= \frac{1}{p}\left(\int_\Omega |u|^p \ln|u|\,dx + \int_\Omega |v|^p \ln|v|\,dx\right),$$

and

$$\left(\|D^m u\|^2 + \|D^m v\|^2\right) \le \frac{1}{p}\left(\int_\Omega |u|^p \ln|u|\,dx + \int_\Omega |v|^p \ln|v|\,dx\right)$$

$$\le \|u\|_{p+1}^{p+1} + \|v\|_{p+1}^{p+1} \le C_1^{p+1}\left(\|D^m u\|^{p+1} + \|D^m v\|^{p+1}\right) \le C_1^{p+1} \tag{4.27}$$

$$\left(\|D^m u\|^2 + \|D^m v\|^2\right)^{\frac{p-1}{2}} \left(\|D^m u\|^2 + \|D^m v\|^2\right)$$

which means

$$\|D^m u\|^2 + \|D^m v\|^2 \ge \left(\frac{1}{C_1^{p+1}}\right)^{\frac{2}{p-1}}. \tag{4.28}$$

From the definition of d, we have $(u, v) \in N$. By the definition of $J(u, v)$, (4.27), (4.7) and $J(u, v) = 0$ we get

$$J(u,v)=\frac{I(u,v)}{p}+\frac{p-2}{2p}\left(\|D^m u\|^2+\|D^m v\|^2\right)+\frac{p-2\gamma-2}{2\gamma+2}$$

$$\left(\|D^m u\|^2+\|D^m v\|^2\right)^{\gamma+1}+\frac{1}{p^2}\left(\|u\|_p^p+\|v\|_p^p\right)$$

$$\geq\frac{p-2}{2p}\left(\|D^m u\|^2+\|D^m v\|^2\right)\geq\frac{p-2}{2p}\left(\frac{1}{C_1^{p+1}}\right)^{\frac{2}{p-1}},$$

where $2\gamma \leq p-2$. Combining Eq (4.26) and (4.28), we can see clearly that

$$d=\frac{p-2}{2p}\left(\frac{1}{C_1^{p+1}}\right)^{\frac{2}{p-1}}.$$

4.3 Blow Up for Problem for *E* (0) < *d*

Lemma 4.5. If $(u_0,v_0)\in H_0^m(\Omega)\times H_0^m(\Omega),(u_1,v_1)\in(H_0^m(\Omega)\times H_0^m(\Omega)$ *and* $E(0)<d$, $u(x,t),v(x,t)$ is a weak solution the problem of (4.1),

i) then $(u, v) \in W$, if $(u_0,v_0) \in W$, *for* $0 \leq t \leq T$;
ii) then $(u, v) \in V$, if $(u_0,v_0) \in V$, *for* $0 \leq t \leq T$,

where T is defined the maximum existence time of $(u(x, t), v(x, t))$.

Proof. The proof of case *(i)* and case *(ii)* is similar. Because of that we prove only case (*i*). If $(u(x, t), v(x, t))$ is a weak solution the problem of (4.1) and $(u_0, v_0) \in W$, then by (4.11) the energy functional is nonincreasing about t. So that, we denote $d > E(0) > E(u(x, t), v(x, t))$ which makes, $I(u(x, t), v(x, t)) > 0$ *for* $0 < t < T$. By using conradiction method, we assume that; we obtain a $t_1 \in (0, T)$ such that $I(u(x, t_1), v(x, t_1)) < 0$. Thus, there exists a $t_2 \in (0, T)$ that satisfies $I(u(x, t_2), v(x, t_2)) = 0$ because of continuity of $I(u(x, t), v(x, t))$ about t. By definition of the d, it is clear that

$$d \leq J(u(x, t^2), v(x, t^2)) \leq E(u(x, t^2), v(x, t^2)) \leq E(0) < d,$$

which is a contradiction.

Lemma 4.6. Assume that, the condition of Lemma 4.5 *(ii)* is holds. So that, we get

$$d < \frac{p-2}{2p}\left(\|D^m u\|^2 + \|D^m v\|^2\right).$$

Proof. By virtue of $I(u, v) < 0$ and definition of $I(u, v)$ and embedding theorems, we obtain

$$\left(\|D^m u\|^2 + \|D^m v\|^2\right) + \left(\|D^m u\|^2 + \|D^m v\|^2\right)^{\gamma+1}$$

$$< \frac{1}{p}\left(\int_\Omega |u|^p \ln|u|\,dx + \int_\Omega |v|^p \ln|v|\,dx\right),$$

and

$$\left(\|D^m u\|^2 + \|D^m v\|^2\right) \leq \frac{1}{p}\left(\int_\Omega |u|^p \ln|u|\,dx + \int_\Omega |v|^p \ln|v|\,dx\right)$$

$$\leq \|u\|_{p+1}^{p+1} + \|v\|_{p+1}^{p+1} \leq C_1^{p+1}\left(\|D^m u\|^{p+1} + \|D^m v\|^{p+1}\right) \leq C_1^{p+1} \tag{4.29}$$

$$\left(\|D^m u\|^2 + \|D^m v\|^2\right)^{\frac{p-1}{2}}\left(\|D^m u\|^2 + \|D^m v\|^2\right)$$

which means

$$\|D^m u\|^2 + \|D^m v\|^2 \geq \left(\frac{1}{C_1^{p+1}}\right)^{\frac{2}{p-1}} \tag{4.30}$$

By Lemma 4.3, we get

$$d < \frac{p-2}{2p} \| D^m u \|^2 + \| D^m v \|^2 .$$

Theorem 4.7. If (u, v) is a weak solution problem of (4.1) and $(u_0, v_0) \in H_0^m(\Omega) \times H_0^m(\Omega), (u_1, v_1) \in \left(H_0^m(\Omega) \times H_0^m(\Omega)\right.$, then the solutions blow up in finite time as long as $E(0) < d$ and $(u_0, v_0) \in V$.

Proof. Our aim is to show the solution goes to the infinity in finite time. If it is not this case, we suppose existence time $T = \infty$. For any $T_1 > 0$, we define the auxiliary functional

$$B(t) = \| u \|^2 + \| v \|^2 + \int_0^t \left(\| D^m u \|^2 + \| D^m v \|^2 \right) d\tau \tag{4.31}$$
$$(T_1 - t)\left(\| D^m u \|^2 + \| D^m v \|^2 \right).$$

It is readily seen that for all $t \in [0, T_1]$, $B(t) > 0$. In wiev of continuity of $B(t)$ in time, we obtain that there exists a $\xi > 0$ which is independent on T_1 such that

$$B(t) > \xi. \tag{4.32}$$

Then by $t \in [0, T_1]$, we derive

$$\begin{aligned} B'(t) &= 2\left(\int_\Omega u u_t \, dx + \int_\Omega v v_t \, dx \right) + \| D^m u \|^2 + \| D^m v \|^2 \\ &- \| D^m u_0 \|^2 + \| D^m v_0 \|^2 = 2\left(\int_\Omega u u_t \, dx + \int_\Omega v v_t \, dx \right) \\ &+ 2\left(\int_0^t (D^m u(\tau), D^m u_\tau(\tau)) + (D^m v(\tau), D^m v_\tau(\tau)) \right), \end{aligned} \tag{4.33}$$

and

$$B''(t) = 2\,(||u_t||^2 + ||v_t||^2) + 2(u, u_{tt}) + 2(v, v_{tt}) + 2\,(D^m u, D^m u_t)$$
$$+ 2\,(D^m v, D^m v_t) = 2\,(||u_t||^2 + ||v_t||^2) - 2I\,(u, v). \tag{4.34}$$

From Eq (4.33), it implies

$$(B'(t))^2 = 4((u,u_t)^2 + (v,v_t)^2) + 4\left(\int_0^t (D^m u(\tau), D^m u_\tau(\tau)) + (D^m v(\tau), D^m v_\tau(\tau))\right)^2$$
$$+8\left[((u,u_t)+(v,v_t))\left(\int_0^t (D^m u(\tau), D^m u_\tau(\tau)) + (D^m v(\tau), D^m v_\tau(\tau))\right)\right] \tag{4.35}$$

Our aim is to estimate the each terms in Eq (4.35). It follows from Young's inequality and Cauchy-Schwarz inequality that

$$(u, u_t)^2 + (v, v_t)^2 \le (||u||||u_t|| + ||v||||v_t||)^2 \le (||u||^2 + ||v||^2)\,(||u_t||^2 + ||v_t||^2), \tag{4.36}$$

and

$$\left(\int_0^t (D^m u(\tau), D^m u_\tau(\tau)) + (D^m v(\tau), D^m v_\tau(\tau))\right)^2$$
$$\le \left(\int_0^t ||D^m u(\tau)||||D^m u_\tau(\tau)|| + ||D^m v(\tau)||||D^m v_\tau(\tau)||\,d\tau\right)^2$$
$$\le \left(\int_0^t (||D^m u(\tau)||^2 + ||D^m v(\tau)||^2)^{\frac{1}{2}} (||D^m u_\tau(\tau)||^2 + ||D^m v_\tau(\tau)||^2)^{\frac{1}{2}}\,d\tau\right)^2$$
$$\le \int_0^t (||D^m u(\tau)||^2 + ||D^m v(\tau)||^2)^{\frac{1}{2}} (||D^m u_\tau(\tau)||^2 + ||D^m v_\tau(\tau)||^2)^{\frac{1}{2}}\,d\tau. \tag{4.37}$$

For the last term by using same calculations and inequalities, we obtain

$$2\left[((u,u_t)+(v,v_t))\left(\int_0^t (D^m u(\tau),D^m u_\tau(\tau))+(D^m v(\tau),D^m v_\tau(\tau))\right)\right]$$

$$\leq 2\left[(\|u\|^2+\|v\|^2)^{\frac{1}{2}}(\|u_t\|^2+\|v_t\|^2)^{\frac{1}{2}}\right]$$

$$\left(\int_0^t \|D^m u(\tau)\|^2+\|D^m v(\tau)\|^2\, d\tau \int_0^t (\|D^m u_\tau(\tau)\|^2+\|D^m v_\tau(\tau)\|^2\, d\tau)\right)^{\frac{1}{2}}$$

$$\leq (\|u_t\|^2+\|v_t\|^2)\int_0^t \|D^m u(\tau)\|^2+\|D^m v(\tau)\|^2\, d\tau(u^2+v^2)d\tau$$

$$(\|u_t\|^2+\|v_t\|^2)\int_0^t (\|D^m u_\tau(\tau)\|^2+\|D^m v_\tau(\tau)\|^2)d\tau. \tag{4.38}$$

Substituting Eqs. (4.36)-(4.38) into Eq. (4.35) becomes

$$(B'(t))^2 \leq 4B(t)\|u_t\|^2+\|v_t\|^2+\int_0^t (\|D^m u_\tau(\tau)\|^2+\|D^m v_\tau(\tau)\|^2)d\tau. \tag{4.39}$$

Combining Eqs. (4.34) and (4.39) and using Eq. (4.34), we have

$$B''(t)B(t)-\frac{2+p}{4}(B'(t))^2 \geq B(t)\Big[B''(t)-(p+2)\big((\|u_t\|^2+\|v_t\|^2)$$

$$\left(\int_0^t (\|D^m u_\tau(\tau)\|^2+\|D^m v_\tau(\tau)\|^2)d\tau\right)\big)\Big]$$

$$\geq B(t)\Big[2(\|u_t\|^2+\|v_t\|^2)-2I(u,v)-(p+2)\big((\|u_t\|^2+\|v_t\|^2)+$$

$$\left(\int_0^t (\|D^m u_\tau(\tau)\|^2+\|D^m v_\tau(\tau)\|^2)d\tau\right)\big)\Big] \tag{4.40}$$

Let

$$\rho(t) = -p\left(\|u_t\|^2 + \|v_t\|^2\right) - 2I(u, \qquad (4.41)$$
$$-(p+2)\int_0^t \left(\|D^m u_\tau(\tau)\|^2 + \|D^m v_\tau(\tau)\|^2\right) d\tau.$$

By using of the Lemma 4.2, we conclude that

$$\begin{aligned} E(0) &= E(u(t),v(t)) + \left(\int_0^t \|D^m u_\tau(\tau)\|^2\, d\tau + \int_0^t \|D^m v_\tau(\tau)\|^2\, d\tau\right) \\ &= \frac{1}{2}\left(\|u_t\|^2 + \|v_t\|^2\right) + \frac{p-2}{2p}\left(\|D^m u\|^2 + \|D^m v\|^2\right) \\ &+ \frac{p-2\gamma-2}{2\gamma+2}\left(\|D^m u\|^2 + \|D^m v\|^2\right)^{\gamma+1} + \frac{I(u,v)}{p} \\ &+ \frac{1}{p^2}\left(\|u\|_p^p + \|v\|_p^p\right) + \left(\int_0^t \|D^m u_\tau(\tau)\|^2\, d\tau + \int_0^t \|D^m v_\tau(\tau)\|^2\, d\tau\right). \end{aligned} \qquad (4.42)$$

Substituting Eq. (4.42) into Eq. (4.41) yields

$$\begin{aligned} \rho(t) &= (p-2)\left(\|D^m u\|^2 + \|D^m v\|^2\right) - 2pE(0) + \frac{p-2\gamma-2}{2\gamma+2} \\ &\left(\|D^m u\|^2 + \|D^m v\|^2\right)^{\gamma+1} + \frac{2}{p}\left(\|u\|_p^p + \|v\|_p^p\right) + (p-2) \\ &\int_0^t \left(\|D^m u_\tau(\tau)\|^2 + \|D^m v_\tau(\tau)\|^2\right) d\tau. \end{aligned} \qquad (4.43)$$

Further from Lemma 4.6, we see clearly that

$$2pE(0) < 2pd < (p-2)\left(\|D^m u\|^2 + \|D^m v\|^2\right). \qquad (4.44)$$

Therefore, by using Eq. (4.44) into Eq. (4.43), we conclude that

$$\rho(t) > +\frac{p-2\gamma-2}{2\gamma+2}(\| D^m u\|^2 + \| D^m v\|^2)^{\gamma+1} + (p-2)\int_0^t$$
$$\left(\| D^m u_\tau(\tau)\|^2 + \| D^m v_\tau(\tau)\|^2\right)d\tau + \frac{2}{p}\left(\| u\|_p^p + \| v\|_p^p\right) > \varsigma_1 > 0, \tag{4.45}$$

Where $\varsigma_1 > 0$ is a constant. Hence, we deduce that Eq. (4.40) becomes

$$B''(t)B(t) - \frac{p+2}{4}(B'(t))^2 > B(t)\varsigma_1 > 0,$$

For $t \in [0, T_1]$.

Let $f(t) = [B(t)]^{\frac{2-p}{4}}$, then we obtain

$$f''(t) \leq \frac{2-p}{4}\varsigma^{1[f(t)]^{\frac{2+p}{p-2}}}, t \in [0, T^1],$$

That is

$$\lim_{t \to T^*} f(t) = 0,$$

where $T^* < T_1$ and T^* is independent of initial selection of T_1. Therefore we can concludethat

$$\lim_{t \to T^*} B(t) = \infty.$$

4.4 Conclusion

As mentioned earlier in the introduction, equations with logarithmic nonlinearity and their qualitative theory of solutions have received much attention from physicists and mathematicians. This work is related with the blow up result for a higher-order Kirchhoff type system with logarithmic nonlinearities and strong damping term. This result is new for these types of systems, and it generalizes many related problems in the literature. The blow up results can be obtained for different method or different initial energy conditions.

References

1. Bartkowski, K. and Górka, P., One-dimensional Klein-Gordon equation with logarithmic nonlinearities. *J. Phys. A Math. Theor.*, 41, 35, 355201, 2008.
2. Bialynicki-Birula, I. and Mycielski, J., Wave equations with logarithmic nonlinearities. *Bull. Acad. Pol. Sci., Ser. Sci. Math. Astron. Phys.*, 23, 4, 461–466, 1975.
3. Cazenave, T. and Haraux, A., Équations d'évolution avec non linéarité logarithmique. *Annales la Faculté Des. Sci. Toulouse: Mathématiques*, 2, 1, 21–51, 1980.
4. Chen, Y. and Xu, R., Global well-posedness of solutions for fourth order dispersive wave equation with nonlinear weak damping, linear strong damping and logarithmic nonlinearity. *Nonlinear Anal.*, 192, 111664, 2020.
5. De Martino, S., Falanga, M., Godano, C., Lauro, G., Logarithmic Schrödinger-like equation as a model for magma transport. *Europhys. Lett.*, 63, 3, 472, 2003.
6. Di, H. and Shang, Y., Song, Initial boundary value problem for a class of strongly damped semilinear wave equations with logarithmic nonlinearity. *Nonlinear Anal. Real World Appl.*, 51, 102968, 2020.
7. Ferreira, J., Pışkın, E., Irkıl, N., Raposo, C., Blow up results for a viscoelastic Kirchhoff-type equation with logarithmic nonlinearity and strong damping. *Mathematica Moravica*, 25, 2, 125–141, 2021.
8. Irkıl, N. and Pişkin, E., Global existence and decay of solutions for a higher-order Kirchhoff-type systems with logarithmic nonlinearities. *Quaest. Math.*, 45, 4, 523–546, 2021.
9. Irkıl, N., Pişkin, E., Agarwal, P., Global existence and decay of solutions for a system of viscoelastic wave equations of Kirchhoff type with logarithmic nonlinearity. *Math. Methods Appl. Sci.*, 45, 5, 2921–2948, 2022.
10. Królikowski, W., Edmundson, D., Bang, O., Unified model for partially coherent solitons in logarithmically nonlinear media. *Phys. Rev. E*, 61, 3, 3122, 2000.
11. Lian, W., Ahmed, M.S., Xu, R., Global existence and blow up of solution for semilinear hyperbolic equation with logarithmic nonlinearity. *Nonlinear Anal.*, 184, 239–257, 2019.
12. Lian, W. and Xu, R., Global well-posedness of nonlinear wave equation with weak and strong damping terms and logarithmic source term. *Adv. Nonlinear Anal.*, 9, 1, 613–632, 2019.

13. Lin, G. and Hu, L., The global attractor for a class of higher-order coupled Kirchhoff-type equations with strong linear damping. *Eur. J. Mathematics Comput. Sci.*, 4, 1, 63–77, 2017.
14. Lin, G. and Yang, S., The global attractor for the higher-order coupled Kirchhoff-type equations with nonlinear strong dampened and source term. *Eur. J. Mathematics Comput. Sci.*, 4, 2, 66–78, 2017.
15. Liu, G., The existence, general decay and blow-up for a plate equation with nonlinear damping and a logarithmic source term. *Electron. Res. Arch.*, 28, 263–289, 20202020.
16. Ma, L. and Fang, Z.B., Energy decay estimates and infinite blow-up phenomena for a strongly damped semilinear wave equation with logarithmic nonlinear source. *Math. Methods Appl. Sci.*, 41, 7, 2639–2653, 2018.
17. Peyravi, A., Blow up solutions to a system of higher-order Kirchhoff-type equations with positive initial energy. *Taiwan. J. Math.*, 21, 4, 767–789, 2017.
18. Pişkin, E. and Harman, E., Energy decay of solutions for a system of higher-order Kirchhoff type equations. *J. New Theory*, 29, 89–100, 2019.
19. Pişkin, E., Blow up of solutions for a system of nonlinear higher-order Kirchhoff-type equations. *Mathematics Stat*, 2, 6, 219–229, 2014.
20. Pişkin, E. and Polat, N., Global existence and exponential decay of solutions for a class of system of nonlinear higher-order wave equations with strong damping. *J. Adv. Res. Appl. Math.*, 4, 4, 26–36, 2012.
21. Pişkin, E. and Polat, N., Uniform decay and blow up of solutions for a system of nonlinear higher-order Kirchhoff-type equations with damping and source term. *Contemp. Anal. Appl. Mathematics*, 1, 2, 181–199, 2013.
22. Pişkin, E. and Irkıl, N., Well-posedness results for a sixth-order logarithmic Boussinesq equation. *Filomat*, 33, 13, 3985–4000, 2019.
23. Pişkin, E. and Irkıl, N., Local existence and blow up for p-Laplacian equation with logarithmic nonlinearity. *Miskolc Math. Notes*, 23, 1, 231–251, 2022.
24. Pişkin, E., Boulaaras, S., Irkil, N., Qualitative analysis of solutions for the p-Laplacian hyperbolic equation with logarithmic nonlinearity. *Math. Methods Appl. Sci.*, 44, 6, 4654–4672, 2021.
25. Piskin, E. and Irkıl, N., Blow up of the solution for hyperbolic type equation with logarithmic nonlinearity. *Aligarh Bull. Math.*, 39, 1-2, 19–29, 2020.
26. Piskin, E. and Irkıl, N., Existence and decay of solutions for a higher-order viscoelastic wave equation with logarithmic nonlinearity. *Commun. Fac. Sci. Univ. Ank. Ser. A1 Math. Stat.*, 70, 1, 300–319, 2021.
27. Pişkin, E. and Irkıl, N., Mathematical behavior of solutions of p-Laplacian equation with logarithmic source term. *Sigma J. Eng. Natural Sci.*, 10, 2, 213–220, 2019.
28. Ye, Y., Global solution and blow-up of logarithmic Klein-Gordon equation. *Bull. Korean Math. Soc.*, 57, 2, 281–294, 2020.
29. Ye, Y., Global existence and asymptotic behavior of solutions for a system of higher-order Kirchhoff-type equations. *Electron. J. Qual. Theory Differ. Equ.*, 2015, 20, 1–12, 2015.

30. Wang, X., Chen, Y., Yang, Y., Li, J., Xu, R., Kirchhoff-type system with linear weak damping and logarithmic nonlinearities. *Nonlinear Anal.*, 188, 475–499, 2019.

5

Developments in Post-Quantum Cryptography

Srijita Sarkar[1]*, Saranya Kumar[2], Anaranya Bose[3] and Tiyash Mukherjee[1]

[1]Computer Science and Engineering, Institute of Engineering and Management, Kolkata, India
[2]Information Technology, Institute of Engineering and Management, Kolkata, India
[3]Computer Science (Internet of Things), Institute of Engineering and Management, Kolkata, India

Abstract
Cryptography deals with the design of mechanisms based on mathematical algorithms that provide fundamental information security services for computers and communication systems. Modern-day cryptography makes use of binary bit sequences and relies on publicly known algorithms such as number theory, computational complexity theory, and probability theory that are practically impossible for conventional computers to compute. Quantum computers use quantum bits or Qubits and can get us "a quantum speedup," which will take seconds to solve these complex computations. This quantum supremacy, which scientists hope to attain by the next two decades, would pose a serious threat to the existing cryptographic systems. The main algorithms that have been proposed for post-quantum primitives are based on lattices, multivariate polynomials, codes, hash functions, supersingular elliptical curves and so on. This paper focuses on the developments that have so far been made in the field of post-quantum cryptography and its future prospects.

Keywords: Information security, modern-day cryptography, quantum computers, qubits, quantum supremacy, cryptographic systems, post-quantum cryptography

**Corresponding author*: srijita.sarkar2015@gmail.com

Sharmistha Ghosh, M. Niranjanamurthy, Krishanu Deyasi, Biswadip Basu Mallik, and Santanu Das (eds.) Mathematics and Computer Science Volume 1, (89–110) © 2023 Scrivener Publishing LLC

5.1 Introduction

The method of protecting our information and communication through the use of codes is called cryptography. This ensures that the person who sends the information and the one who receives it can only access it [12]. For millennia, any written matter has conveyed mankind's significance, under conditions where spoken interaction is either not reasonable or impractical. The human need originates before the composed language itself. Definitely, the need for protection made it necessary for the senders of confidential messages to infer techniques for camouflaging their writings not long after the advancement of the written language. Codes were made to cloud the importance of messages, delivering their substance pointless to any gathering other than the message's expected beneficiary. Encryption forms the basis of digital communication and ensures its security. The need to build up secure and dependable techniques for correspondence is at any rate as significant now as it was the point at which the principal endeavors to encode texts were made. A high expense is put on encryption when data with respect to state insider facts, war procedures, strategies, monetary exchanges, and clinical records should be relayed or communicated. Since its inception, the Internet has enabled unprecedented levels of communication. The sender of a message can now type on a machine on the other side of the world, and the recipient of that message can receive it astonishingly quickly on the other side of the world. The Internet is an advantageous and viable correspondence channel, yet it is anything but a protected one. Present-day cryptosystems give a ground-breaking strategy for guaranteeing the protection, respectability, and legitimacy of messages that are sent across the Internet. However, with changing times, the algorithms deployed today might become insecure, especially when powerful quantum computers come into use.

5.2 Modern-Day Cryptography

The two different types of cryptosystems that are being used today are symmetric key and asymmetric key. These rely on the use of keys and sound key management methods to maintain their security. An attacker can easily obtain a copy of the key that was used to encrypt the message and use it to decrypt it. This is because it is computationally infeasible for classical computers to break the code due to the potency of modern cryptosystems.

5.2.1 Symmetric Cryptosystems

In order to decrypt the cipher text, the symmetric key systems depend on the same key, which most importantly ensures that the secret key is not being compromised. Different systems have various lengths of keys. Data is encrypted bit by bit or in blocks in a stream (usually 64 bits at a time). Because they use shorter key lengths, symmetric key systems allow faster data manipulation than asymmetric key systems [2].

The following are the few major symmetric systems are described briefly:

Data encryption standard (DES):

- Developed so that it can provide a standard method for protecting sensitive commercial and unclassified data.
- Its procedure can be broken in a matter of seconds, hence not considered secure anymore.
- It was replaced by triple DES.

Advanced encryption standard (AES):

- It is remarkably secure with high performance.
- The performance of AES encryption is much faster than DES and triple DES.

Triple DES:

- Block cipher is used for encryption.
- It is necessary to use different types of keys during the encryption process.
- The security in Triple DES is much greater than AES.

Some other symmetric systems are namely Blowfish, IDEA, RC4, and Skipjack.

5.2.2 Asymmetric Cryptosystems

For encrypting and decrypting cipher evenly a private and public key is used in asymmetric key systems, (also known as public key systems) [2].

- It is possible to distribute the public keys freely.
- Once the public keys are distributed between two parties, encrypted communication can be established.

- These happen at a much lower rate than symmetric systems.
- Digital signatures and strong key management are used to provide better security.

The major asymmetric systems are described briefly below:

RSA:

- This cryptosystem provides both encryption and authentication.
- Used in various products and platforms and built-in OS by Apple, Microsoft, etc.

Elliptical curves:

- Short key sizes but provide the same security as conventional cryptosystems.
- Its algorithm is not based on integer factorization.

Secure Hash Algorithm (SHA-1):

- It is an ideal hashing method for providing message integrity.

Hashes identify whether a message's contents were changed during transmission. They are a type of info that has been compressed.

Digital certificates:

- Certificates are made available to people across a globe around a range of organizations.
- Prevents imposters from establishing message authenticity.
- A reputable organization will take a user's key pair and encrypt the message with its own private key. The message receiver is supplied with the organization's public key and a certificate to verify that the sender is legitimate.

5.2.3 Attacks on Modern Cryptosystems

Modern cryptosystems, on the other hand, are vulnerable to attacks from both classical and quantum computers. To fend off attacks via traditional computers, organizations had to switch to 112-bit or 128-bit security instead of the standard 80 bit [1]. Classical computers commonly use two types of attacks.

5.2.3.1 Known Attacks

Cryptographic attacks that are faster than a brute-force attack are referred to as "break" by cryptographers. This entails running one try to decrypt each possible key in the correct order. In a break with modern technology, impossible results can be included. Although this is impracticable, theoretical breaks can sometimes be useful in revealing vulnerability patterns.

5.2.3.2 Side-Channel Attacks

The cipher is not targeted as a black box by the side-channel attack and therefore is not related to cipher security. They are, nonetheless, crucial in practice. They target various encryption implementations on hardware and software systems that result in data leaking unintentionally. There have been several known attacks on various AES implementations. AES hardware instructions are incorporated into many current CPUs, protecting against timing-related side-channel attacks [14].

Apart from this, the most serious threat are attacks from quantum computers, once they come into widespread use. To secure our data against this, we need to analyze and understand where quantum computing stands today. Accordingly, attempts are being made to deploy post-quantum cryptography algorithms to ensure security.

5.3 Quantum Computing

To execute sophisticated computations, quantum computing makes use of the quantum mechanical phenomena of superposition and entanglement. The two bits in a normal computer can be either on or off, whereas in a quantum computer, the fundamental information-carrying components are qubits, which can be either 0 or 1 or a superposition of 0 and 1 Multiple qubits might be entangled together, existing as individual particles but making up an inseparable whole. The quantum state of one of these affects the other in a predictable manner. The probabilities of the outcome will, however, depend on the quantum states of the qubits immediately before measuring. These give rise to the complexity frontier or the entanglement frontier that is based on the principles of quantum complexity and quantum error correction. Quantum computers would help us analyze the properties of complex molecules by diving deep at the atomic or subatomic levels to offer explanations for fundamental physics like the quantum behavior of a black hole, elementary particles, their properties and simulation, and so on [6, 9].

5.3.1 The Main Aspects of Quantum Computing

Computability Theory: Quantum algorithms can be used for computing problems that are intractable for classical computers and are believed to have a much lesser time complexity than their conventional counterparts even though they do not have any extra edge in terms of computational abilities. Though they are unable to solve few undecidable issues like the halting problem, they can solve complex computations very quickly, for example running Peter Shor's algorithm to find the prime factorization of very large integers, computing Pell's Equation, solving discrete logarithms, and so on that would have been practically unsolvable for classical computers in a feasible amount of time. This resulting polynomial speedup is termed as Quantum Supremacy.

Complexity Theory: Quantum states that could be attained by quantum computers would have superclassical properties. Though quantum computers can perform complex computations much faster, their capacity to quicken classical algorithms has rigid upper bounds. Problems, thus easily and quickly solvable by quantum computers with bounded error are termed as "Bounded Error, Quantum, Polynomial Time" or BQP, which has an error probability of 1/3. BPP stands for "Bounded Error, Probabilistic, Polynomial Time" and is a subset of BQP.

A quantum computer does not violate the Church-Turing thesis, thus it could be theoretically simulated by an ideal Turing machine [10]. But the existing classical algorithms would have had exponential space and time complexities to implement quantum bits on their systems. For example, to simulate a quantum system that makes use of n qubits, 2^n bits of information will have to be stored on the classical computer at any instant. This establishes the fact that quantum computers would be supremely powerful.

Quantum gate: Quantum logic gates are the fundamental building elements of quantum circuits. They are reversible. For example, the Toffoli gate, which has a direct quantum equivalent often makes use of ancilla bits to implement Boolean functions [9].

Quantum error correction (QEC): We require quantum error correction (QEC) to preserve quantum information from errors caused by decoherence and quantum noise. This is essential in the process of attaining fault-tolerant quantum computers that would focus on noise reduction in stored quantum information, as well as on problematic quantum preparation and faulty measurements and is to be attained by encoding the quantum system in a highly entangled state. Scientists are thus appreciating the noisy intermediate-scale quantum (NISQ) technology and believing it to be a vital step in building power quantum systems.

It has been observed that to store and process information reliably, uncontrollable disturbances are produced by quantum systems which is why they need to be isolated completely from the surroundings. However, one should be good enough to control the process from the outside to achieve the result of the computation. This is because qubits require to strongly interact with each other to function. The loss of coherence by electromagnetic vibrations, temperature fluctuations, and so on destroy the quantum properties of the computer, making it extremely difficult to engineer quantum systems. The principle of quantum error correction to remove quantum decoherence could be effectively used to optimize the encoding of data in a highly entangled state but the additional use of qubits for this would greatly increase the overhead charges. Also, they require very low temperatures, close to 0K to operate, thus making it clear that we still have a long way to go in order to achieve the optimal materials, control, and fabrication to design quantum computers on a large scale. However, intermediate scale quantum computers ranging from 50 to 200 cubits are expected to be available in the near future.

5.3.2 Shor's Algorithm

To secure information, the public-key system RSA depends on the difficulty of finding the two prime factors of a large positive integer N, where N=pq (for example if N is the product of two 400-digit prime numbers which conventional computers would take an infinite amount of time to solve). The fast quantum algorithm to compute this was proposed by Peter Shor of Bell Laboratories in the year 1994 [1, 19]. Shor's algorithm is still being studied to see whether it can be improved in terms of cost, the number of cubits, and the number of operations. If N=pq fits into n number of bits, Beauregard's variant uses $O(n^3 \log n)$ operations on 2n+3 qubits, which can be further lowered to $n^2+O(1)$ at the cost of extra qubits. Shor's algorithm internally computes an application of quantum Fourier transform, the periodic function $a^e \bmod N$, a being an integer coprime to N. Harnessing these algorithms to deploy the modern-day public-key RSA and ECC systems would, however, require billions of operations to be performed on thousands of qubits.

5.3.3 Grover's Algorithm

The algorithm proposed by Grover in the year 1996 forms the basis of several quantum algorithms. It involves finding an unordered database of size N using $N^{1/2}$ number of quantum queries. Since this does not make use of an ordered database for which O(log N) complexity would have been sufficient, Grover's algorithm is better described as searching the solutions to an equation $f(x)=0$. If one of the N inputs is a root of the function f, the result must be computed using $N^{1/2}$ quantum operations. If a tiny circuit can assess f, quantum assessments of f would not require a large number of qubits.

5.3.4 The Need for Post-Quantum Cryptography

"Integer factorization," "discrete logarithmic problems," or "elliptic-curve discrete logarithm problems" are used to secure public-key cryptography systems. Though classical computers will be unable to solve these problems, quantum computers using Shor's technique will be able to efficiently break public-key ciphers now in use. As a result, the entire electronic and internet security algorithms, including "RSA, Diffie-Hellman, and elliptic curve Diffie-Hellman," which are used to protect web pages, online transactions, sending and receiving encrypted mails, transmitting confidential data, and so on, would be vulnerable to significant security breaches. Even though today's experimental quantum computers lack the processing power to break real-world cryptosystems, once quantum supremacy is achieved, the widespread availability of sufficiently powerful quantum computers (hundreds or thousands of bits) would put today's cryptosystems in grave jeopardy [1, 3]. Huge investments by tech behemoths such as Google, Intel, IBM, and Microsoft resulted in the March 2018 delivery of a general-purpose quantum computer with 72 physical qubits. Quantum computers are predicted to achieve the computational dominance required to break the current 2000-bit RSA cryptosystem by 2031, after which traditional measures, such as raising key size would no longer be able to restore data safety and privacy. The goal of post-quantum cryptography is to create algorithms that are quantum robust or secure against quantum computer assaults. Quantum computing is likely to reach greater heights in the future decades, thus cryptographers are working on new quantum-proof techniques to protect our data if quantum computing becomes a serious threat. Organizations, such as the National Institute of Standards and Technology (NIST), the European Telecommunications Standards

Institute (ETSI), and the Institute for Quantum Computing, among others, have begun to hold conferences and workshops in the field of Post-quantum Cryptography to encourage studies and research [13]. Current symmetric cryptography methods and hash functions have been found to be relatively more robust against quantum computing assaults. Due to the use of quantum Grover's technique, doubling the key size effectively blocks assaults against symmetric cyphers led by speed increases. As a result, post-quantum symmetric cryptosystems do not need to be drastically different from symmetric cryptosystems currently in use. There is additional research underway to see if existing cryptography methods can be modified to cope with and withstand attacks by quantum computers. Because the no-cloning theorem asserts that copying a state in a quantum setting is not always achievable, further variants of the rewinding technique would be required to develop zero-knowledge proof systems that are safe from quantum adversaries.

5.4 Algorithms Proposed for Post-Quantum Cryptography

Public key cryptographic algorithms have been the focus of research into algorithms that are claimed to be immune to attacks from quantum computers. These are based mostly on lattices, codes, and multivariate polynomials, etc.

5.4.1 Code-Based Cryptography

When it comes to public-key encryption or encapsulation, code-based cryptography corrects errors using coded texts and adopts a measured approach. It is based on a well-researched concern that has been around for more than four decades. Large keys are required for these algorithms, and attempts to lower the size of the keys have rendered them susceptible to threats.

Close to the beginning of public-key cryptography, Mc Eliece presented an approach in the year 1978 that employed a "generator matrix" as a public key to encrypt or encapsulate a coded text by introducing a fixed proportion of errors. McEliece's paradigm is a unique "Goppa code" that can successfully fix the amount of problems recorded.

In terms of reliability, highly reliable computer equipment stores "64 bits of data in 72 bits of physical memory" using an "error-correcting code." Each of the 72 physical bits is specified as a "sum, modulo 2, of part of the 64 logical bits by a 72×64 generator matrix G." "GF 64 2 is a 64-dimensional subspace of the vector space F 72 2 with F2 = (0,1)." Any possible error in the 72 bits may be efficiently repaired, and any repetitive faults (by altering two bits) can be efficiently discovered and reported, thanks to the code's design. All of the main approaches for locating the confidential "Goppa-code" structure are often time demanding [5].

5.4.2 Lattice-Based Cryptography

Lattice-based cryptography algorithms are the smallest amount conservative among all, though they are fairly well-liked and perform the best. Lattice cryptography is predicated on the complexity of the "shortest vector downside (SVP)" that involves calculating the "lowest geometric size of a lattice vector." Even with the employment of a quantum computer, SVP has been portrayed to be a polynomial in n. Different lattice-based cryptographic approaches are supported by "Short number Solutions (SIS)" [15]. These are safe and acceptable within the average-case complexity if the SVP here is tough in the worst case.

In the 1990s, Hoffstein, Pipher, and Silverman delivered a brand new encryption device "NTRU" that has shorter keys compared to McEliece's device. This has not been broken till date. The public key is a "p-coefficient polynomial h = h0 +h1x+hp-1* xp-1, for each coefficient in the set {0, 1, ..., q − 1}." "The public key has 743 multiplied by 11 = 8173 bits if p and q having values 743 and 2048" respectively are selected. Some other polynomial in that range, c, will be used as encrypted data. The transmitter selects two unique polynomials "d and e having low values of coefficients (for example, 1, 0, 1) and calculates c = hd + 7 e mod x p 1 mod q." The symbol "mod x p 1" denotes that x raised to the power p is substituted by 1, "xp+1" by x, and so on, whereas "mod q" denotes that every coefficient is substituted by its remainder when divided by variable q.

We do, however, have Lattice-based signatures that employ a "hash function [H : Znq × {−1, 0, 1}*→ {0, 1}k]," in which the resulting vectors also fulfil that "κ entries are non-zero." It is significantly simple to generate H using a standard "hash function h" by suitably encrypting the variables. To begin, the signer must select y from an "m-dimensional distribution," which is a "discrete Gaussian distribution." (A "discrete Gaussian distribution" is one in which only integer values of the "regular Gaussian distribution" are considered and normalized). The signer then calculates "c = H(Ay mod q,)," "z = Sc + y" and μ is the text. The couple is represented by the signature (c, z). Lyubashevsky employs a "rejection sampling technique" to induce an "S-independent distribution" in order to prevent disclosing details about the secret key S via the distribution of (c, z). This indicates that, depending on the circumstances, the entire process is restarted (c, z). If c and z have low values, "H (Az T c mod q,) = c," the signature will be recognised as legitimate. It is true for "valid signatures" since "Az − T c ≡ A(Sc + y) − ASc ≡ Ay mod q [5]."

5.4.3 Multivariate Cryptography

The difficulties of solving the multivariate polynomial (MVP) problem over finite fields are used in "multivariate polynomial cryptography." "MVP" problems are NP-complete in any field and NP-hard in any field if all equations are quadratic. MVP signing systems, such as Rainbow, are chosen because they provide the shortest signatures [15, 16].

This public key is a set of polynomials p1, ..., pm in the n-variable polynomial ring "F2[x1, ..., xn]" over the "field F2, with m<= n." The polynomials are limited to quadratics and contain no squared terms: each polynomial pi has the form" ai + Σj bi,j xj + Σj<k ci,j,k xjxk "with "ai , bi, j, ci,j,k € F2." There are no likely public structures in the coefficients [5].

5.4.4 Hash-Based Cryptography

Digital signatures based on hashes are resistant to quantum-computer assaults. These approaches are based on the underlying cryptographic hash algorithms' security features (which are collision resistance and second pre-image resistance) [12, 15]. They include systems, such as "Lamport signatures" and "Merkle signatures," among others, "SPHINCS and XMSS."

Finding a preimage for a particular output string is computationally challenging, which is one of the design goals for hash functions. In 1975, Lamport found that it might be used to establish a one-time signature system. The one-time signatures of Lamport work as follows: for generating the key pair, Max chooses two random strings which are x0 and x1; both these constitute his secret key. His public key is "(h(x0), h(x1))," where h is a known hash function. If he wants to sign 0 he reveals the string x0; the verifier recomputes the value of h(x0) and checks the result against the first half of the public key [18]. To sign 1, Max reveals x1. More generally, to sign a message which is m-bit, Max takes 2m strings as the secret key "X = (x10, x11, x20, x21, ..., xm0, xm1)" and their hash values as public key "Y = (h(x10), h(x11), h(x20), h(x21), ..., h(xm0), h(xm1))." The signature of, e.g., "10110 ... is (x11, x20, x31, x41, x50, ...)." If Max signs more than one message with the same key, the system's security quickly deteriorates. The number of signatures can be increased indefinitely, however, this increases the signature's size [5].

5.4.5 Supersingular Elliptic Curve Isogeny Cryptography

This cryptosystem uses the features of "supersingular elliptic curves and supersingular isogeny graphs" to produce a "Diffie-Hellman substitute with forward secrecy." They are quantum-resistant and can be used to replace the current Diffie-Hellman algorithm [11]. Because it provides advanced secrecy, it aids in the prevention of government bulk surveillance. Based on "supersingular elliptic curve isogenies," the "Chinese State Key Lab Laboratories for Integrated Service Networks and Xidian University" extended the work of Feo, Jao, and Plut to establish secure "digital signatures" that are quantum-resistant in 2012 [5].

5.4.6 Quantum-Resistant Symmetric Key Cryptography

Symmetric key cryptosystems like "AES and SNOW 3G" would become quantum-resistant if keys of huge sizes could be employed. The key management systems "Kerberos and 3GPP Mobile Network Authentication Structure" are both resistant to quantum computer attacks. Extended versions of "Kerberos-like cryptosystems" can be effective algorithms because they are already in use.

5.5 Launching of the Project Called "Open Quantum Safe"

The "Open Quantum Safe or OQS project" began in the year 2016, aimed at the development and prototyping of quantum-resilient cryptographic systems. It integrates the currently used post-quantum cryptographic algorithms under the open source C library: liboqs. It stores different implementations and includes test standards and benchmarks to compare the performances of these algorithms. It has a common Application Programming Interface that is suited for post-quantum key exchange. Currently, "BCNS15, NewHope, Frodo, NTRU, SIDH, McBits" are the key exchange algorithms that are supported by liboqs.

5.6 Algorithms Proposed During the NIST Standardization Procedure for Post-Quantum Cryptography

The "National Institute of Standards and Technology," commonly known as NIST had conducted a competition-like process to select and standardize public-key cryptographic algorithms that would be resistant and worthy of replacing the modern-day cryptosystems [8]. In November 2017, NIST evaluated 82 algorithm entries and selected 69 of them that met the required acceptance criteria to form the First-Round Candidates on December 20, 2017. Among these, 26 candidates were shortlisted on January 30, 2019, 17 of them being based on "public-key cryptography" and 9 being "digital signature encryption algorithms [4]."

5.7 Hardware Requirements of Post-Quantum Cryptographic Algorithms

Multiprocessor system on chip (MPSoC) would speed up the NTRUEncypt, i.e., one of the most well-known post-quantum cryptography (PCQ) algorithms when software or hardware is designed onXilinx Zynq Ultrascale+. However, the speed in embedded software programs is limited because its primary operation that is the polynomial multiplication takes a long time for execution. The two feasible approaches that can be used in the implementation of polynomial multiplication in the computable Logic of Zynq that would help in the speeding up process are, namely (1) traditional register-transfer level (RTL) and (2) high-level synthesis (HLS).

The remaining NTRUEncrypt operations are executed on the Processing System of Zynq in software using the mode of bare metal.

5.7.1 NTRUEncrypt

The four main parameters of NTRUEncrypt algorithm are: N, p, q and d. The polynomial multiplication in is given by " $R=Z_q[X]/X^N-1$."

When the key is being generated, the two secret polynomials F and g are chosen randomly from R. These have small coefficients $d = 1$ or -1 or 0. The computation of the private key, having small coefficients is done according to the equation $f=1+p$. F and the public key is computed following the equation "$h=f^{-1} *g$. p in R [17]." The encrypting process that consumes the longest time is given by "*t=r*h.*" Here, *r* is a polynomial having small coefficients from *R* and *h* being a public key has large coefficients.

If *F* is a private key (has small coefficients) and belongs to *R* and *c* is a ciphertext (has large coefficients). The key operation performed in the phase 1 decryption (DEC-1) is:

"m" = f * c = (1+3.F) * c"

The key operation performed in the phase 2 decryption (DEC-2) is:

" *t*"=*r*h.*" [7]

5.7.1.1 Polynomial Multiplication

The Poly Mult algorithm is given as:

STEP 1: START
STEP 2: A for loop is considered that would run from $i=0$ to $i=(2d-1)$
STEP 2.1: If the value of i is less than d,
STEP 2.1.1: Another for loop is considered which runs from $j=0$ to $(N-1)$
STEP 2.1.1.1: The value of c_j is incremented by $c_j=c_j+a_{j-bi \bmod N}$
STEP 2.2: Else if the value of i is not less than d,
STEP 2.2.1: Another for loop is considered which runs from $j=0$ to $(N-1)$
STEP 2.2.1.1: The value of c_j is incremented by $c_j=c_j-a_{j-bi \bmod N}$
STEP 3: END

Here, the inputs are:

- Polynomial "*a(X)*" with "*N*" coefficients ranging between *0* and *q-1* (both inclusive)
- Polynomial "*b(X)*" with "*d*" coefficients 1 at locations "b_0" to "b_{d-1}" and *d* coefficients -1 at locations "b_d " to "b_{2d-1}" where "b_i" belongs to [*0*, *N-1*].

Output:

"c(X)=a(X)*b(X) mod X^N -1" [7]

5.7.1.2 Hardware to Accelerate NTRUEncrypt

A fast convolution algorithm was introduced by Bailey in 2001 that exploited the sparsity of polynomials to multiply. Kamal targeted Virtex-E family of FPGAs to compute polynomial multiplication using the ternary nature of polynomials and a Barrel shifter that has been chosen empirically in NTRUEncrypt. In 2016, to apply the truncated polynomial ring multiplier Liu made use of an extended form of Linear Feedback Shift Register.

5.7.2 Hardware-Software Design to Implement PCQ Algorithms

Software-hardware codesign has been applied to improve the algorithms used for post-quantum cryptography. Ghosh implemented a PicoBlaze softcore coprocessor and a number of acceleration units in parallel on SPARTAN-3AN FPGAs for the McEliece Cryptosystem. A very rapid application of lattice-based digital signatures that use the software-hardware codesign was built by Aysu. This design comprised of a NIOS II processor, a polynomial multiplier and a hash unit and its target was the Cyclone IV FPGA family. For the lattice-based Fan-Vercauteren (FV) homomorphic encryption, Migliore presented another software-hardware codesign.

5.7.3 Implementation of Cryptographic Algorithms Using HLS

Approaches based on RTL and HLS to implement AES have been reported. A reduction by a factor 1.19 was found since HLS requires 20% more number of clock cycles.

Points to be noted regarding the Systemization of Knowledge: In this Systemization of Knowledge study, a customary design framework and a common target FPGA platform was used to implement PQC algorithms. The key points regarding this are:

1. In terms of latency and LAP, NTRU-HRSS is superior among all the KEM algorithms.
2. While the CRYSTALS-Dilithium is superior for Signature, for Signature verification purposes, qTesla is superior within the Signature Algorithms that have security level 1. In terms of latency and LAP, SPHINCS+ is the costliest.
3. For high security that is level 3, Saber that has low latency for server applications is ideal.

4. NTRU-HRSS and NewHope are the most rapid ASICs and thus are really commendable for servers having level 1 security. SPHINCS+ ASICS and FrodoKEM serve well in IoT devices as they consume a low amount of power and come in small decapsulation modules.
5. NTRU-HRSS and SPHINCS+ FPGAs are very useful in security level 1 low power IoT devices that need compact designs and low area. Crystals-KYBER FPGA having low latency acts as a good security level 1 server.
6. Loop-pipelining is efficient for Signature Algorithms of level 1 security and makes CRYSTALS-Kyber perform the fastest. To reduce latency in KEM algorithms, high levels of security like 3 and 5, loop unrolling is more efficient. Here NRTU-HRSS is the fastest.
7. When loop pipelined, qTESLA and CRYSTALS-Dilithium had the smallest LAP, whereas when loop unrolled, MQDSS had the smallest.
8. Various algorithm components can be optimized by modifying and re-running the code using HLS for low area. Optimization of different versions of the Keccak function reduces the overall area though this decrease is not much in the case of LAPs, still higher for pipelining than unrolling.
9. Implementations of PQC hardware are not optimised for resistance of side-channel. Both low latency and low area of PQC algorithms can be implemented using Virtex or Artix. The use of HLS can help in optimizing these designs further. Research is still on to implement hardware security to analyze fault-attack side channels, timing, power, and so on.

5.8 Challenges on the Way of Post-Quantum Cryptography

It is still not known when or under what circumstances we will need to switch to post-quantum cryptographic systems. Cryptographic systems like the RSA use a 4000-bit key. They are not resilient to attacks by quantum computers but can counter powerful classical computers. On the other hand, McEliece's system of encryption, having a four-million-bit key, is supposedly able to withstand attacks by both classical and quantum computers. The three most important challenges that are to be overcome, in order to widely implement post-quantum cryptography are improving the efficiency, building confidence and improving the usability.

- **Improving Efficiency**
  A web server that handles millions of users per minute, should have an implementation of a cryptographic system that is efficient, as well as fast. Time and space constraints have always been a huge challenge to cryptographers. However, increased research in these fields have introduced several "speedups." A similar situation is also expected for post-quantum cryptography, and some progress has already been made in this regard.
- **Building Confidence**
  Though Merkle's and McEliece's methods of public-key encryptions remain unbroken today, some proposals for hash-based, lattice-based and multivariate-based systems of cryptography have already been broken. With time, users need to switch to newer systems of cryptography that are faster, smaller and more reliable than classical systems. Rigorous research is needed by cryptanalysts on attacks by quantum computers and the algorithms that have been proposed for post-quantum cryptography, so as to build confidence in these newer systems.
- **Improving Usability**
  A one-function system like the RSA is not capable enough to ensure secure encryption. A secure and efficient encryption system that has been well-defined and standardized should have both software and hardware implementations. Timing leaks and other types of side-channel leaks are to be strictly avoided and correctness and speed has to be ensured. To make them high-speed and leak-resistant, some research on randomization and padding techniques for post-quantum cryptographic systems is still going on.

5.9 Post-Quantum Cryptography Versus Quantum Cryptography

Quantum cryptography manipulates the properties of quantum mechanics to encrypt messages. Quantum key distribution or QKD is an implementation of this, in which the process of sending encryption keys is theoretically resistant to attacks. It is based on the transfer of photons through a fiberoptic line. If one tries to break into the system, the polarization of the photons is affected, thus making it clear to the recipient that the message is not secure. Thus, when one attempts to read the encoded data, wave function collapse will lead to a change in the quantum state. QKD can securely travel through noisy channels over a considerably long distance and involves the use of quantum repeaters to efficiently resolve quantum communication errors. This system is, however, time consuming, and the sending and receiving of photons require the use of specialized and costly equipment. The differences between post-quantum cryptography and quantum cryptography are:

- Quantum cryptography focuses on enlarging a short, shared input stream into a long, shared output stream, whereas post-quantum cryptography handles several secure-communication tasks like public-key signatures, secret-key encapsulation, public-key encryptions, secure electronic voting, and so on.
- Quantum cryptography permits the completion of proven or conjectured cryptographic tasks but post-quantum cryptography contains secure systems that are proven as well as systems conjectured to be secure having lower costs.
- Quantum cryptography requires the use of hardware that is too expensive for most users to afford. Post-quantum cryptography, however, uses methods which can be put into use for a considerable part of present-day Internet communication.

5.10 Future Prospects of Post-Quantum Cryptography

Once quantum computers come into use, which is expected by 2030, a serious threat would be posed to the modern-day cryptosystems. This calls for transitioning on an urgent basis to post-quantum cryptosystems to secure information. Purging the current cryptosystem and integrating new algorithms in the real world would, however, involve huge amounts of time, energy, and money. Today's cryptographic system fits the user's budget, their capabilities and the attacker's capabilities. The transition made earlier from weaker to stronger cryptosystems involved switching to 112 or 128 bits, based on the time complexity of a possible attack by a classical computer. However, the transition to quantum-resilient cryptosystems would not involve the paradigm of bits-of-security. "Grover's algorithm" attains a quadratic speed up and this can be recompensated by doubling the key sizes in symmetric key systems. However, the exact key size needed to resist attacks by a sophisticated quantum system is not known yet. NIST is taking steps to standardize post-quantum cryptographic algorithms. Since 2017, it has been evaluating proposed algorithms for key exchange, digital signatures and public key encryption systems that are quantum resistant. This might help in standardizing some algorithms from among these entries that provide security and are cost-effective. Hash-based signatures have meanwhile almost been standardized by IETF's research branch IRTF. Some other organizations working to standardize post-quantum cryptography algorithms are ETSI that has a "Quantum-Safe working group, OASIS" that works on the KMIP standards and ISO that works with "SC27 and WG2," and the "European EU-H2020 PQCRYPTO project." Agencies should be prepared for a transition by the end of this decade while research and analysis in this field goes on and different organizations continue to evaluate and assess the degree of threat quantum system attacks could pose to standards that exist today and thereby standardize algorithms that inspire confidence and are widely deployable in the years to come.

References

1. Chen, L., Jordan, S., Liu, Y.K., Moody, D., Peralta, R., Perlner, R., Smith-tone, D., *Report on Post-Quantum Cryptography, NISTIR 8105*, p. 10, National Institute of Standards and Technology, Gaithersburg, Maryland, April 2016, https://nvlpubs.nist.gov/nistpubs/ir/2016/nist.ir.8105.pdf.
2. Zwicke, A., An introduction to modern cryptosystems, in: *Global Information Assurance Certification Paper, GIAC Version 1.4b, Option 1*, SANS Institute 2003, https://www.giac.org/paper/gsec/2604/introduction-modern-cryptosystems/104482.
3. Bernstein, D.J. and Lange, T., *Post-Quantum Cryptography*, vol. 549, 7 Macmillan Publishers Limited, part of Springer Nature, London, England, September 2017, https://doi.org/10.1038/nature23461.
4. Alagic, G., Alperin-Sheriff, J., Apon, D., Cooper, D., Dang, Q., Liu, Y.-K., Miller, C., Moody, D., Peralta, R., Perlner, R., Robinson, A., Smith-Tone, D., *Status Report on the First Round of the NIST Post-Quantum Cryptography Standardization Process NISTIR 8240*, National Institute of Standards and Technology, Gaithersburg, Maryland, United States of America, January 2019, https://doi.org/10.6028/NIST.IR.8240.

5. Stebila, D. and Mosca, M., Post-quantum key exchange for the internet and the open quantum safe project, in: *Selected Areas in Cryptography–SAC 2016, SAC 2016. Lecture Notes in Computer Science*, vol. 10532 Springer, Cham, 2017, https://doi.org/10.1007/978-3-319-69453-5_2.
6. Preskill, J., *Quantum Computing in the NISQ Era and Beyond*, Institute for Quantum Information and Matter and Walter Burke Institute for Theoretical Physics, California Institute of Technology, Pasadena CA 91125, USA, 2018, https://doi.org/10.22331/q-2018-08-06-79.
7. Farahmand, F., Nguyen, D.T., Dang, V.B., Ferozpuri, A., Gaj, K., Software/hardware codesign of the post quantum cryptography algorithm NTRU encrypt using high-level synthesis and register-transfer level design methodologies. *2019 29th International Conference on Field Programmable Logic and Applications (FPL)*, pp. 225–231, 2019, doi: 10.1109/FPL.2019.00042.
8. Koziel, B., Azarderakhsh, R., Mozaffari Kermani, M., Jao, D., Post-quantum cryptography on FPGA based on isogenies on elliptic curves. *IEEE Trans. Circuits Syst. I Regul. Pap.*, 64, 1, 86–99, Jan. 2017. doi: 10.1109/TCSI.2016.2611561.
9. Steane, A., *Quantum Computing*, Department of Atomic and Laser Physics, University of Oxford, Clarendon Laboratory, Parks Road, Oxford OX1 3PU, UK, August 13, 1997, https://doi.org/10.1088/0034-4885/61/2/002.
10. Bennett, C.H., Bernstein, E., Brassard, G., Vazirani, U., Strengths and weaknesses of quantum computing. *SIAM J. Comput.*, 26, 5, 1510–1523, 1997. https://doi.org/10.1137/S0097539796300933.
11. Diffie, W. and Hellman, M., New directions in cryptography. *IEEE Trans. Inf. Theory*, 22, 6, 644–654, November 1976. doi: 10.1109/TIT.1976.1055638.
12. Stinson, D.R. and Paterson, M., *Cryptography: Theory and Practice*, CRC Press, Taylor and Francis Group, Boka Raton, Florida, Aug-2018, https://www.ic.unicamp.br/~rdahab/cursos/mo421-mc889/Welcome_files/Stinson-Paterson_CryptographyTheoryAndPractice-CRC%20Press%20%282019%29.pdf
13. Mosca, M., Cybersecurity in an era with quantum computers: Will we be ready? *IEEE Secur. Priv.*, 16, 5, 38–41, September/October 2018. doi: 10.1109/MSP.2018.3761723.
14. Bernstein, D.J., *Cache-Timing Attacks on AES*, Department of Mathematics, Statistics, and Computer Science, The University of Illinois, Chicago, 2005, https://citeseerx.ist.psu.edu/viewdoc/download?doi=10.1.1.140.2835&rep=rep1&type=pdf.
15. Basu, K., Soni, D., Nabeel, M., Karri, R., NIST post-quantum cryptography—A hardware evaluation study. *Proc. IACR Cryptol. ePrint Arch.*, pp. 47, 2019. https://eprint.iacr.org/2019/047.pdf.
16. Gaborit, P., Post-quantum cryptography. *5th International Workshop, PQCrypto 2013, Proceedings*, Limoges, France, June 4-7, 2013, *Lecture Notes in Computer Science*, vol. 7932, 2013. doi: 10.1007/978-3-642-38616-9.

17. Song, F., A note on quantum security for post-quantum cryptography, in: *Post-Quantum Cryptography, PQCrypto 2014, Lecture Notes in Computer Science, vol 8772,* M. Mosca, (Ed.), Springer, Cham, 2014, https://doi.org/10.1007/978-3-319-11659-4_15.
18. Bernstein, D.J., *Introduction to Post-Quantum Cryptography.* Department of Computer Science, University of Illinois at Chicago, 2014, https://doi.org/10.1007/978-3-319-11659-4_15.
19. Buchmann, J.A., Butin, D., Göpfert, F., Petzoldt, A., Post-quantum cryptography: State of the art, in: *The New Codebreakers, Lecture Notes in Computer Science, vol 9100,* P. Ryan, D. Naccache, J.J. Quisquater (Eds.), Springer, Berlin, Heidelberg, 2016, https://doi.org/10.1007/978-3-662-49301-4_6.

6

A Statistical Characterization of MCX Crude Oil Price with Regard to Persistence Behavior and Seasonal Anomaly

Anindita Bhattacharjee[1]*, Jaya Mamta Prosad[2] and M.K. Das[3]

[1]*Symbiosis Centre for Management Studies NOIDA, Symbiosis International (Deemed University), Pune, India*
[2]*Amity College of Commerce and Finance, Amity University, Noida, U.P., India*
[3]*Institute of Informatics and Communication, University of Delhi, South Campus, New Delhi, India*

Abstract

The chapter investigates long-term persistence behavior followed by seasonal anomaly in MCX crude oil returns. Daily return data for the period of 2009 to 2018 is considered for this study. Hurst exponent is computed and used as an indicator of persistence behavior. The data is found to have long range correlation. Return distribution of asset price is leptokurtic which recovers Gaussianity with a lag of 550 days. Additionally, on analyzing the cumulative monthly returns, distinct seasonal patterns were observed for summer and winter months indicating seasonal anomaly that contradicts efficient market hypothesis. This finding has been further utilized to devise suitable investment strategies that may guide investors to gain better return in the crude oil market.

Keywords: Persistence, MCX crude oil, Hurst exponent, power law, seasonal anomaly, efficient market hypothesis

6.1 Introduction

Volatility and time series characteristics are two defining features of stock market data. While, a lot of research work has been done in order to

**Corresponding author*: anindita.bhattacharjee@scmsnoida.ac.in; b.anindita10@gmail.com

Sharmistha Ghosh, M. Niranjanamurthy, Krishanu Deyasi, Biswadip Basu Mallik, and Santanu Das (eds.) Mathematics and Computer Science Volume 1, (111–132) © 2023 Scrivener Publishing LLC

understand the volatile nature of stock market data and model it for future purpose [1–5], a significant amount of these works investigate about the temporal characteristics of price addressing whether the prices are historically dependent or are they random [6–8]. This time dependence of stock market variables imparts significant information about their predictability as well as the market efficiency. The temporal characteristics include a short term as well as long term correlation of price with its past values. A positive correlation of current and historical prices is termed as persistence. A long term correlation due to herding or any other economic or political reason may result in pricing bubble [9]. On the other hand, a short term correlation may exist in a time series as a result of seasonal effect. In financial terms, this pattern (positive short term or long term correlation) can be related to momentum anomaly and is a deviation from the random walk theory of efficient market hypothesis(hereafter referred to as EMH) [10–12]. Simplifying further, this theory states that, in an efficient market the prices or returns follow random walk. It means that there is no correlation between the past and the present prices. However, over a period of time, several studies on stock prices have emerged that defy the efficient market hypothesis [13–15].

The implications of theories that defy EMH are extremely significant and can't be ignored. The temporal correlations, if validated on the real time data, suggest that, it is possible to beat the market by predicting its trend and values. The presence of historical correlation (both positive and negative) makes the markets inefficient. Further, it provides a scope to the investors to develop strategies accordingly and gain supernormal returns. Therefore, it is imperative to investigate the randomness of stock market values.

There are numerous methods to investigate the correlation between past and present market values. One of the concrete methods to detect the same was given by H E Hurst [16] where the author explains the correlation of asset price with time efficiently using a statistical measure known as Hurst exponent (H). Hurst exponent tries to identify whether a time series is self-similar (persistent) or random. Initially, this idea was applied to describe the long range correlation between the water level in rivers and reservoirs [17]. Later on it was also applied to characterize a system based on randomness such as stock market data, commodity market data, forex data etc. [18–20].

In this context, our first aim is to investigate the efficiency of the crude oil market by detecting persistence behavior in the returns. Crude oil is traded as a stock on the commodity exchanges that are an integral component of financial markets. It is a crucial commodity whose price fluctuations can have significant implications on the economy of a country. For instance, increase in stock prices of crude oil have a positive impact on country's economy [21]. Although there are a large number of studies investigating temporal correlation on stock prices [6–8, 22, 23], very few researches have focussed on the detection of this anomaly in the crude oil data. Considering the dearth of researches on this area and to fill this gap in the literature, this chapter attempts to understand the behavior of this commodity.

Additionally, persistence anomaly has significant implications on the prediction of oil prices. If persistence in oil prices prevails in cyclical pattern, then it is possible to predict the price movement and devise a buy/sell strategy accordingly. Therefore, another aim of this chapter is to investigate the occurrence of seasonal patterns in oil prices. The chapter further discusses the implications of these findings on investment behavior.

6.2 Related Literature

Persistence in oil return is an anomaly that defies the norm of random walk theory and weakens market efficiency. Existence of persistence or long term behavior has been studied by [24, 25]. A similar pattern of persistence has been reported by [26] wherein the authors used multifractal detrended moving average method to investigate for market inefficiency and concluded that persistence behavior has made the crude oil futures market inefficient in the long run. Various other researchers have also investigated this anomaly in oil prices [27–29].

The above literature provides a glimpse of presence of persistence in crude oil. Further, it can be noted that the nature of persistence can be closely linked to momentum anomaly wherein positive correlation between present and past stock market data is seen. Momentum in financial market generally follows a pattern. This pattern could be seasonal in nature. Seasonal patterns and calendar anomalies in stock markets have been dealt in wide amount of financial literature historically [30–34]. However, very few researches have paid attention to this phenomenon in the oil market [35]. Auer studies the effect of weekdays in oil prices and found that the price volatility is higher on Mondays [36]. Borowski analyzes seasonality effect in future contracts of various commodities including crude oil and Brent oil [37]. The author indicates monthly effect in November for the crude oil and Brent oil future contracts. This study reports the presence of negative returns in November. Arendas *et al.* detect oil prices yield better return during summer months (May-October) as compared to winter months (November-April) [35, 38]. Quayyoum *et al.* investigate the day of the week effect and month of the year effect in oil price returns [39]. These studies reveal that the crude oil market indeed has certain significant seasonalities that can provide important information for the agents to device investment strategies. Additionally, while a few researches capture seasonal anomaly in crude oil data of developed countries, as mentioned above, there has been a dearth of such studies in Indian context.

In this regard, the present chapter has two pronged focus. First, the chapter investigates the presence of short- and long-term persistence in daily returns of MCX crude oil. Second, it also aims to detect and analyze the seasonal anomaly in MCX crude oil returns.

6.3 Data Description and Methodology

6.3.1 Data

Historical data of daily traded contract (in lots) and value (in Lac.) of crude oil are taken from Energy sector under the head of FUTCOM (www.mcx-india.com). Daily turnover per lot is computed from the above mentioned historical data, and this has been considered as price of crude oil per lot in view of online trading of crude oil (Figure 6.1).

The daily crude oil price (in Lacs) per lot is displayed for 10 years (from January 1, 2009, to January 1, 2019) in the above figure. The price is ranging from 211234.83 on January 1, 2009, to 318217.60 on January 1, 2019. The price has increased gradually from 2009 to 2013 followed by a fall in price from 2014 to 2015. Thereafter, it has again increased till 2018 and fall again during the end of 2018. Hence, the price is found to be changing from high to low and then from low to high recurrently. However, small fluctuations are always observed to exist for a shorter period in the overall span of investigation.

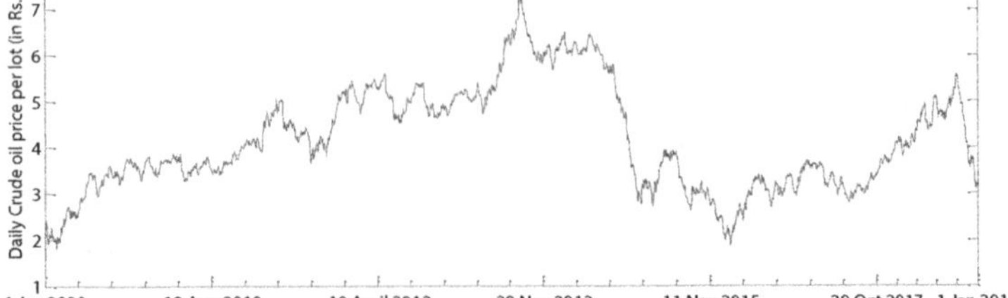

Figure 6.1 Daily crude oil price per lot from January 1, 2009, to January 1, 2019.

6.3.2 Methodology

The methodology has been divided into parts. The first part deals with investigation of persistence in crude oil data through Hurst Exponent. The second part provides the details for detecting seasonal anomaly in crude oil returns.

6.3.2.1 *Characterizing Persistence Behavior of Crude Oil Return Time Series Using Hurst Exponent*

Hurst exponent is an excellent measure of characterizing a time series on the basis of persistence behavior. This technique was first developed to measure the long term storage capacity of reservoir, later on applied to predict the direction of flow of River Nile, in view of making a better dam in Egypt [25]. It is efficient in the sense that it provides all essential information of a complex system in a single parameter (H). Studies suggests that in an independent and identically distributed asset price follows Normal distribution where the standard deviation of the process $\sigma(t) \approx t^H$, t is the increment in time. H is known as Hurst Exponent [16]. The exponent H indicates the scaling relationship of the asset price variation with the change in time. Scaling property is a much discussed topic in econophysics where a variety of techniques of physics are applied to analyze the economic and financial data. The exponent H characterizes the time series as self-similar or random. At times the series is also characterized as monofractal (having single scaling property) or multifractal (different scaling exponents for different part of the series). This technique is widely applied for studying the behavior of financial time series in recent times [40–44]. In the forgoing chapter, Hurst exponent is measured by rescaled–range method [25], which works as the following:

For any time series $\{x(t)\}$, it is divided into n subparts, where $n = 2, 2^2, 2^3, 2^4, \ldots 2^k$ Mean is computed for each subpart as

$$\mu_i = \frac{1}{k}\sum_{t=1}^{k} x_t, i = 1, 2, \ldots, n \tag{6.1}$$

Then, the mean μ is subtracted from each observation of $\{x(t)\}$ and thus a new series $\{y(t)\}$ is formed as:

$$y_t = x_t - \mu \tag{6.2}$$

The cumulative sum of y_t is computed as

$$Y_t = \sum_{i=1}^{t} y_t \tag{6.3}$$

Range of Y_t is computed as follows:

$$Range = Max(Y_1, Y_2, \ldots, Y_t) - Min(Y_1, Y_2, \ldots, Y_t) \tag{6.4}$$

Standard deviation of each range is computed:

$$\sigma = \frac{1}{k}\sqrt{\sum_{t=1}^{k} (x_t - \mu)^2} \tag{6.5}$$

$$\text{Rescaled range } R_i = \left(\frac{Range}{\sigma}\right) \tag{6.6}$$

$$\text{Average value of } R_i \text{ is computed as } R = \frac{1}{n}\sum_{i=1}^{n} R_i \tag{6.7}$$

Next, (log n, log R) are plotted and joined by a line. The slope of the line is referred as Hurst exponent (H).

H=0.5 indicates a random series, $0<H<0.5$ refers to a anti persistence behavior. On the contrary, a persistent behavior is observed for $0.5<H<1$.

In the present work, the time series corresponds to log return of the asset price series $\{p(t)\}$:

$$x(t) = \ln(p_{t+1}) - \ln(p_t) \tag{6.8}$$

6.3.2.2 *Zipf Plot*

Scaling behavior of return is discussed here in terms of Zipf plot (a method of rank-frequency plot). The absolute values of return are used incorporating the case of high risk involving large fluctuation (both direction) in the asset price. Absolute values of returns are arranged in descending order and let these absolute returns be denoted by $\{x_1, x_2, x_3, \ldots, x_N\}$ and the cumulative distribution function (CDF) be denoted by $F(x_i) = P(x_i \leq x_k)$, where $i = 1, 2, \ldots N$. The rank of each such return is obtained as follows [45]:

$$\text{rank } (x_i) = N \times (1 - F(x_i))$$

$$\text{or, } \log[\text{rank}(x_i)] = \log[(1 - F(x_i)] + \log(N) \tag{6.9}$$

The Zipf plot can be obtained by plotting rank (x_i) against $|x_i|$

6.3.2.3 Seasonal Anomaly in Oil Returns

Seasonal anomaly has been checked by testing the significance of the difference between average monthly returns of oil price data for summer and winter. Average monthly returns of oil price data for each year starting from 2009 to 2018 is computed (c.f. Table 6.1). Further, the data is segregated into specific periods based on the return patterns of different months (subperiods) in order to verify seasonality. The difference between the returns of these subperiods (or seasons) are tested through two-sample t-test which is a parametric test for testing the statistical significance of the difference between sample means of two samples [46].

6.4 Analysis and Findings

Analysis of this work is presented in two parts. Persistence behavior of crude oil stock price is discussed at first followed by an analysis of short-term seasonal variation.

6.4.1 Persistence Behavior of Daily Oil Stock Price

The $\{x(t)\}$ computed using Eqn. 6.8 is plotted against time in Figure 6.2 showing fluctuation around a mean value = 0.000146 with a standard deviation of 0.0161. Fluctuation around the mean value is very less at sometimes indicating low risk, on the other hand, it is found to be more at some other places indicating a very high risk [47]. It can be said that the data $\{x(t)\}$ is highly volatile in nature, and the volatility is randomly distributed over the sample period.

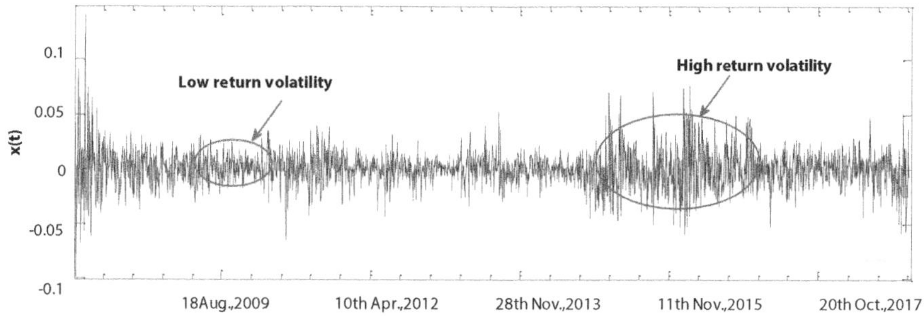

Figure 6.2 Logarithmic return of crude oil price plotted with time.

Volatile nature of the data makes it difficult to forecast, and thus, it is important to know if the volatility is random or following any aperiodic trend with time. Financial data may be mean reverting (randomness) or reinforcing (having long range correlation or nonperiodic trend) in nature. In this analysis, the entire set of observations is considered for characterization in the context of its asymptotic behavior. Hurst exponent is computed for the data following the method discussed earlier. Our computation shown in Figure 6.3 gives, H=0.6920 which lies between (0.5, 1).

This indicates the existence of long term correlation in the data. Since computed H>0.5, this long-term correlation essentially leads to persistence behavior.

In the next part of the analysis, power law behavior of the return distribution has been discussed. Power law phenomenon characterizes the scaling behavior of a time series, which emerges out of a complex system of many interactive units. It is a common phenomenon observed in many economic and financial systems [45, 48]. In this study, the power law behavior of the daily return of crude oil price is discussed with the help of rank frequency plot. Normalized values of {x(t)} is considered for this purpose. The cumulative frequency distribution of the daily absolute return of crude oil price is constructed in order to find the probability of occurrence of a return less than a specific return i.e. $P(|x_t| > X) \approx X^{-\gamma}$ where X is some threshold value. Absolute values of returns $|x_t|$ are considered for ranking so as to investigate the nature of the high return (in both direction). Ranking is done according to decreasing order of their magnitude. A rank frequency relation is thus obtained and displayed in Figure 6.4.

The distribution follows power law with γ = 0.1373 (by considering X > 1.3890) which is determined by least square fit. More precisely, the

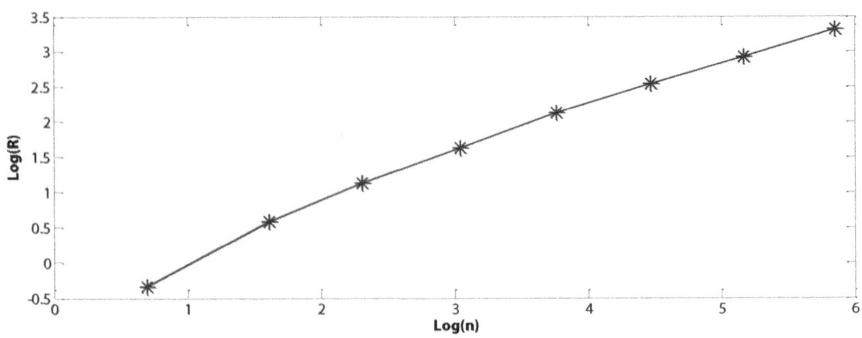

Figure 6.3 Hurst exponent computed as a slope of Log(R) Vs Log(n).

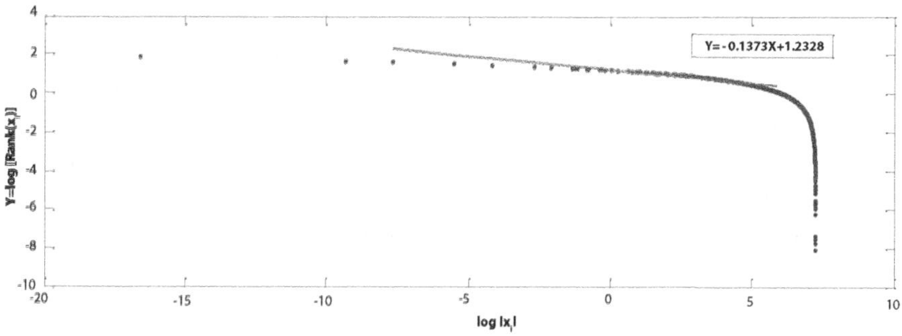

Figure 6.4 Rank frequency plot of logarithm of absolute return.

distribution follows a power law with $\gamma = 0.20$, $\gamma = 0.11$, and $\gamma = 0.04$ for absolute values of return more than 1σ, 2σ, or 3σ limits of the mean value of normalized $|x(t)|$. This describes the fat tail behavior of large returns. It also shows that the distribution follows power law for large returns. Figure 6.4 clearly shows that the lower values of γ corresponds to higher returns. Furthermore, empirical distribution of return is superimposed with the Gaussian distribution of same mean and standard distribution (c.f., Figure 6.5).

The figure shows that the central part of the empirical distribution is more peaked and tails much fatter than the Gaussian distribution superimposed on it. Therefore, it is evident that the daily return distribution is not Gaussian. This triggers a question whether Gaussianity can be recovered for weekly or monthly returns? This question is answered by computing kurtosis of multi period cumulative returns of the asset price with various time lags L. Since kurtosis of a Gaussian distribution is 3, we try to find out lag for which L for which kurtosis of the cumulative probability

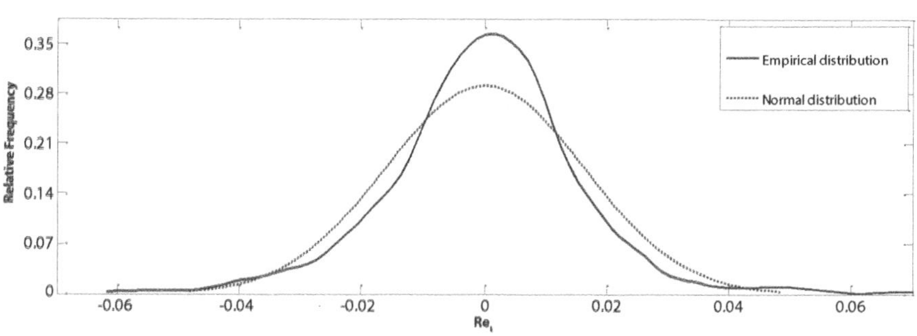

Figure 6.5 Probability distribution of daily logarithmic return and normal distribution with same mean and variance.

distribution becomes 3. This method has been first applied by Mantegena and Stanley, wherein successive variation of S & P index was computed by varying the time interval. In this, time required for price variation to converge to a Gaussian process was shown to be of the order of 1 month [49]. Similar research has also been conducted on gold and stock exchange data to recover Gaussianity [50, 51]. In the foregoing work, time lag has been varied for computing cumulative return as following:

$$Re_t(L) = \frac{1}{L}[\ln(x_{t+L}) - \ln(x_t)] \tag{6.10}$$

where $L = \frac{N}{k}, N$ is the total number of observations and k is the number of non overlapping subparts. More precisely, $L = 1$ refers to daily return, $L = 2$ refers to cumulative return for two days, and so on. Thereafter, for each value of L, we have computed kurtosis of the return set as

$$K(L) = \frac{\langle Re_t^4(L) \rangle}{\langle Re_t^2(L) \rangle^2} \tag{6.11}$$

Table 6.1 presents computed result for few cases.

It is evident that the increase in time lag is reducing the kurtosis and hence the height of the probability distribution is reducing (c.f. Figure 6.5a). The fat tail of the probability distribution is gradually becoming shorter and it is approaching to be a Gaussian Distribution. Computation is thus extended varying the time lag L.

The above mentioned figure (Figure 6.5a) depicts the probability distribution of cumulative return with various time lag showing the transformation in the shape wherein the fat tail gradually becomes shorter for increased time lag. The resulting $K(L)$ when plotted against L clearly

Table 6.1 Kurtosis of cumulative returns.

L	Cumulative return	Kurtosis
1 day	Daily return	7.74
3 days	Return after 3 days	5.54
5 days	Return after 5 days	5.33
10 days	Return after 10 days	3.55

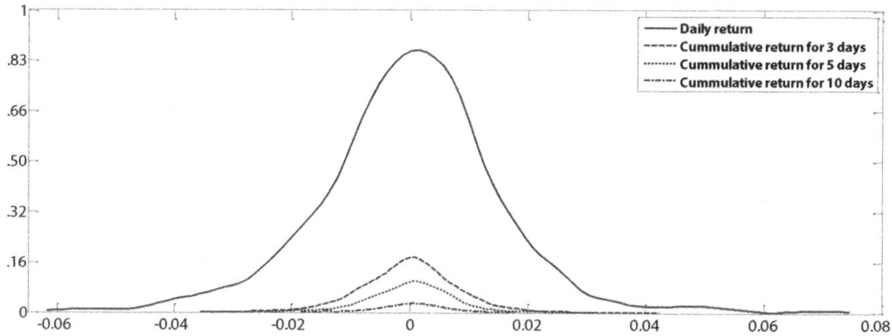

Figure 6.5a Probability distribution of cumulative return with various time lag.

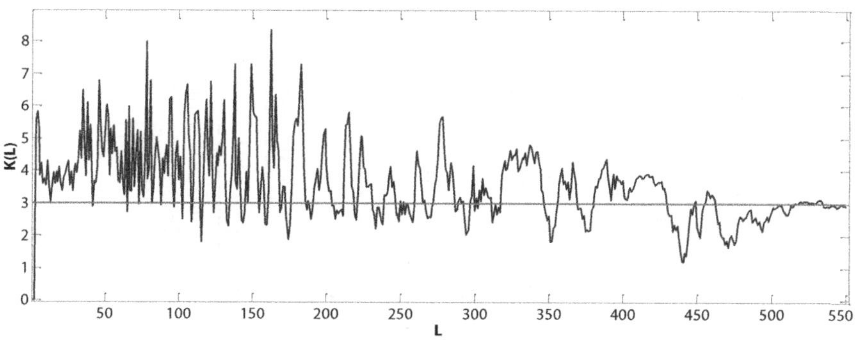

Figure 6.6 Kurtosis K (L) computed by varying L for recovery of Gaussianity of the return distribution.

depicts the convergence of $K(L)$ to 3 for L=550 days (c.f., Figure 6.6). This result indicates that Gaussianity is recovered by the cumulative return with a time lag of 550 days.

6.4.2 Detecting Seasonal Pattern in Oil Prices

Above mentioned results reveal the existence of persistence behavior in the oil returns for the duration of 2009 to 2018. However, researches on seasonal pattern and calendar anomalies reveal that there is a distinct pattern in the oil prices. This section carries forward the research on the same dataset in view of finding seasonal pattern. The result in Table 6.2 depicts average monthly return of oil price data for each year starting from 2009 to 2018. It exhibits an interesting fact that the average monthly returns for 10 years are mostly positive during summer months while negative during winter. More precisely, the oil market shows positive returns from the

Table 6.2 Month-wise return of each year over the period of 2009–2018 positive returns are in white while the negative returns are highlighted in grey.

	Jan	Feb	Mar	Apr	May	June	July	Aug	Sep	Oct	Nov	Dec
2009	-0.0005795	0.0038737	0.0038737	0.0007554	0.008243	0.003436	-0.0021465	0.0027644	-0.0022588	0.0042194	-0.0010387	0.001922
2010	-0.0037463	0.0030929	0.0006808	0.0008522	-0.0037778	0.0008516	0.002031	-0.0018315	0.0008017	0.0008903	0.0029877	0.0011016
2011	0.0010728	0.0028716	0.0024877	0.002758	-0.0034767	-0.0032187	-4.808E.05	-0.0015701	-0.0005559	0.0044272	0.0056674	0.001011
2012	-0.0025544	0.0021858	0.000145	0.0019351	-0.0040802	-0.001863	0.0021517	0.0026533	-0.0034283	-0.0017015	0.0011723	0.001394
2013	0.001565	-0.0011941	0.0028657	-0.002728	0.001627	0.0035624	0.0034802	0.0050716	-0.0052165	-0.0033177	-0.0010427	0.0024239
2014	0.0001264	0.0015782	-0.0018419	-0.0005281	0.0004173	0.0019	-0.0021795	-0.0020385	-3.076E-05	-0.007489	-0.007197	-0.0107405
2015	-0.008781	0.0041073	-0.0007325	0.0105742	5.234E-06	-0.0002103	-0.0086277	-0.0002793	-0.001141	0.0007439	-0.0033726	-0.0063224
2016	-0.0033176	-0.0003041	0.0050481	0.0088972	0.0038097	-0.0003541	-0.0087803	0.0047213	0.0018693	4.546E-05	0.0013826	0.0047987
2017	-0.0009275	4.456E-05	-0.0041165	-0.0014114	-0.0005813	-0.0029168	0.0037913	-0.0028001	0.005637	0.0018949	0.0025124	0.0017776
2018	0.0028016	-193E-05	0.0013935	0.0034828	0.0002675	0.0049628	-0.0028492	0.0020592	0.0030615	-0.0032743	-0.0147326	-0.0052847
Monthly _Average	-0.001434	0.0016236	0.0009803	0.0024587	0.0002454	0.000615	-0.0013177	0.000875	-0.0001262	-0.0003561	-0.0013661	-0.0007919

months of February to June and August while it shows negative returns in July followed by September to January. Investigation is carried out to see if the phenomenon is due to any seasonal impact or is it just a matter of chance. 10 years' average value of each month is computed. These average returns refer to average monthly return of each month for the period of 2009 to 2018. It may be noted that the average monthly return is positive during February to June and in August, whereas it is negative for September to January and in July. This further leads to find out whether there is a significant difference between the returns of summer and winter months. Therefore, following hypothesis is formed in view of testing the significance of the difference of average returns of summer months and winter months of 2009 to 2018.

H0: There is no significant difference between the average returns of summer and winter months for the period of 2009–2018.

Data has been segregated into summer months and winter months. However, as the pattern is slightly disrupted, two separate subperiods for summer and winter months are considered. These subperiods are:

I. February to June compared with July to January

II. February to August compared with September to January.

The comparison of the average returns for these subperiods has been conducted separately using independent sample t-test. In both cases there is significant difference between the subperiods of summer and winter months. For the first subperiod, this difference is significant at 1% level while in the second subperiod the same is significant at 5% level. Therefore, null hypothesis can be rejected (detailed results for both subperiods are presented in Tables 6.3 and 6.4 in Appendix.). These results are found to be in line with the previous researches [35, 38, 52] that can be utilized to develop investment strategies that are discussed in the implication section.

6.5 Conclusion and Implications

This chapter investigates the statistical behavior of the daily crude oil price fluctuation in the context of investment in multicommodity exchange (MCX). The price fluctuation is measured by logarithmic return. It is found that the daily return has a persistent behavior, which is further characterized by the power law behavior. It is therefore evident that the large return is followed by large return in the long run and *vice-versa.* Findings confirm the presence of long term persistence in daily crude oil returns. However, instead of daily returns (where the time lag is one day), if the cumulative return is computed for a duration of 550 days, Gaussianity is found to be recovered. As the duration progresses from daily to weekly to monthly returns, the probability distribution curve flattens more and it finally recovers gaussianity (or normal distribution) by 550 days. Thus, it can be inferred that as the lags increase, the memory of the returns time series fades away. This results in a transition of return series from persistence to achieving market efficiency.

It has also been established that there is a pattern of seasonality in the oil prices such that its performance is better in summer as compared to the winter months, and this information can be utilized to develop a feasible investment strategy. This finding has significant implications for the investors that are discussed in the next section.

Implications

Predictability of asset values like price or returns, allows the investors understand the market beforehand and take advantage to gain better returns. This is applicable in the present case of crude oil data as well. Its predictable pattern gives a scope for developing investment strategies for two distinct periods in Indian context. These strategies aim to exploit the seasonal tendencies of summer and winter months of MCX crude oil data that may provide better returns to the Indian investors. The proposed strategies are as follows:

- Strategy 1: investing in oil during the months of February to June while investing in government securities (like 6-month treasury bills) from July to January
- Strategy 2: investing in oil during the months of February to August while investing in government securities (like 6-month treasury bills) from September to January

The crux of these strategies lies in exploiting the positive returns of summer months while avoiding the negative returns of winter months to the maximum extent possible [35]. The negativity of winter months is balanced by reinvesting the proceeds of summer months in safer options like T-bills. Subsequently, the return generated in winter months is again invested in the oil market for the summer months of next period, and so on.

These strategies are similar to the ones used by Swagerman and Novakovic [35, 53, 54]. These researches reveal that the investment strategies that take advantage of the seasonal patterns in financial markets are able to generate superior returns. Further, these strategies also outperform the normal buy and hold strategies for the same period.

References

1. Bhattacharjee, A. and Kaur, G., Statistical behaviour of daily crude oil price data. *Int. J. Adv. Innov. Res.*, 6, 2, 71–75, 2019.
2. Bollerslev, T., Generalised autoregressive conditional heteroskedasticity. *J. Econ.*, 31, 3, 307–327, 1986.
3. Engle, R.F., Autoregressive conditional heteroskedasticity with estimates of the variance of United Kingdom inflation. *Econometrica*, 50, 4, 987–1007, 1982.
4. Fleming, J., Ostdiek, B., Whaley, R., Predicting stock market volatility: A new measure. *J. Futures Markets*, 15, 3, 265–302, 1995.
5. Schwert, W.G., Why does stock market volatility change over time? *J. Finance*, 44, 5, 1115–1151, 1989.
6. Chordia, T., Roll, R., Subrahmanyam, A., Commonality in liquidity. *J. Financ. Econ.*, 56, 3–28, 2000.
7. Hasbrouck, J. and Seppi, D.J., Common factors in prices, order flows, and liquidity. *J. Financ. Econ.*, 59, 383–411, 2001.
8. Kaur, H., Time varying volatility in the Indian stock market. *Vikalpa*, 29, 4, 25–42, 2004.
9. Prosad, J.M., Kapoor, S., Sengupta, J., Asset pricing bubbles, in: *Handbook of Research on Financial and Banking Crisis Prediction through Early Warning Systems*, IGI Global, Hershey, Pennsylvania, U.S., 2016.
10. Fama, E., Random walks in stock market prices. *Financ. Anal. J.*, 21, 5, 55–59, 1965.
11. Fama, E., The behavior of stock-market prices. *J. Bus.*, 34, 1, 34–105, 1965.
12. Fama, E., Efficient capital markets: A review of theory and empirical work. *J. Finance*, 25, 2, 383, 1970.
13. Shiller, R.J., *Irrational Exuberance*, 3rd ed, Princeton University Press, Princeton, NJ, 2015.
14. Shleifer, A., *Inefficient Markets: An Introduction To Behavioral Finance*, Oxford University Press, Oxford, U.K., 2002.
15. Campbell, J.Y. and Shiller, R.J., Stock prices, earnings and expected dividend. *J. Finance*, 43, 661–76, 1988.
16. Coulliard, M. and Davison, M., A comment on measuring the hurst exponent of financial time series. *Physica A*, 348, 404–418, 2005.
17. Di Matteo, T., Aste, T., Dacorogna, M.M., Scaling behaviors in di erently developed markets. *Physica A*, 324, 183–188, 2003. doi: 10.1016/S0378-4371(02)01996-9.
18. Caraiani, P., Evidence of multifractality from emerging European stock markets. *PloS One*, 7, 7, 1–9, 2012. doi: 10.1371/journal.pone.0040693.
19. Mali, P. and Mukhopadhyay, A., Multifractal characterization of gold market: A multifractal detrended fluctuation analysis. *Phys. A Stat. Mech. Appl.*, 413, 361–372, November 2014. doi: 10.1016/j.physa.2014.06.076.

20. Yuan, Y., Zhuang, X.T., Jin, X., Measuring multifractality of stock price fluctuation using multifractal detrended fluctuation analysis. *Phys. A Stat. Mech. Appl.*, 388, 11, 2189–2197, 2009. doi: 10.1016/j.physa.2009.02.026.
21. Cheema, M.A. and Frank, S., Oil prices and stock market anomalies. *Energy Econ.*, 83, 578–587, 2019.
22. Kogan, L. and Papanikolaou, D., Firm characteristics and stock returns: The role of investment-specific shocks. *Rev. Financ. Stud.*, 26, 11, 2718–2759, 2013.
23. Kwapień, J. and Drozdz, S., Components of multifractality in high-frequency stock returns. *Phys. A Stat. Mech. Appl.*, 350, 466–474, 2005.
24. Menkveld, A.J., High frequency trading and the new market makers. *J. Financial Mark.*, 16, 4, 712–740, 2013. doi: 10.1016/j.finmar.2013.06.006.
25. Voss, J., Rescaled range analysis: A method for detecting randomness, persistence or mean reversion in financial markets. *Enterprising Investor*, 1–35, 2013. [Online blog]. Available: http://http//: blogs.cfainstitute.org/investor/2013/01/30/rescaled-range-analysis-a-method-of-detecting-persistence-randomness-or-mean-reversion-in-financial-markets.
26. Wang, Y. and Wu, C., Efficiency of crude oil futures markets: New evidence from multifractal detrending moving average analysis. *Comput. Econ.*, 42, 393–414, 2013.
27. Alvarez-Ramirez, J., Alvarez, J., Rodriguez, E., Short-term predictability of crude oil markets: A detrended fluctuation analysis approach. *Energy Econ.*, 30, 2645–2656, 2008.
28. Barros, C.P. and Gil-Alana, L.A., Oil prices: Persistence and breaks. *Energy Sources, Part B Econ. Plan. Policy*, 11, 2, 137–149, 2016.
29. Jalali-Naini, A.R. and Asali, M., Cyclical behaviour and shock-persistence: Crude oil prices. *OPEC Rev.*, 28, 2, 107–131, 2004.
30. Ariel, R.A., High stock returns before holidays: Existence and evidence on possible causes. *J. Finance*, 45, 5, 1611–1626, 1990. [Online]. Available: https://doi.org/10.1111/j.1540-6261.1990.tb03731.x.
31. Bouman, S. and Jacobsen, B., The halloween indicator, 'sell in May and go away': Another puzzle. *Am. Econ. Rev.*, 92, 5, 1618–1635, 2002. doi: 10.1257/000282802762024683.
32. Choudry, T., Month of the year effect pre-WWI stock returns: Evidence from a non-linear GARCH model. *Int. J. Financ. Econ.*, 6, 1, 1-11, 2001.
33. Giovanis, E., Calendar effects in fifty-five stock market indices. *Glob. J. Financ. Manage.*, 1, 2, 75–98, 2009.
34. Lakonishok, J. and Smidt, S., Are seasonal anomalies real? A ninety-year perspective. *Rev. Financ. Stud.*, 1, 4, 403–425, 1988. [Online]. Available: https://doi.org/10.1093/rfs/1.4.403.
35. Arendas, P., Tkacova, D., Bukoven, J., Seasonal patterns in oil prices and their implications for investors. *J. Int. Stud.*, 11, 2, 180–192, 2018.
36. Auer, B., Daily seasonality in crude oil returns and volatilities. *Energy Econ.*, 43(C), 82–88, 2014. doi: 43. 10.1016/j.eneco.2014.02.005.

37. Borowski, K., Analysis of selected seasonality effects in markets of futures contracts with the following underlying instruments: Crude oil, brent oil, heating oil, gas oil, natural gas, feeder cattle, live cattle, lean hogs and lumber. *J. Manage. Financ. Sci.*, 9, 27–44, 2016.
38. Arendas, P., "The halloween effect on the agricultural commodities markets. *Agric. Econ.*, 63, 10, 441–448, 2017. [Online]. Available: https://doi.org/10.17221/45/2016-AGRICECON.
39. Quayyoum, S. *et al.*, Seasonality in crude oil returns. *Soft Comput.*, 24, 18, 1–10, 2020.
40. Carbone, A., Castelli, G., Stanley, H.E., Time-dependent Hurst exponent in financial time series. *Phys. A Stat. Mech. Appl.*, 344, 1–2, 267–271, 2004. doi: 10.1016/j.physa.2004.06.130.
41. Eom, C., Choi, S., Oh, G., Jung, W.-S., Hurst exponent and prediction based on weak-form efficient market hypothesis of stock markets. *Physica A*, 387, 4630–4636, 2008. doi: 10.1016/j.physa.2008.03.035.
42. Kroha, P. and Skoula, M., Hurst exponent and trading signals derived from market time series, in: *Proceedings of the 20th International Conference on Enterprise Information Systems, ICEIS*, vol. 1, pp. 371–378, 2018.
43. Mckelvey, B., A comparison of three hurst exponent approaches to predict nascent bubbles. 25, 1, 1–10, 2017. doi: 10.1142/S0218348X17500062.AU: Please provide journal title.
44. Raimundo, M.S. and Okamoto Jr., J., Application of hurst exponent (h) and the r/s analysis in the classification of FOREX securities. *Int. J. Model. Optim.*, 8, 2, 116–124, 2018. doi: 10.7763/IJMO.2018.V8.635.
45. Si, W., Wang, J., Che, Y., Fei, X., Spike trains in Hodgkin-Huxley model and ISIs of acupuncture manipulations. *Chaos Solitons Fractals*, 36, 890–900, 2008.
46. Levin, R., II and Rubin, D.S., *Statistics for Management*, 7th ed, Pearson, London, U.K., 1998.
47. Aas, K. and Dimakos, X.K., Statistical modelling of financial time series: An introduction. (Report No.SAMBA 08/04), *Norwegian Computing Center*, Oslo, Norway, March 2004. [Online]. Available: https://www.uni-ulm.de/fileadmin/website_uni_ulm/mawi.inst.050/Courses/Winter0708/risk/Statistical_modelling_of_financial_time_series_01.pdf.
48. Coudouel, A., Dani, A.A., Paternostro, S., *Poverty and Social Impact Analysis of Reforms: Lessons and Examples from Implementation*, World Bank, Washington, DC, 2006.
49. Mantegena, R.S. and Stanley, H.E., Scaling behaviour in the dynamics of economic index. *Nature*, 376, 46–49, 1995.
50. Mills, T.C., Statistical analysis of daily gold price data. *Phys. A Stat. Mech. Appl.*, 338, 3–4, 559–566, 2004. doi: 10.1016/j.physa.2004.03.003.
51. Pasquini, M. and Serva, M., Multiscaling and clustering of volatility. *Physica A*, 269, 140–147, 1999.

52. Arendas, P., *The Halloween Effect on Financial Markets*, Prague, Wolters Kluwer, 2017.
53. Andrade, S.C., Chhaochharia, V., Fuerst, M.E., Sell in may and go away' just won't go away. *Financial Anal. J.*, 69, 4, 94–105, 2013.
54. Swagerman, D. and Novakovic, I., Multi-national evidence on calendar patterns in stock returns: An empirical case study on investment strategy and the halloween effect. *Int. J. Bus. Financ. Res.*, 4, 4, 23–42, 2010.

Appendix

Table 6.3 Result of t-test: Feb to June vs July to Jan.

Table 6.3a Group statistics

Season		**N**	**Mean**	**Std. deviation**	**Std. error mean**
Average monthly return	Winter	7	-.00064528	.000844573	.000319219
	Summer	5	.00118461	.000875293	.000391443

Table 6.3 Result of t-test: Feb to June vs July to Jan.

Table 6.3b Independent samples test

		Levene's test for equality of variances		t-test for equality of means					95% confidence interval of the difference	
		F	**Sig.**	**t**	**df**	**Sig. (2-tailed)**	**Mean difference**	**Std. error difference**	**Lower**	**Upper**
Average monthly return	Equal variances assumed	.006	.939	-3.647	10	.004	-.001829894	.000501804	-.002947983	-.000711805
	Equal variances not assumed			-3.623	8.564	.006	-.001829894	.000505102	-.002981440	-.000678348

Table 6.4 Result of t-test: Feb to Aug vs Sept to Jan.

Table 6.4a: Group statistics					
Season		**N**	**Mean**	**Std. deviation**	**Std. error mean**
Average monthly return	winter	5	-.00081486	.000585774	.000261966
	summer	7	.00078292	.001175614	.000444340

Table 6.4 Result of t-test: Feb to Aug vs Sept to Jan.

Table 6.4b Independent samples test.										
		Levene's test for equality of variances		**t-test for equality of means**						
									95% Confidence interval of the difference	
		F	**Sig.**	**t**	**df**	**Sig. (2-tailed)**	**Mean difference**	**Std. error difference**	**Lower**	**Upper**
Average monthly return	Equal variances assumed	.798	.393	-2.776	10	.020	-.001597777	.000575647	-.002880398	-.000315156
	Equal variances not assumed			-3.098	9.224	.012	-.001597777	.000515814	-.002760321	-.000435234

7

Some Fixed Point and Coincidence Point Results Involving G_α-Type Weakly Commuting Mappings

Krishna Kanta Sarkar[1]*, Krishnapada Das[2] and Abhijit Pramanik[1]

[1]*Department of Mathematics, Brainware University, Ramkrishnapur Road, Barasat, Kolkata, West Bengal, India*
[2]*Department of Mathematics, Sarojini Naidu College for Women, Jessore Road, Dum Dum, Kolkata, West Bengal, India*

Abstract
G-metric spaces appeared in analysis as a generalization of metric spaces. D-metric and 2-metric are also generalized by G-metric. G-fuzzy metric spaces may be treated as fuzzy version of G-metric. Commuting mapping and weakly commuting mappings are very much important for the study of fixed points. We have established some results for coincidence point in G-fuzzy metric spaces and some results for common fixed point. Here we have deduced some corollaries to our theorems and our theorems are also supported by examples.

Keywords: G-fuzzy metric, G-Cauchy sequence, altering distance function

AMS classification of subject: 54*H*25, 54*E*70, 47*H*10

7.1 Introduction

S. Banach [4] initiated the study of fixed point (fp) theory by his seminal work in 1922. This theory of fp generalized by researcher in many directions. With the help of commuting mappings Jungck [17] established a result on common fixed point (cfp) in the year 1976. This notion further

**Corresponding author*: krishnakanta1505@gmail.com

Sharmistha Ghosh, M. Niranjanamurthy, Krishanu Deyasi, Biswadip Basu Mallik, and Santanu Das (eds.) Mathematics and Computer Science Volume 1, (133–172) © 2023 Scrivener Publishing LLC

generalized to mappings with weakly, R-weakly type, compatible, weakly compatible, R-weakly compatible type and many more. Some authors also interested to find coincidence point (cp) results of these types of mappings. Some related works may be seen in the works like [9, 10, 18, 25, 26, 31].

Khan *et al.* [20], introduced a new type of function which is named as "altering distance function" (adf) in metric space and twisted the study of fp theory to another new direction. After this introduction a huge research works appeared in the literature. This notion has been further generalized to many directions. Choudhury and Das [5] introduced this concept to probabilistic metric spaces in 2008. [1, 2, 6, 7, 9, 10, 12, 13, 21] are some examples of use of adf for the results of fp in the spaces with probabilistic and fuzzy metric.

To overcome some difficulties of D-metric, which was introduced by Dhage [11], Mustafa and Sims [24] introduced G-metric in the year 2006. In [24], they have proved some results for fixed points (fps) which satisfies some contractive inequalities. The notion of G-metric spaces (G-ms) further generalized to G-Menger and G-fuzzy metric spaces (G-fms) etc.

Commuting mapping of G_α-type which are G-weakly and G-R weakly introduced by Mustafa, Aydi and Karapinar [22] in G-ms. These types of mappings further generalized to G-fms by some authors. The works like [3, 8, 14, 15, 23, 27, 30] are relating to the spaces with G-ms and G-fms.

Two cfp results and two cp results have been proved in our work. We have used Hadzic type (H-type) t-norms in two of our results and other two results involving Φ-functions. This work will be an instance of use of Ψ-function and Φ-functions. We have deduced few corollaries of our main works and we have given examples which satisfy our theorems.

7.2 Definitions and Mathematical Preliminaries

We now give some existing definition and lemmas which will be needed for our discussion. Also, we have given some definition and lemmas in *G*-fms for the need of main results.

7.2.1 Definition: G-metric Space (G-ms) [24]

Let us take M be a nonempty set and $Q: M^3 \to [0, \infty)$ be a function satisfying the conditions as in the following $\forall\ u, r, w, e$ in M

(i) $Q(u, r, w) = 0$ if $r = u = w$,

(ii) $0 < Q(u, u, r)$ with $u \neq r$,

(iii) $Q(u, u, r) \leq Q(u, r, w)$ with $r \neq w$,
(iv) $Q(u, r, w) = Q(u, w, r) = Q(r, w, u) = \ldots$(symmetry),
(v) $Q(u, r, w) \leq Q(u, e, e) + Q(e, r, w)$. The Q will be called G-metric or a generalized metric on M and (M, Q) will be called a G-ms.

Throughout the rest of this paper we have taken, $I = [0, 1]$.

7.2.2 Definition: t-norm [16, 28]

A function $\Delta: I^2 \to I$, will be called a t-norm if

(i) $\Delta(1, \rho) = \rho$,
(ii) $\Delta(\rho, \varphi) = \Delta(\varphi, \rho)$,
(iii) $\Delta(r, s) \geq \Delta(\rho, \varphi)$ whenever $r \geq \rho$ and $s \geq \varphi$,
(iv) $\Delta(\Delta(\rho, \varphi), r) = \Delta(\rho, \Delta(\varphi, r))$, where $\rho, \varphi, r, s \in I$.

7.2.3 Definition: t-norm of Hadžić type (H-type) [16, 28]

The t-norm Δ will be called a Hadžić type t-norm if $0 < u, \mu < 1$ with $1 - \mu < \rho \leq 1 \Rightarrow \Delta^n(\rho) > 1 - u$ $(n \geq 1)$ where $\Delta^n(\rho)$ is defined recurrently by $\Delta^1(\rho) = \Delta(\rho, \rho)$ and

$$\Delta^n(\rho) = \Delta(\Delta^{n-1}(\rho), \rho) \qquad (n \geq 2).$$

Minimum t-norm is an trivial H-type t-norm but there also exists H-type t-norm which are weaker than Δ_M. We have used it in our main theorems.

7.2.4 Definition: G-fuzzy metric space (G-fms) [30]

Let M be an arbitrary set, $Q: M^3 \times [0, \infty) \to [0, 1]$, is fuzzy set and Δ is a t-norm which is continuous. (M, Q, Δ) will be called a G-fms if $\forall u, r, w, e \in M$

(i) $Q(u, u, r, t) > 0$ with $u \neq r$,
(ii) $Q(u, u, r, t) \geq Q(u, r, w, t)$ with $r \neq w$,
(iii) $Q(u, r, w, t) = 1$ iff $u = r = w$,
(iv) $Q(u, r, w, t) = Q(P(u, w, r), t)$, where P is permutation on u, r, w,
(v) $Q(u, r, w, t_1 + t_2) \geq \Delta(Q(u, e, e, t_1), Q(e, r, w, t_2))$ where $t_1, t_2 > 0$,
(vi) $Q(u, r, w, .): [0, \infty) \to I$ is left continuous.

7.2.5 Definition [30]

Let (M, Q, Δ) be a G-fms. (M, Q, Δ) will be called symmetric if $Q(u, u, r, t) = Q(u, r, r, t)$ for $u, r \in M$ and for each $t > 0$.

7.2.6 Lemma [30]

Let (M, Q, Δ) be a G-fms, then (M, Q, Δ) is nondecreasing for t.

7.2.7 Lemma [30]

Let (M, Q, Δ) be a G-fms, then the metric Q is continuous on $M \times M \times M \times [0, \infty)$.

Throughout the rest of this paper, we assumed the condition

$$\lim_{t \to \infty} Q(u,r,w,t) = 1 \text{ for all } u, r, w \in M.$$

7.2.8 Definition [30]

Let (M, Q, Δ) be a G-fms. Now a sequence $\{\rho_n\}$ in M will be called G-convergent to $u \in M$ if $Q(\rho_n, \rho_n, u, t) \to 1$ for every $t > 0$, whenever $n \to \infty$.

7.2.9 Definition [30]

Let (M, Q, Δ) be a G-fms. Now $\{\rho_n\}$ in M will be called G-Cauchy sequence (GCs) if $Q(\rho_n, \rho_n, \rho_{n+r}, s) \to 1$ for each $s > 0$, $r = 1, 2, 3, 4, ...$, as $n \to \infty$. (M, Q, Δ) will be called G-compete if every GCs in M is G-convergent.

7.2.10 Definition: Φ-Function [5]

A function f: $[0, \infty) \to [0, \infty)$ will be called a Φ-function if

(i) $f(t) = 0$ iff $t = 0$,
(ii) $f(t)$ is strictly increasing and it tends to infinity as $t \to \infty$,
(iii) the function f is left continuous in $(0, \infty)$ and
(iv) f is continuous at $t = 0$.

We now give the definition of Ψ-functions. This type of functions will be used in our result.

7.2.11 Definition: Ψ-Function [7]

A four variable function g: $I^4 \rightarrow I$, will be called a Ψ-function if

(i) the function g will be monotone increasing and continuous,
(ii) $g(\rho, \rho, \rho, \rho) > \rho$, whenever $0 < \rho < 1$,
(iii) $g(0, 0, 0, 0) = 0$ and $g(1, 1, 1, 1) = 1$.

We now give an example of Ψ-function

$$g(x,v,w,r) = \frac{(a_1\sqrt{x} + a_2\sqrt{v} + a_3\sqrt{w} + a_4\sqrt{u})}{a_1 + a_2 + a_3 + a_4}.$$

The real numbers a_1, a_2, a_3 a_4 are all positive.

With the help of Lemma 7.2.7 and by the property of Φ-function, the following lemma can be proved.

7.2.12 Lemma

Let (M, Q, Δ) be a G-fms. If 0 < k < 1 with $Q(u,r,w,f(t)) \geq Q\left(u,r,w,f\left(\frac{t}{k}\right)\right), \forall u,r,w \in M$ and $t > 0$, then $u = r = w$.

Proof. By the given condition we see that

$$Q(u,r,w,f(t)) \geq Q\left(u,r,w,f\left(\frac{t}{k}\right),\right) \geq Q\left(u,r,w,f\left(\frac{t}{k^2}\right)\right)$$

$$\geq Q\left(u,r,w,f\left(\frac{t}{k^n}\right)\right).$$

Taking $n \rightarrow \infty$ we have from the above

$$\lim_{n\to\infty} Q(u,r,w,f(t)) \geq \lim_{n\to\infty} Q\left(u,r,w,f\left(\frac{t}{k^n}\right)\right).$$

As $n \to \infty$ we have $\frac{t}{k^n} \to \infty$ and hence with the help of Φ-function we get $f\left(\frac{t}{k^n}\right) \to \infty$ as $n \to \infty$. Hence we have $Q(u,r,w,$ $f(t)) \geq \lim_{n\to\infty} Q\left(u,r,w,f\left(\frac{t}{k^n}\right)\right) = 1$. From the properties of Φ-function it is immediate that for given $E > 0$ we can find a positive t such that $E > f(t)$. Hence we have $Q(u,r,w,E) \geq Q(u,r,w,f(t)) \geq \lim_{n\to\infty} Q\left(u,r,w,f\left(\frac{t}{k^n}\right)\right) = 1$, which gives us that $u = r = w$. Hence the lemma is established.

7.2.13 Definition [29]

Let (M, Q, Δ) be G-fms and $\alpha, \beta: M \to M$ be two self-maps. A cp y of the maps α and β in M, *iff* $\alpha y = \beta y$.

7.2.14 Definition [29]

Let (M, Q, Δ) be G-fms and two self-maps $\alpha, \beta: M \to M$ are such that $\alpha y = \beta y = y$ for $y \in M$ then the point y will be called a cfp of α and β.

7.2.15 Definition [19]

Let (M, Q, Δ) be G-fms and $\alpha, \beta: M \to M$ are two self-maps. The maps α, β will be called weakly compatible if they commute at their cp, that is, if $\alpha y = \beta y$ then $\alpha\beta y = \beta\alpha y$.

In 2012, Mustafa *et al.* [22] introduced the notion of G_α-type G-weakly commuting (G-wc) and G_α-type G-R-weakly commuting (GR-wc) maps in G-ms. These two type of mappings are further generalized in the context of G-fms in the year 2014 by Gupta *et al.* [15]. For the requirement of our results we also carried out some modifications as follows:

7.2.16 Definition

Let $\alpha, \beta: M \to M$ be two self-maps in the (M, Q, Δ) where M is a non-empty set, Q be the generalized fuzzy metric on M and Δ is a t-norm. The two mapping α, β will be called G_α-type G-wc if

$$Q(\alpha\beta u, \beta\alpha u, \alpha\alpha u, t) \geq Q\left(\alpha u, \beta u, \alpha u, \frac{t}{k}\right),$$

for all u in M with t is positive and $0 < k < 1$.

7.2.17 Definition

Let α, β: $M \rightarrow M$ be two self-mapping in the (M, Q, Δ) where M be a non-empty set, Q be the generalized fuzzy metric on M and Δ is a t-norm. The above pair of mappings α and β will be called GR-wc of G_α-type if we have some $R > 0$ with $t > 0$, the following is satisfied $\forall u \in M$

$$Q(\alpha\beta u, \beta\alpha u, \alpha\alpha u, \mathrm{t}) \geq Q\left(\alpha u, \beta u, \alpha u, \frac{t}{R}\right).$$

7.2.18 Remarks

Interchanging the mappings α, β in definitions 7.2.16 and 7.2.17 we get G-weakly as well as GR-wc mappings of G_β-type, respectively.

7.2.19 Lemma

Let α, β: $M \rightarrow M$ be two maps on the G-fms (M, Q, Δ). If maps α, β be G-wc and GR-wc of G_α-type, then α, β will be weakly compatible mappings.

Proof. Let $a \in M$ be a coincidence point for the mappings α, β that is, $\alpha a = \beta a$. If the mappings (α, β) be G_α-type G-wc mappings then we have,

$$Q(\alpha\beta a, \beta\alpha a, \alpha\alpha a, \mathrm{t}) \geq Q\left(\alpha a, \beta a, \alpha a, \frac{t}{k}\right).$$

As $\alpha a = \beta a$, then $Q\left(\alpha a, \beta a, \alpha a, \frac{t}{k}\right) = 1$, for all positive t.

Hence, $\alpha\beta a = \beta\alpha a$, whenever $\alpha a = \beta a$, that is α and β commute at their coincidence point. Therefore it is proved that the mapping α, β are weakly compatible if they are of G-wc type.

Now let the mappings α and β are G_α-type GR-wc then for all $a \in M$ and

$$t > 0, Q(\alpha\beta a, \beta\alpha a, \alpha\alpha a, \mathrm{t}) \geq Q\left(\alpha a, \beta a, \alpha a, \frac{t}{R}\right).$$

As $\alpha a = \beta a$, we have $Q\left(\alpha a, \beta a, \alpha a \frac{t}{R}\right) = 1$ for all positive t.

Hence $\alpha\beta a = \beta\alpha a$, whenever $\alpha a = \beta a$, that is α and β commute at their cp. So if the mappings α, β are of G_α type GR-wc, then they are weakly compatible.

Hence proof of the lemma is completed.

7.3 Main Results

7.3.1 Theorem

Let (M, Q, Δ) be a symmetric G-fms with Δ continuous H-type and

$\alpha, \beta: M \to M$ be two self-maps satisfying

(A1) $\beta(M)$ *is a* G-complete sub-space of M,

(A2) $\alpha(M) \subseteq \beta(M)$.

$$(A3)\ Q(\alpha\rho,\alpha\zeta,\alpha r,t) \geq g\left(Q\left(\beta\rho,\beta\zeta,\beta r,\frac{t}{k}\right),Q\left(\alpha\rho,\beta\zeta,\beta r,\frac{t}{k}\right),\left(Q\left(\beta\rho,\alpha\rho,\beta r,\frac{t}{k}\right),Q\left(\beta\zeta,\beta r,\alpha\zeta,\frac{t}{k}\right)\right)\right)$$
$$\forall \rho,\zeta,r \in M \quad \text{with } t > 0 \text{ and } 0 < k < 1.$$

(A4) g is Ψ-function. (7.1)

Then α and β have a cp in M.

Proof. Let ρ_0 in M be an arbitrary point. As α $(M) \subseteq$ β(M), we may choose a point $\rho_1 \in M$ so that $\alpha\rho_0 = \beta\rho_1$. We define the sequence $\{\rho_n\}$ in general inductive process such that $\alpha\rho_n = \beta\rho_{n+1}$. We may assumed that $\beta\rho_n \neq \beta\rho_{n+1}$ for each n, otherwise we have a fp of β.

Let $\rho = \rho_n$, $\zeta = \rho_{n+1}$, $r = \rho_{n+1}$, we can write from inequality (7.1) for all $t > 0$,

$$Q(\alpha\rho_n,\alpha\rho_{n+1},\alpha\rho_{n+1},t) \geq g\begin{pmatrix} Q\left(\beta\rho_n,\beta\rho_{n+1},\beta\rho_{n+1},\frac{t}{k}\right),Q\left(\alpha\rho_n,\beta\rho_{n+1},\beta\rho_{n+1},\frac{t}{k}\right), \\ Q\left(\beta\rho_n,\alpha\rho_n,\beta\rho_{n+1},\frac{t}{k}\right),Q\left(\beta\rho_{n+1},\beta\rho_{n+1},\alpha\rho_{n+1},\frac{t}{k}\right) \end{pmatrix}.$$

Let $\zeta_n = \alpha\rho_n = \beta\rho_{n+1}$, so we get from the above inequality

$$Q(\zeta_n,\zeta_{n+1},\zeta_{n+1},t)\geq g\begin{pmatrix} Q\left(\zeta_{n-1},\zeta_n,\zeta_n,\frac{t}{k}\right),Q\left(\zeta_n,\zeta_n,\zeta_n,\frac{t}{k}\right), \\ Q\left(\zeta_{n-1},\zeta_n,\zeta_n,\frac{t}{k}\right),Q\left(\zeta_n,\zeta_n,\zeta_{n+1},\frac{t}{k}\right) \end{pmatrix}.$$

which implies

$$Q(\zeta_n,\zeta_{n+1},\zeta_{n+1},t)\geq g\begin{pmatrix} Q\left(\zeta_{n-1},\zeta_n,\zeta_n,\frac{t}{k}\right),Q\left(\zeta_n,\zeta_n,\zeta_n,\frac{t}{k}\right), \\ Q\left(\zeta_{n-1},\zeta_n,\zeta_n,\frac{t}{k}\right),Q\left(\zeta_n,\zeta_{n+1},\zeta_{n+1},\frac{t}{k}\right) \end{pmatrix}. \tag{7.2}$$

$\forall\, t > 0$, $n \in$ N (N, natural number set), we claim that

$$Q\left(\zeta_{n-1},\zeta_n,\zeta_n,\frac{t}{k}\right)\leq Q\left(\zeta_n,\zeta_{n+1},\zeta_{n+1},\frac{t}{k}\right). \tag{7.3}$$

If possible let for some $t_1 > 0$,

$$Q\left(\zeta_{n-1},\zeta_n,\zeta_n,\frac{t_1}{k}\right)> Q\left(\zeta_n,\zeta_{n+1},\zeta_{n+1},\frac{t_1}{k}\right).$$

Then we have from inequality (7.2),

$$\begin{aligned} Q(\zeta_n,\zeta_{n+1},\zeta_{n+1},t_1)&\geq g\begin{pmatrix} Q\left(\zeta_n,\zeta_{n+1},\zeta_{n+1},\frac{t_1}{k}\right),Q\left(\zeta_n,\zeta_{n+1},\zeta_{n+1},\frac{t_1}{k}\right), \\ Q\left(\zeta_n,\zeta_{n+1},\zeta_{n+1},\frac{t_1}{k}\right),Q\left(\zeta_n,\zeta_{n+1},\zeta_{n+1},\frac{t_1}{k}\right) \end{pmatrix} \\ &> Q\left(\zeta_n,\zeta_{n+1},\zeta_{n+1},\frac{t_1}{k}\right) \text{[by the property of } g\text{-function.]} \end{aligned}$$

As $0 < k < 1$, we have

$$Q(\zeta_n,\zeta_{n+1},\zeta_{n+1},t_1) > Q\left(\zeta_n,\zeta_{n+1},\zeta_{n+1},\frac{t_1}{k}\right) \geq Q(\zeta_n,\zeta_{n+1},\zeta_{n+1},t_1)$$

which is a self-contradiction

We can say that for all natural number n, (7.3) holds for all $t > 0$.

Using (7.3) in (7.2), for all positive t and $n \in N$.

$$Q(\zeta_n,\zeta_{n+1},\zeta_{n+1},t) \geq g\begin{pmatrix} Q\left(\zeta_{n-1},\zeta_n,\zeta_n,\frac{t}{k}\right), Q\left(\zeta_{n-1},\zeta_n,\zeta_n,\frac{t}{k}\right), \\ Q\left(\zeta_{n-1},\zeta_n,\zeta_n,\frac{t}{k}\right), Q\left(\zeta_{n-1},\zeta_n,\zeta_n,\frac{t}{k}\right) \end{pmatrix}$$

$$> Q\left(\zeta_{n-1},\zeta_n,\zeta_n,\frac{t}{k}\right) [\text{by the property of } g\text{-function.}]$$

$$\therefore Q(\zeta_n,\zeta_{n+1},\zeta_{n+1},t) > Q\left(\zeta_{n-1},\zeta_n,\zeta_n,\frac{t}{k}\right) > Q\left(\zeta_{n-2},\zeta_{n-1},\zeta_{n-1},\frac{t}{k^2}\right)$$

$$> Q\left(\zeta_0,\zeta_1,\zeta_1,\frac{t}{k^n}\right), \text{ for all } t > 0 \text{ and } n \in N.$$

This implies for all positive t,

$$\lim_{n\to\infty} Q(\zeta_n,\zeta_{n+1},\zeta_{n+1},t) = 1 [\text{as } 0 < k < 1]. \tag{7.4}$$

Now let $\lambda \in (0, 1)$ and $t > 0$ be given and $\delta \in (0, 1)$. As Δ is of H-type t-norm we have $\Delta^m (1 - \delta) > 1 - \lambda$ for all $m \in N$.

Let $\delta_1 \in (k, 1)$. Then $\left(\frac{\delta_1}{k}\right)^n \to \infty$ as $n \to \infty$. Now a natural number n_1 can be find so that $Q\left(\zeta_0,\zeta_1,\zeta_1,\frac{\delta_1^{\,n}}{k^n}\right) > 1-\delta$ for all $n \geq n_1$.

As $\delta_1 \in (0,1)$, there is $n_2 \in N$ such that $\sum_{i=n_2}^{\infty} \delta_1^i \leq t$.

Then for all $n \geq N_1 = \max\{n_1, n_2\}$ and $m \in N$,

$$Q(\zeta_n,\zeta_{n+m},\zeta_{n+m},t) \geq Q\left(\zeta_n,\zeta_{n+m},\zeta_{n+m},\Sigma_i^{n+m-1}\delta_1^{\,i}\right)$$

$$\geq \Delta\left(\Delta\left(\ldots\Delta\left(Q\left(\zeta_n,\zeta_{n+1},\zeta_{n+1},\delta_1^{\,n}\right), Q\left(\zeta_{n+1},\zeta_{n+2},\zeta_{n+2},\delta_1^{\,n+1}\right),\ldots,\right.\right.\right.$$

$$Q\left(\zeta_{n+m-1},\zeta_{n+m},\zeta_{n+m},\delta_1^{\,n+m-1}\right)\Big)\ldots\ldots\Big)\Big)$$

$$\geq \Delta\ \Delta\ \ldots\Delta\ Q\ \ \zeta_0,\zeta_1,\zeta_1,\frac{\delta_1^{\,n}}{k^n}\ ,Q\ \ \zeta_0,\zeta_1,\zeta_1,\frac{\delta_1^{\,n+1}}{k^{n+1}}\ \ ,\ldots\ldots,$$

$$Q\ \ \zeta_0,\zeta_1,\zeta_1,\frac{\overset{n+m-1}{1}}{k^{n+m-1}}\quad \ldots$$

$$\geq \Delta^m(1-\delta)>1-\lambda.$$

$\therefore$ The sequence $\{\zeta_n\}$ will be a GCs.

Since, $\zeta_n = \beta\rho_{n+1}$, therefore $\{\beta\rho_{n+1}\}$ is a GCs in $\beta(M)$.

By hypothesis (A1), $\beta(M)$ is G-complete then $\exists\ u \in \beta(M)$ such that,

$$\lim_{n\to\infty}\beta\rho_{n+1} = u = \lim_{n\to\infty}\alpha\rho_n.$$

$\therefore \exists\ x \in M$ such that $u = \beta x$ (as $u \in \beta\ (M)$).

So

$$\lim_{n\to\infty}\beta\rho_n = u = \beta x = \lim_{n\to\infty}\alpha\rho_n. \tag{7.5}$$

We try to prove $\alpha x = \beta x$.

Let $\alpha x \neq \beta x$. Taking $\rho = x$, $\zeta = r = \rho_n$ in inequality (7.1), it turns out

$$Q(\alpha x,\alpha\rho_n,\alpha\rho_n,t) \geq g\left(Q\left(\beta x,\beta\rho_n,\beta\rho_n,\frac{t}{k}\right),Q\left(\alpha x,\beta\rho_n,\beta\rho_n,\frac{t}{k}\right),\right.$$

$$\left.Q\left(\beta x,\alpha x,\beta\rho_n,\frac{t}{k}\right),Q\left(\beta\rho_n,\beta\rho_n,\alpha\rho_n,\frac{t}{k}\right)\right).$$

Taking n → ∞ and with help of (7.5) we can write from the above inequality,

$$Q(\alpha x,\beta x,\beta x,t)\geq g\left(Q\left(\beta x,\beta x,\beta x,\frac{t}{k}\right),Q\left(\alpha x,\beta x,\beta x,\frac{t}{k}\right),\right.$$

$$\left. Q\ \ \beta x,\alpha x,\beta x,\frac{t}{k}\ \ ,Q\ \ \beta x,\beta x,\beta x,\frac{t}{k}\ \ \right).$$

Therefore

$$Q(\alpha x,\beta x,\beta x,t)\geq g\left(Q\left(\alpha x,\beta x,\beta x,\frac{t}{k}\right),Q\left(\alpha x,\beta x,\beta x,\frac{t}{k}\right),\right.$$

$$\left. Q\left(\alpha x,\beta x,\beta x,\frac{t}{k}\right),Q\left(\alpha x,\beta x,\beta x,\frac{t}{k}\right)\right).$$

So by the property of g-function we have

$$Q(\alpha x,\beta x,\beta x,t)t>Q\left(\alpha x,\beta x,\beta x,\frac{t}{k}\right)\geq Q(\alpha x,\beta x,\beta x,t)$$

which is self-contradictory as $0<k<1$.

$$\therefore \alpha x=\beta x.$$

Hence x is a cp of α, β in M.

Hence the theorem is proved.

In the next theorem we will show that if the self-maps α, β are G-wc mappings of G_α-type, ∃ a unique cfp.

7.3.2 Theorem

Let (M, Q, Δ) be a symmetric G-fms with Δ continuous H-type and $\alpha, \beta: M \to M$ be two self-maps satisfying

(B1) $\beta(M)$ is a G-complete sub-space of M,

(B2) $\alpha(M) \subseteq \beta(M)$,

(B3) $$Q(\alpha\rho,\alpha\zeta,\alpha r,t) \geq g\left(Q\left(\beta\rho,\beta\zeta,\beta r,\frac{t}{k}\right), Q\left(\alpha\rho,\beta\zeta,\beta r,\frac{t}{k}\right), Q\left(\beta\rho,\alpha\rho,\beta r,\frac{t}{k}\right), Q\left(\beta\zeta,\beta r,\alpha\zeta,\frac{t}{k}\right)\right) \quad \forall \rho,\zeta,r \in M \text{ and } t > 0. \tag{7.6}$$

(B4) α and β are G-wc of type G_α,

(B5) $k \in (0,1)$ and g is Ψ-function.

Then the two maps α and β have a unique cfp in M.

Proof. Let $\rho_0 \in M$ be arbitrary. We can choose $\rho_1 \in M$ such that $\alpha\rho_0 = \beta\rho_1$ as $\alpha(M) \subseteq \beta(M)$. The sequence $\{\rho_n\}$ may be defined in general inductive process such that $\alpha\rho_n = \beta\rho_{n+1}$. We may assumed that $\beta\rho_n \neq \beta\rho_{n+1}$ for each n, otherwise we have a fp of β.

Let $\rho = \rho_n$, $\zeta = \rho_{n+1}$, $r = \rho_{n+1}$ we can write from inequality (7.6), for all positive t,

$$Q(\alpha\rho_n,\alpha\rho_{n+1},\alpha\rho_{n+1},t) \geq g$$

$$\begin{pmatrix} Q\left(\beta\rho_n,\beta\rho_{n+1},\beta\rho_{n+1}\frac{t}{k}\right), Q\left(\alpha\rho_n,\beta\rho_{n+1},\beta\rho_{n+1},\frac{t}{k}\right), \\ Q\left(\beta\rho_n,\alpha\rho_n,\beta\rho_{n+1},\frac{t}{k}\right), Q\left(\beta\rho_{n+1},\beta\rho_{n+1},\alpha\rho_{n+1},\frac{t}{k}\right) \end{pmatrix}.$$

Taking $\zeta_n = \alpha\rho_n = \beta\rho_{n+1}$, we have from above inequality,

$$Q(\zeta_n,\zeta_{n+1},\zeta_{n+1},t) \geq g\begin{pmatrix} Q\left(\zeta_n,\zeta_n,\zeta_n,\frac{t}{k}\right), Q\left(\zeta_{n-1},\zeta_n,\zeta_n,\frac{t}{k}\right), \\ Q\left(\zeta_{n-1},\zeta_n,\zeta_n,\frac{t}{k}\right), Q\left(\zeta_n,\zeta_n,\zeta_{n+1},\frac{t}{k}\right) \end{pmatrix}.$$

By the symmetric property of G-fms we can write

$$Q(\zeta_n,\zeta_{n+1},\zeta_{n+1},t)\geq g\begin{pmatrix} Q\left(\zeta_{n-1},\zeta_n,\zeta_n,\frac{t}{k}\right),Q\left(\zeta_n,\zeta_n,\zeta_n,\frac{t}{k}\right), \\ Q\left(\zeta_{n-1},\zeta_n,\zeta_n,\frac{t}{k}\right),Q\left(\zeta_n,\zeta_{n+1},\zeta_{n+1},\frac{t}{k}\right) \end{pmatrix}. \tag{7.7}$$

Taking $n \in N$, the set of natural numbers for positive t, we claim that

$$Q\left(\zeta_{n-1},\zeta_n,\zeta_n,\frac{t}{k}\right)\leq Q\left(\zeta_n,\zeta_{n+1},\zeta_{n+1},\frac{t}{k}\right). \tag{7.8}$$

If possible let for some $t_1 > 0$,

$$Q\left(\zeta_{n-1},\zeta_n,\zeta_n,\frac{t_1}{k}\right)> Q\left(\zeta_n,\zeta_{n+1},\zeta_{n+1},\frac{t_1}{k}\right).$$

Then we have from inequality (7.7),

$$Q(\zeta_n,\zeta_{n+1},\zeta_{n+1},t_1)\geq g\begin{pmatrix} Q\left(\zeta_n,\zeta_{n+1},\zeta_{n+1},\frac{t_1}{k}\right),Q\left(\zeta_n,\zeta_{n+1},\zeta_{n+1},\frac{t_1}{k}\right), \\ Q\left(\zeta_n,\zeta_{n+1},\zeta_{n+1},\frac{t_1}{k}\right),Q\left(\zeta_n,\zeta_{n+1},\zeta_{n+1},\frac{t_1}{k}\right) \end{pmatrix}.$$

So, by the property of g-function we have

$$Q(\zeta_n,\zeta_{n+1},\zeta_{n+1},t_1)> Q\left(\zeta_n,\zeta_{n+1},\zeta_{n+1},\frac{t_1}{k}\right)\geq Q(\zeta_n,\zeta_{n+1},\zeta_{n+1},t_1),$$

which is a self-contradiction as $0 < k < 1$.

$$\therefore \quad \forall\, t > 0 \text{ and } n \in N, \text{(7.8) holds.}$$

Using (7.8) in (7.7) we get, $\forall$ t>0 and $n \in N$,

$$Q(\zeta_n,\zeta_{n+1},\zeta_{n+1},t) \geq g\begin{pmatrix} Q\left(\zeta_{n-1},\zeta_n,\zeta_n,\frac{t}{k}\right), Q\left(\zeta_{n-1},\zeta_n,\zeta_n,\frac{t}{k}\right), \\ Q\left(\zeta_{n-1},\zeta_n,\zeta_n,\frac{t}{k}\right), Q\left(\zeta_{n-1},\zeta_n,\zeta_n,\frac{t}{k}\right) \end{pmatrix}$$

$$> Q\left(\zeta_{n-1},\zeta_n,\zeta_n,\frac{t}{k}\right) \text{[by the property of } g\text{-function.]}$$

Therefore

$$Q(\zeta_n,\zeta_{n+1},\zeta_{n+1},t) > Q\left(\zeta_{n-1},\zeta_n,\zeta_n,\frac{t}{k}\right) > Q\left(\zeta_{n-2},\zeta_{n-1},\zeta_{n-1},\frac{t}{k^2}\right)$$

$$> Q\left(\zeta_0,\zeta_1,\zeta_1,\frac{t}{k^n}\right), \text{ for all } t > 0 \text{ and } n \in \mathrm{N}.$$

Taking limit on both the sides with n $\rightarrow \infty$ we get from above inequality for all positive real number t,

$$\lim_{n\to\infty} Q(\zeta_n,\zeta_{n+1},\zeta_{n+1},t) = 1 \text{ [as } 0<k<1]. \tag{7.9}$$

Now let $\lambda \in (0, 1)$ and $t > 0$ be given and $\delta \in (0, 1)$. As Δ is of H-type we have

$$\Delta^m (1 - \delta) > 1 - \lambda \text{ for all } m \in \mathrm{N}.$$

Let $\delta_1 \in (k, 1)$. Then $\left(\frac{\delta_1}{k}\right)^n \rightarrow \infty$ as $n \rightarrow \infty$.

So we may take a natural number $n_1 \in \mathrm{N}$ for which $Q\left(\zeta_0,\zeta_1,\zeta_1\frac{\delta_1^n}{k^n}\right) > 1-\delta$ for all $n \geq n_1$.

As $\delta_1 \in (0, 1)$, there is $n_2 \in \mathrm{N}$ such that $\sum_{i=n_2}^{\infty} \delta_1^i \leq t$.

Then, for all $n \geq N_1 = \max\{n_1, n_2\}$ and $m \in N$,

$$Q(\zeta_n,\zeta_{n+m},\zeta_{n+m},t) \geq Q\left(\zeta_n,\zeta_{n+m},\zeta_{n+m},\sum_i^{n+m-1}\delta_1^i\right)$$

$$\geq \Delta\Big(\Delta\Big(K\Delta\Big(Q\big(\zeta_n,\zeta_{n+1},\zeta_{n+1},\delta_1^n\big),Q\big(\zeta_{n+1},\zeta_{n+2},\zeta_{n+2},\delta_1^{n+1}\big),\ldots,$$

$$Q\big(\zeta_{n+m-1},\zeta_{n+m},\zeta_{n+m},\delta_1^{n+m-1}\big)\Big)\ldots\ldots\Big)\Big)$$

$$\geq \Delta\left(\Delta\left(\ldots\Delta\left(Q\left(\zeta_0,\zeta_1,\zeta_1,\frac{\delta_1^n}{k^n}\right),Q\left(\zeta_0,\zeta_1,\zeta_1,\frac{\delta_1^{n+1}}{k^n+1}\right),\ldots\ldots,\right.\right.\right.$$

$$\left.\left.Q\left(\zeta_0,\zeta_1,\zeta_1,\frac{\delta_1^{n+m-1}}{k^{n+m-1}}\right)\right)\ldots\ldots\right)\Big)$$

$$\geq \Delta^m(1-\delta) > 1-\lambda.$$

$$\therefore \qquad \{\zeta_n\} \text{ is GCs.}$$

Since $\zeta_n = \beta\rho_{n+1}$, therefore $\{\beta\rho_{n+1}\}$ is a GCs in $\beta(M)$ and also $\zeta_n = \beta\rho_{n+1} = \alpha\rho_n$ so $\{\alpha\rho_n\}$ is also GCs.

By the hypothesis ($B1$), $\beta(M)$ is G-complete then $\exists\, \omega \in \beta(M)$ such that,

$$\lim_{n\to\infty}\beta\rho_{n+1} = \omega = \lim_{n\to\infty}\alpha\rho_n.$$

Here $\omega \in \beta(M)$ so in $x \in M$ such that $\omega = \beta x$.

$$\therefore \quad \lim_{n\to\infty}\beta\rho_n = \omega = \beta x = \lim_{n\to\infty}\alpha\,\rho_n. \tag{7.10}$$

Now we want to prove, $\alpha x = \beta x$. Let $\alpha x \neq \beta x$.

Putting, $\rho = x$, $\zeta = r = \rho_n$ in inequality (7.6),

$$Q(\alpha x,\alpha\rho_n,\alpha\rho_n,t) \geq g\left(Q\left(\beta x,\beta\rho_n,\beta\rho_n,\frac{t}{k}\right),Q\left(\alpha x,\beta\rho_n,\beta\rho_n,\frac{t}{k}\right),\right.$$

$$\left.Q\left(\beta x,\alpha x,\beta\rho_n,\frac{t}{k}\right),Q\left(\beta\rho_n,\beta\rho_n,\alpha\rho_n,\frac{t}{k}\right)\right).$$

Whenever $n \to \infty$, on above inequality we have for all positive t,

$$Q(\alpha x,\beta x,\beta x,t) \geq g\left(Q\left(\beta x,\beta x,\beta x,\frac{t}{k}\right),Q\left(\alpha x,\beta x,\beta x,\frac{t}{k}\right),\right.$$
$$\left.Q\left(\beta x,\alpha x,\beta x,\frac{t}{k}\right),Q\left(\beta x,\beta x,\beta x,\frac{t}{k}\right)\right).$$

Therefore

$$Q(\alpha x,\beta x,\beta x,t) \geq g\left(Q\left(\alpha x,\beta x,\beta x,\frac{t}{k}\right),Q\left(\alpha x,\beta x,\beta x,\frac{t}{k}\right),\right.$$
$$\left.Q\left(\alpha x,\beta x,\beta x,\frac{t}{k}\right),Q\left(\alpha x,\beta x,\beta x,\frac{t}{k}\right)\right).$$

So using the property of g-function we have,

$$Q(\alpha x,\beta x,\beta x,t) > Q\left(\alpha x,\beta x,\beta x,\frac{t}{k}\right) \geq Q(\alpha x,\beta x,\beta x,t),$$

which is a contradiction as $0 < k < 1$.

$$\therefore \alpha x = \beta x. \tag{7.11}$$

Again the pair (α, β) are G_α-type G-wc mapping, so $t > 0$ gives us

$$Q(\alpha\beta x,\beta\alpha x,\alpha\alpha x,t) \geq Q\left(\alpha x,\beta x,\alpha x,\frac{t}{k}\right) = 1 \Rightarrow \alpha\beta x = \beta\alpha x = \alpha\alpha x. \tag{7.12}$$

Now taking $\alpha x = \beta x = \omega$, we have from (7.11) and (7.12),

$$\alpha\beta x = \alpha\omega = \beta\alpha x = \beta\omega = \alpha\alpha x. \tag{7.13}$$

Now we show that, the point ω will be a cfp of the maps α, β.
Let $\alpha\omega \neq \omega$, then from (7.6) we can write by putting $\rho = \omega$, $\zeta = r = x$,

$$Q(\alpha\omega,\alpha x,\alpha x,t)\geq g\left(Q\left(\beta\omega,\beta x,\beta x,\frac{t}{k}\right),Q\left(\alpha\omega,\beta x,\beta x,\frac{t}{k}\right),\right.$$

$$\left.Q\left(\beta\omega,\alpha\omega,\beta x,\frac{t}{k}\right),Q\left(\beta x,\beta x,\alpha x\frac{t}{k}\right)\right).$$

By (7.13), we can write.

$$Q(\alpha\omega,\omega,\omega,t)\geq g\left(Q\left(\alpha\omega,\omega,\omega,\frac{t}{k}\right),Q\left(\alpha\omega,\omega,\omega,\frac{t}{k}\right),\right.$$

$$\left.Q\left(\alpha\omega,\alpha\omega,\omega,\frac{t}{k}\right),Q\left(\omega,\omega,\omega,\frac{t}{k}\right)\right).$$

$$Q(\alpha\omega,\omega,\omega,t)\geq g\left(Q\left(\alpha\omega,\omega,\omega,\frac{t}{k}\right),Q\left(\alpha\omega,\omega,\omega,\frac{t}{k}\right),\right.$$

$$\left.Q\left(\alpha\omega,\omega,\omega,\frac{t}{k}\right),Q\left(\alpha\omega,\omega,\omega,\frac{t}{k}\right)\right).$$

So using the property of g-function we have

$$Q(\alpha\omega,\omega,\omega,t)>Q\left(\alpha\omega,\omega,\omega,\frac{t}{k}\right)\geq Q(\alpha\omega,\omega,\omega,t)$$

which is self-contradictory as $0<k<1$.

$$\therefore \alpha\omega = \omega.$$

Using (7.13) and above relation we can write.

$$\alpha\omega=\beta\omega=\omega. \tag{7.14}$$

Next we show, the fp is unique. If not let ν ($\nu \neq \omega$) be the another one, that is,

$$\alpha\nu=\beta\nu=\nu.$$

Putting $\rho=\omega$, $\zeta=r=\nu$ in inequality (7.6) we get.

$$Q(\alpha\omega,\alpha v,\alpha v,t)\geq g\left(Q\left(\beta\omega,\beta v,\beta v,\frac{t}{k}\right),Q\left(\alpha\omega,\beta v,\beta v,\frac{t}{k}\right),\right.$$

$$\left.Q\left(\beta\omega,\alpha\omega,\beta v,\frac{t}{k}\right),Q\left(\beta v,\beta v,\alpha v,\frac{t}{k}\right)\right).$$

which implies,

$$Q(\omega,v,v,t)\geq g\left(Q\left(\omega,v,v,\frac{t}{k}\right),Q\left(\omega,v,v,\frac{t}{k}\right),Q\left(\omega,\omega,v,\frac{t}{k}\right),\right.$$

$$\left.Q\left(v,v,v,\frac{t}{k}\right)\right).$$

that is,

$$Q(\omega,v,v,t)\geq g\left(Q\left(\omega,v,v,\frac{t}{k}\right),Q\left(\omega,v,v,\frac{t}{k}\right),Q\left(\omega,v,v,\frac{t}{k}\right),\right.$$

$$\left.Q\left(\omega,v,v,\frac{t}{k}\right)\right).$$

$\therefore$ Using the property of g-function we have $Q(\omega,v,v,t)>Q\left(\omega,v,v,\frac{t}{k}\right)\geq Q(\omega,v,v,t)$ again we arrived at a contradiction as $0<k<1$.

Hence $\omega = v$.

Therefore the fp ω is the unique for the self-maps α, β, which proves the theorem.

Now, we define another form of Ψ-function which will be known as Ψ-function of sixth order that will be required to prove our next theorems.

7.3.3 Definition Ψ-Function [7]

A six variable function $g: I^6 \to I$ will be called a Ψ-function if,

(i) the function g is monotone increasing and continuous,

(ii) $g(a,a,a,a,a,a) > a$ for all $0 < a < 1$, (iii) $g(0,0,0,0,0,0) = 0$ and $g(1,1,1,1,1,1) = 1$.

We now give an example of Ψ-function

$$g(m,n,r,p,q,h) = \frac{a_1\sqrt{m} + a_2\sqrt{n} + a_3\sqrt{r} + a_4\sqrt{p} + a_5\sqrt{q} + a_6\sqrt{h}}{a_1 + a_2 + a_3 + a_4 + a_5 + a_6}.$$

The real numbers a_1, a_2, a_3, a_4, a_5 and a_6 are all positive.

7.3.4 Theorem

Let (M, Q, Δ) be a symmetric G-fms with Δ continuous and $\alpha, \beta : M \to M$ be two self-maps with

(C1) $\beta(M)$ contains $\alpha(M)$,

(C2) $\beta(M) \subseteq M$ and $\beta(M)$ *is* G-complete

(C3) $Q(\alpha\rho, \alpha\zeta, \alpha r, f(t)) \geq g$

$$\left(\begin{array}{l} Q\left(\beta\rho,\beta\zeta,\beta r, f\left(\frac{t}{k}\right)\right), Q\left(\beta\rho,\beta\rho,\alpha\rho, f\left(\frac{t}{k}\right)\right), Q\left(\beta\zeta,\beta\zeta,\alpha\zeta, f\left(\frac{t}{k}\right)\right), \\ Q\left(\beta r,\beta r,\alpha r, f\left(\frac{t}{k}\right)\right), Q\left(\beta\zeta,\beta\zeta,\alpha r, f\left(\frac{t}{k}\right)\right), Q\left(\alpha\rho,\beta r,\beta r, f\left(\frac{t}{k}\right)\right) \end{array} \right) \tag{7.15}$$

where $0 < k < 1$.

(C4) g is Ψ-function of sixth order and f is a Φ-function, for all $\rho, \zeta, r \in M$.

Then α, β have a cp in M.

Proof. Let $\rho_0 \in M$ be arbitrary. As $\alpha(M) \subseteq \beta(M)$ we can take $\rho_1 \in M$ such that $\alpha\rho_0 = \beta\rho_1$. We define the sequence $\{\rho_n\}$ in general inductive process such that $\alpha\rho_n = \beta\rho_{n+1}$. We may assumed that $\beta\rho_n \neq \beta\rho_{n+1}$ for each n, otherwise we have a fp of β. We are supposing, $\zeta_n = \alpha\rho_n = \beta\rho_{n+1}$, n being a natural number.

Putting $\rho = \rho_n, \zeta = \rho_{n+1}, r = \rho_{n+1}$, in inequality (7.15), for all $t > 0$,

$$Q(\alpha\rho_n,\alpha\rho_{n+1},\alpha\rho_{n+1},f(t))\geq$$

$$g\begin{pmatrix} Q\left(\beta\rho_n,\beta\rho_{n+1},\beta\rho_{n+1},f\left(\frac{t}{k}\right)\right),Q\left(\beta\rho_n,\beta\rho_n,\alpha\rho_n,f\left(\frac{t}{k}\right)\right), \\ Q\left(\beta\rho_{n+1},\beta\rho_{n+1},\alpha\rho_{n+1},f\left(\frac{t}{k}\right)\right),Q\left(\beta\rho_{n+1},\beta\rho_{n+1},\alpha\rho_{n+1},f\left(\frac{t}{k}\right)\right), \\ Q\left(\beta\rho_{n+1},\beta\rho_{n+1},\alpha\rho_{n+1},f\left(\frac{t}{k}\right)\right),Q\left(\alpha\rho_n,\beta\rho_{n+1},\beta\rho_{n+1},f\left(\frac{t}{k}\right)\right) \end{pmatrix}.$$

Therefore

$$Q(\zeta_n,\zeta_{n+1},\zeta_{n+1},f(t))\geq g$$

$$\begin{pmatrix} Q\left(\zeta_{n-1},\zeta_n,\zeta_n,f\left(\frac{t}{k}\right)\right),Q\left(\zeta_{n-1},\zeta_{n-1},\zeta_n,f\left(\frac{t}{k}\right)\right),Q\left(\zeta_n,\zeta_n,\zeta_{n+1},f\left(\frac{t}{k}\right)\right), \\ Q\left(\zeta_n,\zeta_n,\zeta_{n+1},f\left(\frac{t}{k}\right)\right),Q\left(\zeta_n,\zeta_n,\zeta_{n+1},f\left(\frac{t}{k}\right)\right),Q\left(\zeta_n,\zeta_n,\zeta_n,f\left(\frac{t}{k}\right)\right) \end{pmatrix}.$$

Or we can write

$$Q(\zeta_n,\zeta_{n+1},\zeta_{n+1},f(t))\geq$$

$$g\begin{pmatrix} Q\left(\zeta_{n-1},\zeta_n,\zeta_n,f\left(\frac{t}{k}\right)\right),Q\left(\zeta_n,\zeta_{n+1},\zeta_n,f\left(\frac{t}{k}\right)\right),Q\left(\zeta_n,\zeta_{n+1},\zeta_{n+1},f\left(\frac{t}{k}\right)\right), \\ Q\left(\zeta_n,\zeta_{n+1},\zeta_{n+1},f\left(\frac{t}{k}\right)\right),Q\left(\zeta_n,\zeta_n,\zeta_{n+1},f\left(\frac{t}{k}\right)\right),Q\left(\zeta_n,\zeta_n,\zeta_n,f\left(\frac{t}{k}\right)\right) \end{pmatrix}. \tag{7.16}$$

For all natural number n and positive t we claim the following

$$Q\left(\zeta_n,\zeta_{n+1},\zeta_{n+1},f\left(\frac{t}{k}\right)\right)\geq Q\left(\zeta_{n-1},\zeta_n,\zeta_n,f\left(\frac{t}{k}\right)\right). \tag{7.17}$$

If possible, let for some $t_1 > 0, n \in \mathrm{N}$,

$$Q\left(\zeta_n, \zeta_{n+1}, \zeta_{n+1}, f\left(\frac{t_1}{k}\right)\right) < Q\left(\zeta_{n-1}, \zeta_n, \zeta_n, f\left(\frac{t_1}{k}\right)\right),$$

then we can write from (7.16),

$$Q(\zeta_n, \zeta_{n+1}, \zeta_{n+1}, f(t_1)) \geq$$

$$g\begin{pmatrix} Q\left(\zeta_n, \zeta_{n+1}, \zeta_{n+1}, f\left(\frac{t_1}{k}\right)\right), Q\left(\zeta_n, \zeta_{n+1}, \zeta_{n+1}, f\left(\frac{t_1}{k}\right)\right), Q\left(\zeta_n, \zeta_{n+1}, \zeta_{n+1}, f\left(\frac{t_1}{k}\right)\right), \\ Q\left(\zeta_n, \zeta_{n+1}, \zeta_{n+1}, f\left(\frac{t_1}{k}\right)\right), Q\left(\zeta_n, \zeta_{n+1}, \zeta_{n+1}, f\left(\frac{t_1}{k}\right)\right), Q\left(\zeta_n, \zeta_{n+1}, \zeta_{n+1}, f\left(\frac{t_1}{k}\right)\right), \end{pmatrix}$$

$$> Q\left(\zeta_n, \zeta_{n+1}, \zeta_{n+1}, f\left(\frac{t_1}{k}\right)\right), \text{[by the property of } g\text{-function]}$$

$$> Q(\zeta_n, \zeta_{n+1}, \zeta_{n+1}, f(t_1)), \text{ which is a contradiction as } 0 < k < 1.$$

Therefore for any n in N and positive t, (7.17) holds.
Using (7.17) in (7.16) we have $\forall$ positive t and n in N.

$$Q(\zeta_n, \zeta_{n+1}, \zeta_{n+1}, f(t)) \geq$$

$$g\begin{pmatrix} Q\left(\zeta_{n-1}, \zeta_n, \zeta_n, f\left(\frac{t}{k}\right)\right), Q\left(\zeta_{n-1}, \zeta_n, \zeta_n, f\left(\frac{t}{k}\right)\right), Q\left(\zeta_{n-1}, \zeta_n, \zeta_n, f\left(\frac{t}{k}\right)\right), \\ Q\left(\zeta_{n-1}, \zeta_n, \zeta_n, f\left(\frac{t}{k}\right)\right), Q\left(\zeta_{n-1}, \zeta_n, \zeta_n, f\left(\frac{t}{k}\right)\right), Q\left(\zeta_{n-1}, \zeta_n, \zeta_n, f\left(\frac{t}{k}\right)\right), \end{pmatrix}$$

$$> Q\left(\zeta_{n-1}, \zeta_n, \zeta_n, f\left(\frac{t}{k}\right)\right), \text{[by the properties of } g\text{-function].}$$

Therefore

$$Q(\zeta_n, \zeta_{n+1}, \zeta_{n+1}, f(t)) > Q\left(\zeta_{n-1}, \zeta_n, \zeta_n, f\left(\frac{t}{k}\right)\right)$$

$$> Q\left(\zeta_{n-2}, \zeta_{n-1}, \zeta_{n-1}, f\left(\frac{t}{k^2}\right)\right)$$

$$> Q\left(\zeta_0, \zeta_1, \zeta_1, f\left(\frac{t}{k^n}\right)\right),$$

that is

$$Q(\zeta_n, \zeta_{n+1}, \zeta_{n+1}, f(t)) > Q\left(\zeta_0, \zeta_1, \zeta_1, f\left(\frac{t}{k^n}\right)\right).$$

Taking limit on both the sides with $n \to \infty$ we get from above inequality for all positive real number t,

$$\lim_{n\to\infty} Q(\zeta_n, \zeta_{n+1}, \zeta_{n+1}, f(t)) = 1. \tag{7.19}$$

For given $E > 0$, it may be found $t > 0$ with $E > f(t)$. This is possible by virtue of the property of the function f. Hence for all $E > 0$, (7.19) implies that

$$\lim_{n\to\infty} Q(\zeta_n, \zeta_{n+1}, \zeta_{n+1}, E) = 1. \tag{7.20}$$

We now try to show that $\{\zeta_n\}$ be a GCs, if not then we can find positive E along with $\upsilon \in (0, 1)$ for which $\{\zeta_{m(h)}\}$ and $\{\zeta_{n(h)}\}$ subsequences of $\{\zeta_n\}$ can be found with $n(h) > m(h) > h$ such that

$$Q(\zeta_{m(h)}, \zeta_{n(h)}, \zeta_{n(h)}, E) < 1 - \upsilon. \tag{7.21}$$

We may take the integer $n(h)$ which is smaller to the corresponding $m(h)$ and satisfying (7.21) and we have,

$$Q(\zeta_{m(h)}, \zeta_{n(h)-1}, \zeta_{n(h)-1}, E) \geq 1 - \upsilon. \tag{7.22}$$

If $E_1 < E$, then we have

$$Q(\zeta m(h), \zeta n(h), \zeta n(h), E1) \leq Q(\zeta m(h), \zeta n(h), \zeta n(h), E).$$

It may be conclude that, construction of $\{\zeta_{m(h)}\}$ and $\{\zeta_{n(h)}\}$ is possible with the condition $n(h) > m(h) > h$ and satisfying the equations (7.21) and (7.22) with some positive number which is less than E. Also since the function f is continuous at 0 with $f(0) = 0$ and f is strictly monotone increasing function we can find a $E_2 > 0$ with $f(E_2) < E$.

With the help of above argument, we can find increasing sequence $\{\zeta_{m(h)}\}$ and $\{\zeta_{n(h)}\}$ with $n(h) > m(h) > h$ such that

$$Q(\zeta_{m(h)}, \zeta_{n(h)}, \zeta_{n(h)}, f(E_2)) < 1 - \upsilon. \tag{7.23}$$

and

$$Q(\zeta_{m(h)}, \zeta_{n(h)-1}, \zeta_{n(h)-1}, f(E_2)) \geq 1 - \upsilon. \tag{7.24}$$

By (7.23) we can write

$$1-\upsilon > Q(\zeta_{m(h)},\zeta_{n(h)},\zeta_{n(h)},f(t)) = Q(\alpha\rho_{m(h)},\alpha\rho_{n(h)},\alpha\rho_{n(h)},f(t))$$

$$\geq g\left(\begin{array}{l} Q\left(\beta\rho_{m(h)},\beta\rho_{n(h)},\beta\rho_{n(h)},f\left(\frac{t}{k}\right)\right),Q\left(\beta\rho_{m(h)},\beta\rho_{m(h)},\alpha\rho_{m(h)},f\left(\frac{t}{k}\right)\right), \\ Q\left(\beta\rho_{n(h)},\beta\rho_{n(h)},\alpha\rho_{n(h)},f\left(\frac{t}{k}\right)\right),Q\left(\beta\rho_{n(h)},\beta\rho_{n(h)},\alpha\rho_{n(h)},f\left(\frac{t}{k}\right)\right), \\ Q\left(\beta\rho_{n(h)},\beta\rho_{n(h)},\alpha\rho_{n(h)},f\left(\frac{t}{k}\right)\right),Q\left(\alpha\rho_{m(h)},\beta\rho_{n(h)},\beta\rho_{n(h)},f\left(\frac{t}{k}\right)\right) \end{array}\right)$$

$$= g\left(\begin{array}{l} Q\left(\zeta_{m(h)-1},\zeta_{n(h)-1},\zeta_{n(h)-1},f\left(\frac{t}{k}\right)\right),Q\left(\zeta_{m(h)-1},\zeta_{m(h)-1},\zeta_{m(h)},f\left(\frac{t}{k}\right)\right), \\ Q\left(\zeta_{n(h)-1},\zeta_{n(h)-1},\zeta_{n(h)},f\left(\frac{t}{k}\right)\right),Q\left(\zeta_{n(h)-1},\zeta_{n(h)-1},\zeta_{n(h)},f\left(\frac{t}{k}\right)\right), \\ Q\left(\zeta_{n(h)-1},\zeta_{n(h)-1},\zeta_{n(h)},f\left(\frac{t}{k}\right)\right),Q\left(\zeta_{m(h)},\zeta_{n(h)-1},\zeta_{n(h)-1},f\left(\frac{t}{k}\right)\right) \end{array}\right). \tag{7.25}$$

By the equation (7.19) we can find $N_1 \in \mathrm{N}$ for $\upsilon_1 < \upsilon < 1$ and we have for $h > N_1$,

$$Q\left(\zeta_{m(h)-1},\zeta_{m(h)-1},\zeta_{m(h)},f\left(\frac{t}{k}\right)\right) \geq 1-\upsilon_1. \tag{7.26}$$

$$Q\left(\zeta_{n(h)-1},\zeta_{n(h)-1},\zeta_{n(h)},f\left(\frac{t}{k}\right)\right) \geq 1-\upsilon_1. \tag{7.27}$$

As $0 < k < 1$, by the property of f for all positive $t, f\left(\frac{t}{k}\right) > f(t)$.

A positive number η may be chosen such that $\eta < f\left(\frac{t}{k}\right) - f(t)$, that is

$$f\left(\frac{t}{k}\right) - \eta > f(t). \tag{7.28}$$

Now,

$$Q\left(\zeta_{m(h)-1},\zeta_{n(h)-1}\zeta_{n(h)-1},f\left(\frac{t}{k}\right)\right)$$

$$\geq \Delta\left(Q(\zeta_{m(h)-1},\zeta_{m(h)},\zeta_{m(h)},\eta),Q\left(\zeta_{m(h)},\zeta_{n(h)-1},\zeta_{n(h)-1},f\left(\frac{t}{k}\right)-\eta\right)\right)$$
$$\geq \Delta(Q(\zeta_{m(h)-1},\zeta_{m(h)},\zeta_{m(h)},\eta),Q(\zeta_{m(h)},\zeta_{n(h)-1},\zeta_{n(h)-1},f(t)). \tag{7.29}$$

Let $\upsilon_2 \in (0, 1)$ be an arbitrary. By (7.19) a positive integer N_2 can be found with $h > N_2$. We have,

$$Q(\zeta_{m(h)-1},\zeta_{m(h)-1},\zeta_{m(h)},\eta) = Q(\zeta_{m(h)-1},\zeta_{m(h)},\zeta_{m(h)},\eta) > 1-\upsilon_2. \tag{7.30}$$

Using (7.24) and (7.30) in (7.29), for all $h > \max\{N_1, N_2\}$, we can write.

$$Q\left(\zeta_{m(h)-1},\zeta_{n(h)-1},\zeta_{n(h)-1},f\left(\frac{t}{k}\right)\right)\geq \Delta(1-\upsilon_2,1-\upsilon).$$

So for arbitrary υ_2 and continuous Δ, we have

$$Q\left(\zeta_{m(h)-1},\zeta_{n(h)-1},\zeta_{n(h)-1},f\left(\frac{t}{k}\right)\right)\geq \Delta(1,1-\upsilon)=1-\upsilon). \tag{7.31}$$

Using (7.31), (7.26), (7.27) and (7.24) in (7.25), we can write,

$$1-\upsilon > g(1-\upsilon, 1-\upsilon_1, 1-\upsilon_1, 1-\upsilon_1, 1-\upsilon_1, 1-\upsilon), [\text{ as } 0<\upsilon_1<\upsilon].$$

$> 1 - \upsilon$ [by the properties of g-function] and we arrived at a contradiction. Therefore the sequence $\{\zeta_n\}$ turns out to be a Cauchy sequence. Since, $\zeta_n = \beta\rho_{n+1}$, the sequence $\{\beta\rho_{n+1}\}$ will be a GCs in $\beta(M)$.

By hypothesis (C2), β (*M*) is G-complete then there exist $\omega \in \beta$ (*M*) such that

$$\lim_{n\to\infty} \beta\rho_{n+1} = \omega = \lim_{n\to\infty} \alpha\rho_n.$$

As $\omega \in \beta(M)$, $\xi \in M$ exists such that $\omega = \beta\xi$.

We can write

$$\lim_{n\to\infty} \beta\rho_{n+1} = \omega = \beta\xi = \lim_{n\to\infty} \alpha\rho_n. \tag{7.32}$$

Now we have to show that $\alpha\xi = \beta\xi$. Let $\alpha\xi \neq \beta\xi$.

Putting, $\rho = \xi, \zeta = r = \rho_n$ in inequality (7.15).

$$Q(\alpha\xi, \alpha\rho_n, \alpha\rho_n, f(t)) \geq$$

$$g\begin{pmatrix} Q\left(\beta\xi, \beta\rho_n, \beta\rho_n, f\left(\frac{t}{k}\right)\right), Q\left(\beta\xi, \beta\xi, \alpha\xi, f\left(\frac{t}{k}\right)\right), Q\left(\beta\rho_n, \beta\rho_n, \alpha\rho_n, f\left(\frac{t}{k}\right)\right), \\ Q\left(\beta\rho_n, \beta\rho_n, \alpha\rho_n, f\left(\frac{t}{k}\right)\right), Q\left(\beta\rho_n, \beta\rho_n, \alpha\rho_n, f\left(\frac{t}{k}\right)\right), Q\left(\alpha\xi, \beta\rho_n, \beta\rho_n, f\left(\frac{t}{k}\right)\right) \end{pmatrix}.$$

Taking limit $n \to \infty$, and using (7.32) we can write from the above inequality

$$Q(\alpha\xi, \beta\xi, \beta\xi, f(t)) \geq$$

$$g\begin{pmatrix} Q\left(\beta\xi, \beta\xi, \beta\xi, f\left(\frac{t}{k}\right)\right), Q\left(\beta\xi, \beta\xi, \alpha\xi, f\left(\frac{t}{k}\right)\right), Q\left(\beta\xi, \beta\xi, \beta\xi, f\left(\frac{t}{k}\right)\right), \\ Q\left(\beta\xi, \beta\xi, \beta\xi, f\left(\frac{t}{k}\right)\right), Q\left(\beta\xi, \beta\xi, \beta\xi, f\left(\frac{t}{k}\right)\right), Q\left(\alpha\xi, \beta\xi, \beta\xi, f\left(\frac{t}{k}\right)\right) \end{pmatrix}.$$

So

$$Q(\alpha\xi, \beta\xi, \beta\xi, f(t)) \geq$$

$$g\begin{pmatrix} Q\left(\alpha\xi, \beta\xi, \beta\xi, f\left(\frac{t}{k}\right)\right), Q\left(\alpha\xi, \beta\xi, \beta\xi, f\left(\frac{t}{k}\right)\right), Q\left(\alpha\xi, \beta\xi, \beta\xi, f\left(\frac{t}{k}\right)\right), \\ Q\left(\alpha\xi, \beta\xi, \beta\xi, f\left(\frac{t}{k}\right)\right), Q\left(\alpha\xi, \beta\xi, \beta\xi, f\left(\frac{t}{k}\right)\right), Q\left(\alpha\xi, \beta\xi, \beta\xi, f\left(\frac{t}{k}\right)\right) \end{pmatrix}.$$

$\therefore$ By the property of *g*-function we have

$$Q(\alpha\xi, \beta\xi, \beta\xi, f(t)) > Q\left(\alpha\xi, \beta\xi, \beta\xi, f\left(\frac{t}{k}\right)\right).$$

By the Lemma (7.2.12) we can write.

$$\alpha\xi = \beta\xi.$$

$\therefore \xi \in M$ is a cp of α and β. Hence proof of the theorem is completed. In the next, we show that G_α-type G-wc maps α, β have a unique cfp.

7.3.5 Theorem

Let (M, Q, Δ) be a symmetric G-fms with continuous Δ and α, β : $M \rightarrow M$ be two self-maps with

(D1) $\alpha(M) \subseteq \beta(M)$,

(D2) $\beta(M) \subseteq M$ is G-complete.

(D3) $Q(\alpha\rho, \alpha\zeta, \alpha r, f(t)) \geq$

$$g\begin{pmatrix} Q\left(\beta\rho, \beta\zeta, \beta r, f\left(\frac{t}{k}\right)\right), Q\left(\beta\rho, \beta\rho, \alpha\rho, f\left(\frac{t}{k}\right)\right), Q\left(\beta\zeta, \beta\zeta, \alpha\zeta, f\left(\frac{t}{k}\right)\right), \\ Q\left(\beta r, \beta r, \alpha r, f\left(\frac{t}{k}\right)\right), Q\left(\beta\zeta, \beta\zeta, \alpha r, f\left(\frac{t}{k}\right)\right), Q\left(\alpha\rho, \beta r, \beta r, f\left(\frac{t}{k}\right)\right) \end{pmatrix} \tag{7.33}$$

$$\forall \rho, \zeta, r \in M \text{ with positive } t \text{ and } k \in (0,1).$$

(D4) g is Ψ-function of sixth order and f is a Φ-function,

(D5) α and β are of $G_\alpha$type G-wc.

Then the maps α, β have a unique cfp in M.

Proof. Let $\rho_0 \in M$ be arbitrary. As $\alpha(M) \subseteq \beta(M)$, $\rho_1 \in M$ such that $\alpha\rho_0 = \beta\rho_1$. We define the sequence $\{\rho_n\}$ in general inductive process such that $\alpha\rho_n = \beta\rho_{n+1}$. We may assumed that $\beta\rho_n \neq \beta\rho_{n+1}$ for each n, otherwise we have a fp of β. We are supposing, $\zeta_n = \alpha\rho_n = \beta\rho_{n+1}$, $n \in \mathrm{N}$, set of natural numbers.

Putting $\rho = \rho_n, \zeta = \rho_{n+1}, r = \rho_{n+1}$ in inequality (7.33), for all $t > 0$,

$$Q(\alpha\rho_n, \alpha\rho_{n+1}, \alpha\rho_{n+1}, f(t)) \geq$$

$$g\begin{pmatrix} Q\left(\beta\rho_n, \beta\rho_{n+1}, \beta\rho_{n+1}, f\left(\frac{t}{k}\right)\right), Q\left(\beta\rho_n, \beta\rho_n, \alpha\rho_n, f\left(\frac{t}{k}\right)\right), \\ Q\left(\beta\rho_{n+1}, \beta\rho_{n+1}, \alpha\rho_{n+1}, f\left(\frac{t}{k}\right)\right), Q\left(\beta\rho_{n+1}, \beta\rho_{n+1}, \alpha\rho_{n+1}, f\left(\frac{t}{k}\right)\right), \\ Q\left(\beta\rho_{n+1}, \beta\rho_{n+1}, \alpha\rho_{n+1}, f\left(\frac{t}{k}\right)\right), Q\left(\alpha\rho_n, \beta\rho_{n+1}, \beta\rho_{n+1}, f\left(\frac{t}{k}\right)\right) \end{pmatrix}.$$

Therefore,

$$Q(\zeta_n,\zeta_{n+1},\zeta_{n+1},f(t)) \geq g\begin{pmatrix} Q\left(\zeta_{n-1},\zeta_n,\zeta_n,f\left(\frac{t}{k}\right)\right),Q\left(\zeta_{n-1},\zeta_{n-1},\zeta_n,f\left(\frac{t}{k}\right)\right), \\ Q\left(\zeta_n,\zeta_n,\zeta_{n+1},f\left(\frac{t}{k}\right)\right),Q\left(\zeta_n,\zeta_n,\zeta_{n+1},f\left(\frac{t}{k}\right)\right), \\ Q\left(\zeta_n,\zeta_n,\zeta_{n+1},f\left(\frac{t}{k}\right)\right),Q\left(\zeta_n,\zeta_n,\zeta_n,f\left(\frac{t}{k}\right)\right) \end{pmatrix}.$$

So

$$Q(\zeta_n,\zeta_{n+1},\zeta_{n+1},f(t)) \geq g\begin{pmatrix} Q\left(\zeta_{n-1},\zeta_n,\zeta_n,f\left(\frac{t}{k}\right)\right),Q\left(\zeta_{n-1},\zeta_n,\zeta_n,f\left(\frac{t}{k}\right)\right), \\ Q\left(\zeta_n,\zeta_{n+1},\zeta_{n+1},f\left(\frac{t}{k}\right)\right),Q\left(\zeta_n,\zeta_{n+1},\zeta_{n+1},f\left(\frac{t}{k}\right)\right), \\ Q\left(\zeta_n,\zeta_{n+1},\zeta_{n+1},f\left(\frac{t}{k}\right)\right),Q\left(\zeta_n,\zeta_n,\zeta_n,f\left(\frac{t}{k}\right)\right) \end{pmatrix}. \tag{7.34}$$

We claim that for all $t > 0,\ n \in \mathrm{N}$,

$$Q\left(\zeta_n,\zeta_{n+1},\zeta_{n+1},f\left(\frac{t}{k}\right)\right) \geq Q\left(\zeta_{n-1},\zeta_n,\zeta_n,f\left(\frac{t}{k}\right)\right). \tag{7.35}$$

If possible, let for some $t_1 > 0$ and $n \in \mathrm{N}$,

$$Q\left(\zeta_n,\zeta_{n+1},\zeta_{n+1},f\left(\frac{t_1}{k}\right)\right) < Q\left(\zeta_{n-1},\zeta_n,\zeta_n,f\left(\frac{t_1}{k}\right)\right). \tag{7.36}$$

Then we can write from (7.34),

$$Q(\zeta_n,\zeta_{n+1},\zeta_{n+1},f(t_1))\geq g\begin{pmatrix} Q\left(\zeta_n,\zeta_{n+1},\zeta_{n+1},f\left(\frac{t_1}{k}\right)\right),Q\left(\zeta_n,\zeta_{n+1},\zeta_{n+1},f\left(\frac{t_1}{k}\right)\right),\\ Q\left(\zeta_n,\zeta_{n+1},\zeta_{n+1},f\left(\frac{t_1}{k}\right)\right),Q\left(\zeta_n,\zeta_{n+1},\zeta_{n+1},f\left(\frac{t_1}{k}\right)\right),\\ Q\left(\zeta_n,\zeta_{n+1},\zeta_{n+1},f\left(\frac{t_1}{k}\right)\right),Q\left(\zeta_n,\zeta_{n+1},\zeta_{n+1},f\left(\frac{t_1}{k}\right)\right) \end{pmatrix}$$

$$>Q\left(\zeta_n,\zeta_{n+1},\zeta_{n+1},f\left(\frac{t_1}{k}\right)\right) \text{[by the property of } g\text{-function]}$$

$$\geq Q(\zeta_n,\zeta_{n+1},\zeta_{n+1},f(t_1)) \text{ which is a contradiction as } 0<k<1.$$

$\therefore$ $\forall t>0$ and $n\in N$, (7.35) holds.

Now, using (7.35) in (7.34) $\forall t>0$, $n\in N$, we have

$$Q(\zeta_n,\zeta_{n+1},\zeta_{n+1},f(t))\geq g\begin{pmatrix} Q\left(\zeta_{n-1},\zeta_n,\zeta_n,f\left(\frac{t}{k}\right)\right),Q\left(\zeta_{n-1},\zeta_n,\zeta_n,f\left(\frac{t}{k}\right)\right),\\ Q\left(\zeta_{n-1},\zeta_n,\zeta_n,f\left(\frac{t}{k}\right)\right),Q\left(\zeta_{n-1},\zeta_n,\zeta_n,f\left(\frac{t}{k}\right)\right),\\ Q\left(\zeta_{n-1},\zeta_n,\zeta_n,f\left(\frac{t}{k}\right)\right),Q\left(\zeta_{n-1},\zeta_n,\zeta_n,f\left(\frac{t}{k}\right)\right) \end{pmatrix}$$

$$>Q\left(\zeta_{n-1},\zeta_n,\zeta_n,f\left(\frac{t}{k}\right)\right) \text{[by the property of } g\text{-function].}$$

So

$$\begin{aligned} Q(\zeta_n,\zeta_{n+1},\zeta_{n+1}f(t)) &> Q\left(\zeta_{n-1},\zeta_n,\zeta_n,f\left(\frac{t}{k}\right)\right)\\ &> Q\left(\zeta_{n-2},\zeta_{n-1},\zeta_{n-1},f\left(\frac{t}{k^2}\right)\right)\\ &> Q\left(\zeta_0,\zeta_1,\zeta_1,f\left(\frac{t}{k^n}\right)\right), \end{aligned} \tag{7.37}$$

that is

$$Q(\zeta_n,\zeta_{n+1},\zeta_{n+1}f(t))>Q\left(\zeta_0,\zeta_1,\zeta_1,f\left(\frac{t}{k^n}\right)\right).$$

Taking limit on both the sides with $n \to \infty$ we get from above inequality for all positive real number t,

$$\lim_{n\to\infty} Q(\zeta_n, \zeta_{n+1}, \zeta_{n+1}, f(t)) = 1. \tag{7.38}$$

For given $E > 0$, it may be found $t > 0$ with $E > f(t)$. This is possible by virtue of the property of the function f. Hence for all $E > 0$, (7.38) implies that

$$\lim_{n\to\infty} Q(\zeta_n, \zeta_{n+1}, \zeta_{n+1}, E) = 1. \tag{7.39}$$

We now try to show that $\{\zeta_n\}$ be a Cauchy sequence. If not, we can find positive E along with $\upsilon \in (0, 1)$ for which $\{\zeta_{m(h)}\}$ and $\{\zeta_{n(h)}\}$ of $\{\zeta_n\}$ can be found with $n(h) > m(h) > h$ we have

$$Q(\zeta_{m(h)}, \zeta_{n(h)}, \zeta_{n(h)}, E) < 1 - \upsilon. \tag{7.40}$$

We may take the integer $n(h)$ which is smaller to the corresponding $m(h)$ and satisfying (7.40) and we have,

$$Q(\zeta_{m(h)}, \zeta_{n(h)-1}, \zeta_{n(h)-1}, E) \geq 1 - \upsilon. \tag{7.41}$$

If $E_1 < E$, then we have

$$Q(\zeta m(h), \zeta n(h), \zeta n(h), E1) \leq Q(\zeta m(h), \zeta n(h), \zeta n(h), E).$$

It may be conclude that, construction $\{\zeta m(h)\}$ and $\{\zeta n(h)\}$ is possible with the condition $n(h) > m(h) > h$ and satisfying the equations (7.40) and (7.41) some smaller positive value of E. Also since the function f is continuous at 0 with $f(0) = 0$ and f is strictly monotone increasing function, we can find a $E_2 > 0$ with $f(E_2) < E$.

By the above argument, we can find increasing sequences $\{\zeta m(h)\}$ and $\{\zeta n(h)\}$ with $n(h) > m(h) > h$ such that

$$Q(\zeta_{m(h)}, \zeta_{n(h)}, \zeta_{n(h)}, f(E_2)) < 1 - \upsilon. \tag{7.42}$$

and

$$Q(\zeta_{m(h)}, \zeta_{n(h)-1}, \zeta_{n(h)-1}, f(E_2)) \geq 1 - \upsilon. \tag{7.43}$$

By (7.42), we can write.

$$1-\upsilon > Q(\zeta_{m(h)},\zeta_{n(h)},\zeta_{n(h)},f(t)) = Q(\alpha\rho_{m(h)},\alpha\rho_{n(h)},\alpha\rho_{n(h)},f(t))$$

$$\geq g\begin{pmatrix} Q\left(\beta\rho_{m(h)},\beta\rho_{n(h)},\beta\rho_{n(h)},f\left(\frac{t}{k}\right)\right),Q\left(\beta\rho_{m(h)},\beta\rho_{m(h)},\alpha\rho_{m(h)},f\left(\frac{t}{k}\right)\right), \\ Q\left(\beta\rho_{n(h)},\beta\rho_{n(h)},\alpha\rho_{n(h)},f\left(\frac{t}{k}\right)\right),Q\left(\beta\rho_{n(h)},\beta\rho_{n(h)},\alpha\rho_{n(h)},f\left(\frac{t}{k}\right)\right), \\ Q\left(\beta\rho_{n(h)},\beta\rho_{n(h)},\alpha\rho_{n(h)},f\left(\frac{t}{k}\right)\right),Q\left(\alpha\rho_{m(h)},\beta\rho_{n(h)},\beta\rho_{n(h)},f\left(\frac{t}{k}\right)\right) \end{pmatrix}$$

$$= g\begin{pmatrix} Q\left(\zeta_{m(h)-1},\zeta_{n(h)-1},\zeta_{n(h)-1},f\left(\frac{t}{k}\right)\right),Q\left(\zeta_{m(h)-1},\zeta_{m(h)-1},\zeta_{m(h)},f\left(\frac{t}{k}\right)\right), \\ Q\left(\zeta_{n(h)-1},\zeta_{n(h)-1},\zeta_{n(h)},f\left(\frac{t}{k}\right)\right),Q\left(\zeta_{n(h)-1},\zeta_{n(h)-1},\zeta_{n(h)},f\left(\frac{t}{k}\right)\right), \\ Q\left(\zeta_{n(h)-1},\zeta_{n(h)-1},\zeta_{n(h)},f\left(\frac{t}{k}\right)\right),Q\left(\zeta_{m(h)},\zeta_{n(h)-1},\zeta_{n(h)-1},f\left(\frac{t}{k}\right)\right) \end{pmatrix}. \tag{7.44}$$

By the equation (7.39) for $\upsilon_1 \in (0,1)$ and υ_1 less than υ, we can find a positive integer

$$N_1 \text{ with } h > N_1,$$

$$Q\left(\zeta_{m(h)-1},\zeta_{m(h)-1},\zeta_{m(h)},f\left(\frac{t}{k}\right)\right) \geq 1-\upsilon_1. \tag{7.45}$$

$$Q\left(\zeta_{n(h)-1},\zeta_{n(h)-1},\zeta_{n(h)},f\left(\frac{t}{k}\right)\right) \geq 1-\upsilon_1. \tag{7.46}$$

Here t is positive and $k \in (0,1)$ with strictly monotone increasing f.

$$\therefore \quad f\left(\frac{t}{k}\right) > f(t),$$

We can choose $\eta > 0$ such that $\eta < f\left(\frac{t}{k}\right) - f(t)$ that is $f\left(\frac{t}{k}\right) - \eta > f(t)$,

Now

$$Q\left(\zeta_{m(h)-1},\zeta_{n(h)-1},\zeta_{n(h)-1},f\left(\frac{t}{k}\right)\right)$$
$$\geq \Delta\left(Q(\zeta_{m(h)-1},\zeta_{m(h)},\zeta_{m(h)},\eta),Q\left(\zeta_{m(h)},\zeta_{n(h)-1},\zeta_{n(h)-1},f\left(\frac{t}{k}\right)-\eta\right)\right)$$
$$\geq \Delta(Q(\zeta_{m(h)-1},\zeta_{m(h)},\zeta_{m(h)},\eta),Q(\zeta_{m(h)},\zeta_{n(h)-1},\zeta_{n(h)-1},f(t))) \quad (7.47)$$

υ_2 being arbitrary and by (7.39) a positive integer N_2 can be found such that for all $h > N_2$

$$Q(\zeta_{m(h)-1},\zeta_{m(h)-1},\zeta_{m(h)},\eta) = Q(\zeta_{m(h)-1},\zeta_{m(h)},\zeta_{m(h)},\eta) > 1-\upsilon_2. \quad (7.48)$$

Using (7.48) and (7.43) in (7.47) for all $h > \max\{N_1, N_2\}$, we can get.

$$Q\left(\zeta_{m(h)-1},\zeta_{n(h)-1},\zeta_{n(h)-1},f\left(\frac{t}{k}\right)\right)\geq \Delta(1-\upsilon_2,1-\upsilon).$$

As Δ is continuous and υ_2 being arbitrary, we have

$$Q\left(\zeta_{m(h)-1},\zeta_{n(h)-1},\zeta_{n(h)-1},f\left(\frac{t}{k}\right)\right)\geq \Delta\{1,1-\upsilon\}=1-\upsilon. \quad (7.49)$$

Using (7.49), (7.45), (7.46) and (7.43) in (7.44), we can write.

$$1-\upsilon > g(1-\upsilon, 1-\upsilon_1, 1-\upsilon_1, 1-\upsilon_1, 1-\upsilon_1, 1-\upsilon), \text{[where } 0<\upsilon_1<\upsilon].$$

$$> 1-\upsilon \text{ [by the property of g-function]}$$

and we arrived at a contradiction.

$\therefore \{\zeta n\}$ is a GCs.

By hypothesis (D2), $\beta(M)$ is G-complete, so $\omega \in \beta(M)$ such that

$$\lim_{n\to\infty} \beta\rho_{n+1} = \omega = \lim_{n\to\infty} \alpha\rho_n.$$

As $\omega \in \beta(M)$, $\exists\, \xi \in M$ such that $\omega = \beta\xi$.

We can write

$$\lim_{n\to\infty}\beta\rho_n=\omega=\beta\xi=\lim_{n\to\infty}\alpha\rho_n. \tag{7.50}$$

Now we have to show that $\alpha\xi=\beta\xi$. Let $\alpha\xi\neq\beta\xi$.
Putting, $\rho=\xi$, $\zeta=r=\rho_n$ in inequality (7.33).

$$Q(\alpha\xi,\alpha\rho_n,\alpha\rho_n,f(t))\geq g\left(\begin{array}{l}Q\left(\beta\xi,\beta\rho_n,\beta\rho_n,f\left(\frac{t}{k}\right)\right),Q\left(\beta\xi,\beta\xi,\alpha\xi,f\left(\frac{t}{k}\right)\right),Q\left(\beta\rho_n,\beta\rho_n,\alpha\rho_n,f\left(\frac{t}{k}\right)\right),\\Q\left(\beta\rho_n,\beta\rho_n,\alpha\rho_n,f\left(\frac{t}{k}\right)\right),Q\left(\beta\rho_n,\beta\rho_n,\alpha\rho_n,f\left(\frac{t}{k}\right)\right),Q\left(\alpha\xi,\beta\rho_n,\beta\rho_n,f\left(\frac{t}{k}\right)\right)\end{array}\right).$$

Taking limit $n\to\infty$, for all $t>0$ and $0<k<1$ and by using (7.50) we can obtain from the above inequality

$$Q(\alpha\xi,\beta\xi,\beta\xi,f(t))\geq g\left(\begin{array}{l}Q\left(\beta\xi,\beta\xi,\beta\xi,f\left(\frac{t}{k}\right)\right),Q\left(\beta\xi,\beta\xi,\alpha\xi,f\left(\frac{t}{k}\right)\right),Q\left(\beta\xi,\beta\xi,\beta\xi,f\left(\frac{t}{k}\right)\right),\\Q\left(\beta\xi,\beta\xi,\beta\xi,f\left(\frac{t}{k}\right)\right),Q\left(\beta\xi,\beta\xi,\beta\xi,f\left(\frac{t}{k}\right)\right),Q\left(\alpha\xi,\beta\xi,\beta\xi,f\left(\frac{t}{k}\right)\right)\end{array}\right).$$

So

$$Q(\alpha\xi,\beta\xi,\beta\xi,f(t))\geq g\left(\begin{array}{l}Q\left(\alpha\xi,\beta\xi,\beta\xi,f\left(\frac{t}{k}\right)\right),Q\left(\alpha\xi,\beta\xi,\beta\xi,f\left(\frac{t}{k}\right)\right),Q\left(\alpha\xi,\beta\xi,\beta\xi,f\left(\frac{t}{k}\right)\right),\\Q\left(\alpha\xi,\beta\xi,\beta\xi,f\left(\frac{t}{k}\right)\right),Q\left(\alpha\xi,\beta\xi,\beta\xi,f\left(\frac{t}{k}\right)\right),Q\left(\alpha\xi,\beta\xi,\beta\xi,f\left(\frac{t}{k}\right)\right)\end{array}\right).$$

∴ By the property of g-function we have

$$Q\left(\alpha\xi,\ \beta\xi,\ \beta\xi,\ f(t)\right)>Q\left(\alpha\xi,\ \beta\xi,\ \beta\xi,\left(\frac{t}{k}\right)\right).$$

By the Lemma (7.2.12) we can write from above inequality

$$\alpha\xi=\beta\xi. \tag{7.51}$$

By the condition (D5) the pair (α,β) are G_α type Gwc, then by using (7.51), we have $Q(\alpha\beta\xi,\beta\alpha\xi,\alpha\alpha\xi,f(t))\geq Q\left(\alpha\xi,\beta\xi,\alpha\xi,f\left(\frac{t}{k}\right)\right)=1$, which implies that

$$\alpha\beta\xi = \beta\alpha\xi = \alpha\alpha\xi. \tag{7.52}$$

Now taking $\omega = \alpha\xi = \beta\xi$, we can write from (7.51) and (7.52),

$$\alpha\beta\xi = \alpha\omega = \beta\alpha\xi = \beta\omega = \alpha\alpha\xi. \tag{7.53}$$

Here we try to prove that α and β have a cfp.

Putting $\rho = \omega$, $\zeta = r = \xi$, in inequality (7.33) we get.

$$Q(\alpha\omega,\alpha\xi,\alpha\xi,f(t)) \geq g\begin{pmatrix} Q\left(\beta\omega,\beta\xi,\beta\xi,f\left(\frac{t}{k}\right)\right), Q\left(\beta\omega,\beta\omega,\alpha\omega,f\left(\frac{t}{k}\right)\right), Q\left(\beta\xi,\beta\xi,\alpha\xi,f\left(\frac{t}{k}\right)\right), \\ Q\left(\beta\xi,\beta\xi,\alpha\xi,f\left(\frac{t}{k}\right)\right), Q\left(\beta\xi,\beta\xi,\alpha\xi,f\left(\frac{t}{k}\right)\right), Q\left(\alpha\omega,\beta\xi,\beta\xi,f\left(\frac{t}{k}\right)\right) \end{pmatrix}.$$

So

$$Q(\alpha\omega,\omega,\omega,f(t)) \geq g\begin{pmatrix} Q\left(\alpha\omega,\omega,\omega,f\left(\frac{t}{k}\right)\right), Q\left(\alpha\omega,\alpha\omega,\alpha\omega,f\left(\frac{t}{k}\right)\right), Q\left(\omega,\omega,\omega,f\left(\frac{t}{k}\right)\right), \\ Q\left(\omega,\omega,\omega,f\left(\frac{t}{k}\right)\right), Q\left(\omega,\omega,\omega,f\left(\frac{t}{k}\right)\right), Q\left(\alpha\omega,\omega,\omega,f\left(\frac{t}{k}\right)\right) \end{pmatrix},$$

that is

$$Q(\alpha\omega,\omega,\omega,f(t)) \geq g\begin{pmatrix} Q\left(\alpha\omega,\omega,\omega,f\left(\frac{t}{k}\right)\right), Q\left(\alpha\omega,\omega,\omega,f\left(\frac{t}{k}\right)\right), Q\left(\alpha\omega,\omega,\omega,f\left(\frac{t}{k}\right)\right), \\ Q\left(\alpha\omega,\omega,\omega,f\left(\frac{t}{k}\right)\right), Q\left(\alpha\omega,\omega,\omega,f\left(\frac{t}{k}\right)\right), Q\left(\alpha\omega,\omega,\omega,f\left(\frac{t}{k}\right)\right), \end{pmatrix}$$

$$> Q\left(\alpha\omega,\omega,\omega,f\left(\frac{t}{k}\right)\right) \text{[by the property of } g\text{]}.$$

By the Lemma (7.2.12) we can write from above

$$\alpha\omega = \omega.$$

So, with help of (7.53) and the above relation we get.

$$\alpha\omega = \beta\omega = \omega. \tag{7.54}$$

For uniqueness, let $\nu(\nu \neq \omega)$ be the another fp.

Now from inequality (7.33) we can write by putting $\rho = \omega$, $\zeta = r = \nu$,

$$Q(\alpha\omega,\alpha\nu,\alpha\nu,f(t))\geq$$
$$g\begin{pmatrix} Q\left(\beta\omega,\beta\nu,\beta\nu,f\left(\frac{t}{k}\right)\right),Q\left(\beta\omega,\beta\omega,\alpha\omega,f\left(\frac{t}{k}\right)\right),Q\left(\beta\nu,\beta\nu,\alpha\nu,f\left(\frac{t}{k}\right)\right), \\ Q\left(\beta\nu,\beta\nu,\alpha\nu,f\left(\frac{t}{k}\right)\right),Q\left(\beta\nu,\beta\nu,\alpha\nu,f\left(\frac{t}{k}\right)\right),Q\left(\alpha\omega,\beta\nu,\alpha\nu,f\left(\frac{t}{k}\right)\right) \end{pmatrix},$$

that is

$$Q(\omega,\nu,\nu,f(t))\geq$$
$$g\begin{pmatrix} Q\left(\omega,\nu,\nu,f\left(\frac{t}{k}\right)\right),Q\left(\omega,\omega,\omega,f\left(\frac{t}{k}\right)\right),Q\left(\nu,\nu,\nu,f\left(\frac{t}{k}\right)\right), \\ Q\left(\nu,\nu,\nu,f\left(\frac{t}{k}\right)\right),Q\left(\nu,\nu,\nu,f\left(\frac{t}{k}\right)\right),Q\left(\omega,\nu,\nu,f\left(\frac{t}{k}\right)\right) \end{pmatrix},$$

Hence we arrived at

$$Q(\omega,\nu,\nu,f(t))\geq$$
$$g\begin{pmatrix} Q\left(\omega,\nu,\nu,f\left(\frac{t}{k}\right)\right),Q\left(\omega,\nu,\nu,f\left(\frac{t}{k}\right)\right),Q\left(\omega,\nu,\nu,f\left(\frac{t}{k}\right)\right), \\ Q\left(\omega,\nu,\nu,f\left(\frac{t}{k}\right)\right),Q\left(\omega,\nu,\nu,f\left(\frac{t}{k}\right)\right),Q\left(\omega,\nu,\nu,f\left(\frac{t}{k}\right)\right) \end{pmatrix}.$$

Therefore, by the property of g-function,

$$Q(\omega,\nu,\nu,f(t))>Q\left(\omega,\nu,\nu,f\left(\frac{t}{k}\right)\right).$$

By the Lemma (7.2.12) we can write

$$\omega=\nu.$$

Thus, uniqueness is proved and the theorem is completed.

Now putting $f(t) = t$ we get two corollaries for the Theorems 7.3.4 and 7.3.5, respectively.

7.3.6 Corollary

Let (M, Q, Δ) be a symmetric G-fms with continuous H-type t-norm Δ and α,β: $M \to M$ be two self-maps satisfying

(E1) $\alpha(M) \subseteq \beta(M)$,

(E2) $\beta(M) \subseteq M$ is G-complete,

$$\text{(E3) } Q(\alpha\rho,\alpha\zeta,\alpha r,t) \geq g\begin{pmatrix} Q\left(\beta\rho,\beta\zeta,\beta r,\frac{t}{k}\right),Q\left(\beta\rho,\beta\rho,\alpha\rho,\frac{t}{k}\right), \\ Q\left(\beta\zeta,\beta\zeta,\alpha\zeta,\frac{t}{k}\right),Q\left(\beta r,\beta r,\alpha r,\frac{t}{k}\right), \\ Q\left(\beta\zeta,\beta\zeta,\alpha r,\frac{t}{k}\right),Q\left(\alpha\rho,\beta r,\beta r,\frac{t}{k}\right) \end{pmatrix}. \tag{7.55}$$

$\forall \rho,\zeta,r \in M$ with $t > 0$ and $k \in (0,1)$.

(E4) g is Ψ-function of sixth order.

The maps (α,β) have acp in M.

7.3.7 Corollary

Let (M,Q,Δ) be a symmetric G-fms with continuous H-type t-norm Δ and $\alpha,\beta : M \to M$ be two self-maps satisfying

(F1) $\alpha(M) \subseteq \beta(M)$,

(F2) $\beta(M) \subseteq M$ is G-complete,

$$\text{(F3) } Q(\alpha\rho,\alpha\zeta,\alpha r,t) \geq g\begin{pmatrix} Q\left(\beta\rho,\beta\zeta,\beta r,\frac{t}{k}\right),Q\left(\beta\rho,\beta\rho,\alpha\rho,\frac{t}{k}\right), \\ Q\left(\beta\zeta,\beta\zeta,\alpha\zeta,\frac{t}{k}\right),Q\left(\beta r,\beta r,\alpha r,\frac{t}{k}\right), \\ Q\left(\beta\zeta,\beta\zeta,\alpha r,\frac{t}{k}\right),Q\left(\beta r,\beta r,\alpha\rho,\frac{t}{k}\right) \end{pmatrix}. \tag{7.56}$$

$\forall \rho,\zeta,r \in M$ with $t > 0$ and $k \in (0,1)$.

(F4) g is Ψ-function of sixth order.

(F5) α and β are G_α type $G-$wc.

The maps (α, β) have a unique cfp in M.

Now we give some examples to validate Theorems: 7.3.1., 7.3.2., 7.3.4. and 7.3.5, respectively.

7.3.8 Example

Let $M = [0, 1]$, Q be defined as $Q(\rho,\zeta,r,t)=e^{-\frac{|\rho-\zeta|+|\zeta-r|+|r-\rho|}{t}}$ where all ρ, $\zeta, r \in M$, $t > 0$, and Δ be the minimum t-norm. Then (M, Q, Δ) be a complete G-fms. Let us take the self-maps α and β on (M, Q, Δ) defined by $\alpha\rho = \frac{\rho^2}{32}$ and $\beta\rho = \frac{\rho^2}{4}$.

Taking $r=\zeta, \frac{1}{2}<k<1$ and $g(\rho_1,\rho_2,\rho_3,\rho_4)=\frac{mini\{\rho_1,\rho_2,\rho_4\}+\sqrt{\rho_3}}{2}$, the functions α and β satisfied each condition of Theorem: 7.3.1 and we have $\rho = 0$ is a cp of α, β.

7.3.9 Example

Let M = [0, 1], Q be defined as $Q(\rho,\zeta,r,t)=e^{-\frac{|\rho-\zeta|+|\zeta-r|+|r-\rho|}{t}}$ where all ρ, ζ, $r \in M, t > 0$, and Δ be the minimum t-norm. Then (M, Q, Δ) be a complete G-fms. Let us take the self-maps α and β on (M, Q, Δ) defined by $\alpha\rho = \frac{\rho}{40}$ and $\beta\rho = \frac{\rho}{2}$.

Taking $r = \zeta$, $\frac{1}{5}<k<\frac{1}{2}$ and $g(\rho_1,\rho_2,\rho_3,\rho_4)=\frac{\min i\{\rho_1,\rho_2,\rho_4\}+\sqrt{\rho_3}}{2}$, the functions α and β satisfied each condition of Theorem: 7.3.2 and we have $\rho = 0$ is the unique cfp of α, β.

We also found that,

$Q(\alpha\beta\rho,\beta\alpha\rho,\alpha\alpha\rho,t) \geq \left(\alpha\rho,\beta\rho,\alpha\rho,\frac{t}{k}\right)$ but $Q(\beta\alpha\rho,\alpha\beta\rho,\beta\beta\rho,t) \ngeq$ $\left(\beta\rho,\alpha\rho,\beta\rho,\frac{t}{k}\right)$ for all $\rho \in M, t > 0$ with $\frac{1}{5} < k < \frac{1}{2}$. So, the pair of (α, β) is G_α type G-wc but not G_β type G-wc.

7.3.10 Example

Let M = [0, 1], Q be defined as $Q(\rho,\zeta,r,t)=e^{-\frac{|\rho-\zeta|+|\zeta-r|+|r-\rho|}{t}}$ where all ρ, ζ, r ∈ M, t > 0 and Δ be the minimum t-norm. Then (M, Q, Δ) be a complete G-fms. Let us take the self-maps α and β on (M, Q, Δ) defined by $\alpha\rho=\frac{\rho^2}{27}$ and $\beta\rho=\frac{\rho^2}{3}$. Taking $f(t)=\sqrt{t}, \zeta=\rho, \frac{1}{4}<k<1$ and $g(\rho_1,\rho_2,\rho_3,\rho_4,\rho_5,\rho_6)=\frac{\min i\{\rho_1,\rho_3,\rho_5,\rho_6\}+\sqrt{\rho_{2\times}\sqrt[4]{\rho_4}}}{2}$, the functions α and β satisfied every condition of Theorem: 7.3.4 and we have ρ = 0 is a cp of α, β.

7.3.11 Example

Let M = [0, 1], Q be defined as $Q(\rho,\zeta,r,t)=e^{-\frac{|\rho-\zeta|+|\zeta-r|+|r-\rho|}{t}}$ where all ρ, ζ, *r* ∈ *M*, *t* > 0 and Δ be the minimum t-norm. Then (*M*, *Q*, Δ) be a complete G-fms. Let us take the self-maps α and β on (M, Q, Δ) defined by $\alpha\rho=\frac{\rho}{32}$ and $\beta\rho=\frac{\rho}{2}$. Taking f(t) = t^2, ζ = ρ, (0.52 < *k* < 0.70) and $g(\rho_1,\rho_2,\rho_3,\rho_4,\rho_5,\rho_6)=\frac{\min i\{\rho_1,\rho_3,\rho_5,\rho_6\}+\sqrt{\rho_{2\times}\sqrt[4]{\rho_4}}}{2}$, the functions α and β satisfied each condition of Theorem: 7.3.5 and we have ρ = 0 is the unique cfp of α, β.

We also noticed that,

$$Q(\alpha\beta\rho,\beta\alpha\rho,\alpha\alpha\rho,f(t))\geq Q\left(\alpha\rho,\beta\rho,\alpha\rho,f\left(\frac{t}{k}\right)\right)$$

but $Q(\beta\alpha\rho,\alpha\beta\rho,\beta\beta\rho,f(t))\ngeq Q\left(\beta\rho,\alpha\rho,\beta\rho,f\left(\frac{t}{k}\right)\right)$ for all ρ ∈ *M*, with 0.52 < *k* < 0.70.

So, the pair of (α,β) is G_α type G-wc but not G_β type G-wc.

7.4 Conclusion

In this paper, we have proved two fp results for G_α-type G-wc maps. Two of our results are instance of use of Φ-functions. Our results extend and generalize some existing cp results and cfp results in the literature. These results may be applied to other spaces, such as *b*-metric spaces, intuitionistic fuzzy metric spaces, Menger spaces, Cone metric spaces, and many more.

7.5 Open Question

In Theorem: 7.3.1 and Theorem: 7.3.2, we have used Hadzic type t-norm. Is it possible to replace Hadzic type t-norm? If so, what will be the minimum condition of the theorem?

References

1. Aage, C.T., Choudhury, B.S., Das, K., Some fixed point results in fuzzy metric spaces using a control function. *Surv. Math. Appl.*, 12, 23–34, 2017.
2. Ali, R., Batul, S., Farique, M., Shehwar, D., φ-contractive mappings on generalized probabilistic metric spaces. *J. Math. Anal.*, 3, 1–8, 2018.
3. An, T.V., Dung, N.V., Le Hang, V.T., A new approach to fixed point theorems on G-metric spacs. *Topol. Appl.*, 160, 1486–1493, 2013.
4. Banach, S., Sur les operations dans les ensembles abstraits el leur application aux equations integrals. *Fundam. Math.*, 3, 133–181, 1922.
5. Choudhury, B.S. and Das, K., A new contraction principle in Menger spaces. *Acta Math. Sin. Engl. Ser.*, 24, 1379–1386, 2008.
6. Choudhury, B.S. and Das, K., A coincidence point result in menger spaces using a control function. *Chaos Solitons Fractals*, 42, 3058–3063, 2009.
7. Choudhury, B.S. and Das, K., Fixed points of generalized Kannan type mappings in generalized Menger spaces. *Commun. Korean Math. Soc.*, 24, 529–537, 2009.
8. Choudhury, B.S., Kumar, S., Asha, Das, K., Some fixed point theorems in G-metric spaces. *Math. Sci. Lett.*, 1, 25–31, 2012.
9. Choudhury, B.S. and Kundu, A., A coupled coincidence point result in partial order metric spaces for compatible mappings. *Nonlinear Anal.*, 73, 2524–2531, 2010.
10. Das, K., Common fixed point theorem for four compatible mappings in Menger spaces involving altering distance function. *Bull. Cal. Math. Soc.*, 110, 473–492, 2018.
11. Dhage, B.C., Generalized metric spaces and mappings with fixed points. *Bull. Cal. Math. Soc.*, 84, 329–336, 1992.
12. Dutta, P.N. and Choudhury, B.S., A generalized contraction principle in Menger spaces using a control function. *Anal. Theory Appl.*, 26, 110–121, 2010.
13. Dutta, P.N., Choudhury, B.S., Das, K., Some fixed point results in Menger spaces using a control function. *Surv. Math. Appl.*, 4, 41–52, 2009.
14. Gaba, Y.U., Fixed point theorems in G-metric spaces. *J. Math. Anal. Appl.*, 455, 528–537, 2017.
15. Gupta, V., Manro, S., Verma, M., A new approach to symmetric generalized fuzzy metric space. *Ann. Fuzzy Math. Inform.*, X, 1–XX, 2014.

16. Hazic, O. and Pap, E., *Fixed Point Theory in Probabilistic Metric Spaces*, Kluwer Academic Publisher, Amsterdam, Holand, 2001.
17. Jungck, G., Commuting maps and fixed points. *Am. Math. Mon.*, 83, 261–263, 1976.
18. Jungck, G., Compatible mappings and common fixed points. *Int. J. Math. Math. Sci.*, 9, 771–779, 1986.
19. Jungck, G. and Rhoades, B.E., Fixed points for set valued functions without continuity. *Indian J. Pure Appl. Math.*, 29, 227–238, 1998.
20. Khan, M.S., Swaleh, M., Sessa, S., Fixed point theorems by altering distances between the points. *Bull. Aust. Math. Soc*, 30, 1–9, 1984.
21. Ma, P., Guan, J., Tang, Y., Xu, Y., Su, Y., ψ-contraction and $(\varphi-\psi)$-contraction in Menger probabilistic metric space. *Springer Plus*, 5, 1–12, 2016.
22. Mustafa, Z., Aydi, H., Karapinar, E., On common fixed points G-metric spaces using (E.A.) property. *Comput. Math. Appl.*, 64, 1944–1956, 2012.
23. Mustafa, Z., Obiedat, H., Awawdeh, F., Some fixed point theorem for mapping on complete G-metric spaces. *Fixed Point Theory Appl.*, 2008, Article ID 189870, 12, 2008.
24. Mustafa, Z. and Sims, B., A new approach to generalized metric space. *J. Nonlinear Convex Anal.*, 7, 289–297, 2006.
25. Pant, R.P., Common fixed points of two pairs of commuting mappings. *Indian J. Pure Appl. Math.*, 17, 187–192, 1986.
26. Pant, R.P., R-weakly commutativity and common fixed points. *Soochow J. Math.*, 25, 37–42, 1999.
27. Popa, V. and Patriciu, A.M., On some common fixed point theorems in G-metric spaces using G-(E.A.) property. *PJM*, 4, 65–72, 2015.
28. Schweizer, B. and Sklar, A., *Probabilistic Metric Spaces*, Elsevier, North-Holand, 1983.
29. Sukanya, K.P. and Magie, J., Common fixed point theorems in generalized fuzzy metric spaces. *Bull. Kerala Math. Assoc.*, 13, 1, 75–79, 2016.
30. Sun, G. and Yang, K., Generalized fuzzy metric spaces with properties. *Res. J. Appl. Sci.*, 2, 673–678, 2010.
31. Xu, S., Chen, S., Radenovic, S., Some notes on common fixed point of two r-weakly commuting mappings in b-metric spaces. *Int. J. Nonlinear Anal. Appl.*, 9, 161–167, 2018.

8

Grobner Basis and Its Application in Motion of Robot Arm

Anjan Samanta

Department of Basic Science and Humanities, University of Engineering and Management, Kolkata, India

Abstract

We are going to introduce Grobner basis, a topic of interest in algebraic geometry, which has numerous applications, as well as in the domain of applied mathematics and computer science. The upcoming discussion will focus on an application of Grobner basis in computer science—motion of robot arm.

***Keywords*:** Grobner basis, Buchberger's algorithm, Macaulay2, robot arm, forward kinematic problem, inverse kinematic problem

8.1 Introduction

Consider any polynomial ring $K[y_1, y_2, ..., y_n]$; K being a field. If we are given some linear equations to solve, we can find it by Gauss elimination method. But what about nonlinear polynomial equations? The answer is Grobner basis. It was introduced by Austrian Mathematician, Bruno Buchberger, in his Ph.D thesis (1965).

It has numerous applications in different field of mathematics. One immediate application is Ideal Membership Problem, to conclude if a given polynomial is in the Ideal or not. We are going to discuss some basics of Grobner bases and then we apply this tool to solve system of nonlinear equations appearing in motion of Robot arm.

Email: anjansamanta1997@gmail.com

Sharmistha Ghosh, M. Niranjanamurthy, Krishanu Deyasi, Biswadip Basu Mallik, and Santanu Das (eds.) Mathematics and Computer Science Volume 1, (173–190) © 2023 Scrivener Publishing LLC

8.1.1 Define Orderings in $K[y_1, ..., y_n]$

In order to reach up to Grobner basis, we have to talk about division algorithm in several variable polynomial ring. Consider $K[y]$, one variable polynomial ring. It is Euclidian domain, so we can talk about division of two polynomials. In the division procedure, we basically try to remove the leading terms one by one to get the quotient and remainder. But how to find out which is leading term in a given polynomial of $K[y_1, ..., y_n]$? in $K[y]$, it is possible because there is a well-ordering between monomials. So, in order to talk about division in several variable ring, we have to first define some suitable order there.

Lexicographic Order:
Let $t = (t_1, ..., t_n)$ and $u = (u_1, ..., u_n)$ be in $\mathbb{N}^n \cup \{0\}$. We define $t >_{lex} u$ if the nonzero entry in the leftmost position of the order tuple difference $t - u \in \mathbb{N}^n$ is positive. We say $y^t >_{lex} y^u$, if $t >_{lex} u$.
Note: In order to construct ordered pair α, first we have to consider some order in between $y_1, y_2, ..., y_n$. Generally, we assume $y_1 > y_2 > ... > y_n$. Hence, there are $n!$ many lex orders in case of n variables.

Other orderings can be also defined, such as Graded Lex Order, Graded Reverse Lex Order.

Let $h = \Sigma a_r y^r$ be a polynomial in multi-variable polynomial ring and let $>$ be a Lex order.

- The **Multidegree** of h is, $mult\ (h) = Max\ (m \in \mathbb{N}^n \cup \{0\}; a_r \neq 0)$
- The **Leading coefficient** of h, $LC\ (h) = a_{mult(h)} \in k$
- The **Leading monomial** of h is, $LM\ (h) = y^{mult\ (h)}$
- The **Leading term** of h is, $LT\ (h) = LC\ (h).\ LM\ (h)$

Let $h = 4x^3 yz + 4z^2 - 13x^4 + 15xz^3$ and $>$ being Lex order. Then

$$mult\ (h) = (4, 0, 0), LC(h) = -13, LM\ (h) = x^4, LT\ (h) = -13x^4$$

8.1.2 Introducing Division Rule in $K[y_1, ..., y_n]$

Suppose our goal is to divide h by $h_1, h_2, ..., h_s$ in $K[y_1, ..., y_n]$. Since $K[y_1, ..., y_n]$ for $n > 1$ can never be an Euclidean domain, the notion of division applicable on E.D. does not work in this case. But we can think of a similar approach here. Since we have already defined the monomial orderings on multivariable polynomial rings, we can consider leading term of a given polynomial. Then the basic idea of division is same as that of single variable case.

Let us see an example:

Divide $h = xy^3 + 1$ by $h_1 = xy^2 + y$ and $h_2 = y^2 + 4$, under **Lex** order $(x > y)$

$$xy^2 + y)\ xy^3 + 1\ (y$$

$$\frac{xy^3 + y^2}{y^2 + 4) - y^2 + 1(-1}$$

$$\frac{y^2 - 4)}{5}$$

Hence $h = q_1h_1 + q_2h_2 + r$ where $q_1 = y$, $q_2 = -1$ and $r = 5$.

8.2 Hilbert Basis Theorem and Grobner Basis

(Hilbert Basis) [5] *Each ideal H in* $K[y_1, \ldots, y_n]$ *(K being a field) is generated by a suitable finite set. Mathematically,* $H = \langle h_1, \ldots, h_t \rangle$ *where* $h_1, \ldots, h_t$ *some polynomials in H.* We call this generating sets of ideal as basis. So each ideal in $K[y_1, \ldots, y_n]$ has a basis. **Grobner basis** $B = \{b_1, \ldots, b_t\}$ is a basis of ideal H which satisfies the following property:

$$\langle LT(b_1), \ldots, LT(b_t))\rangle = \langle LT(H)\rangle$$

8.3 Properties of Grobner Basis

Proposition: We consider B, a Grobner basis of an ideal H in several variable polynomial ring $K[y_1, \ldots, y_n]$. Take any polynomial h from ring, and apply division algorithm on h with B. We get a unique polynomial $r \in K[y_1, s, y_n]$ satisfying the following:

(i) r is not divisible by the leading terms of the polynomials in Grobner basis,

(ii) There exist some $f \in H$ such that $h = f + r$.

So the above is telling us that division of any polynomial h with Grobner basis gives us unique remainder.

We define **S-polynomial** of h and f as follows:

$$S(h,f)=\frac{y^{\alpha}}{LT(h)}\cdot h-\frac{y^{\alpha}}{LT(f)}\cdot f$$

where y^{α} = l.c.m of leading monomials $LM(h)$, $LM(f)$

Buchberger's Criterion: [5] *We assume H be an ideal. A set of polynomials* $B = \{b_1, \ldots, b_t\} \subset H$ *is a Grobner basis of H iff for any pair u, v, the remainder term when dividing* $S(b_u, b_v)$ *by B is zero.*

This criteria gives rise to an algorithm by which a Grobner basis for H can be generated from any basis of that ideal. This algorithm is known as Buchberger's algorithm. Hence, we can say, in $K[y_1, \ldots, y_n]$, for every ideal, we can have a Grobner Basis.

Now comes the question of minimality. A Grobner basis can be found by using above algorithm, but the minimal no of polynomials required to construct whole H may be less than the no of polynomials present in obtained Grobner basis. A Grobner basis with these minimal no of generators is called reduced Grobner basis.

Definition: A reduced Grobner basis for H is a Grobner basis B satisfying the following:

(i) For any polynomial b in B, The leading coefficient of b must be 1.
(ii) b is not anyhow dependent on the other polynomials in B i.e. no term of b can be produced by the leading terms of other polynomials.

8.4 Applications of Grobner Basis

8.4.1 Ideal Membership Problem

Grobner basis together with the division algorithm gives us the following result: Let $H = \langle h_1,\ldots,h_t\rangle$ be a polynomial ideal. It is to be determined whether a given polynomial h lies in H or not. At first, look for a Grobner basis B of H. Then uniqueness of the remainder implies that $h \in H$ *iff remainder after division with Grobner basis is 0.*

Example: Consider the ideal $H = \langle x^3 - z^2, x^2 y^2 - z^3 \rangle \subset \mathbb{C}\ [x, y, z]$ and a polynomial $h = x^2 y^2 + y^2 z^2 - x z^3 - x^3 z$. We want to see whether h is a member of the given ideal. First, we find Grobner basis of above ideal and then we divide given polynomial with the obtained basis. Now by hand it is tedious to perform Buchberger's algorithm to find the Grobner basis. For computation we will use a software called **Macaulay2**. The codes are as follows:

```
i1 : R = CC[x,y,z]
i2 : H=ideal(x^3-z^2, x^2*y^2-z^3)
                3   2   2  2    3
o2 = ideal (x - z , x  y  - z)
i3 : gens gb H
                3   2   2 2    3   2 2     3
o3 = ideal (x - z , x y  - z , y z - x z)
i4 : h =  x^2*y^2+y^2*z^2-x*z^3-x^3*z
        2 2   3     2 2    3
o4 = x y - x z + y z - x z
i5 : h % gb H
o5 =  0
```

Since the remainder is coming out to be zero, we can conclude that given polynomial h belongs to the ideal $H = (x^3 - z^2, x^2 y^2 - z^3)$.

8.4.2 Solving Polynomial Equations

As we have mentioned earlier, Grobner basis is widely used to solve system of nonlinear polynomial equations. Let us understand this with the help of the following:

Let's assume the following equations:

$$2x^2 + y^2 + 2z^2 = 2$$

$$3x^2 + 4z^2 = y$$

$$2x = z$$

Take the ideal $H = (2x^2 + y^2 + 2z^2, - 2, 3x^2 + 4z^2 - y, 2x - z) \subset \mathbb{C}\ [x, y, z]$. The above equations will be solved by computing Grobner basis of H. Here are the codes to find the basis of above ideal in Macaulay2:

```
i1 : R=CC[x,y,z]
i2 : H=ideal(2*x^2 + y^2 + 2*z^2-2, 3*x^2 + 4*z^2
- y, 2*x-z)
                       2   2     2         2    2
o2 : ideal ( 2 x + y + 2 z - 2, 3 x +4 z - y, 2
x - z )
i3 : gens gb H
                        2               2
o3 : (x - .5z  z  - 0.210526y  y + .536316y - 2)
```

Hence reduced Grobner basis is $\{x - .5z, z^2 - 0.210526y, y^2 + .536316y - 2\}$. Since the last polynomial depends on y alone, we can find the roots. Then by back substitution we can find values of other variables.

8.5 Application of Grobner Basis in Motion of Robot Arm

In this section, we will try to see one application of Grobner basis in the field of computer science—configurations of robot arms. We utilize its capability in solving nonlinear equations to realize how Robot arm moves in 2D.

8.5.1 Geometric Elucidation of Robots

Robot arm are constructed using rigid segments, connected by various kinds of joints. One end will always be fixed and at the other end one 'hand' is attached, the final segment of the robot. Our main goal is to describe the position and the orientation of this "hand."

Here, we consider only the following joints:

1. **Planar revolute joints:** this kind of joint permits rotation of one segment with respect to other restricted on a plane. Hence this kind of rotation can be identified by measuring angle $\theta \in [0, 2\pi]$ between segments, represented by circle S^1.
2. **Prismatic joints:** this joint permits robot arms a translation about an axis. This can be described by the amount of translation occurred in a segment i.e. by an interval of finite length.

8.5.2 Mathematical Representation

The motion of the hand can be described if we know the joint setting, i.e., position-orientation of each joint. If we denote space of all possible joint configuration by ζ and space through which hand moves by C, then the function

$$f : \zeta \rightarrow C$$

represents different hand configuration corresponding to different joint setting. Here C can be visualized as $C = U \times V$, where U denotes space of possible position of hand and V represents possible hand orientation.

We will try to answer the following questions:

- **(Forward Kinematic Problem)** Is it possible to give a detailed explanation for f in terms of joint setting and the dimension of segment of robot arm?
- **(Inverse Kinematic Problem)** Provided $l \in C$, is it possible to describe one or all $z \in \zeta$ for which $f(z) = l$?

8.5.3 Forward Kinematic Problem [2]

In order to describe position of the hand, we have to fix a coordinate system (x_1, y_1) such that origin is placed at joint 1. At each revolute joint i, we set new coordinate system (x_{i+1}, y_{i+1}) to describe the relative positions of hand. Coordinate has been taken as follows: Origin at i, x_{i+1} axis alongside $(i + 1)$ segment, y_{i+1} axis taken in normal direction. We aim for finding position of hand w.r.t (x_1, y_1) system, hence we will establish a relation between ith and $(i + 1)$th coordinate system via traslation and rotation.

Suppose A be any point in the system whose coordinate w.r.t $(i + 1)$th system is (a_{i+1}, b_{i+1}). To obtain its coordinate w.r.t ith system, we perform a rotation of angle θ_i and translation by $(l_i, 0)$.

Problem 1. Let us solve Forward problem for the above diagram (Figure 8.1). Here, three **revolute** joints has been used with angle of rotations shown. Hand is attached with joint 3. We have to describe position of hand w.r.t (x_1, y_1) coordinate system taken at joint 1. Furthermore, we consider (x_2, y_2), (x_3, y_3) coordinate system along segments 2 and 3, respectively, and (x_4, y_4) system along the hand shown in the figure (y_i taken normal to x_i).

Let q be the point at joint 3. Then, hand can be described using q and angle $(\theta_1 + \theta_2 + \theta_3)$. Assume (a_i, b_i) be the coordinate of q w.r.t (x_i, y_i)-coordinate

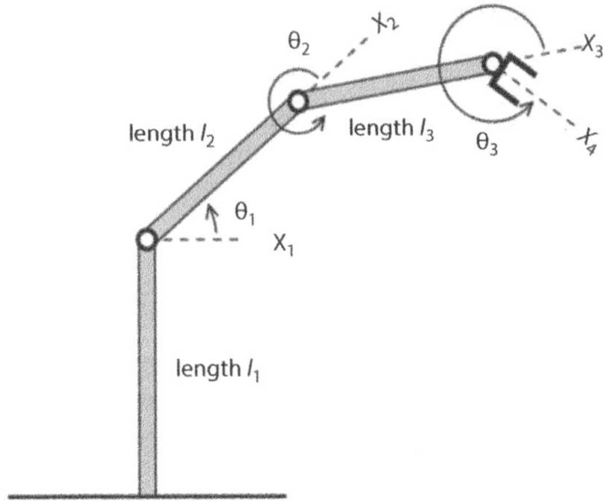

Figure 8.1 Robot arm with three revolute joints [5].

system ($i = 1, \ldots, 4$). Since joint 3 is the origin for (x_4, y_4), $(a_4, b_4) = (0, 0)$. Now, θ_3 angle rotation and l_3 length translation are required to get (a_3, b_3). Means,

$$\begin{pmatrix} a_3 \\ b_3 \end{pmatrix} = \begin{pmatrix} cos\theta_3 & -sin\theta_3 \\ sin\theta_3 & cos\theta_3 \end{pmatrix} \begin{pmatrix} a_4 \\ b_4 \end{pmatrix} + \begin{pmatrix} l_3 \\ 0 \end{pmatrix} = \begin{pmatrix} l_3 \\ 0 \end{pmatrix}$$

And similarly,

$$\begin{pmatrix} a_2 \\ b_2 \end{pmatrix} = \begin{pmatrix} cos\theta_2 & -sin\theta_2 \\ sin\theta_2 & cos\theta_2 \end{pmatrix} \begin{pmatrix} a_3 \\ b_3 \end{pmatrix} + \begin{pmatrix} l_2 \\ 0 \end{pmatrix} = \begin{pmatrix} l_3cos\theta_2 + l_2 \\ l_3sin\theta_2 \end{pmatrix}$$

Finally, θ_1 rotation will give us the hand position w.r.t global coordinate system (x_1, y_1).

$$\begin{pmatrix} a_1 \\ b_1 \end{pmatrix} = \begin{pmatrix} cos\theta_1 & -sin\theta_1 \\ sin\theta_1 & cos\theta_1 \end{pmatrix} \begin{pmatrix} a_2 \\ b_2 \end{pmatrix} = \begin{pmatrix} l_3cos(\theta_1 + \theta_2) + l_2cos\theta_1 \\ l_3sin(\theta_1 + \theta_2) + l_2sin\theta_1 \end{pmatrix}$$

The orientation of the hand can be specified by direction of x_4 axis, and hence, it is given by $(\theta_1 + \theta_2 + \theta_3)$. Therefore, the explicit formula for $f: \zeta \rightarrow C$ is given by

$$f(\theta_1,\theta_2,\theta_3)=\begin{pmatrix} l_3cos(\theta_1+\theta_2)+l_2cos\theta_1 \\ l_3sin(\theta_1+\theta_2)+l_2sin\theta_1 \\ \theta_1+\theta_2+\theta_3 \end{pmatrix}$$

Now, assume $cos\theta_i = c_i$ and $sin\theta_i = s_i$. Then $c_i^2+s_i^2=1$. Substituting these in above, we get the following polynomial representation:

$$f(\theta_1,\theta_2,\theta_3)=\begin{pmatrix} l_3c_1c_2-l_3s_1s_2+l_2c_1 \\ l_3s_1c_2+l_3s_2c_1+l_2s_1 \\ \theta_1+\theta_2+\theta_3 \end{pmatrix}$$

Problem 2. We will do forward problem for Figure 8.2 where now one **prismatic joint** has been added with the joint 3. Prismatic joint allows the hand to move linearly. Hence l_4 is a variable quantity. Now to get the position of the hand, we will proceed as before, only difference is that now $(a_4, b_4) = (l_4, 0)$. After rotations and translations, we get finally:

$$\begin{pmatrix} a_1 \\ b_1 \end{pmatrix} = \begin{pmatrix} l_4cos(\theta_1+\theta_2+\theta_3)+l_3cos(\theta_1+\theta_2)+l_2cos\theta_1 \\ l_4sin(\theta_1+\theta_2+\theta_3)+l_3sin(\theta_1+\theta_2)+l_2sin\theta_1 \end{pmatrix}$$

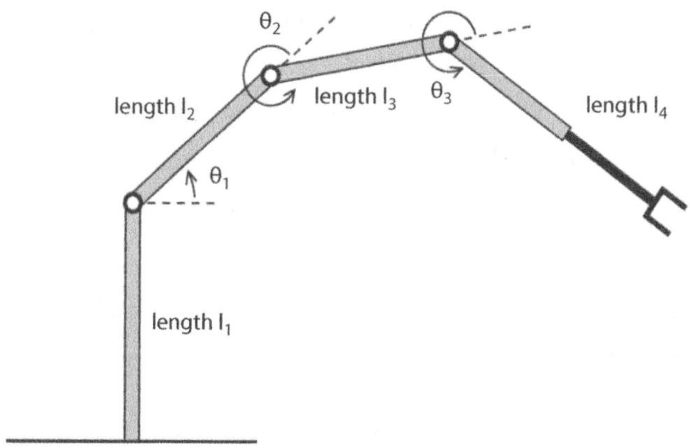

Figure 8.2 Robot arm having one prismatic and three revolute joints [5].

Hence,

$$f(\theta_1,\theta_2,\theta_3,l_4)=\begin{pmatrix} l_4cos(\theta_1+\theta_2+\theta_3)+l_3cos(\theta_1+\theta_2)+l_2cos\theta_1 \\ l_4sin(\theta_1+\theta_2+\theta_3)+l_3sin(\theta_1+\theta_2)+l_2sin\theta_1 \\ \theta_1+\theta_2+\theta_3 \end{pmatrix}$$

Substituting $c_i = cos\theta_i$, $s_i = sin\theta_i$ in above, we get the polynomial representation as:

$$f(\theta_1,\theta_2,\theta_3,l_4)=\begin{pmatrix} l_4(c_1c_2c_3-c_1s_2s_3-s_1c_2s_3-s_1c_3s_2)+l_3(c_1c_2-s_1s_2)+l_2c_1 \\ l_4(s_1c_2c_3-s_1s_2s_3+c_1c_2s_3+c_1c_3s_2)+l_3(s_1c_2+s_2c_1)+l_2s_1 \\ \theta_1+\theta_2+\theta_3 \end{pmatrix}$$

8.5.4 Inverse Kinematic Problem [2]

In the forward problem, we were trying to find the position of hand where configuration of joints are given. Now we do the opposite in inverse problem - for given position and orientation of the hand, we will figure out the configuration of joints required to place the hand. In other words, given $i \in C$ we are interested in finding inverse image of f, $f^{-1}(i)$ in ζ.

Problem 1. We solve inverse problem for the first diagram in previous page. Suppose (a, b) be the given position of the hand with orientation α. Now by forward problem, we have the following description of f:

$$f(\theta_1,\theta_2,\theta_3)=\begin{pmatrix} l_3c_1c_2-l_3s_1s_2+l_2c_1 \\ l_3s_1c_2+l_3s_2c_1+l_2s_1 \\ \theta_1+\theta_2+\theta_3 \end{pmatrix}$$

Now what we want to find is a suitable $(\theta_1, \theta_2, \theta_3)$ such that $f(\theta_1, \theta_2, \theta_3) = (a,b,\alpha)$. Hence,

$$l_3c_1c_2 - l_3s_1s_2 + l_2c_1 = a\ldots \qquad (8.1)$$

$$l_3 s_1 c_2 + l_3 s_2 c_1 + l_2 s_1 = b\ldots \tag{8.2}$$

$$\theta_1 + \theta_2 + \theta_3 = \alpha$$

subject to the constraints

$$c_1^2 + s_1^2 = 1$$
$$c_2^2 + s_2^2 = 1$$

Now the 1st two equations involves only c_i, s_i i.e θ_1, θ_2. So if we solve the four polynomial equations above involving c_1, c_2, s_1, s_2, we get θ_1, θ_2, then $\theta_3 = \alpha - \theta_1 - \theta_2$; means we get all three joint settings. In order to do these steps, we have to solve the above system of polynomial equations. And we can do it by finding **Grobner basis**. The following Macaulay2 codes will do the job:

```
i1 : R= frac (QQ[a,b,l_2,l_3])
o1 = R
o1 : FractionField
i2 : S= R[c_1,s_1,c_2,s_2]
o2 = S
o2 : PolynomialRing

i3      :      H=ideal(l_3*(c_1*c_2-s_1*s_2)+l_2*c_1-
a,l_3*(c_1*s_2+c_2*s_1)
    +l_2*s_1-b,c_1^2+s_1^2-1,c_2^2+s_2^2-1)

o3 = ideal(l c c  - l s s  + l c  - a, l s c  + l
c s  + l s  - b,
            3 1 2    3 1 2    2 1         3 1 2    3
1 2    2 1
             2    2          2    2
            c  + s   - 1, c  + s   - 1)
             1    1          2    2
o3 : Ideal of S
i4 : gens gb H
o5 = | c_2+(-a2-b2+l_2^2+l_3^2)/2l_2l_3
     s_1+al_3/(a2+b2)s_2+(-a2b-b3-bl_2^2+bl_3^2)/
(2a2l_2+2b2l_2)
```

```
     -------------------------------------------
----------------
   c_1+(-bl_3)/(a2+b2)s_2+(-a3-ab2-al_2^2+al_3^2)/
(2a2l_2+2b2l_2)
     -------------------------------------------
-------------------
     s_2^2+(a4+2a2b2+b4-2a2l_2^2-2b2l_2^2+
                         l_2^4-2a2l_3^2-2b2l_3^2-
2l_2^2l_3^2+l_3^4)/4l_2^2l_3^2
             1        4
o5 : Matrix S  <--- S
```

Let us fix the arm lengths $l_2 = l_3 = 5$. After putting these values to the above Grobner basis, it reduces to

$$c_1 - 5bs_2/(a^2+b^2) - a/10$$

$$s_1 + 5as_2/(a^2+b^2) - b/10$$

$$c_2 - (a^2+b^2-50)/50$$

$$s_2^2 + (a^2+b^2)(a^2+b^2-100)/2500$$

When $(a^2 + b^2) \neq 0$, system is consistent. The last equation is a quadratic equation in s_2. Solving that, we get

$$s_2 = \pm\sqrt{(a^2+b^2)(100-(a^2+b^2))}/50$$

When $a^2 + b^2 \leq 100$ we get two real solutions and substituting back to those previous equations we can find values of c_1, s_1, c_2. Hence, two possible configurations are there.

When $(a^2 + b^2) = 0$, means $a = b = 0$. We substitute $a = b = 0$; $l_2 = l_3 = 5$ in (*i*), (*ii*), and compute the Grobner basis as follows:

```
i2 : R=QQ[c_1,c_2,s_1,s_2]
o2 = R
o2 : PolynomialRing
```

```
i3 :H=ideal(5*(c_1*c_2-s_1*s_2)+5*c_1,5*(s_1*c_2+
s_2*c_1)
      +5*s_1,c_1^2+s_1^2-1,c_2^2+s_2^2-1)
o3 = ideal (5c c  - 5s s  + 5c , 5c s  + 5c s  +
5s ,
             1 2      1 2     1    2 1    1 2    1
              2    2       2    2
             c  + s  - 1, c  + s  - 1)
              1    1       2    2

o3 : Ideal of R
i4 : gens gb H
o4 = | s_2 c_2+1 c_1^2+s_1^2-1 |
```

The reduced Grobner basis is $\{s_2, c_2+1, c_1^2+s_1^2-1\}$. Hence, $s_2 = 0$; $c_2 = -1$ implies $\theta_2 = \pi$ and $c_1^2+s_1^2=1$ implies θ_1 is arbitrary. Geometrically this is obvious because to put hand at the origin (where joint 1 also placed) we have to take joint 2 at π angle and no restriction on orientation of joint 1.

Hence, the complete analysis is done when $l_2 = l_3 = 5$. To place hand at (a, b) point, there are

- Two possible joint setting when $(a^2 + b^2) \neq 0$; $a^2 + b^2 \leq 100$,
- Infinitely many possibility of joint 1 and rotation of angle π of joint 2 when $a^2 + b^2 = 0$,
- One possible joint setting when $a^2 + b^2 = 100$,
- No possible configuration when $a^2 + b^2 \geq 100$.

Problem 2. Let us do inverse problem for the 2nd figure. There we have described the function f as follows:

$$f(\theta_1,\theta_2,\theta_3,l_4)=\begin{pmatrix} l_4cos(\theta_1+\theta_2+\theta_3)+l_3cos(\theta_1+\theta_2)+l_2cos\theta_1 \\ l_4sin(\theta_1+\theta_2+\theta_3)+l_3sin(\theta_1+\theta_2)+l_2sin\theta_1 \\ \theta_1+\theta_2+\theta_3 \end{pmatrix}$$

For this case, finding Grobner basis in general would be very tedious. So let us fix the arm lengths $l_2 = l_3 = l_4 = 1$. Assume the hand has been placed at point (a, b) with orientation α. Then we have the following system:

$$l_4 \cos(\theta_1 + \theta_2 + \theta_3) + l_3 \cos(\theta_1 + \theta_2) + l_2 \cos\theta_1 = a$$

$$l_4 \sin(\theta_1 + \theta_2 + \theta_3) + l_3 \sin(\theta_1 + \theta_2) + l_2 \sin\theta_1 = b$$

$$\theta_1 + \theta_2 + \theta_3 = \alpha$$

Substituting $l_2 = l_3 = l_4 = 1$ and using $\theta_1 + \theta_2 + \theta_3 = \alpha$, above can be reduced to:

$$u_1 + (u_1 c_3 + u_2 s_3) + c_1 - a = 0$$

$$u_2 + (u_2 c_3 - u_1 s_3) + s_1 - b = 0$$

$$c_1^2 + s_1^2 - 1 = 0$$

$$c_3^2 + s_3^2 - 1 = 0$$

where $u_1 = \cos\alpha$, $u_2 = \sin\alpha$, $c_i = \cos\theta i$, $s_i = \sin\theta_i$. The Grobner basis computation of above system is the following:

```
i1 : R=frac(QQ[u_1,u_2,a,b])
o1 = R
o1 : FractionField
i2 : S=R[c_1,c_3,s_1,s_3]
o2 = S
o2 : PolynomialRing

i3:H=ideal(u_1+u_1*c_3+u_2*s_3+c_1-a,u_2+u_2*c_3-
u_1*s_3+s_1-b,
    c_1^2+s_1^2-1,c_3^2+s_3^2-1)

o3 = ideal (c  + u c  + u s  + u  - a, u c  + s  -
u s  + u  - b,
            1     1 3    2 3    1          2 3    1
1 3    2
             2    2         2    2
            c  + s  - 1, c  + s  - 1)
             1    1         3    3
o3 : Ideal of S
i4 : gens gb H
o4 =
```

```
|                  s_1+(-u_1^3-u_1u_2^2+u_1^2a+u_2^2a)/
(u_1^2+u_2^2-u_1a-u_2b)s_3
+(-u_2a2-2u_1^2b-2u_2^2b+2u_1ab+u_2b2+u_2)/
(2u_1^2+2u_2^2-2u_1a-2u_2b)
------------------------------------------------------
----------------------

c_3+(-u_2a+u_1b)/(u_1^2+u_2^2-u_1a-u_2b)s_3
+(2u_1^2+2u_2^2-2u_1a+a2-2u_2b+b2-1)/
(2u_1^2+2u_2^2-2u_1a-2u_2b)
------------------------------------------------------
----------------------

c_1+(u_1^2u_2+u_2^3-u_1^2b-u_2^2b)/(u_1^2+u_2^2-
u_1a-u_2b)s_3+
(-2u_1^2a-2u_2^2a+u_1a2+2u_2ab-u_1b2+u_1)/
(2u_1^2+2u_2^2-2u_1a-2u_2b)
------------------------------------------------------
----------------------

s_3^2+(-2u_1^2u_2a-2u_2^3a+2u_1u_2a2-
u_2a3+2u_1^3b+2u_1u_2^2b-2u_1^2ab
+2u_2^2ab+u_1a2b-2u_1u_2b2-u_2ab2+u_1b3+u_2a-
u_1b)
/(u_1^4+2u_1^2u_2^2+u_2^4-2u_1^3a-2u_1u_2^2a+
u_1^2a2+u_2^2a2-2u_1^2u_2b-
2u_2^3b+u_1^2b2+u_2^2b2)s_3+
(4u_1^2a2+4u_2^2a2-4u_1a3+a4-4u_2a2b+4u_1^2b2
+4u_2^2b2-4u_1ab2+2a2b2-4u_2b3+b4-4u_1^2-
4u_2^2+4u_1a-2a2+4u_2b-2b2+1)
/(4u_1^4+8u_1^2u_2^2+4u_2^4-8u_1^3a-
8u_1u_2^2a+4u_1^2a2
+4u_2^2a2-8u_1^2u_2b-8u_2^3b+4u_1^2b2+4u_2^2b2)  |
```

There are four polynomials in the Grobner basis. The denominators appearing in the basis are $u_1^2+u_2^2-u_1a-u_2b=(u_1-a/2)^2+(u_2-b/2)^2$ and $\left(u_1^4+2u_1^2u_2^2+u_2^4-2u_1^3a-2u_1u_2^2a+u_1^2a^2+u_2^2a^2-2u_1^2u_2b-2u_2^3b+u_1^2b^2+u_2^2b^2\right)$ $=\left(u_1^2+u_2^2\right)\left((u_1-a)^2+(u_2-b)^2\right)$, also their constant multiples. When these are nonzero, the system is consistent. One thing to observe that u_1 and u_2 both can't be zero; hence $u_1^2+u_2^2\neq 0$. Since the last polynomial is a quadratic polynomial in s_3, real solution exists only when discriminant $D\geq 0$.

Hence we can conclude that when $u_1 = cos\alpha \neq a$ *or* $u_2 = sin\alpha \neq b$; $u_1 \neq a/2$ *or* $u_2 \neq b/2$ *and* $D \geq 0$, *we can find two joint settings for the hand position* (a, b, α).

For example, take $a=1, b=1, \alpha=\frac{\pi}{4}$. Then the Grobner basis coming out to be:

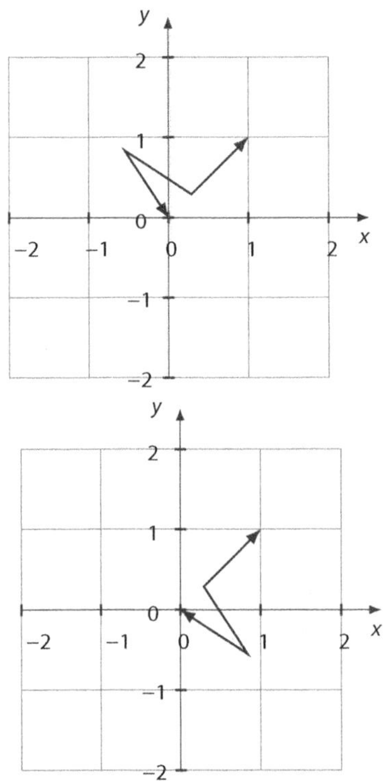

Two distinct joint settings for the hand position (1,1) drawn in LaTeX

```
i10   :   H=ideal(1+c_3+s_3+sqrt(2)*(c_1-1),1+c_3-
s_3+sqrt(2)*(s_1-1),
          c_1^2+s_1^2-1,c_3^2+s_3^2-1)
o10 : Ideal of R
i11 : gens gb H
o11 = | s_1-.707107s_3-.146447 c_3-.207107
          c_1+.707107s_3-.146447 s_3^2-.957107 |
```

Last polynomial in the set of Grobner basis is $s_3^2 - 0.957107$. Solutions are $s_3 = \pm 0.9783$. Now $s_3 = \pm 0.9783$ and $c_3 = 0.207107 \Rightarrow \theta_3 \pm 78.042$ in degree.

Substituting to 1st term we get $s_1 = 0.83815, -0.5485$ and $c_1 = -0.54849, 0.835031 \Rightarrow \theta_1 = 123.26, -33.26$. Hence

$$\theta_1 = 123.26, -33.26$$

$$\theta_3 = 78.042, -78.042$$

$$\theta_2 = 45 - (\theta_1 + \theta_3) = -156.302, 156.302$$

(in degrees) These are the two possible set of angles of rotation of revolute joints to place the hand at (1, 1) with orientation $\frac{\pi}{4}$.

8.6 Conclusion

Grobner basis and its implementation toward solving forward and inverse kinematic problems have been discussed. Forward kinematics involves plotting every joint rotation angle to get the end effector (hand) to where you want it to go, experimentally. Using inverse kinematics for a fixed position-orientation, we can find the joint settings required to reach there. That is why this idea is used to control the trajectory by robot hand. Hence, Grobner basis plays a very crucial role in robotics.

References

1. Eisenbud, D. *et al.*, *Computations in algebraic geometry with Macaulay 2*, vol. 8, Springer Science & Business Media, 2001.
2. Grayson, P., *Robotic Motion Planning*, 2001.
3. Lutes, B.A., *A Survey of Gröbner Bases and Their Applications Departmental Report*, Texas Tech University, 2004.
4. Cox, D.A., Little, J., O'shea, D., *Using Algebraic Geometry*, vol. 185, Springer Science & Business Media, 2006.
5. Cox, D., Little, J., OShea, D., *Ideals, Varieties, and Algorithms: An Introduction to Computational Algebraic Geometry and Commutative Algebra*, Springer Science & Business Media, 2013.

9

A Review on the Formation of Pythagorean Triplets and Expressing an Integer as a Difference of Two Perfect Squares

Souradip Roy[1*], Tapabrata Bhattacharyya[1], Subhadip Roy[1], Souradeep Paul[1] and Arpan Adhikary[2†]

[1]*Department of Computer Science and Engineering, Institute of Engineering & Management (IEM), Kolkata, India*
[2]*Department of Electronics and Communication Engineering, Institute of Engineering & Management (IEM), Kolkata, India*

Abstract
In this paper, we have reviewed [1] the proposition for generalization of formulae that finds the primitive and nonprimitive Pythagorean triplets generated by a given integer. Besides, the procedure for finding complete expressions of the integers as a difference of two perfect squares have been formulated. Finally, the Pythagorean triplets are generated in a range from x to y, programmed with the proposed formulae, which thereby validates them.

***Keywords*:** Pythagorean triplets, primitive and non-primitive, generalization, formularized, difference of two squares

9.1 Introduction

A Pythagorean triple is a set of positive integers (x, y, z) such that

$$x^2 + y^2 = z^2. \tag{9.1}$$

**Corresponding author*: souradiproy2000@gmail.com
†*Corresponding author*: adhikary.arpanjan20@gmail.com

Sharmistha Ghosh, M. Niranjanamurthy, Krishanu Deyasi, Biswadip Basu Mallik, and Santanu Das (eds.) Mathematics and Computer Science Volume 1, (191–212) © 2023 Scrivener Publishing LLC

From the geometrical point of view, (x, y, z) forms a set of sides, where z is the hypotenuse of a right-angled triangle. We have to find the solution to 9.1 when x, y, z are all positive integers (x, y > 2). Equation 9.1 itself ranks among the Diophantine equations. If we find one solution of the equation (primitive), then we can multiply x, y, z with any arbitrary integer resulting in an infinite number of Pythagorean triples (nonprimitive). Say, for an example: $3^2 + 4^2 = 5^2$. If we multiply both sides with 4^2 or 16, we get a new triple (12, 16, 20) from the given triple (3, 4, 5). The triples with x, y, z being co-prime are called primitive triples. Else, they are nonprimitive. Thus, if we find the primitive triples, we can get the non-primitive triples easily.

In algebra, it is well known that $(a - b)^2 + 4ab = (a + b)^2$. Replacing (a, b) by (a^2, b^2) respectively, the formula becomes $(a^2-b^2)^2 + (2ab)^2 = (a^2 + b^2)^2$.

We can represent x, y, and z in terms of a and b as follows:

$$x = 2ab, \ y = a^2 - b^2, \quad z = a^2 + b^2 \quad (a > b) \tag{9.2}$$

Now, to get a primitive solution: x, y, z must be mutually co-prime. If we can find the nature of x, y, z i.e. whether they are even or odd, it will be easier for us to calculate. So, we consider these three possible cases:

- **Case 1:** when both x and y are even
  Let x = 2m and y =2n [m, n > 0 & m, n ∈ Z+]
  then $x^2 + y^2 = (2m)^2 + (2n)^2 = 4(m^2 + n^2) = z^2$
  giving $z = 2\sqrt{m^2 + n^2}$ = an even number.
  In this case x, y, z all will be an even number. Hence, they cannot be coprime as they have a minimum common factor of 2. This case will thus not be considered. However, (x, y, z) might be a non-primitive triple based on the values of x and y.

- **Case 2:** when both x, y are odd
  Let x = 2m + 1 and y = 2n + 1 [m, n > 0 & m, n ∈ Z+].
  Then we have:

$$x^2 + y^2 = (2m + 1)^2 + (2n + 1)^2$$

$$= 4m^2 + 4m + 1 + 4n^2 + 4n + 1$$

$$= 4(m^2 + n^2) + 4(m + n) + 2$$

$$= 4\,[(m + n)^2 - 2mn] + 4(m + n) + 2$$

$$= 4(m + n)^2 + 4(m + n) + (2 - 8mn)$$

$$= 4(m + n)^2 + 4(m + n) + 1 + (1 - 8mn)$$

$$= [2(m + n) + 1]^2 + (1 - 8mn) \tag{9.3}$$

Here we see that the first term is a perfect square while the second one may be or may not be. For $x^2 + y^2$, to be a perfect square, the 2nd term $(1 - 8mn)$ must be a perfect square. So, the minimum value of $(1-8\ mn) = 0$; giving $mn = 1/8$. Hence, at least one of m or n is a fraction. If we take any value other than 0 mn becomes negative which is absolutely not accepted. This case, as well, is thus rejected.

Hence, the only possible case is x, y are of different parity (one is odd and the other is even). We notice that, for a primitive triple, z will always be odd. If we consider equation 9.2, we find that x is even and y is odd. If we just reverse the expressions for x, y, their parities get interchanged but that does not affect the inference. From equation 9.2, we conclude that a, b will be of opposite parity because in that case y will be odd, z will be odd and x is already even $[x = 2ab]$. However, method 9.2 is not that effective to solve this kind of a problem as it generates the same number of triples as the no. of ways in which x can be represented as 2ab (if x is even) or $a^2 - b^2$ (if x is odd).

Following Euclid's method, we get the exact number of primitive triples along with some non-primitive triples. It is quite natural that the number of non-primitive triples will be a lot more than that of primitive triples. Given below, is the detailed calculation for finding the triples and their quantities.

9.2 Calculation of Triples

Considering the fact that both x and y can be even (non-primitive) or of different parity (primitive) we consider different cases and analyze them carefully.

9.2.1 Calculation for Odd Numbers

Let x be an odd number > 1

Then we get an even y and odd z (already discussed above). So $(z - y)$ will be an odd number, i.e., $z - y = 2p + 1$ or, $z = y + (2p + 1)$ where $p \geq 0$ is an integer.

So from 9.1 we obtain:

$$x^2 + [z - (2p + 1)]^2 = z^2$$

$$\text{or, } x^2 + z^2 - 2(2p + 1)z + (2p + 1)^2 = z^2$$

$$\text{or, } x^2 - 2(2p + 1)z + (2p + 1)^2 = 0$$

So,

$$z = \frac{\left[\frac{x^2}{2p+1} + (2p+1)\right]}{2} \tag{9.4}$$

Similarly,

$$y = z - (2p+1) = \frac{\left[\frac{x^2}{2p+1} - (2p+1)\right]}{2} \tag{9.5}$$

For simplicity, we have calculated the values of y and z in terms of x. We will now analyze these two equations and try to find out the solutions. Here, both y and z are integers. Hence, from 9.4 and 9.5 it is clear that (2p + 1) must be a divisor of x^2 so that the quotient is an integer.

We assume $(2p + 1) = d_i$.

Now,

$$z = \frac{\left[\frac{x^2}{d_i} + d_i\right]}{2} \text{ or, } z = \frac{\left[x^2 + d_i^2\right]}{2d_i} \text{ and } y = \frac{\left[x^2 - d_i^2\right]}{2d_i}$$

Here, y will be positive only when $x > d_i$ or $d_i < x$. We choose only those values of d_i which are less than x. Let us find the number of such divisors. In number theory, we know that we can represent any number as a product of primes in the canonical form. So a given integer k can be expressed in the form $k = p_1^{m_1}.p_2^{m_2}.p_3^{m_3}....p_n^{m_n}$ where $p_1, p_2, p_3 \ldots p_n$ are the prime factors of the number k and $m_1, m_2, m_3 \ldots m_n$ are their corresponding powers or indices. If k is a perfect square, then

m_1, m_2, m_3....m_n have to be even. Now, the total number of factors of $k=(m_1+1).(m_2+1).(m_3+1)...(m_n+1)=\prod_{i=1}^{n}(m_i+1)$. Since m_1, m_2, m_3 ... m_n all are even numbers, the product k will definitely be odd. (the number of divisors of any perfect square is always an odd positive integer).

Let us denote the number of divisors of d_i by N_d and the number of Pythagorean triples by N_{tr}. Since the number of divisors N_d is odd, N_{tr} is given by $N_{tr}=\frac{N_d-1}{2}$. This gives the number of Pythagorean triples (when x is an odd number).

Alternative procedure

In the above approach, we expressed y in terms of z. Here, we keep the y-term unaffected and rather substitute z in terms of y. Thus, from equation 9.1, we obtain:

$$x^2+y^2=[y+(2p+1)]^2$$

$$\text{or} \quad x^2+y^2=y^2+2(2p+1)y+(2p+1)^2$$

$$\text{or} \quad x^2-(2p+1)^2=2(2p+1)y$$

$$\text{or} \quad y=\frac{x^2-(2p+1)^2}{2(2p+1)}$$

$$\text{or} \quad y=\frac{\left[\frac{x^2}{2p+1}-(2p+1)\right]}{2}$$

$$\text{So} \quad z=y+(2p+1)=\frac{\left[\frac{x^2}{2p+1}+(2p+1)\right]}{2}$$

Note that we get the same values of y and z. Therefore, we will also get the same result.

9.2.2 Calculation for Even Numbers

Let the number be even and x >2

We already know that x and y can both be even for a nonprimitive triple. In that case, z will also be even. If x is even and y is odd, z will also be odd. We see that y and z will be of the same parity if x is an even number.

Here we can substitute z = y + 2p where p > 0 is an integer. Replacing the value in 9.1 we get:

$$x^2+y^2=[y+(y+2p)]^2$$

or $$x^2+y^2=y^2+4py+4p^2$$

or $$x^2=4py+4p^2$$

or $$x^2-4p^2=4py$$

or $$y=\frac{[x^2-4p^2]}{4p}$$

or $$y=\frac{x^2}{4p}-p \tag{9.6}$$

Substituting the value of y from z = y + 2p we get:

$$z=\frac{x^2}{4p}+p \tag{9.7}$$

From the equations 9.6 and 9.7, we see that p must be a divisor of $\left(\frac{x}{2}\right)^2$. As in the previous case, we take the values of $p<\left(\frac{x}{2}\right)$ for having the positive integral solutions y and z. The number of Pythagorean triples will be $N_{tr}=\frac{N_d-1}{2}$. Now, we find the expressions for N_d and N_{tr} using the canonical expressions. As already discussed above, x can be represented as:

$x=p_1^{f_1}.p_2^{f_2}.p_3^{f_3}\dots.p_n^{f_n}$ where p_1, p_2, p_3 are the distinct prime factors and f_1, f_2, f_3 ... are their corresponding powers or indices.

$\therefore x^2=\left(p_1^{2f_1}\right).\left(p_2^{2f_2}\right).\left(p_3^{2f_3}\right)\dots.\left(p_n^{2f_n}\right)$. Thus, the total number of factors of x^2 is:

$$(2f_1+1).(2f_2+1).(2f_3+1)\dots(2f_n+1)=N_d \tag{9.8}$$

$$\therefore \text{Number of triples } N_{tr}=\frac{(N_d-1)}{2}=\frac{1}{2}[(2f_1+1).(2f_2+1)\dots(2f_n+1)-1] \tag{9.9}$$

This is the total number of Pythagorean triples we have corresponding to any given positive integer x.

For an odd value of x, the solution set will be:

$$\left\{x, \frac{1}{2}\left(\frac{x^2}{d_i} - d_i\right), \frac{1}{2}\left(\frac{x^2}{d_i} + d_i\right)\right\} \tag{9.10}$$

Here, d_i are the divisors of x^2 such that $d_i < x$ and the value of N_{tr} can easily be calculated by the expression (9.9).

When x is even:

For an even value of x, the solution set will be:

$$\left\{x, \frac{\left(\frac{x}{2}\right)^2}{d_i} - d_i, \frac{\left(\frac{x}{2}\right)^2}{d_i} + d_i\right\}$$

$$=> \left\{x, \frac{1}{4}\left(\frac{x^2}{d_i} - 4d_i\right), \frac{1}{4}\left(\frac{x^2}{d_i} - 4d_i\right)\right\}$$

Here d_i are the divisors of $\left(\frac{x}{2}\right)^2$ such that $d_i < \frac{x}{2}$. The value of N_{tr} will be the same and can be calculated from equation 9.9.

Let us consider some examples for illustration.

First, we take an example where x is odd.

Say x = 63.

$$\therefore\ x^2 = 63^2 = (9 \times 7)^2 = 9^2 \times 7^2 = 3^4 \times 7^2$$

As discussed above the total number of divisors

$$N_d = (4 + 1) . (2 + 1) = 5 \times 3 = 15$$

and number of triples $N_{tr} = \frac{1}{2}(15 - 1) = 7$

So, we can have 7 triples using 63. Now the divisors of 63^2 less than 63 are 1, 3, 7, 9, 21, 27 and 49.

∴ The triples using 63 are:

(63, 1984, 1985), (63, 84, 105),

(63, 660, 663), (63, 60, 87)

(63, 280, 287), and

(63, 216, 225), (63, 16, 65).

These are the seven triples that can be constructed from 63.

Similarly, we can take any even value of x and find the triples in the same manner using the same expressions for N_d and N_{tr}.

Two important observations:

(a) When x is a prime: Since all primes excluding 2 are odd, x must be odd. Now, x being prime, has only 2 divisors: 1 and x. So x^2 will have 3 divisors in total. Out of these, only one factor will be less than x and that is 1. So, if x is a prime number, then we can have only one Pythagorean triple.
The triple will be of the form:

$$\left\{x, \frac{x^2-1}{2}, \frac{x^2+1}{2}\right\}$$

Also, the difference between z and y is 1.
Let us take an example. Say x = 37. So $y = \frac{1}{2} \times (1369 - 1) = 684$
Then z = y + 1 = 685
So, the triple will be (37, 684, 685).
Any other prime value of x can be taken and the triples will be found accordingly.

(b) To find a triple, given the value of x such that x, y, z are equidistant i.e., x, y, z are in A.P. Let the common difference be d.

$$\therefore x^2 + (x+d)^2 = (x+2d)^2$$

$$\Rightarrow x^2 + (x+2d)^2 - (x+d)^2$$

$$\Rightarrow x^2 = (2x+3d).d$$

$$\Rightarrow x^2 - 2dx - 3d^2 = 0$$

$$\therefore d = \frac{x}{3}$$

It concludes that when x is a multiple of 3, one possible triple will be $\left(x, \frac{4x}{3}, \frac{5x}{3}\right)$

For example, let us take x = 15. Then, one possible triple will be (15, 20, 25).

9.2.3 Code Snippet

This section contains the brute force implementation (in C language) to generate all Pythagorean triplets, with 10,000 being an upper limit for the longest side. Refer Figure 9.1 for the implementation code.

```
#include<stdio.h>
intmain(){
        int n=20, limit = 10000;
        for(inti=1; i<=limit; i++){
                for(int j=n+1; j<=limit; j++){
                        if(n*n + i*i == j*j)
                                printf("%d %d %d\n", n, i, j);


                }

        }
        return0;
}
```

Figure 9.1 A program in C to display all Pythagorean Triplets considering Hypotenuse length below 10,000.

9.2.4 Observation

This section contains the results of the implementation in section 9.2.3. Refer Figure 9.2 for the obtained result.

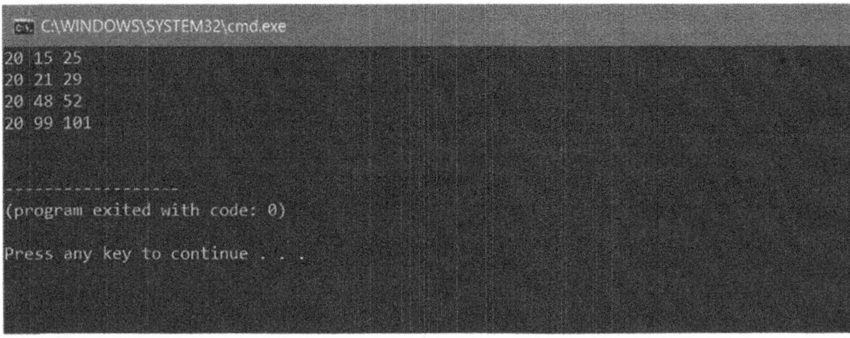

Figure 9.2 Output generated by the code as mentioned in Figure 9.1.

9.3 Computing the Number of Primitive Triples

Here, we will find the number of primitive triples present in N_{tr}.

We consider both the cases - when x is odd and when x is even.

9.3.1 Calculation for Odd Numbers

x is odd:

In the previous section we have already seen that if x is a prime then there will be only one triple (primitive). Now we consider the case when x is not a prime. x can be represented in canonical form as $x = p_1^{s_1}.p_2^{s_2}.p_3^{s_3}....p_n^{s_n}$ where $p_1, p_2, p_3 ... p_n$ are the prime factors and $s_1, s_2, s_3 ... s_n$ are their corresponding powers or indices. For a primitive triple, the solution set (x, y, z) will be such that gcd(x, y, z) = 1. Since x is not a prime and it is an odd number, it is a multiple of some other odd number(s). So we must select the value of y in such a way that x and y have no common factors. Let us analyze it using canonical forms and the equations. Here we can consider x, y and z as three sides of a right-angled triangle with z being the hypotenuse.

We know that,

$$x^2 = p_1^{2s_1}.p_2^{2s_2}.p_3^{2s_3}....p_n^{2s_n}$$

Now, we will choose all the factors in every possible combination, such that it will be less than n. We know that the other side length will be

$\frac{\left\{\frac{x^2}{d_i}-d_i\right\}}{2}$. If the factor d_i is greater than x^2 then $\frac{x^2}{d_i}<x$ and $d_i > x$ so $\frac{\left\{\frac{x^2}{d_i}-d_i\right\}}{2}$ will be less than zero so that will not be a triplet. So we will take d_i as all the probable quantities as well as combinations *viz.* single, double, triple, i.e., all the numbers, which are of the form:

$$p_i^{2s_i}, p_i^{2s_i}.p_j^{2s_j}, \ldots p_i^{2s_i}.p_j^{2s_j} \ldots\ldots p_k^{2s_k} < x \tag{9.11}$$

[where i, j, k … = 1, 2, … q].

When we are dividing x^2 by d_i with a combination of primes as mentioned in type 9.11 then that prime combination will be canceled in the term $\frac{x^2}{d_i}$ but in the next term d_i will contain that prime combination which is cancelled in the first term. So the two terms of the expression $\frac{\frac{x^2}{d_i}-d_i}{2}$ will contain shaperate primes. So, they will not be divided by the same prime divisors. In this way, the triplets will be co-prime to each other and will form a primitive triplet.

Suppose from x^2 is divisible by $p_1^{2s_1}.\ p_2^{2s_2}$ then the other two sides will be:

$$\frac{\left(p_3^{2s_3}p_4^{2s_4}p_5^{2s_5}\ldots p_q^{2s_q}-p_1^{2s_1}p_2^{2s_2}\right)}{2}, \frac{\left(p_3^{2s_3}p_4^{2s_4}p_5^{2s_5}\ldots p_q^{2s_q}+p_1^{2s_1}p_2^{2s_2}\right)}{2}.$$

These two terms are coprime [two parts in the first bracket consist of different primes]. While choosing d_i we have to take one prime with its all power. If we take d_i as p_1^{s1} then the other sides will be:

$$\frac{\left(p_i^{s_i} p_2^{s_2} p_3^{s_3} \cdots p_q^{s_q} - p_i^{s_i}\right)}{2} \text{ and } \frac{\left(p_i^{s_i} p_2^{s_2} p_3^{s_3} \cdots p_q^{s_q} + p_i^{s_i}\right)}{2}$$

We see that $p_1^{s_1}$ is common in them. Now the triplet is not coprime or primitive. So we have to divide x^2 by $p_1^{2s_1}$ [prime term with its entire power].

As there are q different primes, we can write

$$x^2 = \cdot\left(p_1^{2s_1}\right)^1 \cdot \left(p_2^{2s_2}\right)^1 \cdot \left(p_3^{2s_3}\right)^1 \cdots \left(p_q^{2s_q}\right)^1 \quad (9.12)$$

So the number of ways we can choose di is -

$$N_p = \frac{\{(1+1)(1+1)(1+1)\ldots q times\}}{2} = \frac{2^q}{2} = 2^{q-1} \quad (9.13)$$

Now, we take some simple examples to validate our illustration.

First, we take $x = 27 \Rightarrow x^2 = 27^2 = 3^6$.

$$\text{Number of triples } N_{tr} = \frac{6+1-1}{2} = 3$$

The values of divisors: d_i are 1, 3 and 9. Here the possible triples are:-

(27, 364, 365), (27, 120, 123) and (27, 36, 45).

Out of the triples only the first one is primitive. Here we have only one prime 3 and the number of primitive triples = 1 = 2^0 = 2^1 - 1.

We now take another example, x = 45.

Here there will be 7 triples. They are:

(45, 1012, 1013), (45, 60, 75),

(45, 336, 339), (45, 28, 53),

(45, 200, 205), (45, 24, 51) and

(45, 108, 117).

Out of these the triples, (45, 1012, 1013) and (45, 28, 53) are primitive, rest are all non-primitive. Here number of primes = 2 (3 and 5) and the number of primitive triples = 2 = $2^{(2-1)}$.

Similarly, we find that the number of primitive triples $N_{pr} = 2^n - 1$ where n is the number of distinct prime factors in the canonical expression of x provided x is an odd number.

9.3.2 Calculation for Even Numbers

x is even:

When x is even then the other two sides will be $\left\{\frac{\left(\frac{x}{2}\right)^2}{d_i} - d_i\right\}$ and $\left\{\frac{\left(\frac{x}{2}\right)^2}{d_i} + d_i\right\}$.

If $\frac{x}{2}$ is odd then d_i has to be odd because odd term can only be divided by odd terms only and quotient will be odd as well so $\frac{\left(\frac{x}{2}\right)^2}{d_i}$ will be odd

so $\left\{\frac{\left(\frac{x}{2}\right)^2}{d_i} - d_i\right\}$ [odd − odd] and $\left\{\frac{\left(\frac{x}{2}\right)^2}{d_i} - d_i\right\}$ [odd + odd] both will be even so they will not be co-prime and the triplet will not be primitive. But when $\frac{x}{2}$ is even then the canonical form will be $x = 2^{s_1+2}.p_2^{s_2}\ldots p_q^{s_q}\,(s_1 \geq 0)$ [Suppose x can be factored in n prime factors and one of them is 2].

$$\left(\frac{x}{2}\right)^2 = 2^{2s_1+2}.p_2^{2s_2}\cdots p_q^{2s_q}$$

So following the previous theory the number of primitive triplets will be $\frac{(1+1)^q}{2} = 2^{(q-1)}$. In general, we can say the number of primitive triplets for x is $2^{(q-1)}\left[1+(-1)^{\frac{x}{2}}\right]$, q is the total number of distinct primes in the canonical expression of $\frac{x}{2}$.

Now for illustration we take some example:
Let

$$x = 2420 = 2 \times 1210 = 2 \times 121 \times 2 \times 5 = 2^2 \times 11^2 \times 5$$

$$\frac{x}{2} = 121 - = 11^2 \times 2 \times 5.$$

Here, q = number of prime factors in the expression of $\frac{x}{2} = 3$.

So, the total number of primitive triples = $2^{(3-2)}\,[1 + (-1)^{1210}]$

$$= 2^1[1 + 1] = 4.$$

Thus, we can find the number of primitive triples of an even number [2, 3].

A code snippet (in C language) along with its output is hereby attached for confirmation.

9.3.3 Code Snippet

This section contains the brute force implementation (in C language) to generate all Primitive Pythagorean triplets within the range [45, 10,000]. Refer Figure 9.3 for the implementation code.

```
#include<stdio.h>
intgcd(int, int)
intma()in{
    int n = 45, limit = 10000;
    for(inti=1; i<=limit; i++){
        for(intj=n+1; j<=limit; j++){
            if( (n*n + i*i == j*j) && (gcd(n,i)==1) && (gcd(n,j)==1) &&
(gcd(i,j)==1) )
                printf("%d %d %d\n", n, i, j);

        }

    }
    return0;

}
intgcd(int x, int y){
if(y==0)
return x;
else
returngcd(y,x%y);
}
```

Figure 9.3 A program in C to display all Primitive Pythagorean triplets within the range [45, 10,000].

9.3.4 Observation

This section contains the results of the implementation in section 9.3.3. Refer Figure 9.4 for the obtained result.

Figure 9.4 Output generated by the code as mentioned in Figure 9.3.

9.4 Representation of Integers as Difference of Two Perfect Squares

Formula (9.5) is suitable for obtaining all the possible expressions of an integer in the form of a difference of two perfect squares. Suppose x is the number so we have to find all possible values of a and b such that $x = a^2 - b^2$ $[a > b > 0]$.

It is clear from identity 1.2 that $x^2 + (2ab)^2 = (a^2 + b^2)^2$

So if x is a side of a right angle triangle then it will form a Pythagorean triplet. Here too, we have two cases:

9.4.1 Calculation for Odd Numbers

x is odd:

Here, we know the hypotenuse will be:

$$\mathrm{a}^2 + \mathrm{b}^2 = \frac{\left(\dfrac{x^2}{d_i}\right) + d_i}{2} \quad \text{[From 9.5]} \tag{9.14}$$

$$\text{and } \mathrm{a}^2 - \mathrm{b}^2 = \mathrm{x}$$

Adding these two equations we get:

$$2\mathrm{a}^2 = \frac{\left(x^2 + 2xd_i + d_i^2\right)}{2d_i}$$

Giving, $a = \frac{(x + d_i)}{2\sqrt{d_i}}$

Putting the value in any equation we get

$$b = \frac{(x - d_i)}{2\sqrt{d_i}} \tag{9.15}$$

From 9.15 it is clear that a and b both are integers. So, $\sqrt{d_i}$ must be an integer implying, d_i [divisors of x^2] must be a square term itself. As x is odd, all its divisors will also be odd. Thus, x - d_i and x + d_i are both divisible by 2 (even) and x ± d_i will be divisible by $\sqrt{d_i}$ resulting in the final outcome being an integer. So we have to find all the possible divisors which are perfect squares. We know,

$$x^2 = p_1^{2s_1}.p_2^{2s_2}.p_3^{2s_3}\ldots.p_q^{2s_q}$$

$$\Rightarrow x^2 = \left(p_1^2\right)^{s_1} \cdot \left(p_2^2\right)^{s_2} ..\left(p_q^2\right)^{s_q}$$

We have to use the square primes as divisors such that $d_i < x$ [otherwise l will be negative].

So the total possible ways of finding d_i is:

N_d = [if any of s_1, s_2,.....s_q is odd (let s_i), then s_i + 1 will be even. For at least one power of a prime being odd, $x = p_1^{s_1}.p_2^{s_2}\ldots.p_q^{s_q}$ will not be a perfect square].

When all powers (s_1, s_2, ... s_q) are even, then N_d = [as all powers are even so 2 can be taken common from all powers, x will be a perfect square].

9.4.2 Calculation for Even Numbers

x is even:

Here, we know the other two sides are:

$$\left\{\frac{\left(\frac{x}{2}\right)^2}{d_i} + d_i\right\} \text{and} \left\{\frac{\left(\frac{x}{2}\right)^2}{d_i} - d_i\right\}$$

$$\therefore a^2 + b^2 = \left\{ \frac{\left(\frac{x}{2}\right)^2}{d_i} + d_i \right\} \text{[from 9.5]}$$

$$=> a^2 - b^2 = x$$

Adding these two equations, we get $2a^2 = \frac{\left\{\left(\frac{x}{2}\right)^2 + xd_i + d_i^2\right\}}{d_i}$.

$$==> a = \frac{\frac{x}{2} + d_i}{\sqrt{2d_i}}$$

Putting this value in any equation we get $b = \frac{\frac{x}{2} - d_i}{\sqrt{2d_i}}$.

Now $2d_i$ should be a perfect square to get an integral outcome. So d_i must be in the form $d_i = 2^{2a+1} . M^2$, which is obviously even. [$a \geq 0$]. As we know that d_i is a divisor of $(x/2)^2$, so if $\frac{x}{2}$ is odd then all of its divisors will be odd. In that case $\sqrt{2d_i}$ will never be an integer. So n will never be represented as a^2-b^2 for this case.

When $\frac{x}{2}$ is even then we can write the canonical structure as:

$$x = 2^{s_1+1}.p_2^{s_2}.p_3^{s_3}\ldots p_q^{s_q}$$

$$\left(\frac{x}{2}\right)^2 = 2^{2s_1}.p_2^{2s_2}.p_3^{2s_3}\ldots..p_q^{2s_q}$$

$$[s_1 > 0, s_1 \neq 0]$$

$$=> \left(\frac{x}{2}\right)^2 = (2^2)^{(s_1)}.p_2{}^2)^{(s_2)}\cdots(p_q{}^2)^{s_q} \tag{9.16}$$

So total number of divisors = $(s_1 + 1). (s_2 + 1) \ldots (s_q + 1)$.

In the expression the only even portion is 2^{2s_1}, so except that part all combinations made by p_1^2, p_2^2, ... p_q^2 will make odd square divisors. So total odd square divisors are = $(s_2+1).(s_3+1)$ ... (s_q+1).

$$\text{Total even divisors} = \text{total divisors} - \text{odd divisors} = (s_1+1).(s_2+1)\ldots(s_q+1)$$

$$-(s_2+1).(s_3+1)\ldots(s_q+1) = s_1.\prod_{i=2}^{q}(s_i+1) \tag{9.17}$$

We also have to keep in mind that $d_i < \frac{x}{2}$ [otherwise I will be negative]. So total possible ways in which we can represent x as $a^2 - b^2$ is given by the expression

$$N_r = \frac{s_1(s_2+1)(s_3+1)\ldots(s_q+1)}{2} \tag{9.18}$$

[when x is not a square in that case either s_1 is even or at least any one of s_2, s_3, ... s_q is odd]

When s_1 is not even or s_2, s_3, s_4, ... s_q all are even, i.e., x is a perfect square, then [4]

$$N_r = \frac{s_1(s_2+1)(s_3+1)\ldots(s_q+1)-1}{2} \tag{9.19}$$

9.4.3 Corollaries

I. In the expression of $\frac{x}{2}$ if we put $s_1 = 0$ then x will be $p_2^{2s_2}.p_3^{2s_3}\ldots p_q^{2s_q}$ which consists of primes only. It is thus obviously odd. Total ways of representation:

$N_r = \frac{s_1(s_2+1)(s_3+1)\ldots(s_q+1)}{2} = 0$. That is what we observed previously.

For all odd $\frac{x}{2}$, x can't be represented as $a^2 - b^2$.

II. $\frac{x}{2}$ is even, d_i is even. If $\sqrt{2d_i}$ is 2 × even then a = even + even = even and b = even - even = even, else $\sqrt{2d_i}$ is 2 × odd then a = even + odd = odd and b = even - odd = odd. Thus, a and b will always have the same parity.

We know that, if one side of the right angle triangle is $a^2 - b^2$ then the other two sides can be taken as 2ab and $a^2 + b^2$ (hypotenuse). Till now, we were taking the case for n is even and represented as $a^2 - b^2$ and also found the solutions. But 2ab is also even. So, for any even x we can consider it as 2ab as well, i.e., x = 2ab. Then according to 9.5 we know:

$$a^2+b^2=\left\{\frac{\left(\frac{x}{2}\right)^2}{d_i}+d_i\right\} \text{ and } a^2-b^2=\left\{\frac{\left(\frac{x}{2}\right)^2}{d_i}-d_i\right\} \text{[other two sides]}$$

Adding these two equations we get:

$$a^2=\frac{\left(\frac{x}{2}\right)^2}{d_i} \quad \text{and} \quad 2ab=x$$

$$=>a=\frac{x}{2\sqrt{d_i}} \quad \text{and} \quad ==>b=\frac{x}{2a}=\sqrt{d_i}$$

As we know, d_i is the divisor of $\left(\frac{x}{2}\right)^2$ and less than $\frac{x}{2}$ as well d_i has to be a square divisor. So, for $\left(\frac{x}{2}\right)^2=(2^2)^{s_1}\cdot\left(p_2{}^2\right)^{s_2}\cdots\left(p_q{}^2\right)$ the number of square divisors such that $d_i<\frac{x}{2}$ is

$$N_d=\frac{(s_1+1)(s_2+1)\ldots(s_q+1)}{2} \text{ [for non-square x]}$$

$$\text{or, } N_d=\frac{(s_1+1)(s_2+1)\ldots(s_q+1)-1}{2} \text{ [for squared x].}$$

which is exactly the same result as we found earlier.

Straight away we can factorize $\frac{x}{2}=2^{s_1}.p_2^{s_2}.p_3^{s_3}\ldots p_q^{s_q}$. From here, the number of factors we can get which are less than $\frac{x}{2}$ is obtained by:

$$N_d = \frac{(s_1+1)(s_2+1)\ldots(s_q+1)}{2} \text{ [for non-square x]}$$

$$\text{or, } N_d = \frac{(s_1+1)(s_2+1)\ldots(s_q+1)-1}{2} \text{ [for squared x].}$$

If one factor is 'a', then the other factor will be calculated by $\frac{x}{2a}$.

The code snippet and output screenshot are attached herewith like the previous cases.

9.4.4 Code Snippet

This section contains the brute force implementation (in C language) to generate all possible pairs having a difference of squares as 56, keeping the upper limit as 10,000. Refer Figure 9.5 for the implementation code.

```
#include<stdio.h>
intmain(){
    intn;
    printf("Enter the value of n: ");
    scanf("%d", &n);
    printf("\n");
    for(int k=1; k<=10000; k++){
        for(int l=1; l<=10000; l++){
            if(n == k*k - l*l)
                printf("%d\t%d\t%d\n", n, k, l);
        }
    }
    return0;
}
```

Figure 9.5 A program in C to display all possible pairs having a difference of squares as 56, keeping the upper limit as 10,000.

9.4.5 Output

This section contains the results of the implementation in section 9.4.4. Refer Figure 9.6 for the obtained result.

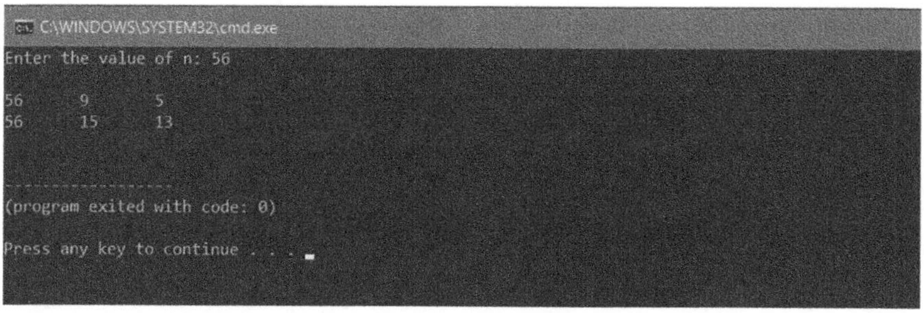

Figure 9.6 Output generated by the code as mentioned in Figure 9.5.

9.5 Conclusion

There exists further ways of generating Pythagorean triples. A variety of special representations of generating numbers are applied, particularly, the representation in terms of Fibonacci numbers [5]. Also, there exist geometrical representations like that of Dickson's method [6] and others [7, 8]. In this paper, we have found out all the possible triples that can be formed using one number and also the possible quantity of triples. Simple algebraic equations and Euclid's formula (9.2) have been used in our derivations. The canonical expression of a number plays a pivotal role here. Using the canonical expressions, we can also represent an integer as a difference of two perfect squares. In each case, suitable examples are given in support of the proposition. The significance of the formulation generalized here, lies in the simplicity with which the triplets have been derived, in contrast with other efforts made earlier.

References

1. Asmaryan, E., On the Generation of Pythagorean triples and representation of integers as a difference of two squares. *Int. J. Math. Comput. Sci.*, 13, arXiv:1711.02379, 2017. https://doi.org/10.48550/arXiv.1711.02379.
2. Robbins, N. and Fibonacci, Q., On the number of primitive pythagorean triangles with a given inradius. *Fibonacci Q.*, 44, 4, 368–369, 2006.
3. Omland, T., How many pythagorean triples with a given inradius? *J. Number Theory*, 170, Supplement C, 1–2, 2017. doi:https://doi.org/10.1016/j.jnt.2016.06.009. http://www.sciencedirect.com/science/article/pii/S0022314X16301603.
4. Nyblom, M.A., On the representation of the integers as a difference of squares. *Fibonacci Q.*, 40, 3, 243–246, 2002.

5. Horadam, A., Fibonacci number triples. *Am. Math. Mon.*, 68, 8, 751–753, 1961.
6. Dickson, L.E., *History of the Theory of Numbers: Diophantine Analysis*, vol. 2, Dover Publications 2005, Mineola, New York (USA), 1920.
7. Amato, R., A characterization of pythagorean triples. *JP J. Algebra Number Theory Appl.*, 39, 2, 221, 2017.
8. Roy, T. and Sonia, F.J., A direct method to generate pythagorean triples and its generalization to pythagorean quadruples and n-tuples, Jadavpur University (Dept. of Physics), Kolkata, 2012, ArXiv e-printsarXiv:1201.2145. https://arxiv.org/ftp/arxiv/papers/1201/1201.2145.pdf.

10

Solution of Matrix Games With Pay-Offs of Single-Valued Neutrosophic Numbers and Its Application to Market Share Problem

Mijanur Rahaman Seikh and Shibaji Dutta*

Department of Mathematics, Kazi Nazrul University, Asansol, India

Abstract
Single-valued neutrosophic numbers (SVNNs) are an immensely convenient tool to counter imprecise circumstances in real-life problems containing indeterminacy. The objective of this chapter is to explore matrix games (MGs) where the pay-offs are assumed as SVNNs. A solution methodology has been derived using the notion of the score function (*SF*) of SVNNs. First, we formulate two neutrosophic mathematical programming problems to obtain optimal strategies. The *SF* of SVNNs is used to transform these problems into two crisp equivalent problems, which are solved to get the optimal strategies. A market share problem is illustrated to express the rationality and applicability of the presented method. A comparative study with an existing work analyzes the superiority of the performed method.

Keywords: Game theory, single-valued neutrosophic numbers, score function, market share problem

10.1 Introduction

Game theory has engaged largely in competitive decision-making systems. In reality, many situations are uncertain due to the imprecision of facts, asymmetric information, and conflict of interest between opponents in the same field of business. However, the pay-offs may not be expressed precisely due to uncertainty in game situations.

**Corresponding author*: dutta.shibaji@gmail.com

Sharmistha Ghosh, M. Niranjanamurthy, Krishanu Deyasi, Biswadip Basu Mallik, and Santanu Das (eds.) Mathematics and Computer Science Volume 1, (213–230) © 2023 Scrivener Publishing LLC

In the literature, fuzzy matrix games (FMGs) have been broadly studied and explored by several researchers. Li [9] developed an effective methodology to solve FMGs. Li [11] evolved several methods to solve matrix games with payoffs as triangular fuzzy numbers. Seikh *et al.* [12] proposed a solution procedure of solving FMG using α-cut of fuzzy numbers (FNs). Seikh *et al.* [15] applied the weighted average operator and the *SF* to design a new approach of solving MGs where triangular hesitant fuzzy elements are considered as pay-offs. Yang [26] solved MG where triangular dual hesitant fuzzy numbers are assumed as pay-offs. Very recently, Seikh *et al.* [14] developed an approach to solve MG where dense fuzzy numbers are utilized as pay-offs.

Fuzzy set (FS) considers only the degree of membership (DOM) $\mu_S(\epsilon) \in U$ (= [0, 1]) for all $\epsilon \in \mathbf{S}$ in a universe. FS cannot anticipate any other direction concerning the imperfect conception of elements. Intuitionistic fuzzy set (IFS) [1] examine the degree of non-membership (DONM) $\nu_S(\epsilon) \in U$ together with the DOM $\mu_S(\epsilon) \in U$, where $0 \leq \mu_S(\epsilon) + \nu_S(\epsilon) \leq 1$. So, IFS expresses impreciseness in the pay-offs in a generalized way. Several researchers worked on MGs [8, 10, 13, 25] where IFSs are presented as pay-offs. Verma and Kumar [20] introduced Ambika Method to solve MGs, where intuitionistic fuzzy numbers (IFNs) are assumed as pay-offs. Seikh *et al.* [16] defined a new method to defuzzify the Type-2 fuzzy variables and used it to solve MG problems. Xing and Qiu [24] implemented the accuracy function method to solve the MG where pay-offs are considered as triangular intuitionistic fuzzy numbers. Karmakar *et al.* [7] analyzed a solution approach of Type-2 intuitionistic fuzzy MGs by using a new distance measure.

The available information in the real world always contains some vague and imprecise data which consists of conflicting, unpredictable, and indeterminate information. The FS can not express the false membership information and the IFS is unable to control the indeterminacy of information. Neutrosophic sets (NSs) [18] assume the degree of indeterminacy (DOIN), i.e., the neutrality $\omega_S(\epsilon) \in U$ together with the DOM and DONM for uncertainty problems. Therefore, NS represents the indeterminate data whereas FS and IFS fail. NSs are described by DOM (μ), DOIN (ω), and DONM (ν), where μ, ω, and ν are independent and $0 \leq \mu + \omega + \nu \leq 3$. NS becomes the classical set when $\omega = 0$, μ, ν either 0 or 1 and $\mu + \omega + \nu = 1$; the FS when $\omega = 0$, $0 \leq \mu, \omega, \nu \leq 1$ and $\mu + \omega + \nu = 1$; the IFS when $0 \leq \mu, \omega, \nu \leq 1$ and $0 < \mu + \omega + \nu < 1$. Therefore, one can decide that the NS is a generalization of the classical set, FS and IFS.

Single-valued neutrosophic set (SVNS) was conceptualized by Wang *et al.* [21] for realistic applications. SVNS is a special form of NS.

Single-valued neutrosophic numbers (SVNNs) are a distinctive case of SVNSs and are the rationalization of FNs and IFNs. Ye [27] defined single-valued trapezoidal neutrosophic set (SVTNS) by combining the concept of trapezoidal fuzzy numbers and SVNS. Ye [28] merged the idea of interval number and SVNN and introduced single-valued neutrosophic interval number. Deli and Subas [3] were first to define the idea of cut sets of SVNNs and applied to SVTNNs and single-valued triangular neutrosophic numbers. Very recently, Garai *et al.* [4] developed an attentive ranking methodology by defining the weighted possibility mean for SVNNs. Garg and Nancy [6] conceptualized new distance measures in the SVNS circumstances and evolved an algorithm for neutrosophic decision-making problems. Some recent works using SVNS are References [5, 17, 19, 22, 23].

In MGs, due to the lack of information in the available data, the DOIN plays a vital role while assessing the pay-off values. Therefore, the idea of FS or IFS cannot depict the components of the pay-off matrix. For example, suppose a medicine manufacturing company is going to dispatch another important item on the market, and the marketing manager of the company wants to estimate the sales amount of the item. The sales amount of the new item depends on various uncertain parameters such as production cost, the demand for the product, the capacity of supply, selling price, etc. However, the company wants to know whether "the guaranteed sales amount would be $1200 per day or less" before starting its production. Then some experts are consulted for their opinions about the guaranteed sales amount. The existence of this uncertain guaranteed sales amount always contains some "neutral" knowledge. This situation cannot be revealed by FNs or IFNs and SVNNs comes into consideration.

By applying neutrosophic estimations, one can feel more versatility and better trustability to counter uncertainty. So in many uncertain situations, players can consider SVNN as pay-offs of the MG problems. This guides us to implement SVNN as pay-offs in the MG problems.

Here we characterized the idea of the solution of MGs with SVNN as payoffs. We used the notion of the *SF* of SVNN to solve the MG problem with SVNN as pay-offs. Two auxiliary neutrosophic mathematical programming problems are formulated at first. Then, the *SF* of SVNNs is used to transform these problems into two crisp equivalent problems and solved using the simplex method on the LINGO platform. An illustration of a market share problem manifests the legitimacy and appropriateness of the performed approach. Present work is also compared with existing work.

The remainder of this chapter is structured as follows. Section 10.2 recalled basic ideas related to SVNNs. Section 10.3 is dedicated to discussing the MG with SVNN as pay-offs and the concept of the solution. Section 10.4 outlines the mathematical model construction of the problem. A market share problem is illustrated and a comparative study is analyzed with existing work in Section 10.5. Section 10.6 concludes the chapter.

10.2 Preliminaries

NSs and their basic operations are outlined here.

Definition 10.2.1 [18] Let $\tilde{M} = \{\langle \xi, (\mu_{\tilde{M}}(\xi), \omega_{\tilde{M}}(\xi), \nu_{\tilde{M}}(\xi))\rangle \mid \xi \in \mathbf{S}, \mu_{\tilde{M}}(\xi),$ $\omega_{\tilde{M}}(\xi)$, and $\nu_{\tilde{M}}(\xi) \in]^{-}0,1^{+}[\}$. The set $\tilde{M}$ is called a NS over a universe **S**, where $\mu_{\tilde{M}}(\xi), \omega_{\tilde{M}}(\xi), \nu_{\tilde{M}}(\xi)$ are respectively the DOM, DOIN, and DONM of ξ with $^{-}0 \le \mu_{\tilde{M}}(\xi) + \omega_{\tilde{M}}(\xi) + \nu_{\tilde{M}}(\xi) \le 3^{+}$.

Definition 10.2.2 [21] Let $\tilde{M} = \{\langle \xi (\mu_{\tilde{M}}(\xi), \omega_{\tilde{M}}(\xi), \nu_{\tilde{M}}(\xi))\rangle \mid \xi \in \mathbf{S}, \mu_{\tilde{M}}(\xi),$ $\omega_{\tilde{M}}(\xi), \nu_{\tilde{M}}(\xi) \in [0,1]\}$. The set $\tilde{M}$ is called a SVNS over a universe **S** where

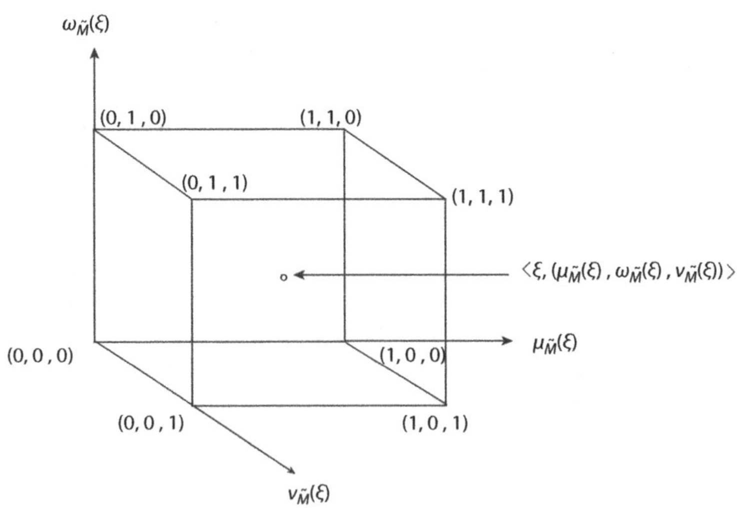

Figure 10.1 Picture representation of SVNS $\tilde{M}$.

$\mu_{\tilde{M}}(\xi),\omega_{\tilde{M}}(\xi),\nu_{\tilde{M}}(\xi)$ are respectively the DOM, DOIN, and DONM of ξ with $0\leq\mu_{\tilde{M}}(\xi)+\omega_{\tilde{M}}(\xi)+\nu_{\tilde{M}}(\xi)\leq 3$.

Figure 10.1 depicts the picture representation of SVNS $\tilde{M}$.

For convenience, $\langle\mu_{\tilde{M}}(\xi),\omega_{\tilde{M}}(\xi),\nu_{\tilde{M}}(\xi)\rangle$ is said to be a SVNN and is usually represented as $\langle\mu, \omega, \nu\rangle$.

Definition 10.2.3 (Basic Arithmetic Operations of SVNNs) [2]
Let $\tilde{W}=\langle\mu,\omega,\nu\rangle,\tilde{W}_1=\langle\mu_1,\omega_1,\nu_1\rangle$, and $\tilde{W}_2=\langle\mu_2,\omega_2,\nu_2\rangle$ be three SVNNs, and $\lambda > 0$ be a scalar. Then, their algebraic operations are outlined as follows:

1. $\tilde{W}_1+\tilde{W}_2=\langle\mu_1+\mu_2-\mu_1\mu_2,\omega_1\omega_2,\nu_1\nu_2\rangle$;
2. $\tilde{W}_1\times\tilde{W}_2=\langle\mu_1\mu_2,\omega_1+\omega_2-\omega_1\omega_2,\nu_1+\nu_2-\nu_1\nu_2\rangle$;
3. $\lambda\tilde{W}=\langle 1-(1-\mu)^{\lambda},\omega^{\lambda},\nu^{\lambda}\rangle$;
4. $\tilde{W}^{\lambda}=\langle\mu^{\lambda},1-(1-\omega)^{\lambda},1-(1-\nu)^{\lambda}\rangle$.

Definition 10.2.4 [2] Let $\tilde{W}=\langle\mu,\omega,\nu\rangle$ be a SVNN. Then, (i) the SF of $\tilde{W}$ *is* $\Omega(\tilde{W})=\dfrac{2+\mu-\omega-\nu}{3}$ (ii) the accuracy function of $\tilde{W}$ is $\Psi(\tilde{W})=\mu-\nu$ and (iii) the certainty function of $\tilde{W}$ is $\Upsilon(\tilde{W})=\mu$.

Let $\tilde{W}_1$ and $\tilde{W}_2$ be two SVNNs. Then, the following is the ranking order relation for $\tilde{W}_1$ and $\tilde{W}_2$.

1. $\Omega(\tilde{W}_1)>\Omega(\tilde{W}_2)\Rightarrow\tilde{W}_1\succ_n\tilde{W}_2$;
2. $\Omega(\tilde{W}_1)=\Omega(\tilde{W}_2)$, and $\Psi(\tilde{W}_1)>\Psi(\tilde{W}_2)\Rightarrow\tilde{W}_1\succ_n\tilde{W}_2$;
3. $\Omega(\tilde{W}_1)=\Omega(\tilde{W}_2),\Psi(\tilde{W}_1)=\Psi(\tilde{W}_2)$, and $\Upsilon(\tilde{W}_1)>\Upsilon(\tilde{W}_2)\Rightarrow\tilde{W}_1\succ_n\tilde{W}_2$;
4. $\Omega(\tilde{W}_1)=\Omega(\tilde{W}_2),\Psi(\tilde{W}_1)=\Psi(\tilde{W}_2)$, and $\Upsilon(\tilde{W}_1)=\Upsilon(\tilde{W}_2)\Rightarrow\tilde{W}_1\simeq_n\tilde{W}_2$.

Here '$>_n$' stands for 'larger than', '$\simeq_n$' stands for 'equal to' and '$\geq_n$', '$\leq_n$' have similar explanations in neutrosophic sense.

10.3 Matrix Games With SVNN Pay-Offs and Concept of Solution

Let Player-I ($\mathbf{P}_1$) and Player-II ($\mathbf{P}_2$) choose the pure strategy ζ_h and η_k with probability γ_h and δ_k, respectively, for $h \in \{1,2,\ldots, p\}$ and $k \in\{1,2,\ldots, q\}$. If $\sum_{h=1}^{p}\gamma_h=1$ and $\sum_{k=1}^{q}\delta_k=1$ for $(\gamma,\delta)\in \Re_+^p\times\Re_+^q$, where $\gamma = (\gamma_1,\gamma_2,\ldots,\gamma_p)$ and $\delta = (\delta_1, \delta_2, \ldots, \delta_q)$, then γ and δ are considered as the mixed strategies. For $\mathbf{P}_1$ and $\mathbf{P}_2$, respectively. Let $\mathbf{M}_1=\left\{\gamma=(\gamma_1,\gamma_2,\ldots,\gamma_p)\in\Re_+^p \middle| \sum_{h=1}^{p}\gamma_h=1\right\}$ and $\mathbf{M}_2=\left\{\delta=(\delta_1,\delta_2,\ldots,\delta_p)\in\Re_+^q \middle| \sum_{k=1}^{q}\delta_k=1\right\}$ are the sets of all mixed strategies for $\mathbf{P}_1$ and $\mathbf{P}_2$, respectively.

If $\mathbf{P}_1$ and $\mathbf{P}_2$ choose their pure strategy ζ_h and η_k respectively then $\mathbf{P}_1$ gains as pay-off represented by a SVNN $\tilde{a}_{hk} = \langle\mu_{hk}, \omega_{hk}, \nu_{hk}\rangle$, while $\mathbf{P}_2$ gains a negation of the SVNN $\tilde{a}_{hk}$. Thus, a MG with pay-offs represented by SVNNs is expressed by $\tilde{A}=(\tilde{a}_{hk})_{p\times q}$, which is generally mentioned as pay-off matrix. Then, the MG with SVNN pay-offs is represented by $\{\mathbf{M}_1,\mathbf{M}_2,\tilde{A}\}$ and is referred as SVNN matrix game (SVNNMG) $\tilde{A}$.

For $(\gamma, \delta) \in \mathbf{M}_1 \times \mathbf{M}_2$, the expected pay-off $(\mathbf{E\tilde{P}})$ for $\mathbf{P}_1$ will be an SVNN, which will be calculated (using Definition 10.2.3) as

$$\mathbf{E\tilde{P}}(\gamma,\delta)=\gamma^T\tilde{A}\delta=\sum_{h=1}^{p}\sum_{k=1}^{q}\tilde{a}_{hk}\gamma_h\delta_k=\sum_{h=1}^{p}\sum_{k=1}^{q}\langle\mu_{hk},\omega_{hk},\nu_{hk}\rangle\gamma_h\delta_k$$

$$=\sum_{h=1}^{p}\sum_{k=1}^{q}\langle 1-(1-\mu_{hk})^{\gamma_h\delta_k},\omega_{hk}{}^{\gamma_h\delta_k},\nu_{hk}{}^{\gamma_h\delta_k}\rangle.$$

As the SVNNMG is two-person zero-sum game, $\mathbf{E\tilde{P}}$ of $\mathbf{P}_2$ is $-\ \mathbf{E\tilde{P}}(\gamma,\delta)$. Irrespective of the use of best strategies of the players, the maximum guaranteed gain (or the minimum possible loss) is the value of the MG for $\mathbf{P}_1$ (or $\mathbf{P}_2$).

Definition 10.3.1 (Gain floor for $\mathbf{P}_1$ and loss ceiling for $\mathbf{P}_2$)

Let $\tilde{G}^* = \max_{\gamma \in \mathbf{M}_1} \min_{\delta \in \mathbf{M}_2} \{\mathbf{E}\tilde{\mathbf{P}}(\gamma,\delta)\}$ and $\tilde{L}^* = \min_{\delta \in \mathbf{M}_2} \max_{\gamma \in \mathbf{M}_1} \{\mathbf{E}\tilde{\mathbf{P}}(\gamma,\delta)\}$. Then G^* and L^* are respectively the gain floor for $\mathbf{P}_1$ and the loss ceiling for $\mathbf{P}_2$.

Theorem 10.3.1 For the SVNNMG $\tilde{\mathrm{A}}$, $\mathbf{P}_1$'s gain floor is less than $\mathbf{P}_2$'s loss ceiling, i.e., $\tilde{G}^* \preceq_n \tilde{L}^*$ holds.

Proof: For any $\gamma \in \mathbf{M}_1, \min_{\delta \in \mathbf{M}_2} \{\mathbf{E}\tilde{\mathbf{P}}(\gamma,\delta)\} \preceq_n \{\mathbf{E}\tilde{\mathbf{P}}(\gamma,\delta)\}$.

Again, for any $\delta \in \mathbf{M}_2$, we have $\mathbf{E}\tilde{\mathbf{P}}(\gamma,\delta) \preceq_n \max_{\gamma \in \mathbf{M}_1} \{\mathbf{E}\tilde{\mathbf{P}}(\gamma,\delta)\}$.

Thus, for any $\gamma \in \mathbf{M}_1$ and $\delta \in \mathbf{M}_2$, we obtain

$$\min_{\delta} \{\mathbf{E}\tilde{\mathbf{P}}(\gamma,\delta)\} \preceq_n \max_{\gamma} \{\mathbf{E}\tilde{\mathbf{P}}(\gamma,\delta)\}.$$

Therefore, $\min_{\delta} \{\mathbf{E}\tilde{\mathbf{P}}(\gamma,\delta)\} \preceq_n \min_{\delta} \max_{\gamma} \{\mathbf{E}\tilde{\mathbf{P}}(\gamma,\delta)\}$.

Hence,

$$\max_{\gamma} \min_{\delta} \{\mathbf{E}\tilde{\mathbf{P}}(\gamma,\delta)\} \preceq_n \min_{\delta} \max_{\gamma} \{\mathbf{E}\tilde{\mathbf{P}}(\gamma,\delta)\},$$

Therefore, $\tilde{G}^* \preceq_n \tilde{L}^*$. □

Definition 10.3.2 (Optimal solutions and value of the game) If for some $(\gamma^0, \delta^0) \in \mathbf{M}_1 \times \mathbf{M}_2$, such that

$$\gamma^{0^T} \tilde{A}\, \delta^0 = \max_{\gamma \in \mathbf{M}_1} \min_{\delta \in \mathbf{M}_2} (\gamma^{\mathrm{T}} \tilde{A}\, \delta) = \min_{\delta \in \mathbf{M}_2} \max_{\gamma \in \mathbf{M}_1} (\gamma^{\mathrm{T}} \tilde{A}\, \delta),$$

then γ^0 and δ^0 are called optimal solutions for $\mathbf{P}_1$ and $\mathbf{P}_2$, respectively and $\gamma^{0^T} \tilde{A} \delta^0$ is said to be the value of the SVNNMG $\tilde{A}$.

10.4 Mathematical Model Construction for SVNNMG

According to Definition 10.3.2, the maximin strategy $\gamma^* \in \mathbf{M}_1$ for $\mathbf{P}_1$ and minimax strategy $\delta^* \in \mathbf{M}_2$ for $\mathbf{P}_2$ can be obtained by solving a pair of neutrosophic programming models (10.1) and (10.2), respectively, as follows:

$$\begin{aligned}
&\max \tilde{G} = \langle \rho, \sigma, \tau \rangle \\
\text{s.t.,}\quad &\sum_{h=1}^{p} \tilde{a}_{hk}\, \gamma_h \succeq_n \tilde{G}, k \in \{1,2,\ldots,q\}, \\
&\sum_{h=1}^{p} \gamma_h = 1, \gamma_h \geq 0, h \in \{1,2,\ldots,p\}, \\
&\rho \geq 0, \sigma \geq 0, \tau \geq 0, \rho + \sigma + \tau \leq 3,
\end{aligned} \tag{10.1}$$

$$\begin{aligned}
&\max \tilde{L} = \langle \in, \psi, \xi \rangle \\
\text{s.t.,}\quad &\sum_{k=1}^{q} \tilde{a}_{hk} \delta_k \preceq_n \tilde{L}, h \in \{1,2,\ldots,p\}, \\
&\sum_{k=1}^{q} \delta_k = 1, \delta_k \geq 0, k \in \{1,2,\ldots,q\}, \\
&\in \geq 0, \psi \geq 0, \xi \geq 0, \in + \psi + \xi \leq 3,
\end{aligned} \tag{10.2}$$

Using Definition 10.2.3, model (10.1) and (10.2) reduce to model (10.3) and (10.4), respectively, as follows.

$$\begin{aligned}
&\max \tilde{G} = \langle \rho, \sigma, \tau \rangle \\
&\text{s.t.,}\quad \left\langle 1 - \prod_{h=1}^{p} \left(1 - \mu_{hk}\right)^{\gamma_h}, \prod_{h=1}^{p} \omega_{hk}{}^{\gamma_h}, \prod_{h=1}^{p} \nu_{hk}{}^{\gamma_h} \right\rangle \succeq_n \langle \rho, \sigma, \tau \rangle, \quad k \in \{1,2,\ldots q\}, \\
&\sum_{h=1}^{p} \gamma_h = 1, \gamma_h \geq 0, h \in \{1,2,\ldots,p\}, \\
&\rho \geq 0, \sigma \geq 0, \tau \geq 0, \rho + \sigma + \tau + \leq 3,
\end{aligned} \tag{10.3}$$

and

$$\min \tilde{L} = \langle \in, \psi, \xi \rangle$$

$$\text{s.t.,} \left\langle 1 - \prod_{k=1}^{q} (1-\mu_{hk})^{\delta_k}, \prod_{k=1}^{q} \omega_{hk}{}^{\delta_k}, \prod_{k=1}^{q} \nu_{hk}{}^{\delta_k} \right\rangle \preceq \langle \in, \psi, \xi \rangle, \quad h \in \{1,2,\ldots p\},$$

$$\sum_{k=1}^{q} \delta_k = 1, \delta_k \geq 0, \quad k \in \{1,2,\ldots,q\},$$

$$\in \geq 0, \psi \geq 0, \xi \geq 0, \in + \psi + \xi \leq 3. \tag{10.4}$$

Theorem 10.4.1 Let (γ^*, δ^*) be an optimal solution of the SVNNMG $\tilde{A} = (\tilde{a}_{\text{hk}})$. Then (γ^*, δ^*) is also the solution of the pay-off matrix after using SF to convert it into a crisp pay-off matrix and vice-versa.

Proof: Let (γ^*, δ^*) be an optimal solution of the SVNNMG $\tilde{A}$. Then,

$$\begin{aligned} \min_{\delta}\left\{\max_{\gamma}\{\mathbf{E}\tilde{\mathbf{P}}(\gamma,\delta)\}\right\} &= \mathbf{E}\tilde{\mathbf{P}}(\gamma^*,\delta^*) = \sum_{h}\sum_{k} \tilde{a}_{hk}\gamma_h\delta_k \\ &= \sum_{k}\left(\sum_{h} \tilde{a}_{hk}\gamma_h\right)\delta_k = \max_{\gamma}\left\{\min_{\delta}\{\mathbf{E}\tilde{\mathbf{P}}(\gamma,\delta)\}\right\} \end{aligned}$$

Applying SF to the above relation, we have the following two equations

$$\begin{aligned} \Omega\left(\min_{\delta}\left\{\max_{\gamma}\{\mathbf{E}\tilde{\mathbf{P}}(\gamma,\delta)\}\right\}\right) &= \Omega(\mathbf{E}\tilde{\mathbf{P}}(\gamma^*,\delta^*)) = \Omega\left(\sum_{h}\sum_{k} \tilde{a}_{hk}\gamma_h\delta_k\right) \\ &= \Omega\left(\sum_{k}\left(\sum_{k} \tilde{a}_{hk}\gamma_h\right)\delta_k\right) = \sum_{h}\sum_{k} \Omega(\tilde{a}_{hk})\gamma_h\delta_k \end{aligned} \tag{10.5}$$

and

$$\begin{aligned} \Omega\left(\max_{\gamma}\left\{\min_{\delta}\{\mathbf{E}\tilde{\mathbf{P}}(\gamma,\delta)\}\right\}\right) &= \Omega(\mathbf{E}\tilde{\mathbf{P}}(\gamma^*,\delta^*)) = \Omega\left(\sum_{h}\sum_{k} \tilde{a}_{hk}\gamma_h\delta_k\right) \\ &= \Omega\left(\sum_{k}\left(\sum_{h} \tilde{a}_{hk}\gamma_h\right)\delta_k\right) = \sum_{h}\sum_{k} \Omega(\tilde{a}_{hk})\gamma_h\delta_k \end{aligned} \tag{10.6}$$

From Equation (10.5) and Equation (10.6), we obtain

$$\Omega\left(\min_{\delta}\left\{\max_{\gamma}\left\{\tilde{\mathbf{EP}}(\gamma,\delta)\right\}\right\}\right)=\Omega\left(\max_{\gamma}\left\{\min_{\delta}\left\{\tilde{\mathbf{EP}}(\gamma,\delta)\right\}\right\}\right)=\sum_{k}\left(\sum_{h}\Omega(\tilde{a}_{hk})\gamma_h\delta_k\right)\delta_k=\tilde{\mathbf{EP}}\left(\Omega(\tilde{a}_{hk})\right).$$

Hence, (γ^*, δ^*) is also the solution for SVNNMG after applying the SF for SVNN. To prove the converse part, let us assume that (γ^*, δ^*) is the solution of the SVNNMG after using the SF , i.e.,

$$\Omega\left(\min_{\delta}\left\{\max_{\gamma}\left\{\tilde{\mathbf{EP}}(\gamma,\delta)\right\}\right\}\right)=\Omega\left(\max_{\gamma}\left\{\min_{\delta}\left\{\tilde{\mathbf{EP}}(\gamma,\delta)\right\}\right\}\right)=\tilde{\mathbf{EP}}\left(\Omega\left(\tilde{a}_{hk}\right)\right).$$

Now, we can write

$$\begin{aligned}\Omega\left(\min_{\delta}\left\{\min_{\gamma}\left\{\tilde{\mathbf{EP}}(\gamma,\delta)\right\}\right\}\right)&=\tilde{\mathbf{EP}}\left(\Omega(\tilde{a}_{hk})\right)=\sum_{h}\sum_{k}\Omega(\tilde{a}_{hk})\gamma_h\delta_k\\&=\Omega\left(\sum_{k}\left(\sum_{h}\tilde{a}_{hk}\gamma_h\right)\delta_k\right)=\Omega\left(\sum_{h}\sum_{k}\tilde{a}_{hk}\gamma_h\delta_k\right)=\Omega\left(\tilde{\mathbf{EP}}(\gamma^*,\delta^*)\right)\end{aligned}$$

and

$$\begin{aligned}\Omega\left(\max_{\gamma}\left\{\min_{\delta}\left\{\tilde{\mathbf{EP}}(\gamma,\delta)\right\}\right\}\right)&=\tilde{\mathbf{EP}}\left(\Omega(\tilde{a}_{hk})\right)=\sum_{h}\sum_{k}\Omega(\tilde{a}_{hk})\gamma_h\delta_k\\&=\Omega\left(\sum_{k}\left(\sum_{h}\tilde{a}_{hk}\gamma_h\right)\delta_k\right)=\Omega\left(\sum_{h}\sum_{k}\tilde{a}_{hk}\gamma_h\delta_k\right)=\Omega\left(\tilde{\mathbf{EP}}(\gamma^*,\delta^*)\right)\end{aligned}$$

Therefore, we can write

$$\Omega\left(\min_{\delta}\left\{\max_{\gamma}\left\{\tilde{\mathbf{EP}}(\gamma,\delta)\right\}\right\}\right)=\Omega\left(\tilde{\mathbf{EP}}(\gamma,\delta)\right)=\Omega\left(\max_{\gamma}\left\{\min_{\delta}\left\{\tilde{\mathbf{EP}}(\gamma,\delta)\right\}\right\}\right).$$

Now, according to Definition 10.2.4, we have

$$\min_{\delta}\left\{\max_{\gamma}\left\{\tilde{\mathbf{EP}}(\gamma,\delta)\right\}\right\}=\tilde{\mathbf{EP}}(\gamma^*,\delta^*)=\max_{\gamma}\left\{\min_{\delta}\left\{\tilde{\mathbf{EP}}(\gamma,\delta)\right\}\right\}.$$

Hence, (γ^*, δ^*) is the optimal solution of the SVNNMG $\tilde{A}$. □

Therefore, applying *SF* as defined in Definition 10.2.4, model (10.3) and (10.4) reduce to the crisp equivalent model (10.7) and (10.8) respectively as follows.

$$\max\left\{\frac{2+\rho-\sigma-\tau}{3}\right\}$$

$$\text{s.t.,}\left(3-\prod_{h=1}^{p}(1-\mu_{hk})^{\gamma_h}-\prod_{h=1}^{p}\omega_{hk}{}^{\gamma_h}-\prod_{h=1}^{p}\nu_{hk}{}^{\gamma_h}\right)\geq(2+\rho-\sigma-\tau),k\in\{1,2,\ldots,q\}$$

$$\sum_{h=1}^{p}\gamma_h=1,\gamma_h\geq 0,\quad h\in\{1,2,\ldots,p\},$$

$$\rho\geq 0,\sigma\geq 0,\tau\geq 0,\rho+\sigma+\tau\leq 3, \tag{10.7}$$

and

$$\min\left\{\frac{2+\in-\psi-\xi}{3}\right\}$$

$$\text{s.t.,}\left(3-\prod_{k=1}^{q}(1-\mu_{hk})^{\delta_k}-\prod_{k=1}^{q}\omega_{hk}{}^{\delta_k}-\prod_{k=1}^{n}\upsilon_{hk}{}^{\delta_k}\right)\leq(2+\in-\psi-\xi),h\in\{1,2,\ldots,p\}$$

$$\sum_{k=1}^{q}\delta_k=1,\delta_k\geq 0,\quad k\in\{1,2,\ldots,q\},$$

$$\in\geq 0,\psi\geq 0,\xi\geq 0,\in+\psi+\xi\leq 3. \tag{10.8}$$

Solving model (10.7) and (10.8), we can get the optimal mixed strategy γ^* for $\mathbf{P}_1$ and δ^* for $\mathbf{P}_2$ respectively. Then, $\gamma^{*T}\tilde{A}\delta^*$ is the expected value for $\mathbf{P}_1$.

10.4.1 Algorithm for Solving SVNNMG

Step-1: Consider a MG with pay-offs of SVNNs.
Step-2: Formulate two neutrosophic mathematical programming models for $\mathbf{P}_1$ and $\mathbf{P}_2$.
Step-3: Applying *SF* in both sides of the constraint and restrictions, construct two mathematical programming models with crisp values.
Step-4: Solve the reduced models as constructed in Step-3 by using LINGO software to obtain optimal strategies for both the players.
Step-5: $\mathbf{E\tilde{P}}$ for $\mathbf{P}_1$ can be computed using optimal strategies.

The corresponding algorithmic flowchart of the proposed approach is presented in Figure 10.2.

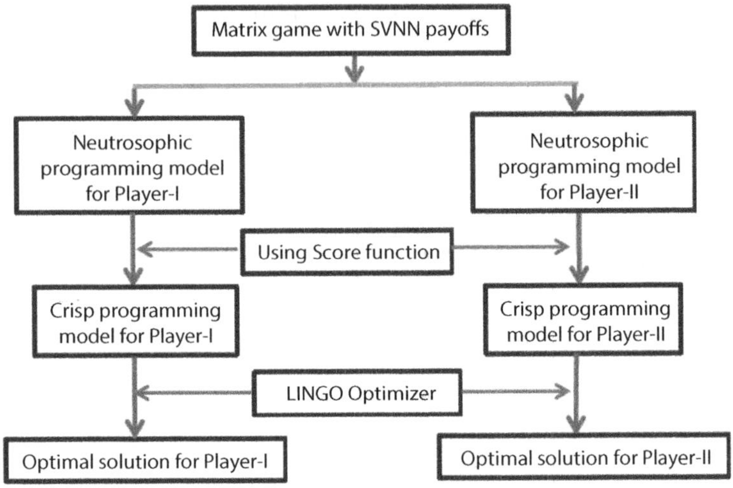

Figure 10.2 Algorithmic flowchart of the prescribed method.

10.5 Numerical Example

Here, a numerical example is solved and results are discussed to demonstrate the solution of an SVNNMG. A comparative study with an existing work is also displayed in this section.

10.5.1 A Market Share Problem

Assume that two different disinfectant manufacturing agencies $\boldsymbol{X}_1$ and $\boldsymbol{X}_2$ want to bid for the supply of the product in hospitals in the state of West Bengal, where the demand for the product is fixed. So they take some strategies mainly, like advertising to get the doctors' attention (strategy-$\boldsymbol{S}_1$), reducing the price (strategy-$\boldsymbol{S}_2$), enhancing in time delivery system (strategy-$\boldsymbol{S}_3$). Here, X_1 and X_2 are assumed as $\mathbf{P}_1$ and $\mathbf{P}_2$, and the order amount from the hospital authority can be considered as the pay-off for the agencies. As there always exists uncertainty in the healthcare equipment supply industry, the order amount can not be forecasted precisely by the marketing research departments (MRDs) of X_1 and X_2. MRD can roughly calculate the sales amount with a DOIN, along with DOM and DONM of the order amount. To come out from this unsettled situation, SVNN is applied to express the order amount of the product.

According to the MRD of X_1, the pay-offs are termed using linguistic terms as follows:

$$\tilde{A} = \begin{matrix} & S_1 & S_2 & S_3 \\ S_1 \\ S_2 \\ S_3 \end{matrix} \begin{pmatrix} VH & H & M \\ L & VH & H \\ M & VL & VH \end{pmatrix}.$$

Table 10.1 depicts the corresponding relations between linguistic terms and SVNNs.

Then, the pay-off matrix $\tilde{A}$ with SVNNs may be transformed according to Table 10.1 as follows.

$$\tilde{A} = \begin{matrix} S_1 \\ S_2 \\ S_3 \end{matrix} \overset{\begin{matrix} S_1 & & & S_2 & & & S_3 \end{matrix}}{\begin{pmatrix} \langle 0.95,0.07,0.05\rangle & \langle 0.70,0.50,0.25\rangle & \langle 0.50,0.40,0.40\rangle \\ \langle 0.25,0.30,0.70\rangle & \langle 0.95,0.07,0.05\rangle & \langle 0.70,0.50,0.25\rangle \\ \langle 0.50,0.40,0.4\rangle & \langle 0.05,0.10,0.95\rangle & \langle 0.95,0.07,0.05\rangle \end{pmatrix}}.$$

Table 10.1 Assigned SVNN corresponding to linguistic terms.

Linguistic terms	VH (Very High)	H (High)	M (Medium)	L (Low)	VL (Very Low)
SVNN	⟨0.95, 0.07, 0.05⟩	⟨0.7, 0.5, 0.25⟩	⟨0.5, 0.4, 0.4⟩	⟨0.25, 0.3, 0.7⟩	⟨0.05, 0.1, 0.95⟩

10.5.2 The Solution Procedure and Result Discussion

Our aim is to find the optimal strategies $\gamma^* = (\gamma_1, \gamma_2, \gamma_3)$ and $\delta^* = (\delta_1, \delta_2, \delta_3)$. According to models (10.7) and (10.8), the following models (10.9) and (10.10) are constructed, respectively, as follows:

$$\begin{aligned}
&\max\left\{\frac{2+\rho-\sigma-\tau}{3}\right\}\\
&\text{s.t.,}(3-0.05^{\gamma_1}0.75^{\gamma_2}0.5^{\gamma_3}-0.07^{\gamma_1}0.3^{\gamma_2}0.4^{\gamma_3}-0.05^{\gamma_1}10.7^{\gamma_2}0.4^{\gamma_3})\geq(2+\rho-\sigma-\tau),\\
&\quad(3-0.3^{\gamma_1}0.05^{\gamma_2}0.95^{\gamma_3}-0.5^{\gamma_1}0.07^{\gamma_2}0.1^{\gamma_3}-0.25^{\gamma_1}0.05^{\gamma_2}0.95^{\gamma_3})\geq(2+\rho-\sigma-\tau),\\
&\quad(3-0.5^{\gamma_1}0.3^{\gamma_2}0.05^{\gamma_3}-0.4^{\gamma_1}0.5^{\gamma_2}0.07^{\gamma_3}-0.4^{\gamma_1}0.25^{\gamma_2}0.05^{\gamma_3})\geq(2+\rho-\sigma-\tau),\\
&\quad\gamma_1+\gamma_2+\gamma_3=1,\gamma_1\geq0,\gamma_2\geq0,\gamma_3\geq0,\rho+\sigma+\tau\leq3,\rho\geq0,\sigma\geq0,\tau\geq0,
\end{aligned}\tag{10.9}$$

and

$$\begin{aligned}
&\min\left\{\frac{2+\epsilon-\psi-\xi}{3}\right\}\\
&\text{s.t.,}(3-0.05^{\delta_1}0.3^{\delta_2}0.5^{\delta_3}-0.07^{\delta_1}0.5^{\delta_2}0.4^{\delta_3}-0.05^{\delta_1}0.25^{\delta_2}0.4^{\delta_3})\leq(2+\epsilon-\psi-\xi),\\
&\quad(3-0.75^{\delta_1}0.05^{\delta_2}0.3^{\delta_3}-0.3^{\delta_1}0.07^{\delta_2}0.5^{\delta_3}-0.7^{\delta_1}0.05^{\delta_2}0.25^{\delta_3})\leq(2+\epsilon-\psi-\xi),\\
&\quad(3-0.5^{\delta_1}0.95^{\delta_2}0.05^{\delta_3}-0.4^{\delta_1}0.1^{\delta_2}0.07^{\delta_3}-0.4^{\delta_1}0.95^{\delta_2}0.05^{\delta_3})\leq(2+\epsilon-\psi-\xi),\\
&\quad\delta_1+\delta_2+\delta_3=1,\delta_1\geq0,\delta_2\geq0,\delta_3\geq0,\\
&\quad\epsilon+\psi+\xi\leq3,\epsilon\geq0,\psi\geq0,\xi\geq0.
\end{aligned}\tag{10.10}$$

Solving model (10.9) and (10.10), we have the optimal solution as $\gamma^* = (0.3802, 0.3101, 0.3097)$ for $\mathbf{P}_1$ and $\delta^* = (0.3037, 0.3286, 0.3677)$ for $\mathbf{P}_2$, respectively. Substituting the optimum solutions γ^* and δ^*, we have the $\mathbf{E\tilde{P}}$ for X_1 as $\langle 0.7711, 0.2002, 0.2066\rangle$, which is a SVNN. This implies that X_1 chooses strategy S_1, S_2, and S_3 with probability 0.3802, 0.3101 and 0.3097 respectively, to get order at a maximum scale, and X_2 chooses strategy S_1, S_2 and S_3 with probability 0.3037, 0.3286, and 0.3677, respectively, to reduce losses. In this circumstances, the share of the company X_1 in the market will be regarded as 'High'.

10.5.3 Analysis and Comparison of Results With Li and Nan's [8] Approach

Li and Nan [8] solved MG with IFS as pay-offs by developing a nonlinear programming approach. Considering the indeterminacy degree as zero, the pay-off matrix $\tilde{A}$ with SVNN becomes a pay- off matrix $\tilde{A}'$ with Atanassov's IFS, where

$$\tilde{A}' = \begin{pmatrix} \langle 0.95, 0.05 \rangle & \langle 0.70, 0.25 \rangle & \langle 0.50, 0.4 \rangle \\ (\langle 0.25, 0.70 \rangle & \langle 0.95, 0.05 \rangle & \langle 0.70, 0.25 \rangle) \\ \langle 0.50, 0.40 \rangle & \langle 0.05, 0.95 \rangle & \langle 0.95, 0.05 \rangle \end{pmatrix}.$$

Here, $\tilde{A}'$ is the pay-off matrix, which is the same as in Li and Nan's [8] work. We used the proposed technique for the pay-off matrix $\tilde{A}'$ to find the optimal solutions for the players.

Table 10.2 shows the results for the pay-off matrix $\tilde{A}'$ with Atanassov's IFSs obtained by the proposed technique discussed in this chapter and Li and Nan's [8] work.

From Table 10.2, we can say that the optimal solution for both the players obtained by the performed approach and by the technique proposed in Li and Nan [8] are approximately the same. This manifests the effectiveness and superiority of the performed technique.

Table 10.2 Results for the pay-off matrix $\tilde{A}'$.

	Li and Nan [8] (for $\lambda = 0.8$)	**Our proposed technique**
γ^*	(0.4034,0.3306,0.2660)	(0.4071,0.3323,0.2605)
δ^*	(0.2752,0.2970,0.4278)	(0.2686,0.2958,0.4355)

10.6 Conclusion

The solution procedure of an SVNNMG is developed in this work. To find the optimal strategies, two neutrosophic mathematical programming models are constructed. The *SF* of SVNNs is used to transform these models into crisp equivalent problems. Solving these problems on the LINGO platform, optimal strategies are computed for each player. A market share problem is solved using the performed method. We also illustrate the proposed method by considering the DOIN as zero and compare the results with the results obtained by Li and Nan [8], which demonstrate the reliability of our approach.

The discussed approach in this chapter does not guarantee the existence of a solution for the SVNNMG. It is also dependent on the *SF* of SVNN and various kinds of *SF* s will give various solutions. Therefore, further study is needed to investigate a more general methodology. The proposed methodology may be applied to counter different problems related to the political voting system, e-commerce, advertising, cybersecurity, etc.

References

1. Atanassov, K.T., *Intuitionistic Fuzzy Sets*, Springer-Verlag, Heidelberg, 1999.
2. Broumi, S., Bakali, A., Talea, M., Smarandache, F., Ulucay, V., Sahin, M., Dey, A., Dhar, M., Tan, R.P., Bahnasse, A., Pramanik, S., Neutrosophic sets: An overview, in: *New Trends in Neutrosophic Theory and Applications*, 2, 403–434, 2018.
3. Deli, I. and Subas, Y., A ranking method of single valued neutrosophic numbers and its application to multiattribute decision making problems. *Int. J. Mach. Learn. Cybern.*, 8, 4, 1309–1322, 2017.
4. Garai, T., Garg, H., Roy, T.K., A ranking method based on possibility mean for multi-attribute decision making with single valued neutrosophic numbers. *J. Ambient Intell. Humaniz. Comput.*, 11, 1, 5245–5258, 2020. https://doi.org/10.1007/s12652-020-01853-y.
5. Garg, H. and Nancy, New logarithmic operational laws and their applications to multiattribute decision making for single-valued neutrosophic numbers. *Cogn. Syst. Res.*, 52, 931–946, 2018.
6. Garg, H. and Nancy, Algorithms for single-valued neutrosophic decision making based on TOPSIS and clustering methods with new distance measure. *AIMS Press*, 5, 3, 2671–2693, 2020.
7. Karmakar, S., Seikh, M.R., Castillo, O., Type-2 intuitionistic fuzzy matrix games based on a new distance measure: Application to biogas-plant

implementation problem. *Appl. Soft Comput.*, 106, 2021. DOI: https://doi.org/10.1016/j.asoc.2021.107357.

8. Li, D.F. and Nan, J.X., A nonlinear programming approach to matrix games with payoffs of Atanassov's intuitionistic fuzzy sets. *Int. J. Uncertain. Fuzz.*, 17, 4, 585–607, 2009.
9. Li, D.F., An effective methodology for solving matrix games with fuzzy payoffs. *IEEE Trans. Cybern.*, 43, 2, 610–621, 2013.
10. Li, D.F. and Liu, J.C., A parameterized non-linear programming approach to solve matrix games with payoffs of I-fuzzy numbers. *IEEE Trans. Fuzzy Syst.*, 23, 4, 885–896, 2015.
11. Li, D.F., *Linear Programming Models and Methods of Matrix Games with Payoffs of Triangular Fuzzy Numbers*, Springer-Verlag, Heidelberg, 2016.
12. Seikh, M.R., Nayak, P.K., Pal, M., An alternative approach for solving fuzzy matrix games. *Int. J. Math. Comput. Sci.*, 5, 1, 79–92, 2015.
13. Seikh, M.R., Nayak, P.K., Pal, M., Matrix game with intuitionistic fuzzy payoffs. *J. Inf. Optim. Sci.*, 36, 1–2, 159–181, 2015.
14. Seikh, M.R., Karmakar, S., Nayak, P.K., Matrix games with dense fuzzy payoffs. *Int. J. Intell. Syst.*, 36, 4, 1770–1799, 2021.
15. Seikh, M.R., Karmakar, S., Xia, M., Solving matrix games with hesitant fuzzy pay-offs. *Iran. J. Fuzzy Syst.*, 17, 4, 25–40, 2020.
16. Seikh, M.R., Karmakar, S., Castillo, O., A novel defuzzification approach of Type-2 fuzzy variable to solving matrix games: An application to plastic ban problem. *Iran. J. Fuzzy Syst.*, 18, 5, 155–172, 2021.
17. Selvachandran, G., Quek, S., Smarandache, F., An extended technique for order preference by similarity to an ideal solution (TOPSIS) with maximizing deviation method based on integrated weight measure for single-valued neutrosophic sets. *Symmetry*, 10, 7, 236, 2018. doi.org/10.3390/sym10070236.
18. Smarandache, F., *A Unifying Field in Logics. Neutrosophy: Neutrosophic Probability, Set and Logic*, American Research Press, Rehoboth, 1998.
19. Sodenkamp, M.A., Tavana, M., Di Caprio, D., An aggregation method for solving group multi-criteria decision-making problems with single-valued neutrosophic sets. *Appl. Soft Comput.*, 71, 715–727, 2018.
20. Verma, T. and Kumar, A., Ambika methods for solving matrix games with Atanassov's intuitionistic fuzzy pay-offs. *IEEE Trans. Fuzzy Syst.*, 26, 1, 270–283, 2018.
21. Wang, H., Smarandache, F., Zhang, Y.Q., Sunderraman, R., Single valued neutrosophic sets. *Multispace Multistructure*, 4, 410–413, 2010.
22. Wei, G. and Zhang, Z., Some single-valued neutrosophic Bonferroni power aggregation operators in multiple attribute decision making. *J. Ambient Intell. Humaniz. Comput.*, 10, 863–882, 2019.
23. Wei, G. and Wei, Y., Some single-valued neutrosophic dombi prioritized weighted aggregation operators in multiple attribute decision making. *J. Intell. Fuzzy Syst.*, 35, 1–13, 2018.

24. Xing, Y. and Qiu, D., Solving traingular intuitionistic fuzzy matrix game by applying the accuracy function method. *Symmetry*, 11, 10, 1258, 2019. doi:10.3390/sym11101258.
25. Xue, W., Xu, Z., Zeng, X.J., Solving matrix games based on ambika method with hesitant fuzzy information and its application in the counterterrorism issue. *Appl. Intell.*, 51, 3, 1227–1243, 2020. https://doi.org/10.1007/s10489-020-01759-4.
26. Yang, Z. and Song, Y., Matrix game with payoffs represented by triangular dual hesitant fuzzy numbers. *Int. J. Comput. Commun. Control*, 15, 3, 1–11, 2020.
27. Ye, J., Trapezoidal neutrosophic set and its application to multiple attribute decision-making. *Neural Comput. Appl.*, 26, 1157–1166, 2015.
28. Ye, J., Multiple attribute group decision-making method with single-valued neutrosophic interval number information. *Int. J. Syst. Sci.*, 50, 1, 152–162, 2019.

11

A Novel Score Function-Based EDAS Method for the Selection of a Vacant Post of a Company with q-Rung Orthopair Fuzzy Data

Utpal Mandal* **and Mijanur Rahaman Seikh**†

Department of Mathematics, Kazi Nazrul University, Asansol, India

Abstract

In this chapter, we introduce an improved score function in the q-rung orthopair fuzzy (q-ROF) environment. Then, we establish several suitable properties of our introduced score function. Next, we compare our introduced score function with some existing score functions under the q-ROF environment, by considering a few numerical examples. The comparison result shows that the introduced score function outperforms the existing ones. Then, we utilize our introduced score function to develop a q-ROF evaluation based on the distance from the average solution (EDAS) method for solving the multiattribute decision-making problem. Further, we present a numerical problem of selection for the vacant post of a company to verify the validity and effectiveness of our proposed model. Finally, we elaborate on the superiority of our proposed model by comparing it with several existing methods.

Keywords: q-rung orthopair fuzzy sets, score function, EDAS method, MADM

11.1 Introduction

Multiattribute decision making (MADM) is the procedure of identifying the best choice from several given alternative choices based on multiple criteria. The purpose of MADM is to facilitate decision makers to deal with

**Corresponding author*: utpalmandal2204@gmail.com

†*Corresponding author*: mrseikh@ymail.com

Sharmistha Ghosh, M. Niranjanamurthy, Krishanu Deyasi, Biswadip Basu Mallik, and Santanu Das (eds.) Mathematics and Computer Science Volume 1, (231–250) © 2023 Scrivener Publishing LLC

such types of problems. This problem does not have a unique optimal solution. So, the decision makers' preference is necessary to determine the best choice from the available collection of alternatives. These types of MADM problems arise in many real-life situations. Several researchers are actively engaged in solving MADM problems with different approaches. The EDAS method [4] is considered as a proficient tool to handle MADM problems. The main concept of the EDAS method is to calculate positive distance from average solution (PDAS) and negative distance from average solution (NDAS) for each of the available alternatives based on the average solution. The preference of an alternative is proportionate to its PDAS value and disproportionate to its NDAS value.

But, to handle real-life MADM problems, the decision makers' face vagueness and uncertainty. Fuzzy set theory [24] is a useful tool to deal with such situations. But, the fuzzy set provides a degree of support only. So, modifying the fuzzy set theory, Atanossov [1] established the concept of the intuitionistic fuzzy set (IFS). The IFS associated with the MADM problems consists of the "degree of support" (DS) and the "degree of against" (DA). The sum of the DS and the DA in IFS always belong to the unit interval. Several researchers are attracted to study the procedures of MADM problems with IFSs and their dynamic applications [6, 16]. Zou *et al.* [27] developed geometric operators under the intuitionistic fuzzy environment and utilized them to MADM. Joshi [7] introduced biparametric exponential information measure for IFSs.

Later, Yager [22] introduced Pythagorean fuzzy set (PFS). The sum of the squares of DS and DA in PFS is bounded by one. Clearly, PFS can handle vagueness of data more efficiently than IFS. MADM with PFS has received a great amount of interest from many researchers [14, 17, 21].

Thereafter, IFS and PFS are considered as two specific subsets of q-rung orthopair fuzzy set (q-ROFS) [23]. The q-ROFS is characterized by DS and DA, satisfying that sum of the qth power of them is bounded by 1. Clearly, the range of applicability of the q-ROFS is much bigger because of the flexibility of the parameter q. Under the abundant progress of our society in the last few years, the research on MADM problems has been succeeded to attain more and more attention for its validity, precision, and accuracy along with its simplicity [9, 18]. Liu and Wang [11] built several aggregation operators with q-ROF data and applied them in MADM. Garg and Chen [3] presented neutrality aggregation operators and applied them in MAGDM under q-ROF fuzzy environment. Liu and Wang [12] introduced Maclaurin symmetric mean operators under q-ROF environment.

In fuzzy MADM problems, collective information obtained for each of the available alternatives needs to be defuzzified to rank them in proper order. Generally, for the defuzzification of the collective information of available alternatives, it was required to calculate the score and accuracy functions for q-ROFSs. Therefore, score function is an efficient tool for handling MADM. But if the score values for distinct alternatives be the same, then the task was little bit difficult for the complexity of calculations. Now, the paper is focused mainly on removing the complexities in calculations of finding the ranking of alternatives in an MADM. For this, we construct a novel score function, which is able to easily distinguish between the alternatives. Also, we develop a new approach for the EDAS method [4] using our proposed score function to handle the MADM problem with q-ROF information. Many researchers extended EDAS method to solve MADM problem in several fuzzy environments, for instance, fuzzy EDAS [5], EDAS with intuitionistic fuzzy environment [8], PF environment-based EDAS [13], EDAS with q-ROF environment [10], etc.

The major augmentations of the present chapter are as follows:

- A novel score function under q-ROF environment is introduced.
- Introduced score function is compared with some existing ones to establish its superiority.
- The desirable features of our introduced score function are disclosed.
- Using the proposed score function, the EDAS method is extended to MADM under q-ROF environment.
- Proposed method is applied for solving a real life numerical problem of selecting the most promising candidate to fill the vacant post of a company.
- The new model, suggested in this paper, is compared with existing well-known methods, which highlight the superiority of our proposed procedure.

The remaining part of the article can be organized in the following sequence:

In section 11.2, the necessary basic definitions are discussed, which are prerequisites to understand the concept of the article. A novel score function of q-ROFN is introduced, as well as compared with the existing score functions in section 11.3. We construct q-ROF EDAS method to solve MADM problem using proposed score function in section 11.4. In section 11.5, we have scrutinized the validity of our introduced procedure with its successful application to a real life problem. Section 11.6 holds the comparison analysis and the results which clarifies the superiority of our method over the existing ones. Finally, we conclude in section 11.7.

11.2 Preliminaries

Here, we have recalled some basic definitions and preliminaries.

Definition 2.1. [1] An IFS I over the universal set $\mathbb{U}$ is defined by

$$I = \{\langle i, \mu_I(i), \nu_I(i)\rangle : i \in \mathbb{U}\}$$

where $\mu_I(i), \nu_I(i) \in [0, 1]$ and satisfies $0 \leq \mu_I(i) + \nu_I(i) \leq 1, \forall i \in \mathbb{U}$. $\mu_I(i)$ and $\nu_I(i)$ represents the DS and DA of $i \in \mathbb{U}$, respectively. The hesitancy degree is denoted by $\pi_I(i) = 1 - \mu_I(i) - \nu_I(i)$.

Definition 2.2. [22] A PFS F over the universal set $\mathbb{U}$ is described as

$$F = \{\langle f, a_F(f), b_F(f)\rangle : f \in \mathbb{U}\}$$

where $a_F(f), b_F(f) \in [0, 1]$ and satisfies $0 \leq (a_F(f))^2 + (b_F(f))^2 \leq 1, \forall f \in \mathbb{U}$. $a_F(f)$ and $b_F(f)$ represents the DS and DA of $f \in \mathbb{U}$ respectively. The hesitancy degree $\psi_F(f)$ is defined by $\psi_F(f) = \sqrt{1-(a_F(f))^2-(b_F(f))^2}$.

Definition 2.3. [23] A q-ROFS H over the universal set $\mathbb{U}$ is defined by

$$H = \{\langle e, a_H(e), b_H(e)\rangle : e \in \mathbb{U}\}$$

where $a_H(e), b_H(e) \in [0, 1]$ and satisfies $0 \leq (a_H(e))^q + (b_H(e))^q \leq 1, \forall e \in \mathbb{U}$. $a_H(e)$ and $b_H(e)$ represents the DS and DA of $e \in \mathbb{U}$, respectively. The hesitancy degree $\psi_H(e)$ is defined by $\psi_H(e) = q\sqrt{1-(a_H(e))^q-(b_H(e))^q}$. For convenience, Yager [23] called the pair $(a_H(e), b_H(e))$ a q-ROFN and denoted by $h = (a_h, b_h)$.

The differences among the spaces of IFSs, PFSs and q-ROFSs are exhibited in Figure 11.1. From Figure 11.1, it is clear that, the area of application increases along with the increment of the value of q. Therefore, in handling the uncertainty of MADM problems q-ROFS is more competent than IFS and PFS.

Now, we recall some basic operations of q-ROFNs which we will be using later.

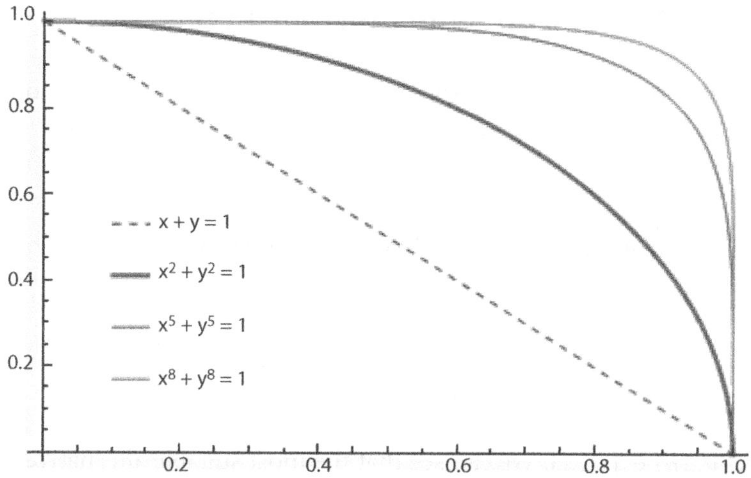

Figure 11.1 Comparison among the spaces of IFSs, PFSs and q-ROFS [q = 5, 8].

Definition 2.4. [23] The q-ROFNs $h_1 = \left(a_{h_1}, b_{h_1}\right)$ and $h_2 = \left(a_{h_2}, b_{h_2}\right)$ satisfies the following relations:

1. $h_1 \oplus h_2 = \left(\sqrt[q]{a_{h_1}^q + a_{h_2}^q - a_{h_1}^q a_{h_2}^q}, b_{h_1} b_{h_2}\right).$
2. $h_1 \otimes h_2 = \left(a_{h_1} a_{h_2}, \sqrt[q]{b_{h_1}^q + b_{h_2}^q - b_{h_1}^q b_{h_2}^q}\right).$
3. $\lambda h_1 = \left(\sqrt[q]{1-\left(1-a_{h_1}^q\right)^\lambda}, b_{h_1}^\lambda\right)$, for $\lambda > 0 \in \mathbb{R}$.
4. $h_1^\lambda = \left(a_{h_1}^\lambda, \sqrt{1-\left(1-b_{h_1}\right)^\lambda}\right)$, for $\lambda > 0 \in \mathbb{R}$.

Definition 2.5. [23] The aggregated value of the q-ROFNs $h_b = \left(a_{h_b}, b_{h_b}\right)$, $b = 1(1)n$ using q-ROF weighted averaging (q-ROFWA) operator is given by

$$q-ROFWA(h_1, h_2, \ldots, h_n) = \bigoplus_{b=1}^{n} h_b \delta_b = \left(\sqrt[q]{1 - \prod_{b=1}^{n}(1-a_{h_b}^q)^{\delta_b}}, \prod_{b=1}^{n} b_{h_b}^{\delta_b}\right) \tag{11.1}$$

where the weight vector corresponding to h_b is δ_b with $\sum_{b=1}^{n} \delta_b = 1$.

11.3 A Novel Score Function of *q*-ROFNs

Along the present section, some existing *q*-ROF score functions are discussed, and a novel score function is constructed under *q*-ROF environment.

11.3.1 Some Existing *q*-ROF Score Functions

For the *q*-ROFN $h = (a_h, b_h)$, Liu [11] defined the score function as:

$$\Phi_1(h) = a_h^q - b_h^q \tag{11.2}$$

where $\Phi_1(h) \in [-1, 1]$. We can see that the most suitable alternative which fulfils the decision makers' expectations is whose corresponding Φ -functional value is maximum.

Wei *et al.* [19] presented a modified score function as follows:

$$\Phi_2(h) = \frac{1 + a_h^q - b_h^q}{2} \tag{11.3}$$

Example 1. *Suppose that* $h_1 = (0.39, 0.39)$ *and* $h_2 = (0.47, 0.47)$ *be two q-ROFNs. If the score functions, which were introduced by Liu [11] and Wei et al. [19], be used to select the best alternative between* h_1 and h_2 *then we have* $\Phi_1(h_1) = \Phi_1(h_2) = 0$ *and* $\Phi_2(h_1) = \Phi_2(h_2) = 0.5$. *Hence, we can not distinguish between* h_1 and h_2. *Therefore, these score functions fail to rank the q-ROFNs* h_1 and h_2 *in this case.*

So, to overcome the above-mentioned disadvantages of Φ_1 *and* Φ_2, *Liu [11] and Wei et al. [19] used the accuracy function* Δ *presented by Liu [11], to distinguish q-ROFNs whose score values are equal. The accuracy function* Δ *is given by:*

$$\Delta(h) = a_h^q + b_h^q \tag{11.4}$$

where $\Delta(h) \in [0, 1]$

Remark 1. *If the score function* $\Phi_1(h)$ *is greater, then corresponding q-ROFN h is larger. When the score values for two or more q-ROFNs are same, then we need to find the accuracy function* $\Delta(h)$. *If* Δ *gives the maximum value for* *more than one alternative, then any of the alternatives, whose corresponding Δ-functional value is maximum. The score function Φ_2 [19] face the same situation. Therefore, these existing score functions fails to rank the alternatives in a proper order.*

From Example 1 and Remark 1, we realize that the existing score functions cannot sufficiently handle every decision-making situations. So, we need to define a more capable score function with higher degree of precision. For this we define a new score function which can easily distinguish among alternatives.

11.3.2 A Novel Score Function of q-ROFNs

The above section proves that the score functions introduced by Liu [11] and Wei *et al.* [19] are not sufficient to distinguish q-ROFNs. If the score value of two distinct q-ROFNs turn to be same then it is necessary to implement accuracy function to rank them. To overcome this complexity of calculation, we define an effective score function to easily discriminate the q-ROFNs.

Definition 3.1. The score function is defined for a q-ROFN h = (a_h, b_h), as

$$\Phi(h) = e^{a_h^q} + e^{-b_h^q} \tag{11.5}$$

Now we give an example to check the validity of our proposed score function.

Example 2. If $h = (0.39, 0.39)$ then for $q = 2$ we obtain $\Phi(h) = 2.023$.

Theorem 3.1. *$\Phi(h)$ is proportionate to a_h and disproportionate to b_h for any q-ROFN $h = (a_h, b_h)$.*

Proof: Differentiating both sides of Equation (11.5) partially with respect to a_h we get:

$$\frac{\partial \Phi(h)}{\partial a_h} = q a_h^{q-1} e^{a_h^q}$$

Here, $\dfrac{\partial \Phi(h)}{\partial a_h} \geq 0$ as $q > 0$ and $a_h^q \geq 0$.

Again, differentiating both sides of Equation (11.5) partially with respect to b_h we obtain:

$$\frac{\partial\Phi(h)}{\partial b_h} = -q b_h^{q-1} e^{-b_h^q}$$

Clearly, $\frac{\partial\Phi(h)}{\partial b_h} \geq 0$ as $q > 0$ and $b_h^q \geq 0$.

Therefore, we have for $h = (a_h, b_h)$, $\Phi(h)$ is proportionate to a_h and disproportionate to b_h. □

Theorem 3.2. For a *q-ROFN* $h = (a_h, b_h)$, our proposed score function $\Phi(h)$ satisfies the following properties:

1. $\Phi(h) = e + 1$ *if and only if* $h = (1, 0)$.
2. $\Phi(h) = \frac{e+1}{e}$ *if and only if* $h = (1, 0)$.
3. $\frac{e+1}{e} \leq \Phi(h) \leq e+1$.

Proof: From Theorem 3.1, we come to the point that $\Phi(h)$ is minimized at $h = (1, 0)$ and maximized at $h = (0, 1)$. So, minimum and maximum values of $\Phi(h)$ are $\frac{e+1}{e}$ and $e + 1$ respectively. Consequently, $\frac{e+1}{e} \leq \Phi(h) \leq e+1$.

Theorem 3.3. *Let* $h_1 = (a_{h_1}, b_{h_1})$ *and* $h_2 = (a_{h_2}, b_{h_2})$ *be two q-ROFNs, where* $a_{h_1} > a_{h_2}$ *and* $b_{h_1} < b_{h_2}$ *then* $\Phi(h_1) > \Phi(h_2)$

Proof: From Theorem 3.1 we have Φ is proportionate to a_h and disproportionate to b_h.

Therefore, $a_{h_1} > a_{h_2}$ and $b_{h_1} < b_{h_2}$ together implies that $\Phi(h_1) > \Phi(h_2)$. □

We have shown the feasibility, superiority and applicability of our introduced score function Φ in ranking q-ROFNs, by comparing it with existing score functions Φ_1 and Φ_2. We exhibit the comparison results in Table 11.1.

Table 11.1 shows that our introduced score function Φ ranks the alternatives successfully while other score functions fail. For example, for $q = 3$, $h_1 = (0.39, 0.39)$ and $h_2 = (0.47, 0.47)$ we have $\Phi_1(h_1) = \Phi_2(h_2) = 0$, $\Phi_2(h_1) = \Phi_2(h_2) = 0$. Therefore, Φ_1 and Φ_2 cannot distinguish h_1 and h_2. But, if we utilize our proposed score function Φ then $\Phi(h_1) = 2.003$ and $\Phi(h_2) = 2.010$. So, $h_1 < h_2$. Therefore, the insufficiencies of the existing score

Table 11.1 Comparison among introduced score function and existing score functions.

Examples	Value of the parameters q	Φ_1 [11]	Ranking	Φ_2 [19]	Ranking	Φ	Ranking
$h_1 = (0.5, 0.5)$	$q = 1$	$\Phi_1(h_1) = 0$, $\Phi_2(h_2) = 0$	$h_1 = h_2$	$\Phi_1(h_1) = 0.5$, $\Phi_2(h_2) = 0.5$	$h_1 = h_2$	$\Phi_1(h_1) = 2.255$, $\Phi_2(h_2) = 2.090$	$h_1 > h_2$
and	$q = 2$	$\Phi_1(h_1) = 0$, $\Phi_2(h_2) = 0$	$h_1 = h_2$	$\Phi_1(h_1) = 0.5$, $\Phi_2(h_2) = 0.5$	$h_1 = h_2$	$\Phi_1(h_1) = 2.062$, $\Phi_2(h_2) = 2.008$	$h_1 > h_2$
$h_2 = (0.3, 0.3)$	$q = 3$	$\Phi_1(h_1) = 0$, $\Phi_2(h_2) = 0$	$h_1 = h_2$	$\Phi_1(h_1) = 0.5$, $\Phi_2(h_2) = 0.5$	$h_1 = h_2$	$\Phi_1(h_1) = 2.015$, $\Phi_2(h_2) = 2$	$h_1 > h_2$
$h_1 = (0.39, 0.39)$	$q = 1$	$\Phi_1(h_1) = 0$, $\Phi_2(h_2) = 0$	$h_1 = h_2$	$\Phi_1(h_1) = 0.5$, $\Phi_2(h_2) = 0.5$	$h_1 = h_2$	$\Phi_1(h_1) = 2.154$, $\Phi_2(h_2) = 2.225$	$h_1 > h_2$
and	$q = 2$	$\Phi_1(h_1) = 0$, $\Phi_2(h_2) = 0$	$h_1 = h_2$	$\Phi_1(h_1) = 0.5$, $\Phi_2(h_2) = 0.5$	$h_1 = h_2$	$\Phi_1(h_1) = 2.023$, $\Phi_2(h_2) = 2.049$	$h_1 > h_2$
$h_2 = (0.47, 0.47)$	$q = 3$	$\Phi_1(h_1) = 0$, $\Phi_2(h_2) = 0$	$h_1 = h_2$	$\Phi_1(h_1) = 0.5$, $\Phi_2(h_2) = 0.5$	$h_1 = h_2$	$\Phi_1(h_1) = 2.003$, $\Phi_2(h_2) = 2.010$	$h_1 > h_2$
$h_1 = (0.45, 0.45)$	$q = 1$	$\Phi_1(h_1) = 0$, $\Phi_2(h_2) = 0$	$h_1 = h_2$	$\Phi_1(h_1) = 0.5$, $\Phi_2(h_2) = 0.5$	$h_1 = h_2$	$\Phi_1(h_1) = 2.205$, $\Phi_2(h_2) = 2.162$	$h_1 > h_2$
and	$q = 2$	$\Phi_1(h_1) = 0$, $\Phi_2(h_2) = 0$	$h_1 = h_2$	$\Phi_1(h_1) = 0.5$, $\Phi_2(h_2) = 0.5$	$h_1 = h_2$	$\Phi_1(h_1) = 2.041$, $\Phi_2(h_2) = 2.025$	$h_1 > h_2$
$h_2 = (0.4, 0.4)$	$q = 3$	$\Phi_1(h_1) = 0$, $\Phi_2(h_2) = 0$	$h_1 = h_2$	$\Phi_1(h_1) = 0.5$, $\Phi_2(h_2) = 0.5$	$h_1 = h_2$	$\Phi_1(h_1) = 2.008$, $\Phi_2(h_2) = 2.004$	$h_1 > h_2$

functions are removed by our proposed score function, which make our score function Φ more powerful. The introduced score function successfully discriminates the alternatives when other score functions be failure.

11.4 EDAS Method for *q*-ROF MADM Problem

Here, we extend the EDAS method [4] under q-ROF environment to solve MADM problems. The EDAS method mainly ranks the alternatives and provides the best alternative of the decision makers while solving MADM problems containing conflicting attribute. EDAS method is very easy to apply for its simplicity of calculations.

For an MADM problem, let the set $D = \{D_1, D_2,\ldots, D_m\}$ contains m alternatives and the set $H = \{H_1, H_2,\ldots, H_n\}$ contains n attributes. The evaluation of the alternative $D_p \in D$ $(p = 1\ (1)m)$ on attribute $H_r \in H$ $(r = 1\ (1)n)$ is expressed with $h_{pr} = \left(a_{h_{pr}}, b_{h_{pr}}\right)$ where $a_{h_{pr}} \in [0,1]$ and $b_{h_{pr}} \in [0,1]$ and $\left(a_{h_{pr}}\right)^q + \left(b_{h_{pr}}\right)^q \leq 1$, where q is any positive integer. Let δ_r $(r = 1\ (1)\ n)$ be the weight of the attribute H_r $(r = 1\ (1)n)$ such that $\sum_{r=1}^{n} \delta_r = 1$. Let $A = (h_{pr})_{m\times n} = \left(\left(a_{h_{pr}}, b_{h_{pr}}\right)\right)_{m\times n}$ be the q-ROF decision matrix, which is exhibited in Table 11.2.

The modification of the conventional EDAS method to deal MADM problems with q-ROFNs is proposed below.

At first, the decision matrix need to be normalized as both of the benefit and cost attribute may be present simultaneously in the decision matrix. These two types of criteria works in opposite direction. The performance of the benefit attribute improves when the attribute value increases. But the performance of the cost attribute declines with the increment of the

Table 11.2 q-ROF decision matrix.

	H_1	H_2	...	H_4
D_1	$\left(a_{h_{11}}, b_{h_{11}}\right)$	$\left(a_{h_{12}}, b_{h_{12}}\right)$	...	$\left(a_{h_{1n}}, b_{h_{1n}}\right)$
D_2	$\left(a_{h_{21}}, b_{h_{21}}\right)$	$\left(a_{h_{22}}, b_{h_{22}}\right)$	...	$\left(a_{h_{2n}}, b_{h_{2n}}\right)$
⋮	⋮	⋮	⋱	⋮
D_3	$\left(a_{h_{m1}}, b_{h_{m1}}\right)$	$\left(a_{h_{m2}}, b_{h_{m2}}\right)$	...	$\left(a_{h_{mn}}, b_{h_{mn}}\right)$

attribute value. So to remove such ambiguity and ensure the compatibility of all the attributes, we make all the attributes of benefit type. We normalize the q-ROF decision matrix A into $A' = (h'_{pr})_{m \times n}$ by Equation (11.6).

$$h'_{pr} = \begin{cases} \left(a_{h_{pr}}, b_{h_{pr}}\right), & \text{when } D_p \text{ be benefit attribute;} \\ \left(b_{h_{pr}}, a_{h_{pr}}\right), & \text{when } D_p \text{ be cost attribute.} \end{cases} \tag{11.6}$$

Next, we determine the average solution $\omega = (\omega_1, \omega_2, \ldots, \omega_m)$ in conformity with all attributes utilizing Definition 2.4, which is exhibited as follows:

$$\omega_p = \frac{1}{n} \bigoplus_{r=1}^{n} h_{pr} = \left(\sqrt[q]{1 - \left(\prod_{r=1}^{n} \left(1 - a_{h_{pr}}^q\right) \right)^{\frac{1}{n}}}, \prod_{r=1}^{n} \left(b_{h_{pr}}\right)^{\frac{1}{n}} \prod_{r=1}^{n} \left(b_{h_{pr}}\right)^{\frac{1}{n}} \right). \tag{11.7}$$

Thereafter, we compute PDAS S^+ with $S^+ = \left(S_{pr}^+\right)_{m \times n}$ and NDAS S^- with $S^- = \left(S_{pr}^-\right)_{m \times n}$ corresponding to different types of attributes. The values of S_{pr}^+ and S_{pr}^- are obtained by using the following equations:

$$S_{pr}^+ = \frac{max\left\{1, e^{\Phi(h_{pr}) - \Phi(\omega_p)}\right\}}{e^{\Phi(\omega_p)}} \tag{11.8}$$

and

$$S_{pr}^- = \frac{max\left\{1, e^{\Phi(\omega_p) - \Phi(h_{pr})}\right\}}{e^{\Phi(\omega_p)}}. \tag{11.9}$$

Let $\mathfrak{R}_p^+$ and $\mathfrak{R}_p^-$ be the weighted sum of S_{pr}^+ and S_{pr}^-, respectively. We compute $\mathfrak{R}_p^+$ and $\mathfrak{R}_p^-$ by Equations (11.10) and (11.11), respectively.

$$\mathfrak{R}_p^+ = \sum_{r=1}^{n} \delta_r S_{pr}^+ \tag{11.10}$$

$$\mathfrak{R}_p^- = \sum_{r=1}^{n} \delta_r S_{pr}^- \tag{11.11}$$

Let $N\mathfrak{R}_p^+$ and $N\mathfrak{R}_p^-$ be the normalized values of $\mathfrak{R}_p^+$ and $\mathfrak{R}_p^-$, respectively, where

$$N\mathfrak{R}_p^+ = \frac{\mathfrak{R}_p^+}{\max_i \{\mathfrak{R}_p^+\}} \tag{11.12}$$

$$N\mathfrak{R}_p^- = 1 - \frac{\mathfrak{R}_p^-}{\max_i \{\mathfrak{R}_p^-\}} \tag{11.13}$$

Finally, we derive the integrative appraisal score $\wp_p$, $\forall p = 1\,(1)m$, which are given by:

$$\wp_p = \frac{1}{2}\left(N\mathfrak{R}_p^+ + N\mathfrak{R}_p^-\right) \tag{11.14}$$

where $0 \leq \wp_p \leq 1$.

As the appraisal score increases, a particular alternative approaches to the ideal solution accordingly. We rank the alternatives depending upon the scale of appraisal score. Higher appraisal score fetches better rank for the alternatives. The alternative corresponding to the greatest appraisal score, is supposed to be the optimal solution of the respective MADM problem.

The above-mentioned procedure of q-ROF EDAS method can be summarized as follows:

Step I: Identify the attributes and alternatives, denoted the set of attributes and set of alternatives by $H = \{H_1, H_2, \ldots, H_n\}$ and $D = \{D_1, D_2, \ldots, D_m\}$ respectively.
Step II: Pull the decision maker's opinion to get the evaluation of each alternative on the attributes that are constructing the decision matrix $A = (h_{pr})_{m\times n} = \left(\left(a_{h_{pr}}, b_{h_{pr}}\right)\right)_{m\times n}$ under q-ROF environment.

Step III: Normalized decision matrix A' is obtained by utilizing Equation (11.6). Note that this step can be omitted when all the attributes are of benefit type.
Step IV: Calculate the average solution $\omega = (\omega_1, \omega_2, \ldots, \omega_m)$ by Equation (11.7).
Step V: Using Equations (11.8) and (11.9) calculate the PDAS S^+ and NDAS S^- respectively.
Step VI: Calculate the weighted sum $\mathfrak{R}_p^+$ and $\mathfrak{R}_p^-$ for every alternatives using Equations (11.10) and (11.11) respectively.
Step VII: Compute normalized values of $\mathfrak{R}_p^+$ and $\mathfrak{R}_p^-$ utilizing Equations (11.12) and (11.13) respectively.
Step VIII: Derive the integrative appraisal score $\wp_p$ for all alternatives by Equation (11.14).
Step IX: Determine the ranking order of the alternatives with the help of appraisal score values $\wp_p$.

For better view, we exhibit the above steps in Figure 11.2.

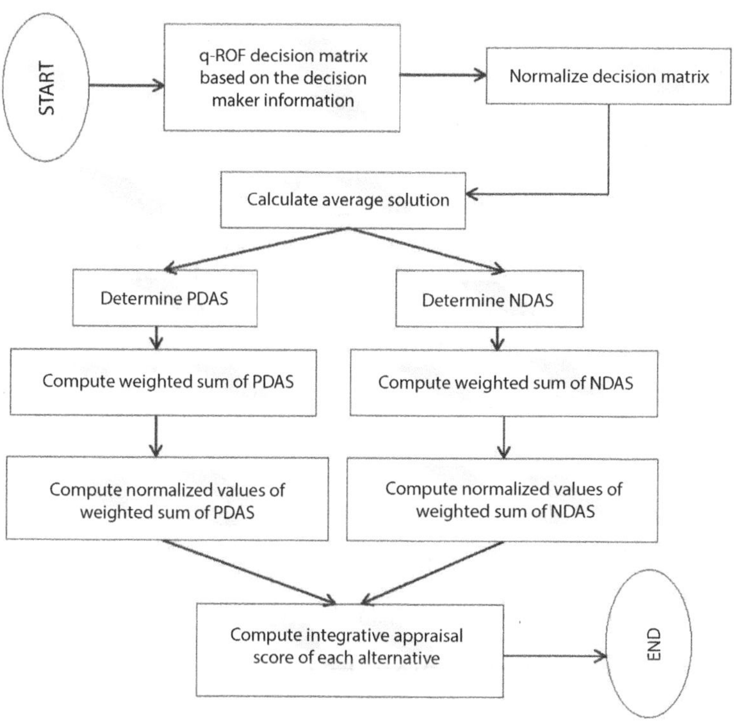

Figure 11.2 Flowchart of the proposed model.

11.5 Numerical Example

Here, we construct a numerical problem with q-ROF data for the selection of a vacant post of a company to illustrate the above mentioned approach.

Suppose ABC Limited is a private company. Recently one faithful aged worker has retired. So the company requires one efficient worker for that vacant post. To choose the worker, they organized a selection procedure with one written test and an interview. Suppose three candidates D_1, D_2, D_3 cracked the written examination and are called for the interview, so now, the company has three feasible choices D_p ($p = 1, 2, 3$) for the post. Now, to choose the most capable one with desirable qualities, the management board of the company selected the following attributes:

- H_1: Educational qualification.
- H_2: Score in the written test.
- H_3: Past job experiences.
- H_4: Physical fitness.

Meanwhile, H_1, H_2, H_3 are benefit attributes, and H_4 is the cost attribute. To avoid conflict between them, the management board set weights for each of the attributes according to their requirements. The management board has set weights of all attributes as 0.30, 0.25, 0.25, 0.20. Based on the decision maker's opinion, the DS and DA of the alternative D_p ($p = 1, 2, 3$) to the attribute H_r ($r = 1, 2, 3, 4$) are exhibited in Table 11.3.

To find the most favorable candidate D_p ($p = 1, 2, 3$), we utilize our proposed q-ROF EDAS method, which can be executed by the following:

Table 11.3 Decision maker's opinion.

	H_1	H_2	H_3	H_4
D_1	(0.9,0.3)	(0.7,0.5)	(0.5,0.6)	(0.4,0.3)
D_2	(0.4,0.7)	(0.8,0.2)	(0.7,0.1)	(0.8,0.5)
D_3	(0.8,0.4)	(0.6,0.5)	(0.8,0.3)	(0.5,0.6)

Step I: The q-ROF decision matrix given by the members of the management board is given in Table 11.3.
Step II: After taking the decision matrix from selection criteria, we normalize the q-ROF decision matrix according to Equation (11.6). The normalized decision matrix is:

$$\begin{pmatrix} (0.9,0.3) & (0.7,0.5) & (0.5,0.6) & (0.3,0.4) \\ (0.4,0.7) & (0.8,0.2) & (0.7,0.1) & (0.5,0.8) \\ (0.8,0.4) & (0.6,0.5) & (0.8,0.3) & (0.6,0.5) \end{pmatrix}.$$

Step III: The average solutions of all the alternatives are given by $\omega_1 = (0.7777, 0.435)$, $\omega_2 = (0.528, 0.325)$ and $\omega_3 = (0.764, 0.416)$.
Step IV: The PDAS and NDAS are exhibited in matrix S^+ and S^- respectively.

$$S^+ = \left(S^+_{pr}\right)_{3\times4} = \begin{pmatrix} 0.1136 & 0.7022 & 0.7022 & 0.7022 \\ 0.1085 & 0.2084 & 0.1620 & 0.1627 \\ 0.0805 & 0.7186 & 0.0857 & 0.7186 \end{pmatrix}$$

$$S^- = \left(S^-_{pr}\right)_{3\times4} = \begin{pmatrix} 0.7022 & 0.0888 & 0.0755 & 0.0753 \\ 0.1675 & 0.1085 & 0.1085 & 0.1627 \\ 0.7186 & 0.1094 & 0.7186 & 0.1094 \end{pmatrix}$$

Step V: Weighted sum $\mathfrak{R}^+_p$ of S^+ and $\mathfrak{R}^-_p$ of S^- for all the alternatives are executed in Table 11.4.
Step VI: For each alternative the values of $\mathfrak{R}^+_p$ and $\mathfrak{R}^-_p$ are normalized through Equation (11.12) and Equation (11.13) respectively and they are given in Table 11.4.
Step VII: Compute the integrative appraisal score $\wp_p$ ($p = 1, 2, 3$) for all candidates utilizing Equation 11.14, which are given by: $\wp_1 = 0.6673$, $\wp_2 = 0.3724$, and $\wp_3 = 0.5$.

From the last step, we have seen that appraisal score of the alternative D_1 is maximum among the alternatives D_1, D_2, and D_3. Hence, D_1 is the best candidate to be selected by the company.

Table 11.4 Weighted sum and normalized weighted sum of S^+ and S^-.

p	$\mathfrak{R}_p^+$	$\mathfrak{R}_p^-$	$N\mathfrak{R}_p^+$	$N\mathfrak{R}_p^-$
1	0.5264	0.2667	0.9371	0.3976
2	0.1567	0.1370	0.2789	0.3724
3	0.5617	0.4444	1	0

11.6 Comparative Analysis

In the literature, Li *et al.* [10] developed EDAS method for solving MADM problems under q-ROF environment. We call this method Li *et al.*'s method. Liu [11] defined q-ROFWA operator to aggregate q-ROF information and develop a model tosolve MADM problems. We call this method Liu's method. Here, we compare our proposed EDAS method with the above-mentioned methods. Li *et al.*'s method and Liu's method are dependent on the score function Φ_1 [11].

Liu's method produces the ranking order of the alternatives as $D_1 > D_3 > D_2$ and D_1 is the best candidate to be selected by the company. The integrative appraisal score $\wp_p$ obtained by utilizing Li *et al.*'s method are $\wp(D_1) = 0.4230$, $\wp(D_2) = 0.5$, $\wp(D_3) = 0.4603$. Based on these integrative appraisal scores $\wp_p$, the ranking order of the three candidates is $D_2 > D_3 > D_1$, so the best candidate is D_2.

Table 11.5 Comparison table.

Methods	Decision values	Ranking	Optimal alternative
q-ROFWA [$r = 3$] [11]	$\Phi_1(D_1) = 0.335$, $\Phi_1(D_2) = 0.247$, $\Phi_1(D_3) = 0.326$	$D_1 > D_3 > D_2$	D_1
EDAS [$r = 3$] [10]	$\wp(D_1) = 0.4230$, $\wp(D_2) = 0.5$, $\wp(D_3) = 0.4603$	$D_2 > D_3 > D_1$	D_2
EDAS (Proposed) [$r = 3$]	$\wp(D_1) = 0.6673$, $\wp(D_2) = 0.372$, $\wp(D_3) = 0.5$	$D_1 > D_3 > D_2$	D_1

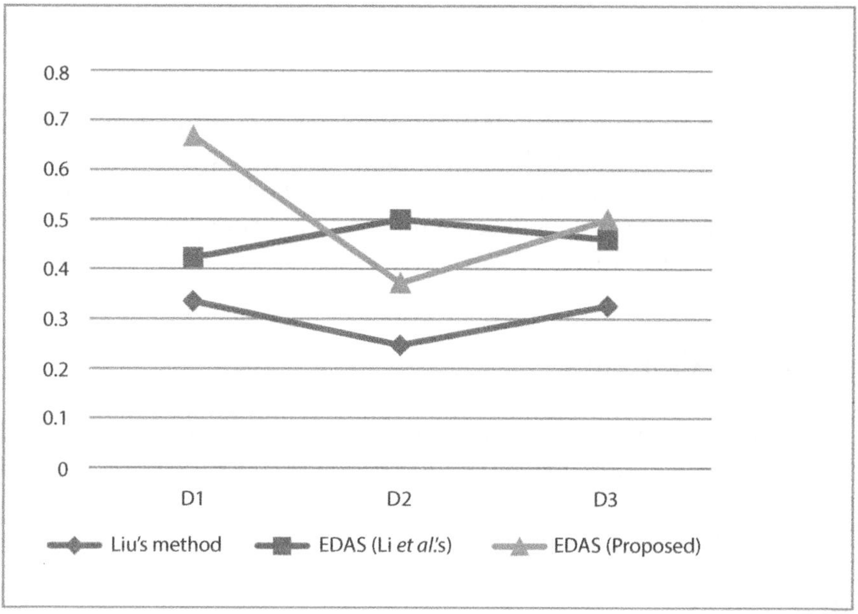

Figure 11.3 Ranking of the alternatives.

Again, the ranking order obtained by utilizing our proposed EDAS method is $D_1 > D_3 > D_2$. Therefore, the best candidate according to our proposed method is D_1.

We have presented the comparison analysis in Table 11.5. Also for better understanding, we have illustrated the comparison results graphically in Figure 11.3. Table 11.5 highlights that Liu's method and our proposed method give the same ranking result, but, Li *et al.*'s method and our proposed method give different ranking order.

11.7 Conclusions

Along this chapter, we have suggested a modified score function of q-ROFSs and thoroughly discussed its suitable properties. Then, we establish the superiority of our introduced score functions by comparing it with some existing score functions. Next, we extend the EDAS method under the q-ROF environment to solve MADM problem by using the proposed score function. Then, a numerical example of candidate selection in a company is given to show the validity and effectiveness of our proposed model. Finally, our proposed method is compared to the existing methods and the comparison analysis points out the validity, as well as the superiority of our proposed method.

In the future, our proposed score function and proposed model can be implemented to solve many MADM problems, e.g., project evaluation, supplier selection, medical diagnosis, etc. Also, we can extend the TOPSIS method [25], VIKOR method [2], COPRAS method [15], and so on, with q-ROFNs using our proposed score function. We can hybridize EDAS to AHP/Entropy/any other methods.

Acknowledgments

The work was financially supported by Council of Scientific and Industrial Research (CSIR), India having grant 09/1269(0001)/2019-EMR I, Dated 02/07/2019.

References

1. Atanassov, K.T., Intuitionistic fuzzy sets. *Fuzzy Sets Syst.*, 20, 1, 87–96, 1986.
2. Chen, C.T., Huang, S.F., Hung, W.Z., Linguistic VIKOR method for project evaluation of ambient intelligence product. *J. Ambient Intell. Humaniz. Comput.*, 2018. DOI: https ://doi.org/10.1007/s1265 2-018-0889-x.
3. Garg, H. and Chen, S.M., Multiattribute group decision making based on neutrality aggregation operators of q-rung orthopair fuzzy sets. *Inf. Sci.*, 517, 427–447, 2020.
4. Ghorabaee, M.K., Zavadskas, E.K., Olfat, L., Turskis, Z., Multi-criteria inventory classification using a new method of evaluation based on distance from average solution (EDAS). *Informatica*, 26, 3, 435–451, 2015.
5. Ghorabaee, M.K., Zavadskas, E.K., Amiri, M., Turskis, Z., Extended EDAS method for fuzzy multi-criteria decision-making: An application to supplier selection. *Int. J. Comput. Commun. Control*, 11, 3, 358–371, 2016.
6. Gumusab, S., Kucukvarc, M., Tataria, O., Intuitionistic fuzzy multi-criteria decision making framework based on life cycle environmental, economic and social impacts: The case of U.S. wind energy. *Sustain. Prod. Consum.*, 8, 78–92, 2016.
7. Joshi, R., A new multi-criteria decision-making method based on intuitionistic fuzzy information and its application to fault detection in a machine. *J. Ambient Intell. Humaniz. Comput.*, 11, 739–753, 2020.
8. Kahraman, C., Ghorabaee, M.K., Zavadskas, E.K., Onar, S.C., Yazdani, M., Oztaysi, B., Intuitionistic fuzzy EDAS method: An application to solid waste disposal site selection. *J. Environ. Eng. Landsc. Manage.*, 25, 1, 1–12, 2017.

9. Khan, M.J., Kumam, P., Shutaywi, M., Knowledge measure for the q-rung orthopair fuzzy sets. *Int. J. Intell. Syst.*, 36, 2, 628–655, 2021.
10. Li, Z., Wei, G., Wang, R., Wu, J., EDAS method for multiple attribute group decision making under *q*-rung orthopair fuzzy environment. *Technol. Econ. Dev. Eco.*, 26, 1, 1–17, 2019.
11. Liu, P. and Wang, P., Some *q*-rung orthopair fuzzy aggregation operators and their applications to multiple attribute decision making. *Int. J. Intell. Syst.*, 33, 2, 259–280, 2018.
12. Liu, P. and Wang, Y., Multiple attribute decision making based on *q*-rung orthopair fuzzy generalized Maclaurin symmetic mean operators. *Inf. Sci.*, 518, 181–210, 2020.
13. Peng, X. and Selvachandran, G., Pythagorean fuzzy set: State of the art and future directions. *Artif. Intell. Rev.*, 52, 1873–1927, 2019.
14. Ren, P.J., Xu, Z.S., Gou, X.J., Pythagorean fuzzy TODIM approach to multi-criteria decision making. *Appl. Soft Comput.*, 42, 246–259, 2016.
15. Roy, J., Sharma, H.K., Kar, S., Zavadskas, E.K., Saparauskas, J., An extended COPRAS model for multi-criteria decision-making problems and its application in web-based hotel evaluation and selection. *Econ. Res. Ekon. Istraz.*, 32, 1, 219–253, 2019.
16. Seikh, M.R. and Mandal, U., Intuitionistic fuzzy Dombi aggregation operators and their application to multiple attribute decision-making. *Granul. Comput.*, 6, 3, 473–488, 2021.
17. Wang, L. and Li, N., Pythagorean fuzzy interaction power Bonferroni mean aggregation operators in multiple attribute decision making. *Int. J. Intell. Syst.*, 35, 1, 150–183, 2020.
18. Wang, J., Wei, G., Wei, C., Wei, Y., MABAC method for multiple attribute group decision making under *q*-rung orthopair fuzzy environment. *Def. Technol.*, 16, 1, 208–216, 2020.
19. Wei, G., Gao, H., Wei, Y., Some *q*-rung orthopair fuzzy Heronian mean operators in multiple attribute decision making. *Int. J. Intell. Syst.*, 33, 7, 1426–1458, 2018.
20. Xue, W., Xu, Z., Zhang, X., Tian, X., Pythagorean fuzzy LINMAP method based on the entropy theory for railway project investment decision making. *Int. J. Intell. Syst.*, 33, 93–125, 2018.
21. Yager, R.R., Pythagorean fuzzy subsets. *Proc Joint IFSA World Congress & NAFIPS Annual Meeting*, pp. 57–61, 2013.
22. Yager, R.R., Generalized orthopair fuzzy sets. *IEEE Trans. Fuzzy Syst.*, 25, 1222–1230, 2017.
23. Zadeh, L.A., Fuzzy sets. *Inf. Control*, 8, 3, 338–352, 1965.
24. Zhang, X.L. and Xu, Z.S., Extension of TOPSIS to multiple criteria decision making with pythagorean fuzzy sets. *Int. J. Intell. Syst.*, 29, 12, 1061–1078, 2014.

25. Zou, X.Y., Chen, S.M., Fan, K.Y., Multiple attribute decision making using improved intuitionistic fuzzy weighted geometric operators of intuitionistic fuzzy values. *Inf. Sci.*, 535, 242–253, 2020.

12

Complete Generalized Soft Lattice

Manju John* **and Susha D.**

Department of Mathematics, Catholicate College, Pathanamthitta, Kerala, India

Abstract
A soft set is a representation of vague data and can be used to study real-life problems containing various types of uncertainty. Complete lattices have many applications in various fields. Here, we introduce the concepts of complete gs-lattice, s-closure system and s-Moore family. Here, we discuss some ways to construct complete gs-lattices and a representation of a complete gs-lattice as an s-closure system. Fixed point theory is a useful tool in applied mathematics. We conclude this study with Tarski's fixed point theorem in generalized soft lattices (gs-lattices).

Keywords: Complete gs lattice, s-closure system, s-Moore family

12.1 Introduction

Theories of probability, fuzzy set, rough set [18], vague set, etc. are very useful for dealing with different kind of uncertainty. Zadeh's fuzzy set theory [20] is most applicable and useful among all these theories, which was proposed in 1965. A soft set is a fairly accurate description of an object, and this description can be made convenient by choosing parametrization as words, sentences, real numbers, functions, etc., which reduces the problems like setting membership functions. Molodstov showed that this theory has various applications in the fields, like theory of games, integration, and the theory of measure. Many researchers like Maji [15], Shabir [19], Cagman [1, 5] and Feng [11] have contributed more to soft set theory.

In [4], G. Birkhoff presented more concepts to the theory of lattice. Many concepts of mathematics were described in terms of soft elements

**Corresponding author*: manjujohn75@gmail.com

Sharmistha Ghosh, M. Niranjanamurthy, Krishanu Deyasi, Biswadip Basu Mallik, and Santanu Das (eds.) Mathematics and Computer Science Volume 1, (251–270) © 2023 Scrivener Publishing LLC

[9] by Das *et al.* (also in [8] and [7]). Based on soft elements, we introduced gs-group in [12] and gs-lattice, soft valuation and soft metric lattice in the study of Manju and Susha [13]. Here, we consider more concepts in the theory of lattice in soft context in soft element terms. In this study, mainly, we define complete generalized soft lattice with some properties. Also, we can see that complete gs-lattices can be constructed many ways like from soft power set, complete lattices, and s-closure systems. All these can be used for modeling vague or uncertain data.

12.2 Soft Sets and Soft Elements—Some Basic Concepts

Here, we consider soft sets over a universal set U and C, D as parameter sets.

1. [16] A pair (G, D), where $G: D \rightarrow P(U)$ is a function, is called a soft set over U.
2. [11] The soft set (G,D) is a soft subset of (H, C) if, $D \subseteq C$ and for each $d \in D$, $G(d) \subseteq H(d)$ and we write $(G,D) \tilde{\subseteq} (H,C)$.
   (b) (G, D) and (H, C) are soft equal if $(G,D) \tilde{\subseteq} (H,C)$ and $(H,C) \tilde{\subseteq} (G,D)$. we write $(G, D) = (H, C)$.
3. [14] Any soft subset (G, D) over U, where $D \subseteq C$ can be considered as the soft set (K, C) by defining

$$K(c) = \begin{cases} G(c) & \text{if } c \in D \\ \phi & c \in C \setminus D \end{cases}$$

Hence, we can consider any soft set over U with a common parameter set.

4. [15] An absolute soft set over U is of the form (G, D), where $G(d) = U, \forall d \in D$ and is null if $G(d) = \phi, \forall d \in D$.
   We denote the absolute and null soft sets, respectively, as $\tilde{U}$ or $(\tilde{U},C)$ and $\tilde{\Phi}$ or $(\tilde{\Phi},C)$.
5. [3] The concept Cartesian product of soft sets (G, C) and (H, D) is the soft set $(K, C \times D)$, where K is a function from $C \times D \rightarrow P(U \times U)$ given by $K(c, d) = G(c) \times H(d)$.

6. [9] A soft element $\tilde{g}$ in $(\tilde{U},D)$ is a function from D to U. We denote it by $\tilde{g}\tilde{\in}\tilde{U}$.
   $\tilde{g}$ is a soft element of (G, D) if $\tilde{g}(d)\in G(d),\forall d\in D$ and we write $\tilde{g}\tilde{\in}(G,D)$.
   Any soft set (G,D) over U with $G(d)\neq\phi,\forall d\in D$ we have $G(d)=\{\tilde{g}(d):\tilde{g}\tilde{\in}(G,D)\}$. Usual notation for soft elements are $\tilde{u},\tilde{v},\tilde{w}$, etc.
7. [8] $S(\tilde{U})=\{(G,D)\tilde{\subseteq}\tilde{U}:G(d)\neq\phi \text{ for all d in D}\}\cup\{(\tilde{\Phi},D)\}$.
   SE (*G*, *D*) denotes the set of all soft elements of (G,D). The soft set SS(B) generated by B, a subset of SE(G,D), is defined by *SS* (*B*) = (*K*, *D*), where $K(d)=\{\tilde{u}(d):\tilde{u}\in B\}$.
8. [6] The operations elementary complement, the elementary union and the elementary intersection of elements of $S(\tilde{U})$ are respectively given by

$$(G,D)^{\Subset}=SS\{\tilde{u}\tilde{\in}(\tilde{U},D):\tilde{u}\tilde{\in}(G,D)^{c}\},\text{where }(G,D)^{c}=(G^{c},D)$$

and $G^{c}(d)=U-G(d),\text{for all d in D}$

$$(G,D)\Cup(H,D)=SS\{\tilde{u}\tilde{\in}(\tilde{U},D):\tilde{u}\tilde{\in}(G,D)\text{ or }\tilde{u}\tilde{\in}(H,D)\}\text{ and}$$

$$(G,D)\Cap(H,D)=SS\{\tilde{u}\tilde{\in}(\tilde{U},D):\tilde{u}\tilde{\in}(G,D)\text{ and }\tilde{u}\tilde{\in}(H,D)\}.$$

12.3 gs-Posets and gs-Chains

The mathematical structure partially ordered sets (posets) are fundamental in various fields. A soft subset (G,D) of $\tilde{U}$ together with $\tilde{\preceq}$, a soft partial ordering on (*G*, *D*) is called a generalized soft poset or gs-poset [see [13]].

Proposition 12.3.1. *Let ((P, ≼ D) be any poset. Then* $(\tilde{P},D)$ *is a gs poset. Also, any* $(G,D)\in S(\tilde{P})$ *is a generalized soft poset where D is any parameter set.*

Proof *Consider* $(\tilde{P},D)$*. Define* $\tilde{\preceq}$ *on* $SE(\tilde{P})$ *by* $\tilde{u}\tilde{\preceq}\tilde{v}$ *if* $\tilde{u}(d)\preceq\tilde{v}(d),\forall d\in D$.

i. *Since (P, ≼) is a poset and* $\tilde{u}(d)\in P$ *for each d in D, we have* $\tilde{u}(d)\preceq\tilde{u}(d)$ *for each d in D.*
   i.e., for each soft element $\tilde{u}$ *in (P, D), we have* $\tilde{u}\tilde{\preceq}\tilde{u}$.
   Hence, $\tilde{\preceq}$ *in* $\tilde{P}$ *is a soft reflexive relation.*

ii. *Also,* $(P, \preccurlyeq)$ *is a poset and* $\tilde{u}(d), \tilde{v}(d) \in P, \forall d \in D$, *we have* $\tilde{u}(d) \preceq \tilde{v}(d)$ *and* $\tilde{v}(d) \preceq \tilde{u}(d)$ *implies* $\tilde{u}(d) = \tilde{v}(d)$ *for each d in D.*
Hence, $\tilde{u} \tilde{\preceq} \tilde{v}$ *and* $\tilde{v} \tilde{\preceq} \tilde{u}$ *implies* $\tilde{u} \tilde{=} \tilde{v}$ *in* $\tilde{P}$.
So $\tilde{\preceq}$ *is a soft antisymmetric relation on* $\tilde{P}$.

iii. *If* $d \in D, \tilde{u}(d) \preceq \tilde{v}(d)$ *and* $\tilde{v}(d) \preceq \tilde{w}(d)$ *implies* $\tilde{u}(d) \preceq \tilde{w}(d)$.
So $\tilde{u} \tilde{\preceq} \tilde{v}$ *and* $\tilde{v} \tilde{\preceq} \tilde{w}$ *and this implies* $\tilde{u} \tilde{\preceq} \tilde{w}$ *and* $\tilde{\preceq}$ *is a soft transitive relation on* $\tilde{P}$.
Hence, $\tilde{P}$ *is a gs poset generated by* $(\tilde{P}, \tilde{\preceq})$.

Since $SE(G,D) \subseteq SE(\tilde{P},D)$ and $G\ (d) \subseteq P$ for each d in D, as in the above we get any $(G,D) \in S(\tilde{P})$ is a gs poset.

In a similar manner, by defining $\tilde{\preceq}$ on $SE(\tilde{U})$ by $\tilde{u} \tilde{\preceq} \tilde{v}$ iff $\tilde{u}(c) \preceq_c \tilde{v}(c)$ for each c in C, we get the following.

Proposition 12.3.2. *The family* $\{\preccurlyeq_c : c \in C\}$, *where* $\preccurlyeq_c$ *is a soft partial ordering on U for each c in C, generate a gs-poset. In general, any collection of partial orderings* $\{\preccurlyeq_c : c \in C\}$ *on any soft set (G, C) in* $S(\tilde{U})$ *generate a gs poset.*

Definition 12.3.3. *A soft subset (G, C) of* $S(\tilde{U})$ *of a gs poset* $((\tilde{U},C), \tilde{\preceq})$ *is a gs subposet of* $\tilde{U}$ *if* $((G,C), \tilde{\preceq})$ *is a gs poset.*

Example 12.3.4. *Consider* $(\tilde{R},C)$, *where R, the set of all real numbers with* $\leq$ *(less than or equal to) on R.*
Define $\tilde{\leq}$ *on* $\tilde{R}$ *by* $\tilde{r} \tilde{\leq} \tilde{s}$ iff $\tilde{r}(c) \leq \tilde{s}(c)$, *for all c in C. Then* $\tilde{R}$ *is a gs poset.*

Theorem 12.3.5. *Let* $(\tilde{U},C)$ *be an absolute gs poset with a soft partial ordering* $\tilde{\preceq}$ *in* $\tilde{U}$ *such that* $\tilde{u} \tilde{\prec} \tilde{u}$ *for no soft element* $\tilde{u}$ of $\tilde{U}$. *Then* $\widetilde{u_1} \tilde{\prec} \widetilde{u_2}$ *and* $\widetilde{u_2} \tilde{\prec} \widetilde{u_3}$ *implies* $\widetilde{u_1} \tilde{\prec} \widetilde{u_3}$ *(Here we consider* $\widetilde{u_1} \tilde{\preceq} \widetilde{u_2}$ *and* $\widetilde{u_1} \neq \widetilde{u_2}$ *as* $\widetilde{u_1} \tilde{\prec} \widetilde{u_2}$*).*

Proof *Consider the absolute gs poset* $(\tilde{U},C)$ *with a soft partial ordering* $\tilde{\preceq}$ *and let* $\tilde{u} \tilde{\prec} \tilde{u}$ *for no soft element* $\tilde{u} \tilde{\in} \tilde{U}$.

Let $\widetilde{u_1} \tilde{\prec} \widetilde{u_2}$. *and* $\widetilde{u_2} \tilde{\prec} \widetilde{u_3}$.

So $\widetilde{u_1} \neq \widetilde{u_2}$ *and* $\widetilde{u_2} \neq \widetilde{u_3}$ *and hence* $\widetilde{u_1} \neq \widetilde{u_3}$.

But $\widetilde{u_1} \tilde{\preceq} \widetilde{u_3}$. *(By soft transitive property)*

i.e., $\widetilde{u_1} \tilde{\prec} \widetilde{u_3}$ or $\widetilde{u_1} \tilde{=} \widetilde{u_3}$

$$\therefore \widetilde{u_1} \tilde{\prec} \widetilde{u_3}.$$

Definition 12.3.6. *Consider a gs poset* $(\tilde{U},C)$ *with the soft partial ordering* $\tilde{\preceq}$. *If* $\widetilde{m_1}\tilde{\preceq}\tilde{u}$ *and* $\tilde{u}\tilde{\preceq}\widetilde{m_2}\forall\tilde{u}\tilde{\in}\tilde{U}$ *for some* $\widetilde{m_1},\widetilde{m_2}\tilde{\in}\tilde{U}$ *then* $\widetilde{m_1}$ *is the least soft element and* $\widetilde{m_1}$ *is called the soft infimum or inf* $\tilde{U}$ *and* $\widetilde{m_2}$ *is the greatest soft element and is caled the soft supremum or sup* $\tilde{U}$ *of* $\tilde{U}$.

Theorem 12.3.7. *A gs poset contain at most one least (greatest) soft element(If inf* $\tilde{U}$ *or sup* $\tilde{U}$ *exists, then it is unique).*

Proof *If possible, let* $\widetilde{q_1}$ *and* $\widetilde{q_2}$ *be two least soft elements of the gs-poset* $\tilde{U}$.

Then $\tilde{u}\preceq\widetilde{q_1}$ *and* $\tilde{u}\preceq\widetilde{q_2},\forall\tilde{u}\tilde{\in}\tilde{U}$.

By taking $\tilde{u}=\widetilde{q_2}$*and* $\tilde{u}=\widetilde{q_1}$, *we get* $\widetilde{q_2}\preceq\widetilde{q_1}$ *and* $\widetilde{q_1}\preceq\widetilde{q_2}$ $\widetilde{q_1}=\widetilde{q_2}$.

Hence, a gs poset contain at most one least soft element.

Similarly, we can prove that a gs poset contain at most one greatest soft element.

Definition 12.3.8. *If the least soft element* $\tilde{0}$ *and greatest soft element* $\tilde{1}$ *exist in a soft poset* $(\tilde{U},C)$, *then they are called universal soft bounds of* $\tilde{U}$.

Example 12.3.9. *Let* I = [0, 1] *and consider the gs poset* $(\tilde{I},C)$ *generated by* $(I, \leq)$ *with a nonempty parameter set* C. *Then* $\tilde{0}$ *and* $\tilde{1}$, *where* $\tilde{0}(c)=0$ *and* $\tilde{1}(c)=1,\forall c\in C$ *are universal soft bounds for the gs poset* $(\tilde{I},C)$.

Definition 12.3.10. *A gs-poset* $((G,D),\tilde{\preceq})$ *is called a soft total ordering on* (G, D) *if* $\tilde{u}\tilde{\preceq}\tilde{v}$ *or* $\tilde{v}\tilde{\preceq}\tilde{u}\forall\tilde{u},\tilde{v}\tilde{\in}(G,D)$.

Example 12.3.11. *Consider the gs poset* $(\tilde{Z},B)$, *where* $B = \{e_1, e_2\}$ *generated by the poset* $(Z, \leq)$.

Let $(G,B)\tilde{\subseteq}(\tilde{Z},B)$ *be given by* $G(e_1) = \{2\}$, $G(e_2) = \{4, 8, 12\}$. *Then* $SE(G,B)=\{\tilde{s},\tilde{p},\tilde{q}\}$, *where* $\tilde{s}(e_1)=2,\tilde{s}(e_2)=4$, $\tilde{p}(e_1)=2,\tilde{p}(e_2)=8$, $\tilde{q}(e_1)=2,\tilde{q}(e_2)=12$.

Here $\tilde{s}\tilde{\leq}\tilde{p}\leq\tilde{q}$.

Hence, (G, B) is a gs-chain.

Note: *All gs posets are not gs chains.*

Example 12.3.12. Consider the gs poset generated by '|' (divides) on Z^+ with the parameter set $T = \{t_1, t_2, t_3\}$. Define the soft partial ordering | on

$\tilde{Z}^+$ by $\tilde{u}|\tilde{v}$ *iff* $\tilde{u}(t)|\tilde{v}(t)$, for all t in T. Here, $\tilde{Z}^+$ *is not a gs chain.*

Proof Consider $\tilde{u}, \tilde{v} \in \widetilde{Z^+}$ with $\tilde{u}(t) = 2$ *and* $\tilde{v}(t) = 3, \forall t \in T$. *Then* $\tilde{u}(t)$ *does not divide* $\tilde{v}(t), \forall t \in T$.

Hence, $\tilde{u}$ does not divide $\tilde{v}$ *and so* $\tilde{Z}^+$ *is not a gs chain.*

Theorem 12.3.13. *(Soft anticircularity)* $\tilde{q}_1 \tilde{\preceq} \tilde{q}_2 \tilde{\preceq} \ldots \tilde{\preceq} \tilde{q}_n \tilde{\preceq} \tilde{q}_1$ *implies* $\tilde{q}_1 = \tilde{q}_2 = \ldots = \tilde{q}_n$, *where* $\tilde{q}_1, \tilde{q}_2, \ldots, \tilde{q}_n$ *are soft elements of any gs poset.*

Proof *Since* $\tilde{q}_1 \tilde{\preceq} \tilde{q}_2 \tilde{\preceq} \tilde{q}_1$, *we get* $\tilde{q}_1 \tilde{\preceq} \tilde{q}_2$ *and* $\tilde{q}_2 \tilde{\preceq} \tilde{q}_1$.

So $\tilde{q}_1 \tilde{=} \tilde{q}_2$.

Similarly, we get $\tilde{q}_1 \tilde{=} \tilde{q}_3, \tilde{q}_1 \tilde{=} \tilde{q}_4, \ldots$ *and* $\tilde{q}_1 \tilde{=} \tilde{q}_n$.

Hence, $\tilde{q}_1 \tilde{=} \tilde{q}_2 \tilde{=} \ldots \tilde{=} \tilde{q}_n$.

Definition 12.3.14. *The soft elements* $\tilde{u}$ *and* $\tilde{v}$ *of a gs poset are called comparable if* $\tilde{u} \leq \tilde{v}$ *or* $\tilde{v} \leq \tilde{u}$ *holds and soft incomparable if neither* $\tilde{u} \leq \tilde{v}$ *or* $\tilde{v} \leq \tilde{u}$.

Example 12.3.15. *Let A* = {1, 2} *and* $(\tilde{R}, \leq, A)$ *be the gs poset generated by* $\leq$ *on R.*

Let $\tilde{u}(a) = a^2$ *and* $\tilde{v}(a) = a^2 + 1, \forall a \in A$.

Here $\tilde{u}$ *and* $\tilde{v}$ *are soft comparable.*

Consider $(G, A) = SS(\{\tilde{u}, \tilde{v}\})$. *Then* $SE(G, A) = \{\tilde{u}, \tilde{v}, \tilde{u}_1, \tilde{u}_2\}$, *with* $\tilde{u}_1(1) = 1$, $\tilde{u}_1(2) = 5$ *and* $\tilde{u}_2(1) = 2, \tilde{u}_2(2) = 4$.

Here $\tilde{u}$ *and* $\tilde{u}_1$ *are soft comparable. Also* $\tilde{u}_2$ *and* $\tilde{v}$ *are soft comparable. But* $\tilde{u}_1$ *and* $\tilde{u}_2$ *are soft incomparable.*

Theorem 12.3.16. *Any soft finite (having a finite number of soft elements) gs-chain has a minimal soft element and a maximal soft element.*

Proof *Consider the gs-chain (G, C) with n soft elements* $\widetilde{u_1}, \widetilde{u_2}, \ldots, \widetilde{u_n}$.

Let $\tilde{x} = \widetilde{u_1}$. *Then for i* = 2, 3, …, *n do the following:*

If $\tilde{x} \tilde{>} \widetilde{u_k}$ *then replace* $\tilde{x}$ *by* $\widetilde{u_k}$ *and no change otherwise.*

Hence, *we get* $\tilde{x} \tilde{=} \widetilde{u_j}$ *for some j and for which* $\tilde{u} \tilde{\leq} \widetilde{u_k}$ *for k* = 1, 2, …, *n*. *So* $\tilde{x} \tilde{=} \widetilde{u_j}$ *becomes the minimal element of (G, C)*

In the same manner, we get (G, C) has a maximal element.

Example 12.3.17. *In example 12.3.11, (G, B) have the soft elements* $\tilde{s}$ *and* $\tilde{q}$ *which are respectively the minimal and maximal soft elements.*

12.4 Soft Isomorphism and Duality of gs-Posets

Here we discuss the concepts soft isomorphism and dual soft isomorphism. Main result is the duality principle related to gs posets.

Definition 12.4.1. *A soft isotone (order preserving) mapping from a gs poset (G, C) to a gs poset (H, D) is an isotone mapping* $\tilde{g}: SE(G,C) \rightarrow SE(H,D)$.

i.e., Let $\tilde{\preceq}$ *and* $\tilde{\precsim}$ *be respectively soft partial orderings on (G, C) and (H, D),* $\tilde{u} \tilde{\in} (G,C)$ *and* $\tilde{v} \tilde{\in} (H,D)$ *then if* $\tilde{u} \tilde{\preceq} \tilde{v} \Rightarrow \tilde{g}(\tilde{u}) \tilde{\precsim} \tilde{g}(\tilde{v})$, *we say that the mapping* $\tilde{g}$ *is soft isotone.*

Definition 12.4.2. *A soft isomorphism between two gs posets (G, C) and (H, D) is a bijective function* $f: SE(G,C) \rightarrow SE(H,D)$ *which is order preserving. Then we called (G, C) and (H, D) are soft isomorphic and we write* $(G,C) \hat{=} (H,D)$.

Theorem 12.4.3. *Every soft finite chain (G, C) of n soft elements is soft isomorphic with the soft chain (N, C) with n soft elements* $\widetilde{u_j}$ *given by* $\widetilde{u_j}(c) = j$ for $j = 1$ *to n*, where $N(c) = \{1, 2, 3, \ldots, n\}$ *with a singleton parameter set* $C = \{c\}$.

Proof *Let the soft chain (G, C) of n soft elements be such that* $\tilde{u}_1 \leq \tilde{u}_2 \leq \ldots \leq \tilde{u}_n$ *and define* $\tilde{g}: SE(G,C) \rightarrow \{\tilde{1}, \tilde{2}, \ldots, \tilde{n}\}$ *by* $\tilde{g}(\tilde{u}_k) = \tilde{k}$, for $k = 1, 2, 3, \ldots, n$.

Then clearly it is a bijection and $\tilde{x}_k \leq \tilde{x}_m$ *iff* $\bar{k} \leq \tilde{m}$ *iff* $\tilde{g}(\tilde{u}_k) \leq \tilde{g}(\tilde{u}_m)$ *for k,* $m = 1, 2, 3, \ldots, n$.

Definition 12.4.4. *Consider a soft relation R. Then the converse of R is the soft relation* $\tilde{R}$, which *is given by* $\tilde{u}\tilde{R}\tilde{v}$ *if and only if* $\tilde{v}R\tilde{u}$.

Theorem 12.4.5. *Converse of any soft partial ordering is itself a soft partial ordering.*

Proof $\tilde{R}$, *the converse of the soft partial ordering R on a gs poset (G, C) is given by* $\tilde{u}\tilde{R}\tilde{v}$ *iff* $\tilde{v}R\tilde{u}$.

(i)For $\tilde{u} \tilde{\in} (G,C), \tilde{u}R\tilde{u}$ *and so we have* $\tilde{u}\tilde{R}\tilde{u}$ *for any* $\tilde{u} \tilde{\in} (G,C)$
(ii)Let $\tilde{u}\tilde{R}\tilde{v}$ *and* $\tilde{v}R\tilde{u}$.
$\Rightarrow \tilde{v}R\tilde{u}$ *and* $\tilde{u}\tilde{R}\tilde{v}$. *(∵ by definition)*
$\tilde{u} \hat{=} \tilde{v}$. *(∵ R is soft antisymmetric.)*
(iii)Let $\tilde{u}\tilde{R}\tilde{v}$ *and* $\tilde{v}\tilde{R}\tilde{w}$.

$\Rightarrow \tilde{v}R\tilde{u}$ *and* $\tilde{w}R\tilde{v}$. ($\because$ *by definition)*
arrow$\tilde{w}R\tilde{u}$ ($\because$ *R is soft transitive)*
$\Rightarrow \tilde{u}\tilde{R}\tilde{w}$. (by definition)

From (i),(ii) and (iii) we get the converse of a soft partial ordering is a soft partial ordering.

Definition 12.4.6. *The dual of a gs poset ((G, C), R) is the gs poset* $((G,C),\tilde{R})$, *where* $\tilde{R}$ *is the converse of the soft partial ordering R.*

Note: *Dual concepts of soft lower bound, soft minimal element, soft infimum etc. in a gs poset become the soft upper bound, soft maximal element, soft supremum etc. in its dual poset. A dual statement of a gs-poset is the statement obtained by replacing all concepts occurring in the statement by its dual concepts. Thus if a statement holds for a gs poset, its dual holds for the dual of the gs poset and we get the following result.*

Theorem 12.4.7. *(Duality principle for gs posets)Dual of a statement is true if the statement is true for all gs posets.*

Note *By duality principle, for proving a theorem and its dual we have enough to prove any one of them.*

Note *Let* $((G,C),\tilde{\preceq})$ *be a gs poset. Then* $(G, C)^*$ *denotes its dual and is the gs poset* $((G,C),\tilde{\succeq})$, *where* $\tilde{\succeq}$ *is the converse of the soft partial ordering* $\tilde{\preceq}$. *Here* $((G, C)^*)^* = (G, C)$.

Definition 12.4.8. *A soft function p from a gs poset (L, C) to a gs poset (H, D) is soft antitone if* $\tilde{l}_1 \tilde{\preceq} \tilde{l}_2$ *implies* $p(\tilde{l}_1) \tilde{\succeq} p(\tilde{l}_2)$ *and* $p(\tilde{l}_1) \tilde{\preceq} p(\tilde{l}_2)$ *implies* $\tilde{l}_1 \tilde{\succeq} \tilde{l}_2$.

Definition 12.4.9. *A dual soft isomorphism between two gs posets (G, C) to (H, D) is a soft bijective antitone function between (G, C) to (H, D).*

Example 12.4.10. *Consider the gs-poset* $((\tilde{R},\tilde{\leq}),D)$ *generated by (R,* $\leq$*) (usual meaning) and* $\tilde{\leq}$ *defined by* $\widetilde{q_1} \tilde{\leq} \widetilde{q_2}$ *iff* $\widetilde{q_1}(d) \leq \widetilde{q_2}(d)$ *for all d in D.*

Here $\tilde{R}$ is self dual. [i.e. there exists a dual soft isomorphism between $(\tilde{R},\tilde{\leq})$ and $(\tilde{R},\tilde{\geq})$.

12.5 gs-Lattices and Complete gs-Lattices

Lattices and complete lattices possess a major role in mathematics and computer science. Here, we study on complete gs-lattice and its main properties. An element of $S(\tilde{U})$ such that its soft elements form a lattice is a generalized soft lattice (see [13]). A soft lattice is defined as follows.

Definition 12.5.1. [17] *Consider a lattice L. Then a soft lattice over L is a soft set (H, D) such that H (d) is a sublattice of L for each d in D.*
Since any lattice generate a gs lattice and any soft lattice is a gs lattice (see [13]), we can say that gs lattice is a generalization of lattice and soft lattice.

Definition 12.5.2. *A gs-lattice (G, D) is complete if each soft subset (H, D) of (G, D) generated by subsets of SE (G, D) has a soft infimum (least soft element) and a soft supremum (greatest soft element) in (G, D).*

Let $\tilde{0}_{(G,D)}$ *denotes the least soft element of (G, D) or infimum of SE (G, D)) and* $\tilde{1}_{(G,D)}$ *represents the greatest soft element of (G, D) or supremum of SE (G, D)).*

Theorem 12.5.3. *Every complete gs-lattice possess a maximal soft element (soft supremum) and a minimal soft element (soft infimum).*

Proof *Consider a complete gs-lattice (G, D). Then by definition, every soft subset of (G, D) generated by any subset of SE (G, D) has a greatest and a least soft element called respectively soft supremum and soft infimum.*

Since (G, D) is same as the softset generated by SE (G, D) [see [6]] it have the greatest and the least soft element.

i.e., $\tilde{\vee}_{\tilde{u}\tilde{\in}(G,D)}\tilde{u} = \tilde{\vee}SE(G,D)$ *and* $\tilde{\wedge}_{\tilde{u}\tilde{\in}(G,D)}\tilde{u} = \tilde{\wedge}SE(G,D)$ *are respectively the maximal soft element and minimal soft element respectively.*

Example 12.5.4. *Consider I = [0, 1], C = {l, m, n}. Then* $(\tilde{I},C)$ *becomes a gs-lattice which is complete. Here the minimal soft element and the maximal soft element of* $(\tilde{I},C)$ *are given by* $\tilde{0}_{(\tilde{I},C)} = (0,0,0)$ *and* $\tilde{1}_{(\tilde{I},C)} = (1,1,1)$.

Theorem 12.5.5. *An absolute soft set over any complete lattice form a complete gs-lattice with any parameter set D.*

Proof
Let $<U, \preccurlyeq, \vee, \wedge>$ *be a complete lattice and consider the absolute soft set* $(\tilde{U},D)$.

Then by definition $\tilde{U}(d)=U$, *for all d in D and* $((\tilde{U},D),\preceq)$ *form a gs-lattice by the following definitions.*

For $\tilde{u},\tilde{v}\tilde{\in}\tilde{U},\tilde{u}\tilde{\preceq}\tilde{v}$ *iff* $\tilde{u}(d)\preceq\tilde{v}(d)$ *for all d in D,* $(\tilde{u}\tilde{\vee}\tilde{v})(d)=\tilde{u}(d)\vee\tilde{v}(d)$ *and* $(\tilde{u}\tilde{\wedge}\tilde{v})(d)=\tilde{u}(d)\wedge\tilde{v}(d)$, *for all d in D. for all d in D.*

Also any soft subset (G, D) of $(\tilde{U},D)$ *generated by a subset of* $SE(\tilde{U})$ *is such that G (d) ⊆ U, for all d in D. Since U is complete, each G (d) has a maximal and a minimal element in U. So (G, D) has a maximal and a minimal soft element in* $(\tilde{U},D)$, *which are respectively* $\tilde{l}$ *and* $\tilde{m}$ *and are given by* $\tilde{l}(d)=\vee_{u_d\in G(d)}u_d\in U$ *and* $\tilde{m}(d)=\wedge_{u_d\in G(d)}u_d\in U$.

Definition 12.5.6. *The complete gs-lattice* $(\tilde{U},D)$ *in the above proof is called the complete absolute generalized soft lattice generated by the complete lattice* $<U, \preccurlyeq, \vee, \wedge>$.

Theorem 12.5.7. *If* $(\tilde{U},K)$ *and* $(\tilde{V},T)$ *be two complete gs-lattices generated by the two lattices* $<U, \preccurlyeq_1, \vee_1, \wedge_1>$ *and* $<U, \preccurlyeq_2, \vee_2, \wedge_2>$. *Then the Cartesian product* $(\tilde{U},K)(\tilde{U},T)$ *becomes a complete gs-lattice with parameter set* $K\times T$.

Proof *Here* $(\tilde{U},K)(\tilde{V},T)=(H,K\times T)$, *where H:* $K\times T\rightarrow P(U\times V)$ *and* $H(k,t)=\tilde{U}(k)\times\tilde{V}(t)=U\times V,\forall(k,t)\in K\times T$.

Also $SE(\tilde{U},K)(\tilde{V},T)=\{\tilde{p}:K\times T\rightarrow U\times V\setminus\tilde{p}(k,t)=(u_{p_k},v_{p_t})$ where $u_{p_k}\in U$ and $v_{p_t}\in V\}$. *For two soft elements* $\tilde{p},\tilde{q}$ of $(\tilde{U},K)(\tilde{V},T)$, *define* $\tilde{p}\tilde{\leq}\tilde{q}$ iff $\tilde{p}(k,t)\leq\tilde{q}(k,t)$, $\forall(k,t)\in K\times T$, *where* $\tilde{p}(k,t)\leq\tilde{q}(k,t)$ *iff* $u_{\tilde{p}(k,t)}\leq_1 u_{\tilde{q}(k,t)}$, *and* $v_{\tilde{p}(k,t)}\leq_2 v_{\tilde{q}(k,t)}$ *where* $\tilde{p}(k,t)=(u_{\tilde{p}(k,t)},v_{\tilde{p}(k,t)})$ *and* $\tilde{q}(k,t)=(u_{\tilde{q}(k,t)},v_{\tilde{q}(k,t)})$.

Here $u_{\tilde{q}(k,t)}\in\tilde{U}(k)=U$ *and* $y_{\tilde{q}(k,t)}\in\tilde{V}(t)=V$.

If we take any subset B of $SE[(\tilde{U},K)(\tilde{V},T)]$, *then SS (B) have a minimal soft element and* a *maximal soft element in* $(\tilde{U},K)(\tilde{V},T)$ *and are respectively* $\tilde{l}$, *where* $\tilde{l}(k,t)=left(\wedge_{\tilde{p}\in B}u_{p_k},\wedge_{\tilde{p}\in B}v_{p_t}$ *and* $\tilde{u}$, *where* $\tilde{u}(k,t)=(\vee_{\tilde{p}\in B}u_{p_k},\vee_{\tilde{p}\in B}v_{p_t})$.

Example 12.5.8. *Let* $T=\{1,2\}$, $W=\{4,5\}$, $C=\{c_1,c_2\}$, $D=\{d_1,d_2\}$. *Here T and W are complete lattices with usual ≤, ∨ and ∧, where t ∨ w and t ∧ w are respectively maximum and minimum among t and w.*

Let $(\tilde{T},C)$ *and* $(\tilde{W},D)$ *be the the complete gs-lattices generated by T and W with* $SE(\tilde{T},C)=\{\tilde{t}_1,\tilde{t}_2,\tilde{t}_3,\tilde{t}_4\}$ *and* $SE(\tilde{W},D)=\{\tilde{w}_1,\tilde{w}_2,\tilde{w}_3,\tilde{w}_4\}$, *where* $\tilde{t}_i:C\rightarrow T$ *and* $\tilde{w}_i:D\rightarrow W$ *for i = 1, 2, 3, 4 which are given by Table 12.1.*

Table 12.1

i	$\tilde{t}_i(c_1)$	$\tilde{t}_i(c_2)$	$\tilde{w}_i(d_1)$	$\tilde{w}_i(d_2)$
1	1	1	4	4
2	1	2	4	4
3	2	1	5	4
4	2	2	5	5

Here the Cartesian product $(\tilde{T},C)(\tilde{W},D)=(H,C\times D)$ *and* $H(c,d)=\tilde{T}(c)\times\tilde{W}(d)=\{1,2\}\times\{4,5\},\forall(c,d)\in C\times D$.

The Cartesian product $(\tilde{T},C)(\tilde{W},D)$ *is a complete gs-lattice and have* 4^4 *soft elements and each soft element* $\tilde{p}$ *is a function such that* $\tilde{p}:C\times D\rightarrow T\times W$, *where* $\tilde{p}(c,d)=(t_{p_c},w_{p_d})$ *defined by* $t_{p_c}\in T$ *and* $w_{p_d}\in W$. *Also we can see that* $(\tilde{T},C)(\tilde{W},D)$ *and the set* $\{(\tilde{t},\tilde{w}):C\times D\rightarrow T\times W$, *where* $(\tilde{t},\tilde{w})(c,d)=(\tilde{t}(c),\tilde{w}(d))\}$ *are not equal.*

But $(\tilde{T},C)(\tilde{W},D)$ *and the set* $SS\{(\tilde{t},\tilde{w}):C\times D\rightarrow T\times W, where (\tilde{t},\tilde{w})(c,d)=(\tilde{t}(c),\tilde{w}(d))\}$ *are equal.*

Theorem 12.5.9. *Consider be a complete gs-lattice (G, D) and any non-null soft subset (H, D) of (G, D) in* $S(\tilde{U})$ *such that*

$$(i)\ \tilde{i}_{(G,D)}\tilde{\in}(H,D)$$

$$(ii)\ a\ non-null\ soft\ set (K,D)\tilde{\subseteq}(H,D)\ in\ S(\tilde{U})\ implies\ \tilde{0}_{(K,D)}\tilde{\in}(H,D).$$

Then the soft set (H, D) is a complete gs-lattice.

Proof *Consider a non-null soft subset (G, D) of (H, D) in* $S(\tilde{U})$, *we have* $\tilde{0}_{(K,D)}\tilde{\in}(H,D)$[∵ *By (ii)*] *and is the greatest lower bound of (K, D) in (H, D).*

Let B ⊆ SE (H, D) be the collection of all upper bounds of (K, D) in (H, D). Then $B\neq\phi[\because \tilde{i}_{(G,D)}\tilde{\in}(H,D)]$. *Now, inf B is also an upper bound of SE (K, D) and is the least upper bound* [∵ *inf B ≤ u, ∀u ∈ B*]. Hence, *(H, D) is a complete gs-lattice.*

Theorem 12.5.10. *$S(\tilde{U})$ is the collection of all soft sets generated by subsets of $SE(\tilde{U})$.*

Proof *Let $(K,D) \in S(\tilde{U})$.*

Then $(K,D) = (\tilde{\Phi},D)$ or $K(d) \neq \phi$ for all d in D.
If $(K,D) = (\tilde{\Phi},D)$ then (K, D) = SS (ϕ), where $\phi \subseteq SE(\tilde{U})$.
If K (d) ≠ ϕ, ∀d ∈ D, then *(K, D) = SS (B), where $B = \{\tilde{u} \in SE(\tilde{U}) : \tilde{u}(d) \in K(d), \text{for all } d \text{ in D}\}$ is a non-null subset of $SE(\tilde{U})$.*
Hence, $S(\tilde{U}) \subseteq \{SS(B) : B \subseteq SE(\tilde{U})\}$.
Conversely, let (K, D) = SS (B), where $B \subseteq SE(\tilde{U})$.
If B = ϕ, then $(K,D) = (\tilde{\Phi},D)$.
If B ≠ ϕ, then $K(d) = \{\tilde{u}(d) : \tilde{u} \in B\} \neq \phi$ *for each d in D.* [*since* B ≠ ϕ].
$\therefore (K,D) \in S(\tilde{U})$ *and hence the result.*

Definition 12.5.11. *The collection of all soft subsets of $(\tilde{U},D)$ generated by subsets of $SE(\tilde{U})$ is the soft power set of $(\tilde{U},D)$.*

Corollary 1. $S(\tilde{U}) = SP(\tilde{U})$. *(From Theorem 12.5.10 and Definition 12.5.11)*

Example 12.5.12. *Let $(\tilde{X},A)$ defined by X = {0, 1}, ∀a ∈ A, where A = {a, b}, a parameter set.*

Then $S(\tilde{X}) = \{(H_i,A) : i = 1,2,3,4,5,6,7,8,9,10\}$, where (H_i, A) are given by Table 12.2.

Here, $SE(\tilde{X},A) = \{(0,0),(0,1),(1,0),(1,1)\}$.

From tables 12.2 and 12.3 we get $S(\tilde{X}) = SP(\tilde{X})$.

Note *Let $(K,D),(T,D) \in S(\tilde{U})$. Then (K, D) = (T, D) iff SE (K, D) = SE (T, D). But if $B_i, B_j \subseteq SE$ (K, D) such that SS (B_i) = SS (B_j) then B_i and B_j need not be same. This is clear from the above table [SS (β_8) = SS (β_9) but $\beta_8 \neq \beta_9$].*

Also, from Table 12.3, we can verify that SS (SE (Hi, A)) = (Hi, A), for any $(H_i A) \in S(\tilde{X})$ [see [6]].

Table 12.2

i	1	2	3	4	5	6	7	8	9	10
$H_i(a)$	ϕ	{0}	{0}	{0}	{1}	{1}	{1}	{0,1}	{0,1}	{0,1}
$H_i(b)$	ϕ	{0}	{1}	{0,1}	{0}	{1}	{0,1}	{0,1}	{0}	{1}

Table 12.3

j	**subsets** $\mathcal{B}_j$ of $SE(\tilde{X},A)$	$SS\ (B_j)$
1	ϕ	$(H_1,A)=(\tilde{\Phi},A)$
2	{(0, 0)}	(H_2, A)
3	{(0, 1)}	(H_3, A)
4	{(1, 0)}	(H_5, A)
5	{(1, 1)}	(H_6, A)
6	{(0, 0),(0, 1)}	(H_4, A)
7	{(0, 0), (1, 0)}	(H_9, A)
8	{(0, 0), (1, 1)}	$(H_8,A)=(\tilde{X},A)$
9	{(0, 1), (1, 0)}	$(H_8,A)=(\tilde{X},A)$
10	{(0, 1), (1, 1)}	(H_{10}, A)
11	{(1, 0), (1, 1)}	(H_7, A)
12	{(0, 0), (0, 1), (1, 0)}	$(H_8,A)=(\tilde{X},A)$
13	{(0, 0), (0, 1), (1, 1)}	$(H_8,A)=(\tilde{X},A)$
14	{(0, 0), (1, 0), (1, 1)}	$(H_8,A)=(\tilde{X},A)$
15	{(0, 1), (1, 0), (1, 1)}	$(H_8,A)=(\tilde{X},A)$
16	{(0, 0), (0, 1), (1, 0), (1, 1)}	$(H_8,A)=(\tilde{X},A)$

Theorem 12.5.13. *Soft power set is a lattice which is complete.*

Proof *Soft power set become a lattice with soft subset relation, elementary soft set union and intersection and is complete.*

Remark *Soft power set is not a distributive lattice since the elementary union and intersection of soft sets are not distributive (see [6]).*

Theorem 12.5.14. *Any absolute soft set* $(\tilde{U},D)$ *over a soft power set is a complete gs-lattice.*

Proof *Any absolute soft set* $(\tilde{U}, D)$ *is a complete gs-lattice over a soft power set with* $\tilde{u}_1 \lessapprox \tilde{u}_2$ iff $\tilde{u}_1(d) \tilde{\subseteq} \tilde{u}_2(d), \forall d \in D$. *[From theorems 12.5.5 and 12.5.13]*

12.6 s-Closure System and s-Moore Family

Most of the complete lattices are characterized in terms of the closure property concept, since it gives more fruitful results and applications in several areas. So, we define s-closure system and s-Moore family and discuss its main properties.

Definition 12.6.1. *Consider a non-null soft set (P, D) be in* $S(\tilde{U})$. *Then an s-closure operation scl on (P, D) is defined as a function scl: SP (P, D) → SP (P, D) such that for any soft sets* (G, D) *and* (H, D) *of* (P, D) *belong to* $S(\tilde{U})$, (*i*) (G, D) $\tilde{\subseteq}$ (G, D), (*ii*) *scl* (*scl* ((G, D)) = *scl* (G, D) and (*iii*) (G, D) $\tilde{\subseteq}$ (H, D) ⇒ *scl* (G, D) $\tilde{\subseteq}$ *scl* (H, D).

Definition 12.6.2. (*G, D*)*, a soft subset of* (*P, D*) *in SP* (*P, D*) *is said to be s-closed under the s-closure operation scl if* (G, D) = *scl* (G, D).

In the following examples *c*, *c′* and *c″* are s-closure operators.

Example 12.6.3. *Consider U* = [0, 4] *and D* = $\{d_1, d_2\}$. *Let G* (d_1) = [3.5, 4] *and G* (d_2) = [3, 4]. *Let c: SP* (*G, D*) → *SP* (*G, D*) *defined by* $c(P,D) = (\overline{P}, D)$, *where* $\overline{P}(d) = \overline{P(d)}$, *the intersection of all closed supersets of P* (*d*).

Example 12.6.4. *Consider* $(\tilde{U}, D)$, *an absolute soft set over U. Then define* $c' : S(\tilde{U}) \to S(\tilde{U})$ *by* $c'((G,D)) = (G,D)$, *for all* (G, D) *in* $S(\tilde{U})$. *Here* $S(\tilde{U})$ *is the corresponding s-closure system.*

Example 12.6.5. *Let* $(\tilde{U}, D)$ *be any absolute soft set. Then define*

$$c'' : S(\tilde{U}) \to S(\tilde{U}) \text{ by } c''((P,D)) = \begin{cases} (\tilde{U}, D) & \text{if } (P,D) \neq (\tilde{\Phi}, D) \\ (\tilde{\Phi}, D) & \text{if } (P,D) = (\tilde{\Phi}, D) \end{cases}.$$

Here $\{(\tilde{U}, D), (\tilde{\Phi}, D)\}$ *is the corresponding s-closure system.*

Definition 12.6.6. *An s-Moore family M of soft subsets of* (*P, C*) *in* $S(\tilde{U})$ *is a collection of soft subsets of* (*P, C*) *with the following properties.*
(*i*) (*P, C*) ∈ *M and* (*ii*) *M is closed under arbitrary elementary intersection.*

Theorem 12.6.7. *The collection of all s-closed subsets of a soft set (P, D) in $S(\tilde{U})$ under any s-closure operation form an s-Moore family, and conversely.*

Proof *Let {(G_α, D): α ∈ I} be the collection of all s-closed subsets of (P, D) under an s-closure operation.*

To prove that it is an s-Moore family.

By definition, $scl: SP(P,D) \rightarrow SP(P,D), (P,D) \tilde{\subseteq} scl(P,D)$ *and* (P,D) *is the maximal soft subset of (P, D).*

So (P, D) = scl (P, D).

∴ (P, D), is s-closed and belongs to the family.

Now, to prove that the above family is closed under arbitrary elementary intersection.

Let (G, D) = ⋒ {(G_b, D)| b ∈ J}, where J ⊆ I.

Then $(G,D) \tilde{\subseteq} (G_b,D)$, *for all b and ($G_b$, D) is s-closed.*

Here scl (G_b, D) = (G_b, D), ∀ b

$$\therefore scl(G,D) \tilde{\subseteq} \Cap (G_b,D) = (G,D).$$

We have $(G,D) \tilde{\subseteq} scl(G,D)$ *(By definition), and so (G, D) = scl (G, D).*

∴ (G, D) is closed.

Hence, *{(G_α, D): α ∈ I} is an s-Moore family.*

Conversely, let M = {(G_α, D): α ∈ I} be an s-Moore family of soft subsets of (P, D) in $S(\tilde{X})$. *To prove that this family form an s-closure system.*

Define scl: SP (P, D) → SP (P, D) such that scl (G, D) is the elementary soft intersection of all $(G_\alpha,D) \tilde{\subseteq} (P,D)$ *in* $S(\tilde{U})$ *which contains (G, D)*

(i) Since (P, D) ∈ M, M is nonempty.

Also, since scl (G, D) is the elementary soft intersection of all $(G_\alpha,D) \tilde{\subseteq} (P,D)$ *in* $S(\tilde{U})$ *which contains* $(G,D), (G,D) \tilde{\subseteq} scl(G,D)$.

(ii) Let (H, D) ∈ M. Then to prove that scl (H, D) = scl (scl (H, D)).

By definition of scl, every (G_α, D) ∈ M which contains (H, D) contains scl (H, D) and conversely, every (G_α, D) ∈ M which contains scl (H, D) contains scl (scl (H, D)).

Hence, scl (H,D) = scl (scl (H, D)).

(iii) Let (H, D), (L, D) be soft subsets of (P, D) in $S(\tilde{U})$ *such that* $(H,D) \tilde{\subseteq} (L,D)$.

Then $scl(H,D) \tilde{\subseteq} scl(L,D)$. *(By definition of scl)*

Theorem 12.6.8. *The collection of all soft closed subsets of a soft topological space [6] form an s-Moore family.*

Proof *For any soft topological space* $(\tilde{U},D),\tilde{U}$ *is soft closed and arbitrary elementary intersection of soft subsets is soft closed. Hence the soft closed subsets of a soft topological space form an s-Moore family [see [6]].*

12.7 Complete gs-Lattices From s-Closure Systems

Consider the set of all closed sets corresponding to a closure operator. They form a complete lattice [see [4]. The set of s-closed sets under an s-closure operator is an ordinary set. It can be make a soft set by identifying it with one parameter absolute soft set over it.

Let Υ be the set consist of all s-closed soft subsets of (P, D) under the s-closure operation scl in $S(\tilde{U})$. Then Υ = (K, B) by taking B = {b}, any singleton parameter set and identifying each element (H, D) of Υ by a soft element $\tilde{k}$ of (K, B) such that $\tilde{k}(b)=(H,D)$. That is there is a one to one crrespondence between the soft elements of (K, B) and Υ.

Hence Υ can be considered as the soft set (K, B) uniquely. Now, (K, B) becomes a complete gs lattice with the following definitions.

For $\widetilde{k_1},\widetilde{k_2} \tilde{\in} (K,B), \widetilde{k_1} \tilde{\leq} \widetilde{k_2}$ if $\widetilde{k_1}(b) \tilde{\subseteq} \widetilde{k_2}(b), (\widetilde{k_1} \tilde{\vee} \widetilde{k_2})(b) = scl(\widetilde{k_1}(b) \tilde{\Cup} \widetilde{k_2}(b))$ and $(\widetilde{k_1} \tilde{\wedge} \widetilde{k_2})(b) = \widetilde{k_1}(b) \tilde{\Cap} \widetilde{k_2}(b)$

Also, any soft subset (L, B) of (K, B) in $S(\tilde{\Upsilon})$, we have the soft infimum, *inf* (K, B) = $\Cap$ (H, D) which is the elementary soft intersection of all soft subsets (H, C) of (K, B) in Υ which contains (L, B) and the soft supremum, *sup* (L, B) = *scl* ($\Cup$ (H, D)) which is the elementary soft union of all soft subsets (H, D) of (K, B) in Υ which contains (L, B).

So (K, B) and hence Υ becomes a complete gs lattice. Hence the following theorem.

Theorem 12.7.1. *Any s-Moore family of a soft set* $(P,D) \in S(\tilde{U})$ *form a complete gs-lattice.*

Corollary 2. *The set of all soft subsets a soft set (P, D) in* $S(\tilde{U})$ *which are s-closed with respect to any s-closure operation on (P, D) form a complete gs-lattice.*

By theorem 12.5.5, 12.7.1 and corollary 2, we get the following theorem.

Theorem 12.7.2. *Any absolute soft set over any family* Υ *of s-closed subsets (s-Moore family) of (P, D) in* $S(\tilde{U})$ *form a complete gs-lattice.*

So we can construct any number of complete gs-lattices by their connection with s-closure systems. Here these lattices may be constructed by means of any topological space, algebraic structures, ordered set etc. in soft terms.

12.8 A Representation Theorem of a Complete gs-Lattice as an s-Closure System

We can represent any complete gs-lattice as an s-closure system as follows.

Take (P, D) as a complete gs-lattice. Define an s-closure operator δ on (P, D) by $\delta((G,D)) = SS(\{\tilde{x} \tilde{\in} (P,D) : \tilde{x} \tilde{\leq} \vee(G,D)\})$, where $V(G,D) = \wedge\{\tilde{x} \tilde{\in} (P,D) : \tilde{x} \tilde{\geq} \tilde{g}, \forall \tilde{g} \tilde{\in} (G,D)\}$.

Then there is a soft isomorphism between (P, D) and C_δ, the set of all s-closed subsets of (P, D) under δ by the soft map ϕ: (P, D) → (δ, B) by $\phi(\tilde{g}) = SS\{\tilde{x} \tilde{\in} (P,D) : \tilde{x} \tilde{\leq} \tilde{g}\}$, where $B = \{b\}$ and the soft set (δ, B) is such that δ (b) = C_δ. [By identifying soft elements of (δ, B) with elements of C_δ we can consider both are identical.]

Here ∨ (G, D) denotes the least upper bound of *SE* (P, D). Since (P, D) is a complete gs-lattice, (G, D) has a greatest soft element and so $\{\tilde{x} \tilde{\in} (P,D) : \tilde{x} \tilde{\geq} \tilde{g}, \forall \tilde{g} \tilde{\in} (G,D)\}$ is nonempty.

So $\wedge\{\tilde{x} \tilde{\in} (P,D) : \tilde{x} \tilde{\geq} \tilde{g}, \forall \tilde{g} \tilde{\in} (G,D)\} = \tilde{l}$ (say) exists.

Also, ϕ is a soft isotone mapping since $\tilde{g} \tilde{\leq} \tilde{s} \Rightarrow \phi(\tilde{g}) \tilde{\subseteq} \phi(\tilde{s})$. $\left[\because SS\{\tilde{x} \tilde{\in} (P,D) : \tilde{x} \tilde{\leq} \tilde{g}\} \tilde{\subseteq} SS\{\tilde{x} \tilde{\in} (P,D) : \tilde{x} \tilde{\leq} \tilde{s}\} \text{ for any } \tilde{g}, \tilde{s} \tilde{\in} (P,D)\right]$.

Hence we get the next theorem.

Theorem 12.8.1. *If (P, D) be complete gs-lattice, then there exists an s-closure operator δ such that the collection of s-closed subsets C_δ is soft isomorphic to (P, D).*

Example 12.8.2. *Let (P, D) be defined by P (d) = {0, 1}, ∀d ∈ D, where D = $\{d_1, d_2\}$ a parameter set.*

Then SP (P, D) = $\{(H_i, D): i = 1, 2, 3, 4, 5, 6, 7, 8, 9, 10\}$, where (H_i, D) are given by Table 12.4.

By the above s-closure operation δ, we have

$\delta((H_1, D)) = (H_8, D) = (P, D)$, $\delta((H_2, D)) = (H_2, D)$, $\delta((H_3, D)) = (H_4, D)$,
$\delta((H_4, D)) = (H_4, D)$, $\delta((H_5, D)) = (H_9, D)$, $\delta((H_6, D)) = (H_8, D) = (P, D)$,
$\delta((H_7, D)) = (H_8, D) = (P, D)$, $\delta((H_8, D)) = (H_8, D) = (P, D)$, $\delta((H_9, D))$
$= (H_9, D))$ *and*
$\delta((H_{10}, D)) = (H_8, D) = (P, D)$.

Table 12.4

i	1	2	3	4	5	6	7	8	9	10
$H_i(d_1)$	ϕ	{0}	{0}	{0}	{1}	{1}	{1}	{0,1}	{0,1}	{0,1}
$H_i(d_2)$	ϕ	{0}	{1}	{0,1}	{0}	{1}	{0,1}	{0,1}	{0}	{1}

So the collection of all s-closed soft subsets of (P, D), $C_\delta = \{(H_2, D), (H_4, D), (H_8, D), (H_9, D)\}$.

We have $SE(P,D)=\{\tilde{t}_1=(0,0),\tilde{t}_2=(0,1),\tilde{t}_3=(1,0),\tilde{t}_4=(1,1)\}$ *and*

$$\phi(\tilde{t}_1)=SS\{\tilde{t}_1\}=(H_2,D),$$
$$\phi(\tilde{t}_2)=SS\{\tilde{t}_1,\tilde{t}_2\}=(H_4,D),$$
$$\phi(\tilde{t}_3)=SS\{\tilde{t}_1,\tilde{t}_3\}=(H_9,D),$$
$$\phi(\tilde{t}_4)=SS\{\tilde{t}_1,\tilde{t}_2,\tilde{t}_3,\tilde{t}_4\}=(H_8,d)=(P,D).$$

That is there exists a soft isomorphism between (P, D) and (δ, B), where $\delta(b) = C_\delta$.

By identifying soft elements of (δ, B) with elements of C_δ, we get a soft isomorphism between (P, D) and C_δ.

12.9 gs-Lattices and Fixed Point Theorem

The fixed point theorem [2] in lattice theory have many applications in various mathematical areas such as the theory of sets, Boolean algebra, general and algebraic topology, nonlinear functional analysis, partial differential equations, and so on. Davis [10] proved that its converse is also true. Here, we consider another form of fixed point theorem of Tarski's in gs-lattices.

Definition 12.9.1. *An isotone (soft order preserving) function on a gs-lattice (P, D) is an order preserving mapping on SE (P, D).*

Theorem 12.9.2. *If a gs-lattice (P, D) is complete then every soft order preserving mapping on (P, D) has a fixed soft element.*

Proof *Let (P, D) be a complete gs-lattice with an order preserving soft mapping $\tilde{l}$ on (P, D).*

Let $B=\left\{\tilde{s}\in(P,D):\tilde{l}(\tilde{s})\tilde{\geq}\tilde{s}\right\}$ *and (G, D) = SS (B).*

Since B ⊊ (P, D), [see [8]], $(G,D)\tilde{\subseteq}(P,D)$ *Since (P, D) is a complete gs-lattice, it has a minimal soft element* $\tilde{0}$ *[By Theorem 12.5.3] and so* $(G,D)\neq(\tilde{\Phi},D)$.

Let $\tilde{k}=\vee_{\tilde{s}\tilde{\in}(G,D)}\tilde{s}$.

Then $\tilde{k}\tilde{\geq}\tilde{s}$ *for all* $\tilde{s}\tilde{\in}(G,D)$ *and* $\tilde{l}$ *is a soft order preserving map, we get* $\tilde{l}(\tilde{k})\tilde{\geq}\tilde{l}(\tilde{s})$ *for all* $\tilde{s}\tilde{\in}(G,D)$.

Hence $\tilde{l}(\tilde{k})\tilde{\geq}\vee_{\tilde{s}\tilde{\in}(G,D)}\tilde{l}(\tilde{s})\tilde{\geq}\vee_{\tilde{s}\tilde{\in}(G,D)}\tilde{s}=\tilde{k}$.

Since $\tilde{l}(\tilde{k})\tilde{\geq}\tilde{k},\tilde{k}\tilde{\in}(G,D)$ *and hence in (P, D).*

Also $\tilde{k}\tilde{\leq}\tilde{l}(\tilde{k})$ *implies* $\tilde{l}(\tilde{k})\tilde{\leq}\tilde{l}(\tilde{l}(\tilde{k}))$

$\Rightarrow\tilde{l}(\tilde{k})\tilde{\in}(G,D)$ *and this implies* $\tilde{l}(\tilde{k})=\tilde{k}$.

Hence, $\tilde{l}(\tilde{k})=\tilde{k}$.

i.e., the soft element $\tilde{k}$ *is fixed.*

We can apply more properties of complete lattices to soft sets, which are representations of vague data.

References

1. Aktas, H. and Cagman, N., Soft sets and soft groups. *Inform. Sci.*, 77, 13, 2726–2735, 2007.
2. Tarski, A., A lattice-theoretical fixed point theorem and its applications. *Pac. J. Math.*, 5, 285–309, 1955.
3. Babitha, K.V. and Sunil, J.J., Soft set relations and functions. *Comput. Math. Appl.*, 60, 1840–1849, 2010.
4. Birkhoff, G., Lattice theory, in: *Colloquium Publications*, vol. 25, 1940.
5. Cagman, N., Karatas, S., Enginoglu, S., Soft topology. *Comput. Math. Appl.*, 62, 1, 351–358, 2011.
6. Chiney, M. and Samanta, S.K., Soft topology redefined, December 26, 2016. http://arxiv.org/abs/1701.00466v1[math.GM].
7. Das, S., Majumdar, P., Samanta, S.K., On soft linear spaces and Soft normed linear spaces. *Ann. Fuzzy Math. Inform.*, 9, 1, 91–109, 2015.
8. Das, S. and Samanta, S.K., On soft metric spaces. *J. Fuzzy Math.*, 21, 3, 707–734, 2013.

9. Das, S. and Samanta, S.K., Soft real sets, soft real numbers and their properties. *J. Fuzzy Math.*, 20, 3, 551–576, 2013.
10. Davis, A.C., A characterization of complete lattices. *Pac. J. Math.*, 5, 311–319, 1955.
11. Feng, F. *et al.*, Soft sets combined with fuzzy sets and rough sets. A tentative approach. *Soft Comput.*, 14, 8999–9911, 2010.
12. Manju, J. and Susha, D., Generalized soft groups. *J. Comput. Math. Sci.*, 10, 6, 1302–1313, 2019.
13. Manju, J. and Susha, D., Soft valuation on a generalized soft lattice. *J. Math. Comput. Sci.*, 10, 2, 339–358, 2020.
14. Ma, Z., Yang, W., Hu, B., Soft set theory based on its extension. *Fuzzy Inf. Eng.*, 2, 4, 423–432, 2010.
15. Maji, P.K., Biswas, R., Roy, R., Soft set theory. *Comput. Math. Appl.*, 45, 555–563, 2003.
16. Molodtsov, D., Soft set theory first results. *Comput. Math. Appl.*, 37, 19–31, 1999.
17. Nagarajan, E.K.R. and Meenambigai, G., An application of soft set to lattices. *Kragujev. J. Math.*, 35, 1, 75–87, 2011.
18. Pawlak, Z., Rough sets. *Int. J. Comput. Inf. Sci.*, 11, 5, 341–356, 1982.
19. Shabir, M. and Naz, M., On soft topological spaces. *Comput. Math. Appl.*, 61, 7, 1786–1799, 2011.
20. Zadeh, L.A., Fuzzy sets. *Inf. Control*, 8, 338–353, 1965.

13

Data Representation and Performance in a Prediction Model

Apurbalal Senapati[1], Soumen Maji[2]* and Arunendu Mondal[3]

[1]*Department of Computer Science & Engineering, Central Institute of Technology Kokrajhar, Kokrajhar, Assam, India*
[2]*Department of Civil Engineering, Central Institute of Technology Kokrajhar, Kokrajhar, Assam, India*
[3]*Department of Chemistry, Central Institute of Technology Kokrajhar, Kokrajhar, Assam, India*

Abstract
Prediction is a fundamental task in science and engineering disciplines, especially in numerical methods, data science, machine learning, weather forecasting, etc. There are several algorithms for predictions that are developed, also known as predictive modeling or predictive analytics. It is a mathematical model that attempts to predict future events or outcomes by analyzing the patterns of past observation. In a nutshell, it can be stated that, based on known past behavior, tried to answer what is most likely to happen in the future? The key element of future prediction is the past observation or past behavior is termed as past data or data. The data must be in a specific format such that it is fitted with a specific predictive model. But in real life, data are available in different forms and formats. Sometimes data may not be fitted into the various mathematical models directly or sometimes gives poor performance. This manuscript gives a comparative study of a predictive model with a different representation of the same data from which it can be chosen as the better representation. In this experiment, the linear regression model is considered as the predictive model with the COVID-19 data set of India.

***Keywords*:** Predictive model, data representation, machine learning, regression model, COVID-19 data

**Corresponding author*: s.maji@cit.ac.in

Sharmistha Ghosh, M. Niranjanamurthy, Krishanu Deyasi, Biswadip Basu Mallik, and Santanu Das (eds.) Mathematics and Computer Science Volume 1, (271–284) © 2023 Scrivener Publishing LLC

13.1 Introduction

The predictive task comes under the multidisciplinary branch of science and engineering, and it is dedicated to the domain of machine learning and deep learning. There are various underlying algorithms for building the predictive models for making predictions using past data or information. Currently, it is being used for various applications like face detection, speech recognition, email filtering, Facebook autotagging, recommender system, weather forecasting, and many more. On the availability of the large data set, it can be solved many complex problems by simple statistical models trained on that massive data sets [1]. The performance of the prediction model depends on the amount of past data, as a large amount of data helps to build a better model that predicts the output more accurately. However, a basic question arises continually, does increasing amounts of training data continue to progress the performance? Everingham *et al.* [2] tried to tackle this question. But, empirically, they got surprise result that, on increasing of training data, performance is decreasing. Another aspect of the performance is the data representation. In real life, data can be represented in various forms and formats. This chapter tried to address this issue empirically with the COVID-19 data set of India [3] with the linear regression predictive model.

13.1.1 Various Methods for Predictive Modeling

Several classical predictive models are used for various purposes and choosing the correct model is a crucial task. For a specific task, the random selection of a model may mislead its interpretation and can degrade its performance. To choose the correct model, it is a prerequisite that you must have a clear idea about the domain knowledge, data representation, the output, and the model. In predictive modeling, the data is the main component. Broadly, we can say that there are four types of data, these are categorical, ordinal, interval, and ratio. These are different from each other concerning the kind of relationship among them, and the mathematical operations, which can be derived from individual elements [4]. The ordinal variables are generally treated as categorical (shown in Figure 13.1), which implies the binary value like yes/no, 0/1, true/false, pass/fail, etc. On the other hand, the interval and ratio are taken as numeric values (shown in Figure 13.2).

Based on the characteristic of availability data, the predictive mechanism is of different types. So, first study the input data, and if it is a labeled

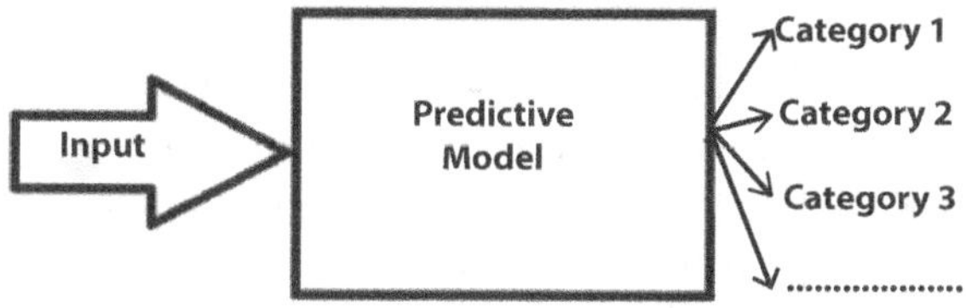

Figure 13.1 Classification.

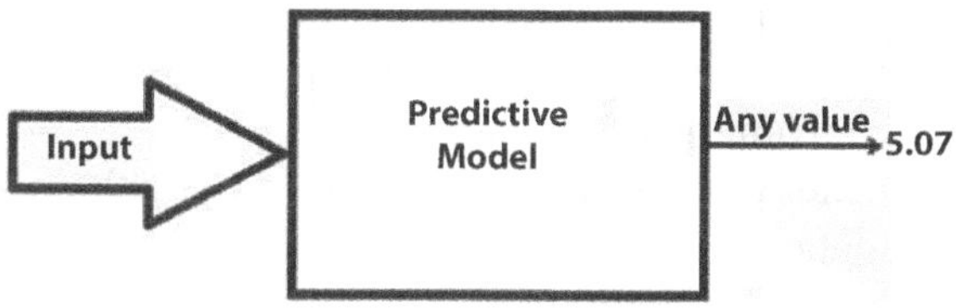

Figure 13.2 Regression.

data, it is a supervised learning problem. If data is unlabeled to find structure, then it is an unsupervised learning problem. If there is an interaction with the environment to improve the performance, it is a reinforcement learning problem. Similarly, the analysis of output is also needed in order to select the model. If the output is a number, it is suitable for the regression model else if the output is a class, it is considered as a classification problem. When the output belongs to a set of input groups, it is a clustering problem (shown in Figure 13.3).

In the earlier section, we have discussed various models based on the nature of the outputs. Figures 13.1 to 13.3 gives their high-level meaning or interpretations. In this section, we tried to express the models as a low-level representation. In each of the predictive models, there is an underline algorithm or in other words, we can say there is a mathematical model.

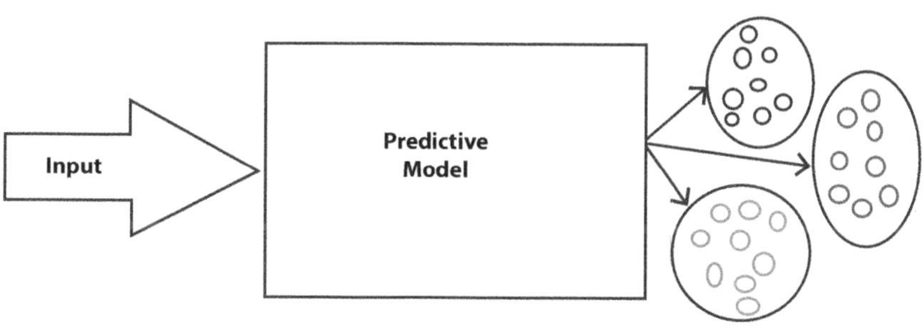

Figure 13.3 Clustering.

Now any mathematical model can be viewed as a set of unknown parameters that need to predict or have to assign a fixed value [4]. This is done by the algorithm using the past data, and this method is known as the training method. Once the model is trained, that means the parameters get the fixed value then the model is ready to serve or ready to test.

For example, the naive Bayes classifiers are a collection of classification algorithms based on Bayes' Theorem. Here, it is assumed that the features are independent of each other and equally contribute to the outcome. In this probabilistic model, the underline mathematical model is the Bayes' theorem shown in Figure 13.4. In the training process, all such probabilities are calculated from the given data set.

In our experiment, we have used the linear regression model with the COVID-19 data set of different forms. The high-level view of a two-dimension linear regression model [5] can be represented by a straight line that can be defined by the model $y = a_0 + a_1x$, and the low-level representation of the model, i.e., particularly in this case, it is equivalent to the parameters a_0 and a_1 or the set of parameters $\{a_0, a_1\}$. To build the model, it needs to find the value of these parameters from the past data. Figure 13.5 shows the pictorial representation of the model. Here x-axis represents the number of days, and the y-axis represents the confirmed COVID-19–positive cases. Each dot in the figure represents the number of COVID-19–positive cases on a particular day, and these are represented as the past data. Particularly, in this case, the data is represented cumulative-wise. From these data, the value of the parameters is calculated through the training process and get a fixed value, i.e., a fixed-line shown in Figure 13.5. There are several learning algorithms to train the model or to find the value of the unknown parameters. The common thing in the learning algorithm is that in each case there is an optimization technique that minimizes the error or tends to a closer approximation [6].

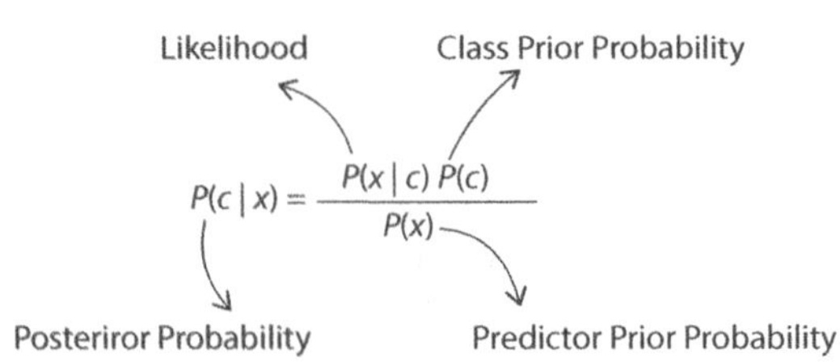

$$P(c \mid X) = P(x_1 \mid c) \times P(x_2 \mid c) \times \cdots \times P(x_n \mid c) \times P(c)$$

Figure 13.4 Bayes' classifier and Bayes' theorem.

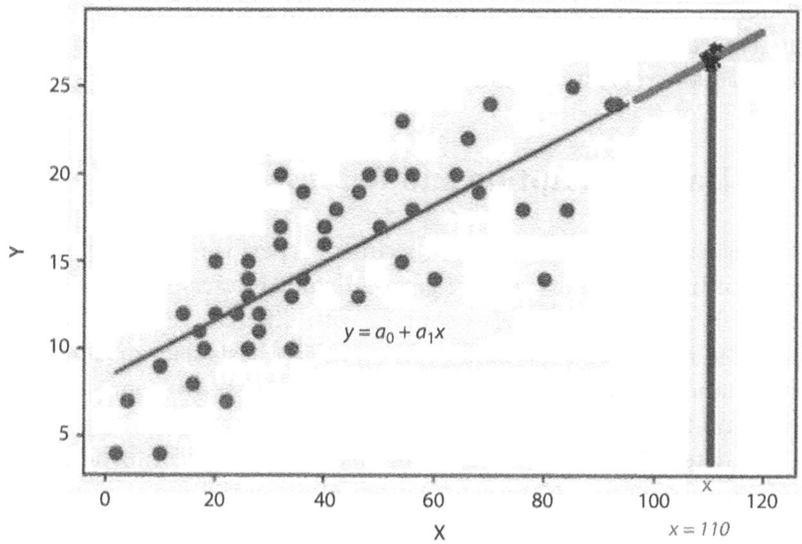

Figure 13.5 Graphical representation of regression model.

13.1.2 Problem Definition

As we mentioned, in real life, same data can be represented in various ways. For example, the COVID-19 data is given on a daily basis, cumulative basis, or some other representation. There is a basic question, which data representation is most suitable in predictive modelling, or which data format gives the most accurate result? In this chapter, we are going to address this issue. Here the issue is addressed in two different ways. First, the different format of data is plotted graphically and from the mathematical knowledge tried to guess the best-suited format and it is discussed along with the geometrical representation. Next, it is done empirically and verified our guess. For this experiment, the COVID-19 data are represented in three different formats and then the prediction is done for each case and accuracy shows the most suitable representation of the data.

In predictive modeling, past data or training data are the major components. Based on the nature of the input data and the output, a suitable model is chosen. In our experiment, we have used the linear regression model for the prediction and used the COVID-19 data with their various representations for the prediction. The experiment has been performed using the same data with various representations and compared the results along with the accuracy, which gives an insight into the better representation of them. Several works can be found for the COVID-19 prediction.

Researchers have used various models, including linear regression, polynomial regression, logistic regression [7–10]. Some of them have used [11] a variant of the regression model.

13.2 Data Description and Representations

As said earlier that, our empirical experiment is performed on the COVID-19 data of India [3]. This section gives a detailed description of the data. The data is available in various sources with different representations. It is available in the form of column format daily basis data or cumulative representation or graphical format. Table 13.1 has shown one such sample column format COVID-19 data of India. It shows the (i) serial number, (ii) date, (iii) time (iv) state/union territory, (v) confirmed Indian National, (vi) confirmed foreign national (vii) cured, (viii) deaths, and (ix) confirmed. The last three columns cured, deaths, and confirmed give the cumulative data statewise. Instead of cumulative representation, the data may be obtained daily from some other representation. Particularly, for our study, we considered the data from a randomly selected state, say, Assam up to November 11, 2020.

As we have mentioned earlier, that the same data can be represented in various ways, and in this experiment, COVID-19 data is represented in three different ways:

- Cumulative: total number of positive cases by successive additions of consecutive days.
- Daily basis: number of positive cases daily basis.
- Gradient representation: a growth rate in a specified time interval. It can be defined in both the cumulative and daily cases.

The graphical representation of each type's data is shown in the four following figures. Figure 13.6 shows the cumulative confirm cases, Figure 13.7 shows the daily confirm cases, whereas Figure 13.8 shows the direction of growth (gradient) of cumulative confirm cases and Figure 13.9 shows the direction of growth (gradient) of daily confirm cases. The gradient is defined as the growth rate of confirmed cases with respect to the date. The growth rate is defined by the formula (Eq. 13.1).

$$\left(Y_{confirm\_cases\_day2} - Y_{confirm_{cases\_day1}}\right) / \left(X_{day2} - X_{day1}\right) \tag{13.1}$$

Table 13.1 Sample COVID-19 data of India in column format.

Sl. no.	Date	Time	State/Union Territory	Confirmed Indian National	Confirmed Foreign National	Cured	Deaths	Confirmed
…	…	…	…	…	…	…	…	…
3683	03-07-20	8:00 AM	Dadra and Nagar Haveli and Daman and Diu	-	-	89	0	230
3684	03-07-20	8:00 AM	Delhi	-	-	63007	2864	92175
3685	03-07-20	8:00 AM	Goa	-	-	734	4	1482
3686	03-07-20	8:00 AM	Gujarat	-	-	24593	1886	33913
3687	03-07-20	8:00 AM	Haryana	-	-	11019	251	15509
3688	03-07-20	8:00 AM	Himachal Pradesh	-	-	628	10	1014
3689	03-07-20	8:00 AM	Jammu and Kashmir	-	-	4974	115	7849
3690	03-07-20	8:00 AM	Jharkhand	-	-	1983	15	2584
3691	03-07-20	8:00 AM	Karnataka	-	-	8334	272	18016
3692	03-07-20	8:00 AM	Kerala	-	-	2640	25	4753

(Continued)

Table 13.1 Sample COVID-19 data of India in column format. *(Continued)*

Sl. no.	Date	Time	State/Union Territory	Confirmed Indian National	Confirmed Foreign National	Cured	Deaths	Confirmed
3693	03-07-20	8:00 AM	Ladakh	-	-	730	1	990
3694	03-07-20	8:00 AM	Madhya Pradesh	-	-	10815	589	14106
3695	03-07-20	8:00 AM	Maharashtra	-	-	101172	8178	186626
3696	03-07-20	8:00 AM	Manipur	-	-	617	0	1279
3697	03-07-20	8:00 AM	Meghalaya	-	-	42	1	56
3698	03-07-20	8:00 AM	Mizoram	-	-	126	0	162
3699	03-07-20	8:00 AM	Nagaland	-	-	182	0	501
...	...	...	...	...	...	...	...	...

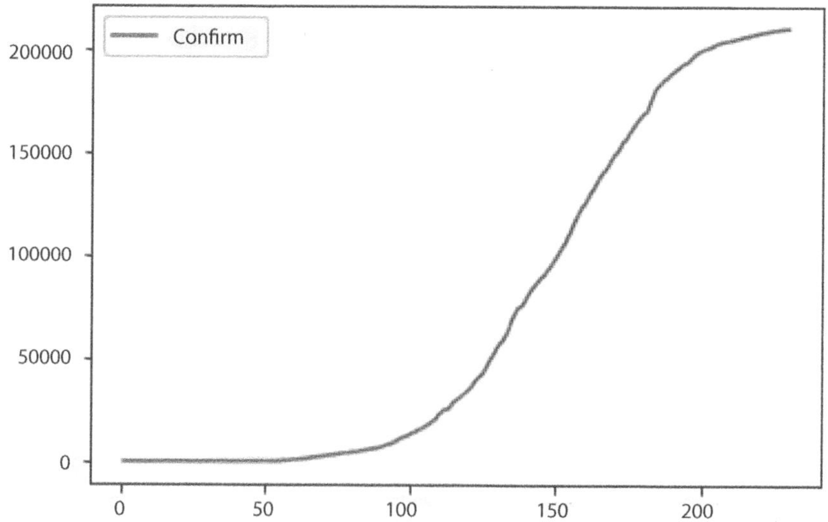

Figure 13.6 Confirmed positive cases (cumulative representation) of state Assam (X axis: number of days and Y axis: number of confirmed positive COVID-19 cases).

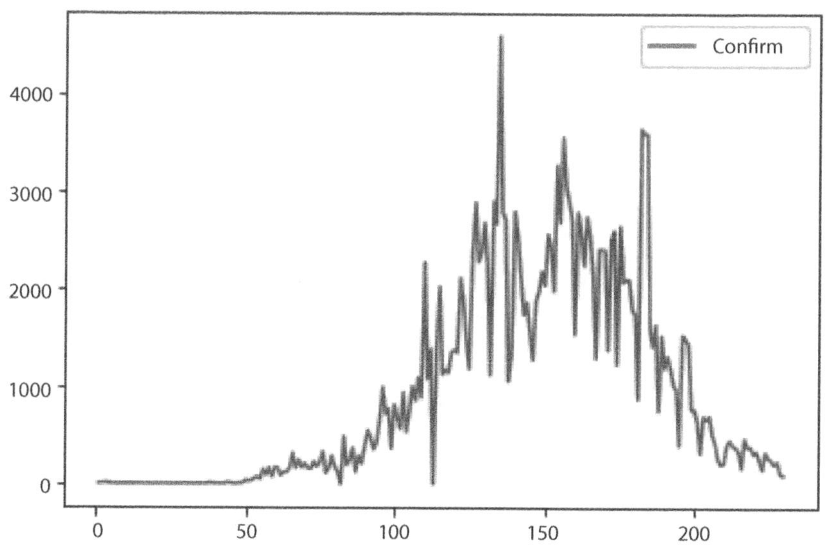

Figure 13.7 Confirmed positive cases (daily representation) of state Assam (X axis: number of days and Y axis: number of confirmed positive COVID-19 cases).

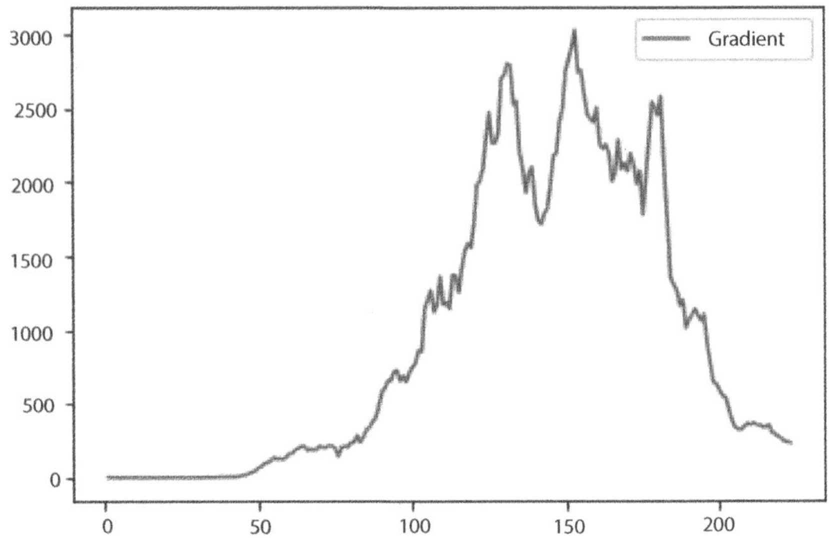

Figure 13.8 Confirmed positive cases (cumulative gradient) of state Assam (X axis: number of days and Y axis: number of confirmed positive COVID-19 cases).

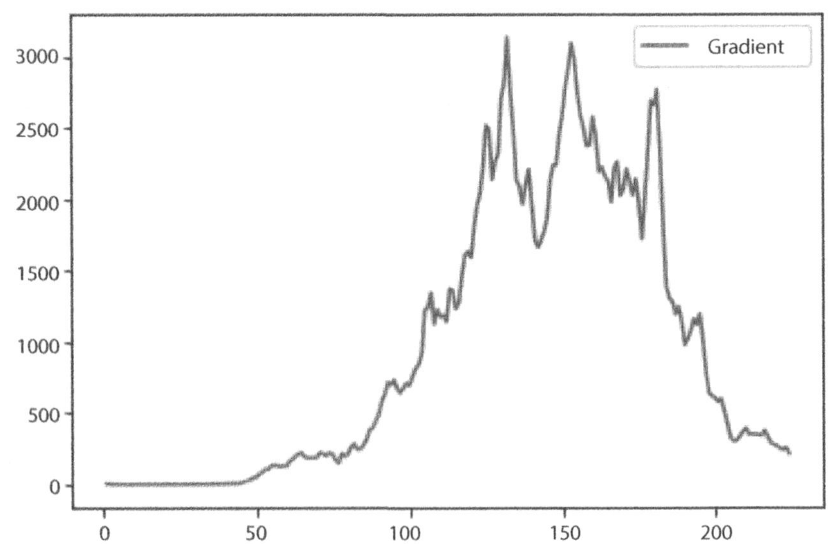

Figure 13.9 Confirmed positive cases (daily gradient) of state Assam (X axis: number of days and Y axis: number of confirmed positive COVID-19 cases).

Since we have considered a window size of 5 days, and hence, the exact formula is defined by (Eq. 13.2).

$$\left(Y_{confirm\_cases\_day\_(i+5)} - Y_{confirm_{cases\_day\_i}}\right)/(X_{day\_(i+5)} - X_{day\_i}) \quad (13.2)$$

where $Y_{confirm_{cases\_day\_i}}$ is the positive confirm cases at the day X_{day_i}. Note that the same data are represented in three different ways (daily, cumulative, and gradient), and it can be converted from one represented to another.

13.3 Experiment and Result

Since we are using the regression model, hence, a best-fitted curve (linear or polynomial) through the points is generated for future prediction. Now, if we consider the linear regression, the curve is a straight line like $y = mx + c$. Particularly, in this case, we have considered the data points like (x_i, y_i), where x_i = related to days and y_i = related to the number of positive confirm cases (as said earlier in data description section). In the above figures (Figures 13.6 to 1.3.9), except Figure 13.6, none others are suitable to fit the data by a straight line. Hence, in this case, only the cumulative data is suitable for linear regression.

Further, note that, if a single straight line is used to fit the entire data it still not suitable (Figure 13.10), rather a set of lines (Figure 13.11) is best suited for the data [11].

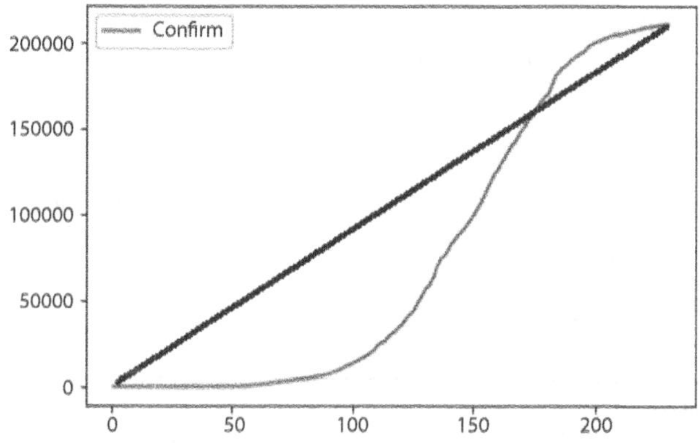

Figure 13.10 Fitted with a single line (X axis: number of days and Y axis: number of confirmed positive COVID-19 cases).

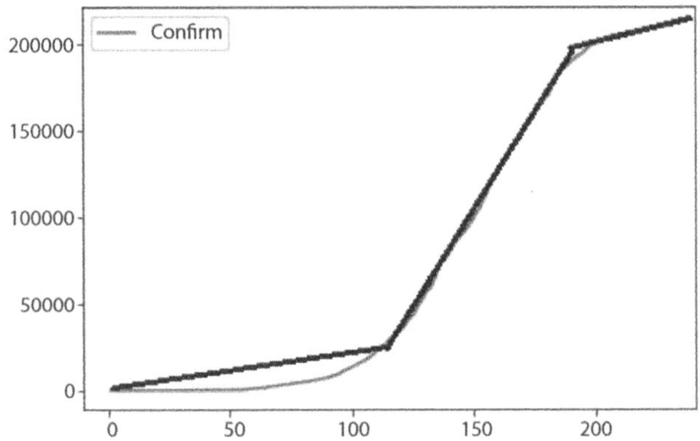

Figure 13.11 Fitted with a multiple line (X axis: number of days and Y axis: number of confirmed positive COVID-19 cases).

So far, the discussion was done from the graphical perspective. Now the experiment can be done on the said data set. In our experiment, we have considered only the COVID-19–confirmed cases, and training data are considered for 180 days (data set up to September 30, 2020). The experiment is done on the Python library. The data are represented in three different ways like cumulative, daily, and gradient representations. The test data are considered for five values corresponding to the date shown in Table 13.2 and the corresponding outputs are shown in respective columns along with the actual value.

13.4 Error Analysis

Table 13.2 shows the actual value and predicted value of three different cases and gives the error. The error is calculated from the predicted–actual

Table 13.2 Predicted and actual value.

Date	Cumulative data	Daily data	Gradient data	Actual
01-10-2020	1,82,393	1,79,565	1,83,327	1,80,811
07-10-2020	1,96,304	1,90,082	1,91,321	1,88,902
15-10-2020	2,14,850	1,99,177	2,11,387	1,98,213
21-10-2020	2,28,761	2,03,119	2,23, 618	2,02,073
30-10-2020	2,49,626	2,07,677	2,32,846	2,05,635

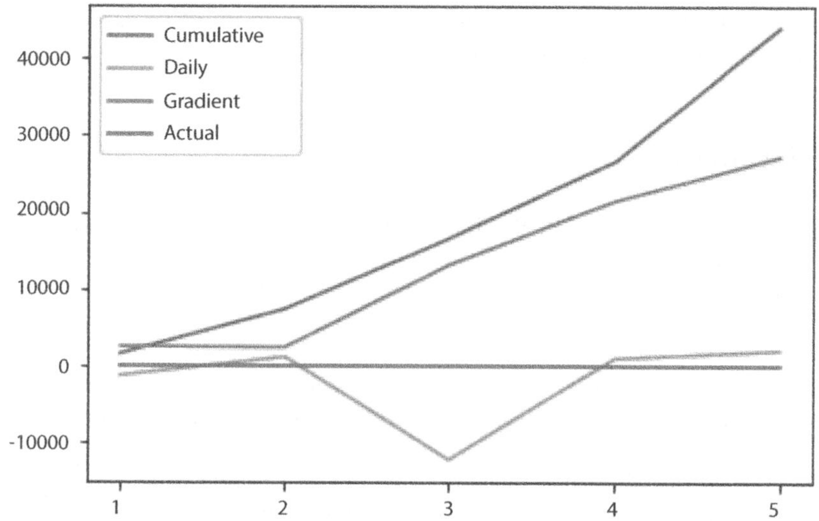

Figure 13.12 Errors in three different representations (X axis: days corresponding to the test date and Y axis: error [predicted–actual]).

in Table 13.2 in each case. Figure 13.12 shows the relative errors that stretch an idea regarding better accuracy. In cumulative representation, the values become increasing over time and hence the outlier values affect a lot compared to others, which can be further investigated by removing the outlier values. Figure 13.12 also shows that in the long run, gradient representation provides better accuracy.

13.5 Conclusion

The issue presented in this manuscript is relatively less explored. From the graphical representation, the conclusion could be made that cumulative representation will give better accuracy. It may be also data-dependent, i.e., the performance of a particular representation varies in data of different domains. The experiment result shows that it contradicts the general perception from graphical representation. To investigate this contradiction, there is a need to study the issue in depth.

Our experiment only focuses on the linear regression, but it is a general issue and hence can be investigated with any predictive model. On the other hand, the experiment is performed with a limited number of test cases (only five points), but it would be better to do with a large number of test cases. It may also be noted that the performance of a particular representation may vary with different domains. A lot of research may be conducted to conclude about the presented issue.

References

1. Halevy, A., Norvig, P., Pereira, F., The unreasonable effectiveness of data. *IEEE Intell. Syst.*, 24, 2, 8–12, March-April 2009.
2. Everingham, M., Van Gool, L., Williams, C.K.I., Winn, J., Zisserman, A., The pascal Visual Object Classes (voc) challenge. *Int. J. Comput. Vision*, 88, 2, 303–338, 2010.
3. COVID-19 in India, dataset on novel corona virus disease 2019 in India, https://www.kaggle.com/sudalairajkumar/covid19-in-india. [Accessed 30 October 2020.]
4. Aishwarya, Ch, Suvarchala, K., Aravind, B., Shashank, G., Anand, M., Krishna Rao, N.V., Prediction of disease using machine learning and deep learning, in: *Energy Systems, Drives and Automations*, pp. 69–79, Springer, Singapore, 2020.
5. Yan, X. and Su, X., *Linear regression analysis: Theory and computing*, World Scientific, Singapore, 2009.
6. Montgomery, D.C., Peck, E.A., Vining, G.G., *Introduction to linear regression analysis*, John Wiley & Sons, Hoboken, NJ, USA, 2021.
7. Senapati, A., Maji, S., Mondal, A., Piece-wise linear regression: A new approach to predict COVID-19 spreading. *IOP Conf. Ser.: Mater. Sci. Eng.*, 1020, 1, 012017, 2021, IOP Publishing.
8. da Rocha, A., *Regression Polynomial Analysis of the COVID 19 Epidemics: Progress Report 1*, 2020, Available at SSRN: https://ssrn.com/abstract=3636824 or http://dx.doi.org/10.2139/ssrn.3636824.
9. Shen, C.Y., Logistic growth modelling of COVID-19 proliferation in China and its international implications. *Int. J. Infect. Dis.*, 96, 582–589, 2020.
10. Pelinovsky, E., Kurkin, A., Kurkina, O., Kokoulina, M., Epifanova, A., Logistic equation and COVID-19. *Chaos, Solitons Fractals*, 140, 110241, 2020.
11. Senapati, A., Nag, A., Mondal, A., Maji, S., A novel framework for COVID-19 case prediction through piecewise regression in India. *Int. J. Inf. Technol.*, 13, 1, 41–48, 2021.

14

Video Watermarking Technique Based on Motion Frames by Using Encryption Method

Praful Saxena* **and Santosh Kumar**

Department of CSE, Maharishi University of Information Technology, Lucknow, India

Abstract

Rapidly growing technologies over the Internet are used to transfer digital media content at a high rate nowadays. The growth in these technologies results in the creation of identical copies of original multimedia data easily. The creation of these identical copies states the false claim of the ownership of digital media by any malicious user. The proposed method suggests protecting these types of the false claim by inserting a watermark into selected motion frames. Before embedding the watermark, it is scrambled by using an algorithm having a secret key. With help of this key, the scrambled watermark will be reconstructed whenever it is required. With the help of DWT method along with using SVD technique watermark is inserted into the selected frame from the original media. The results of the proposed algorithm show that the scheme is secure and robust against different types of attacks whether it is intentional and unintentional. The proposed scheme is found more secure for the protection of false claims of malicious users over digital media.

***Keywords*:** Ownership protection, wavelet transform, motion frames, singular value decomposition, video watermarking

**Corresponding author*: shyam.praful@gmail.com

Sharmistha Ghosh, M. Niranjanamurthy, Krishanu Deyasi, Biswadip Basu Mallik, and Santanu Das (eds.) Mathematics and Computer Science Volume 1, (285–298) © 2023 Scrivener Publishing LLC

14.1 Introduction

The higher growth in the internet bandwidth makes it easily possible to share multimedia data over various devices. This high order sharing reduces the possibility of security of digital media content. The advancement of technology related to multimedia content introduces lots of issues like ownership, authentication of content, copyright protection, and many more. Tempering with digital media is one of the major concerns related to the copyright protection of media. Proper solutions should be developed to protect the media from such unauthorized user access. The use of cryptography enables the protection of media content during transmission. The malicious user cannot access or modify the media content without knowing the encryption key. The scheme associated with the cryptographic approach encrypts the data at the authorized user end and at the end of the receiver the data will be decrypted which results in the original data. Therefore, a secure and robust scheme needs to be developed to avoid illegal media reproduction and false claims of digital media content. The advent of watermarking schemes gives a feasible solution for such problems having some challenges still to be faced.

Digital Watermarking is a feasible scheme in which additional information is hidden into host media content having a relation between both of them. Digital watermarking can be applied to audio, video, and image. Watermark is used proof copyright protection by embedding the watermark into the real source video and after that extraction process is performed which results from the same watermark which was extracted. In the process of digital watermarking proper balanced tradeoff should be maintained among payload, robustness, and perceptibility. In the watermarking method, the extraction process can only be performed by an authorized user only in the case to prove copyright protection. Figure 14.1 shows the scheme used to embed the watermark should maintain robustness so that embedded information should not be destroyed by any malicious user. In case to select the type of watermark we can consider any gray level image, name, logo any timestamp, etc. These types of watermark are frequently used in various watermark implementation techniques.

In the present scenario of the market, various types of video content are available like advertisements, training videos, workshop videos, games, etc. The video objects are categorized as real source video and compressed video [1]. In the case of the real source video, watermarking can be implemented by the means of spatial domain or frequency domain. During the implementation of the same, the balanced trade-off among watermarking features should be balanced. Any malicious user can try to modify the video

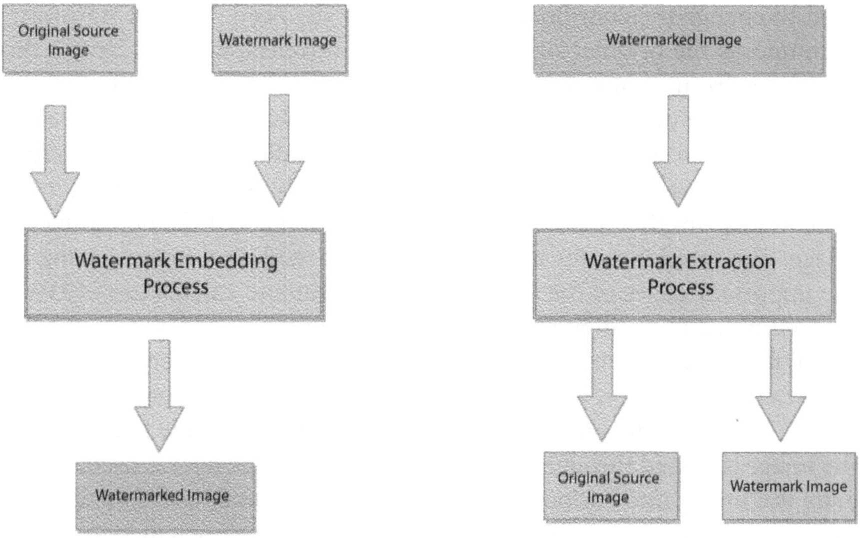

Figure 14.1 Digital watermarking.

content by performing some kind of attack. Attacks can be categorized as geometric attacks, cropping, resize, and rotation attacks [2]. Another category of attacks performed on the image is image enhancement like histogram equalization, histogram processing, noise removal, frame restoration, contrast enhancement, and many more. In the video, media attacks are intentional and unintentional specific attacks. Any malicious user can insert the video frame or some clip of video are cut down from the real source video content [3]. Such kind of attacks comes under the category of unintentional attack. The replacement of frames comes under intentional attacks.

During the implementation of the watermarking scheme, the perceptibility should be maintained up to a minimum threshold value. Another consideration is to maintain the payload capacity which states that the amount of information to be embedded into the host file.

14.2 Methodology Used

14.2.1 Discrete Wavelet Transform

Discrete wavelet transform is the kind of mathematical tool that is used to implement wavelet transform. It uses the discrete set of wavelet scales having some defined rules. The wavelets are the mathematical tool that is used to change the coordinate system as per the need. Wavelet transform decomposes the original signal into a set of wavelets, which is having the difference from continuous wavelet transform. The construction of the wavelet can be done from a scaling function, which describes the scaling properties of an object [4]. Wavelet has an unconditional basis, as a result, the size of the wavelet coefficients drops off rapidly. The layer having more resolution states that it contains more information about the object like an image. The wavelet results in the natural multiresolution image having all the important edges of the image. Its generation and calculation of DWT are well suited to the digital computer [5]. If an image is passed through the DWT function, then it is a sum of the wavelets having different locations and scales. DWT represents the image into high pass and low pass coefficients. In this, the image is passed through the high and low filters. When the image is passed through the DWT then the analysis filter is used to separate the image into various frequency bands. The analysis filters separate the image into four subbands, LL, LH, HL, and HH as shown in Figure 14.2. It is also called the first level of decomposition and it also represent the finer scale coefficients. The various level of decomposition depends on the need of the application implementation. "In most of the watermarking methods the decomposition is done up to three levels. The multi resolution concept is a conceptual concept that shows the signals that will be decomposed into fine details and finer approximations. The finer details subspaces are represented by a coarse and finer approximation. One of the major benefit for using the discrete wavelet transform method is its excellent spatiofrequency localization property [6]. DWT is used by many watermarking algorithm for the copyright protection due to the imperceptibility property also [7]".

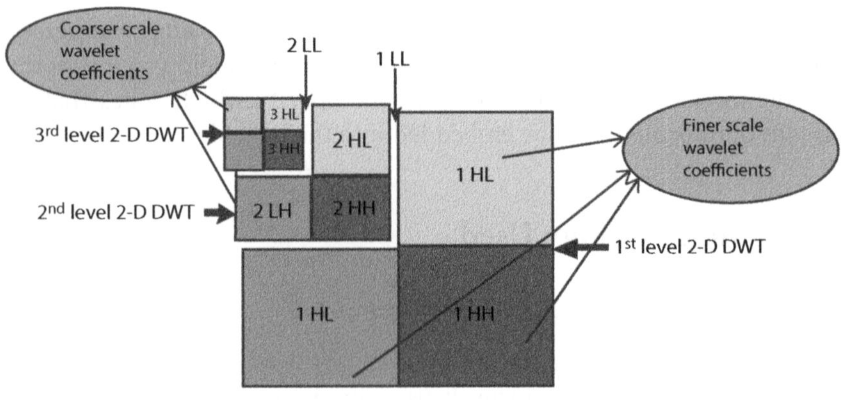

Figure 14.2 Three-level DWT.

14.2.2 Singular-Value Decomposition

Singular-value decomposition is a linear algebra mathematical tool which is used for the purpose of factorization of a real or complex matrix [8]. SVD tool performs the decomposition of square matrix into ant m × n matrix with the extension polar decomposition. When SVD is applied on any m × n complex or real matrix M, then it will be factorized into the form of $U\Sigma V^*$ where U is defined as the real or complex unitary matrix of the order m × m. Σ is defined as the rectangular matrix having non negative numbers on the diagonal and its order is m × n. V is defined as the real or complex unitary matrix having order of n × n. The diagonal entries are called the singular values of matrix M.

14.3 Literature Review

The fast motion frame to embed the watermark into video object. The author converts the fast frame RGB to YCbCr color space, and after that, the frame gets transformed into one-level decomposition of MR SVD. SVD is applied on each block approximation which provides robustness to the algorithm. The algorithm is tested against various attacks and the watermark is extracted from the video object having a satisfied perceptibility [9].

The various applications of watermarking and its design features. Authors describes the implementation schemes which are categorized in spatial domain and frequency domain. A scheme to extract the watermark from the distorted video by using distortion model based on barrel. After applying the attacks the watermark calculated is by means of correlation and mean square error to find the accuracy of watermark signal [2].

The use of scrambled watermark for the implementation of watermarking over video object. Author uses unencoded video to measure the security of the algorithm. Generalized multistage Arnold scheme is used to scramble the watermark. Scrambling the watermark improves the security feature of the algorithm. The author implements the watermarking process by selecting the U component from the YUV model. Discrete wavelet transform and singular value decomposition methods are proposed in the algorithm watermark embedding. The quality of the extracted watermark is measured by SSIM which is having lower quality than the extracted from the luminance channel. Proposed algorithm gives a good quality to the video object after embedding [5].

To avoid accidentally embedding the watermark in all the video frames, we only use the frames with big motion energy to embed it. This method works since the human visual system cannot detect the details of fast moving regions [10].

The algorithm is focused on the classification of blocks and shot segmentation. The watermark is computed as a small image that is proportional to the size of the host image. It is then embedded into the selected frames [11].

14.4 Watermark Encryption

The logic behind the encryption of watermark is to provide the security to embedded video object. The encryption will be done during embedding, and decryption will be done during the extraction of watermark. The secret key is generated for this purpose. The benefit of encryption mechanism is that the cipher image is not logically readable by any user. The watermark will be in a state of readable after decrypting it with the secret key. Appropriate algorithm [2] proposed by author found a strong algorithm for encrypting the watermark. Initially, we decide the key used for scrambling the watermark, which will be as follows:

$$K=\{16,1,14,3,12,5,10,7,8,9,6,11,4,13,2,15\}$$

A key is an arrangement of even and odd numbers, which will be used in scrambling the watermark. The size of the watermark will be same as of the video object. The watermark is portioned and the again rearranged as per the decided value K. This process results the encryption process of watermark. Figure 14.3 shows the original watermark, and Figure 14.4 shows some parts of the scrambled watermark.

After the scrambling of watermark, the next step is to extract the motion frames from real source video object. The motionless area of the video object is not robust, and it is a comparatively easy task to exploit on the motionless frame-based watermarking method. After execution of proposed process, as shown in Figure 14.5, 14 frames are found, which were based on the motion from the coastguard video object. The threshold value is set up to 5000 to extract the frames. The proposed scheme embed the scrambled watermark into the frame, which results that few frames get watermarked instead of all. It is also observed that quality of watermarked video also maintained.

Figure 14.3 Original watermark.

Figure 14.4 Parts of scramble watermark.

Figure 14.5 Extracted frames.

14.5 Proposed Watermarking Scheme

14.5.1 Watermark Embedding

The algorithm for inserting the scrambled watermark is implemented by embedding the encrypted part of watermark into motion frame. If the selected frame as displayed in Figure 14.6 is not a motion frame than no embedding process will be executed. The embedding process done only on motion frames. If the count of extracted motion frames is more than the encrypted parts of watermark than repeat the sequence of encrypted watermark from one. Now transform motion frame into

Figure 14.6 Real source video frame.

various luminance components. After the conversion, discrete wavelet transformation is executed up to 2-Level which results in four sub bands LL, LH, HL, HH. The HH band is used for the insertion of watermark. Apply the SVD method on this band which results the matrix of three components from which diagonal matrix is used insert the watermark. SVD method is also applied on watermark and diagonal matrix is used for watermarking purpose. The embedding of watermark is executed by following equation:

$$SVD_{WD} = SVD_{Y} + \alpha SVD_{W} \tag{14.1}$$

SVD_Y is the diagonal matrix of Y component after the apply SVD operation on it. SVDw represents the diagonal matrix of scrambled watermark. The value of α will be 0.01. After performing the watermarking operation, the resultant diagonal matrix is SVD_{WD} which is watermarked frame. After the insertion of watermark on all frames the watermarked video is generated which leads the final watermarked video as shown in Figure 14.7.

Figure 14.7 Watermarked video frame.

14.5.2 Watermark Extraction

The Extraction process is required when we need to extract the watermark, which is inserted into motion frame as in the Figure 14.8. For this, extract the adjacent RGB frame from the watermarked video and evaluate that if it is motion frame. If it is a motion frame than it must contain a watermark in it. Apply the process which results RGB frame into three different components stated as Y, Cb, and Cr. Now apply DWT of two-level and select HH band. Apply SVD method to find the diagonal matrix which is singular values. The singular values of watermark is generated by given mathematical formula

$$S_{w'} = (S_o - S_w)/\alpha \tag{14.2}$$

Here, Sw and So represent the singular values of the frame, which is extracted from the real source video and watermarked video frame. Now repeat this process to obtain all the watermark and then the final watermark is generated as shown in Figure 14.9.

$$\text{Watermark} = \text{UW x SW' x VW} \tag{14.3}$$

UW and VW are the orthogonal matrix of watermark. These values are referred as the values of original watermark.

Figure 14.8 Original frame.

Figure 14.9 Watermarked frame.

14.6 Experimental Results

After the extraction of watermark, two major characteristics are to be checked one is robustness of algorithm and perceptibility. For the result processing, various attacks are applied on watermarked video, and then, the extraction process is implemented. PSNR is a tool used for the measurement of the perceptibility. The unit of PSNR is dB. The features of video are investigated in order to assess the suggested watermarking scheme's durability. Various attacks are executed on the watermarked video to destroy the watermark inserted in real video object. The proposed is found robust against defined attacks above. The evaluation is carried by analyzing the original watermark and driven out watermark by observing NC. Tables 14.1 and 14.2 shown below is the level of sturdiness of proposed algorithm. The proposed algorithm is applied on two videos foreman.avi and akiyo.

Table 14.1 Proposed algorithm robustness results (foreman.avi).

S. no.	Attack performed	NC
1.	Speckle noise	0.881
2.	Frame deletion (around 10%)	0.908
3.	Rotation	0.893
4.	Gaussian low pass filter	0.811
5.	Cropping	0.898
6.	Salt and pepper	0.882

Table 14.2 Proposed algorithm robustness results (akiyo.avi).

S. no.	Attack performed	NC
1.	Speckle noise	0.902
2.	Frame deletion (around 10%)	0.911
3.	Rotation	0.901
4.	Gaussian low pass filter	0.806
5.	Cropping	0.885
6.	Salt and pepper	0.917

avi. The applied attacks mentioned above measure the robustness of the algorithm.

14.7 Conclusion

The quality of the proposed scheme is that it manages to maintain the robust ness against attacks, video-specific attacks. The security proposed in the algorithm is implemented at two-layer security. The watermark is encrypted before the process of embedding, and further, it is segmented into subimages by extracting the motion frames from real source video object. Figure 14.10 shows total number of frames in video object is 300 but only very few of the frames have gone through the process of embedding, which improves the quality of video after embedding. After implementing the proposed algorithm, it is found that the quality of the watermarked video is satisfactory in terms of imperceptibility as is zero equivalent visual gap between the original source video and watermarked video. Because the real inserted watermark and real source video are provided whenever the extraction phase is executed, such resulting watermarking technique is ideal for applications, which are private in nature. The greater robustness should be tested for composite assaults, collusion attacks, ambiguity attacks, algorithms should also be designed for video watermarking, which are in blind category, according to future studies. In the blind video watermarking, there is no requirement of real source video and original watermark for the extraction of embedded watermark.

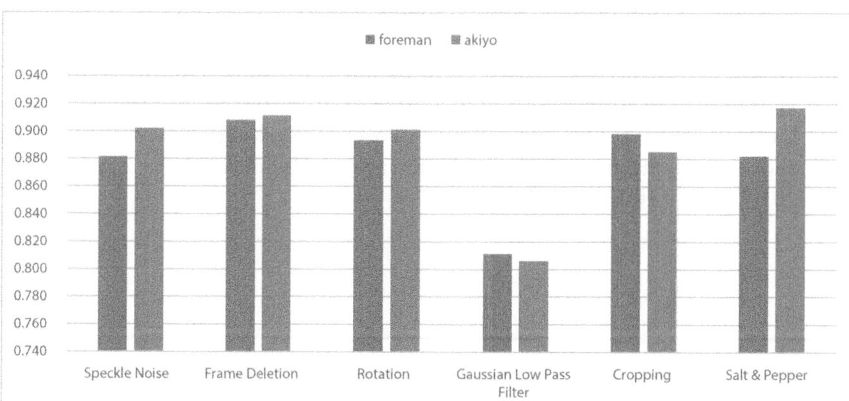

Figure 14.10 Result comparison (NC values) on foreman and akiyo video.

References

1. Li, Q. and Chang, E.C., Zero-knowledge watermark detection resistant to ambiguity attacks, in: *Proceedings of the Multimedia and Security Workshop 2006, MM and Sec'06*, vol. 2006, 2006, doi: 10.1145/1161366.1161395.
2. Ahuja, R., Purnima, Haque, M.J., Tanwar, S., Gautam, N., Rana, A., Secure and robust watermarking scheme based on motion features for video object. *J. Theor. Appl. Inf. Technol.*, 96, 17, 2020, doi: 10.1109/ICRITO48877.2020.9197967.
3. Praful Saxena, D.S.K., Robust video watermarking scheme based on DWT and SVD approach using multiple frequency bands. *Parishodh*, 8, 9, 10, 2019.
4. Kaur, R. and Jindal, S., Semi-blind image watermarking using high frequency band based on DWT-SVD, 2013, doi: 10.1109/ICETET.2013.5.
5. Veličković, Z.S., Milivojević, Z.N., Veličković, M.Z., A secured digital video watermarking in chrominance channel, in: *2018 23rd International Scientific-Professional Conference on Information Technology, IT 2018*, vol. 2018-January, 2018, doi: 10.1109/SPIT.2018.8350858.
6. Mohammed, A.A., Salih, D.A., Saeed, A.M., Kheder, M.Q., An imperceptible semi-blind image watermarking scheme in DWT-SVD domain using a zigzag embedding technique. *Multimed. Tools Appl.*, 79, 43–44, 2020, doi: 10.1007/s11042-020-09694-9.
7. Shanmugam, M. and Chokkalingam, A., Performance analysis of 2 level DWT-SVD based non blind and blind video watermarking using range conversion method. *Microsyst. Technol.*, 24, 12, 2018, doi: 10.1007/s00542-018-3870-x.
8. Abdallah, H.A., Hadhoud, M.M., Shaalan, A.A., SVD-based watermarking scheme in complex wavelet domain for color video, 2009, doi: 10.1109/ICCES.2009.5383220.
9. Nouioua, I., Amardjia, N., Belilita, S., A novel blind and robust video watermarking technique in fast motion frames based on SVD and MR-SVD. *Secur. Commun. Netw.*, 2018, 2018, doi: 10.1155/2018/6712065.
10. C.K.R. and R.K., DWT based blind digital video watermarking scheme for video authentication. *Int. J. Comput. Appl.*, 4, 10, 2010, doi: 10.5120/863-1213.
11. Jiang, X., Liu, Q., Wu, Q., A new video watermarking algorithm based on shot segmentation and block classification. *Multimed. Tools Appl.*, 62, 3, 2013, doi: 10.1007/s11042-011-0857-3.

15

Feature Extraction and Selection for Classification of Brain Tumors

Saswata Das

Department of Radiophysics and Electronics, University of Calcutta, Calcutta, India

Abstract

When malignant masses grow in an uncontrolled way, then it produces obscureness in the boundaries, shape and location. Sometimes it produces imprecise feature values. So selection of proper feature is very necessary to classify the brain tumors and diagonose it correctly. In this case, the tumor image is contrast enhanced and segmented. The segmented image is edged by using canny edge detection algorithm. From this segmented edged image, the proper feature is extracted. After that, the feature is selected using genetic algorithm and particle swarm optimization algorithm. Finally, classifiers like decision tree, support vector machine, K-nearest neighbor, etc., are used to classify the images. Experimental results show that this method reduces the feature redundancy to produce high efficiency in reasonably low time. The aim of our study is to check whether the tumor is benign or malignant.

***Keywords*:** Brain tumor, feature extraction, feature selection, classification, SVM, KNN

15.1 Introduction

The main objective of this work is to analyze the brain MRI for proper detection of lesion characteristics in early stage. Different brain tumors like glioma, oligodendroglioma, astrocytoma, acoustic neuroma are very fertile

Email: saswatadasraja@gmail.com

Sharmistha Ghosh, M. Niranjanamurthy, Krishanu Deyasi, Biswadip Basu Mallik, and Santanu Das (eds.) Mathematics and Computer Science Volume 1, (299–316) © 2023 Scrivener Publishing LLC

to human being if not diagnosed properly. It causes different abnormalities life vision problem, headache, difficulty in balance, speech difficulty, personality and behavior changes, etc. [1]. As the symptoms are very much similar to other diseases in early stage so these are often neglected, so it is very necessary to diagnose properly to encounter the disease. It is observed that the rate of death due to brain tumor is rapidly increasing with respect to other diseases. The survival rate of the people who is suffering with brain or central nervous system tumor is very poor and it is 34% in case of male and 36% in case of female.

The tumor may develop in the brain or it may also develop and spread out in the brain from other parts of the body [2]. Brain tumors are classified into four, starting from grade 1 to grade 4. The higher the grade of the tumor, higher will be its fatality rate.

Monitoring should be done from stage 2 of brain cancerous tumor by means of MRI [3]. From different studies It is seen that the composition of grey and white matter density of brain tissue smooth to complex pattern of brightness [4, 5]. As we know that the composition of memory depends on different factors like different relaxation time, what are contents within the tissues, noise is mainly caused due to field strength variation with time, presence of RF pulses, receiver bandwidth pattern etc. [6].

So the above mentioned factors shows and accuracy regarding the exact location and the boundaries of the tumors. Sometimes the brain MRI is obscure to detect the proper position, boundaries, contrast etc. so it becomes very difficult for a radiologist to diagnose the tumor properly. Routine checkup the computer assisted system is very necessary to detect the tumor properly and analyze the risk factors.

Feature extraction is a process through which we can get relevant shape based; texture based etc. information from an image which is very necessary for proper classification. it is a special form of dimensionality reduction. The main goal of feature extraction is to extract the proper information about an image. It is used for image processing, character recognition, speech processing, signal Processing and many others field. Not only feature extraction but feature selection is also an important process for proper classification of images. In this word in the result section we have shown how the classification accuracy increases with the process of feature selection.

Our work comprises of five different steps (Figure 15.1). These steps are contrast enhancement, segmentation, feature extraction, feature selection and classification. Contrast of an image is defined as the ability to detect different objects in an image very clearly and separately. So, to enhance

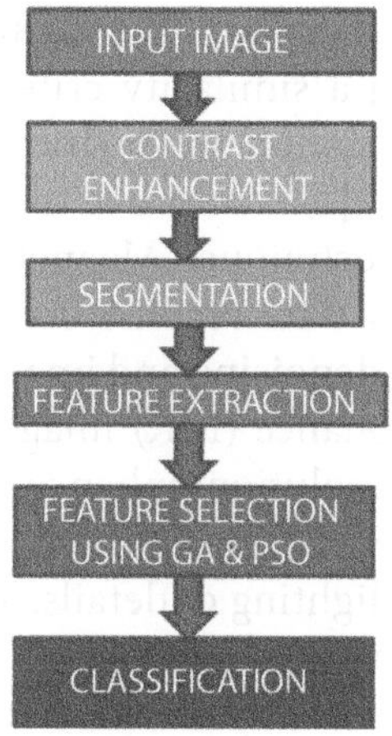

Figure 15.1 The flowchart of the present work.

the contrast of the image, it is very important to detect the proper position brain tumor. The next step after contrast enhancement is segmentation. Segmentation is done by the process of K means clustering algorithm, and the tumor portion is separated. For denoising the image median filter is used. The segmented brain tumor is edged by using canny edge detection algorithm. Then the feature is extracted from the edged images. Now the feature is selected using genetic algorithm and particle Swarm Optimization. Finally, the tumors are classified.

Our input data set comprises of 50 benign tumors and 50 malignant tumors MRI. Images are taken from the whole brain atlas (a unit of Harvard Medical School). These 50 images are classified in this work.

15.2 Related Work

Image segmentation is a low-level image processing task that aims at partitioning an image into homogeneous regions. How region homogeneity is defined depending on the application. A great number of segmentation methods are available in the literature to segment images according to various criteria, such as gray level, color, or texture [7]. Several automated methods have been developed to process the acquired images and identify features of interest, including intensity-based methods, region-growing methods, and deformable contour models. Intensity-based methods identify local features, such as edges and texture, in order to extract regions of interest. Region-growing methods start from a seed-point on the image and perform the segmentation task by clustering neighborhood pixels using a similarity criterion. Recently, researchers have investigated the application of genetic algorithms into the image segmentation problem [8, 9]. To improve the image quality, we can use any one of the filtering technique. Magnetic resonance (MR) image enhancement are mainly used for reconstruction of missing or corrupted parts of MR images, image denoising and image resolution enhancement. While using magnetic resonance (MR) images resolution enhancement face many problems like resolution enhancement of MR images (512 × 512 pixels times more), there is conservation of sharp edges in the image and conservation and highlighting of details. There are two designed and tested methods used for image resolution enhancement: discrete Fourier transform (DFT) and discrete wavelet transform (DWT). Recently, wavelets have been successfully used in a large number of biomedical applications. The multi-resolution framework makes wavelets into very powerful compression and filter tool and the time and frequency localization of wavelets makes it into a powerful tool for feature detection. This chapter, 2D discrete wavelet transform is used for removing noise from MRI brain image. The performance of an image denoising system using discrete wavelet transform (DWT) is experimentally analyzed for four levels of DWT decomposition.

Some works have applied genetic algorithms (GA) to image processing and to segmentation particularly. As segmentation can be seen as a process which finds out the optimal regions partition of an image according to a criterion, GA are well adapted to achieve this goal. Indeed, GA is particularly efficient when the search space is really important and when the criterion to optimize is numerically complicated which is always the case in image processing. The main advantages of using GA for segmentation lie in their ability to determine the optimal number of regions of a segmentation result or to choose some features such as the size of the analysis window or some heuristic thresholds. The GA proposed is a general-purpose global optimization technique based on randomized search. They incorporate some aspects of iterative algorithm.

A genetic algorithm is based on the idea that natural evolution is a search process that optimizes the structures it generates. An interesting characteristic of GA is their high efficiency for difficult search problems without being stuck in local extreme. In a GA, a population of individuals, described by some chromosomes, is iteratively updated by applying operators of selection, mutation and crossover to solve the problem. Each individual is evaluated by a fitness function that controls the population evolution in order to optimize it. GA can be used to find out the optimal label of each pixel to determine the optimal parameters of a segmentation method or to merge regions of a fine segmentation result. Concerning the fitness function, it can be an unsupervised quantitative measure of a segmentation result or a supervised one using some a priori knowledge. In this chapter, we deal with a general scheme for MRI brain tumor image segmentation that involves a GA. GA is used here as an optimization method for the optimal combination of segmentation results whose quality is quantified through an evaluation criterion. We use a general scheme to define segmentation methods by optimization.

15.3 Methodology

15.3.1 Contrast Enhancement

It helps to detect the region of tumor very clearly from the MRI. So, the visibility of the tumor region in the MRI increases. For this, we will take the help of power law (gamma) transformation.

- **Power law Transformation:** The equation describing power law transformation is given below

$$S = C * r^{\gamma}$$

where "s" is the output pixel values and "r" is the input pixel values. "c" is a positive constant, and "gamma" is also a positive constant. For, gamma value greater than one the mapping is weighted towards brighter side where as for gamma less than one the mapping is toward darker side. For gamma value, one linear mapping occurs (Figure 15.2).

Gamma correction, gamma encoding, or gamma compression is the same as above. The mapping is shown in the graph below for different values of gamma.

This gamma correction is applied on a brain tumor MRI (Figure 15.3 and Figure 15.4). The result is shown below (Figure 15.5).

15.3.2 K-Means Clustering

It is used for segmenting the tumor portion from the contrast enhanced image. This step is done after contrast enhancement (Figure 15.6). The steps of K-means clustering [10] is described below (Figure 15.7)

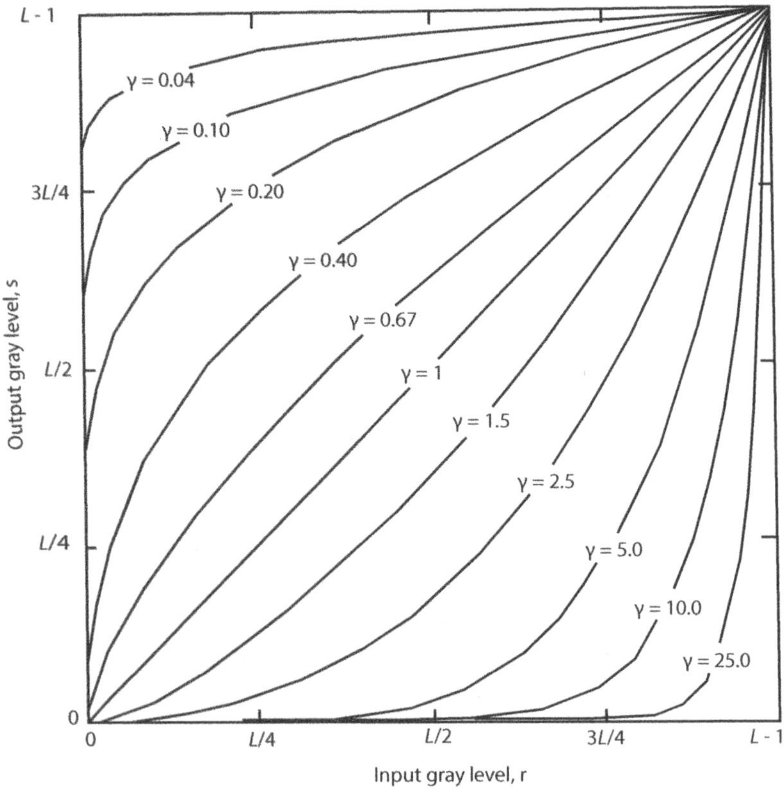

Figure 15.2 Gamma correction curve for different values of gamma.

(1) Determine the value of K where K is the no of clusters.
(2) Centroid is to be determined by first shuffling the dataset and then randomly selecting K data points for the centroids without replacement.
(3) We have to keep iterating until there is no change to the centroids. That is assigning the data points to clusters is not changing.
(4) Distance of each datapoint is to be calculated from the centroid.
(5) Data points are to be assigned in that cluster from which the distance of the point is less.
(6) New centroid is to be calculated by taking the average of present dataset and previous dataset.
(7) The above steps is to be repeated until all data are clustered properly.

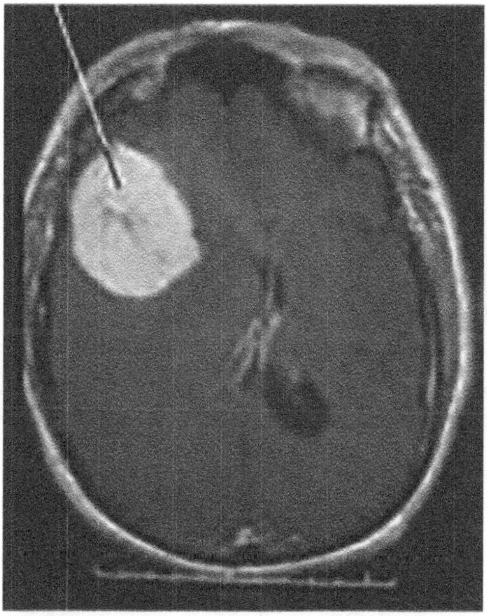

Figure 15.3 Original MRI of tumor.

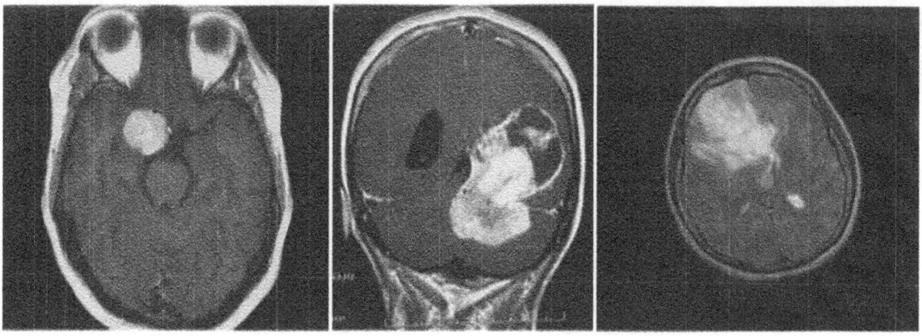

Figure 15.4 Original MRI.

15.3.3 Canny Edge Detection

This is performed after K-Means Clustering (Figure 15.8). This particular edge detection algorithm is discovered by scientist John. F. Canny in 1986. The advantage of this canny edge detection algorithm is low error rate, good localization, and minimal response. The steps of canny edge detection is described below.

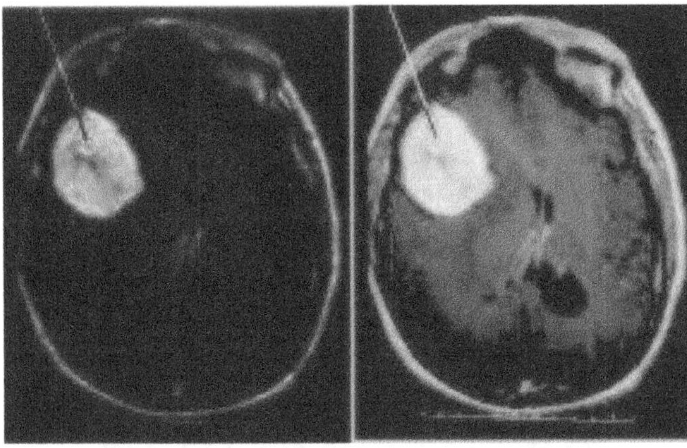

Figure 15.5 Gamma corrected image with gamma value 2.2 (left) and 0.4 (right).

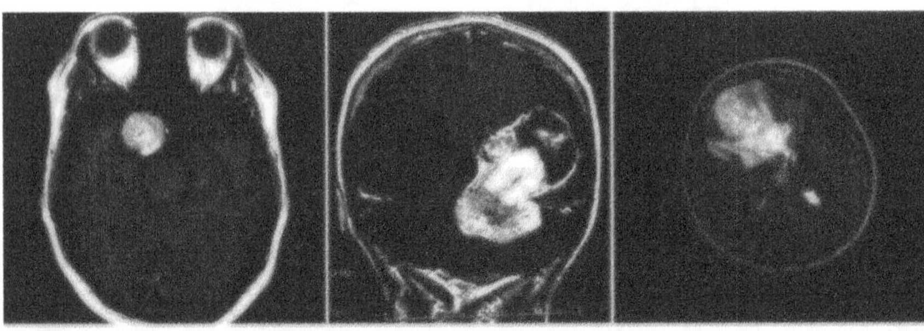

Figure 15.6 Images after contrast enhancement.

I. Any noise is to be filtered from the image. Then we have to try to locate and detect any edges using Gaussian Filter. After finding a suitable mask gaussian smoothing is to be performed.
II. After smoothing the image sobel kernel is to be used in both horizontal and vertical direction. In this way, we can get first derivative in both horizontal and vertical directions.
III. No nonmaximum supressions are to be applied, i.e., the pixels which are not the part of an edge is not considered. Result is a binary image with thin edge. Trace and supress any pixel value (sets it equal to 0) that is not considered to be an edge. This will give a thin line in the output image

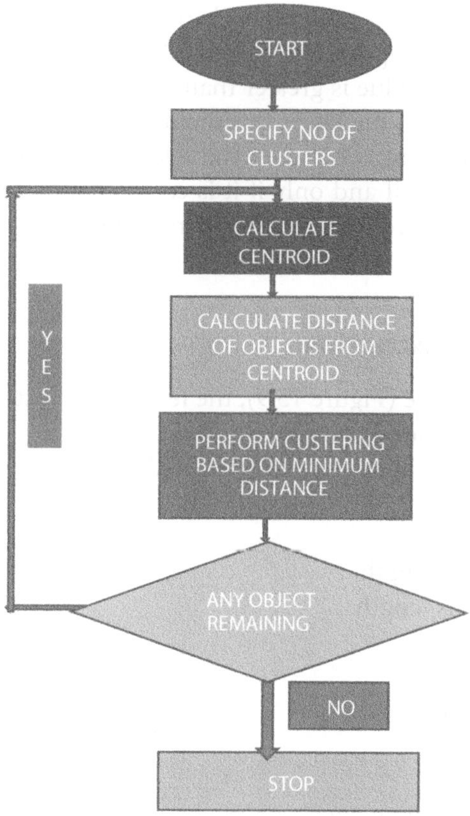

Figure 15.7 The flowchart of K-means clustering algorithm.

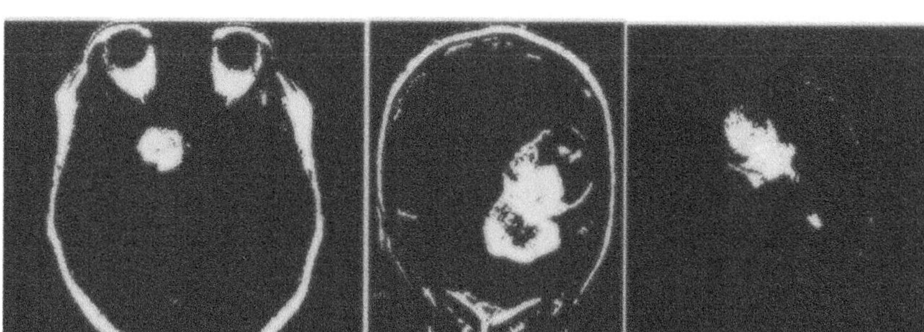

Figure 15.8 Images after segmentation (using K-means clustering).

IV. The final step is hysteresis. Canny edge detection uses two thresholds (upper and lower). A pixel is accepted as an edge if the value is greater than higher threashold and if it is lower than the lower threshold, then it is rejected. If the pixel gradient is between the two thresholds, then it will be accepted if and only if it is connected to a pixel that is higher with respect to upper threshold.

15.3.4 Feature Extraction

From the edged image (Figure 15.9), the features are extracted [11]. These are basically shape-based feature

The shape based feature that are extracted from the above images are:

1. Area
2. Major axis length
3. Minor axis length
4. Eccentricity
5. Orientation
6. ConvexaArea
7. Circularity
8. Filled area
9. Equivalent diameter
10. Solidity
11. Extent
12. Preimeter
13. Perimeter old

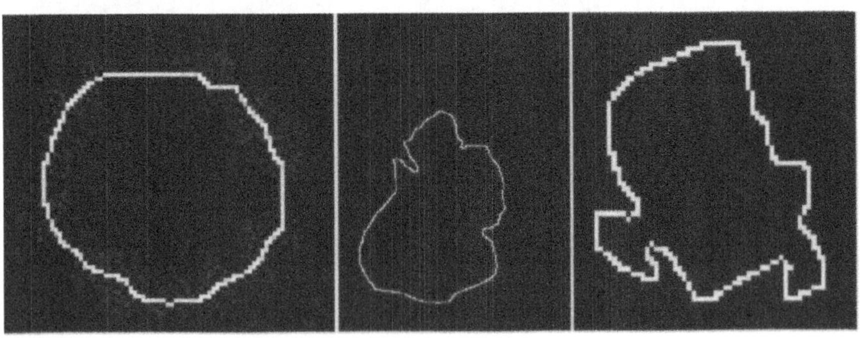

Figure 15.9 Images edged by canny edge detection method.

14. Max Feret diameter
15. Min Feret diameter
16. Max Feret angle
17. Min Feret angle

15.3.5 Feature Selection

Feature selection [11] is very important for classification. Because with all features, the classification accuracy will be very low. So when we classify the two type of tumor images, we will feed the accurate features to the classifier like SVM, KNN such that our classification accuracy increases. This is done by using two optimization technique one is GENETIC ALGORITHM and another is PARTICLE SWARM OPTIMIZATION. The necessity for feature selection can be summarized below as.

- It enables machine learning algorithm to train faster
- It reduces complexity of a model and makes it easier to interpret
- It improves the accuracy of a modelif the right subset is chosen
- It reduces overfitting

15.3.5.1 Genetic Algorithm for Feature Selection

Genetic Algorithm [12] is an evolutionary algorithm which follows the rules of genetics. This algorithm contains by selection, mutation and cross-over this three nature inspired method (Figure 15.10). First of all the features are incorporated as genes in the chromosomes. Eg area, solidity, eccentricity are geans and combination of these three gene forms a chromosome. Among the chromosomes the best chromosomes are selected through optimization of a objective function. Then the crossover and mutation between the chomosomes are done. Thus the best chromosome is selected through optimization (minimization/maximiztion) of the objective function. The objective function used for this selection is given below [13].

$$\alpha \frac{Mw}{M} + \delta \frac{|S|}{|R|}$$

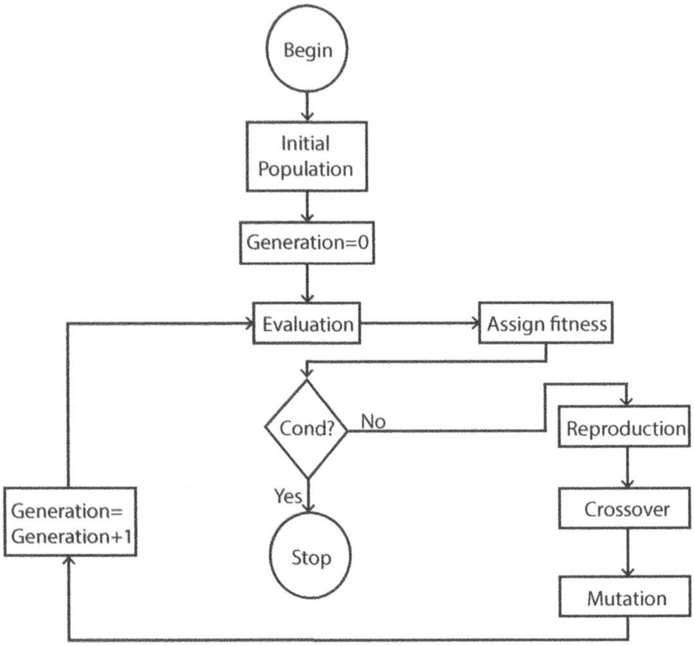

Figure 15.10 The flowchart for genetic algorithm.

- Mw is the no of instances that is predicted uncorrect
- M is the total no of instances in the dataset
- |S| is no of selected features
- |R| is the total no of features
- α and δ are the weight vectors in [0,1], which aims to balance
- the classification accuracy and feature size

The flowchart for genetic algorithm is given below
The selected features using genetic algorithms are

- Minor Axis Length
- Eccentricity
- Orientation
- Circularity
- Equivalent Diameter
- Solidity
- Extent
- Min Feret Diameter
- Min Feret Angle

15.3.5.2 Particle Swarm Optimization for Feature Selection

This is an population based stochastic algorithm. This is inspired from the flock of birds or a class of fish searching for food (Figure 15.11). Here the feature set are selected as particle. For each particle a velocity and a position is defined as per the equation (15.1). The objective function used here is written below [14–16].

$$\alpha \frac{Mw}{M} + \delta \frac{|S|}{|R|}$$

- Mw is the no of instances that is predicted uncorrect
- M is the total no of instances in the dataset
- |S| is no of selected features
- |R| is the total no of features
- α and δ are the weight vectors in [0,1], which aims to balance

the classification accuracy and feature size.

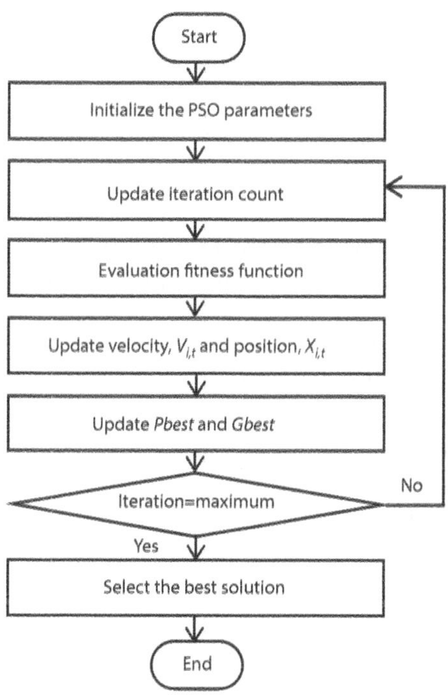

Figure 15.11 The flowchart for particle swarm optimization.

$$V_i(t+1)=wV_i(t)+c_1r_1(pbest_i(t)-x_i(t))+c_2r_2(gbest_i(t)-x_i(t)) \quad (15.1)$$

Here V is the velocity and X is the solution(position). c1 and c2 are accleration factor. r1 and r2 is the random no generated which is between [0,1]. Actually r2 is the velocity which is updated as per equation (15.1).

Then velocity is converted into a sigmoid function as given in equation (15.2). Depending on the sigmoid value the feature is to be selected as per criterion given in equation (15.3).

$$S(v_i(t+1))=\frac{1}{1+exp(-Vi(t+1)} \quad (15.2)$$

X_i={ 1 , if rand< $S(V_i(t+1))$
{ 0 Otherise 15.3
Here 1 means the feature is selected and 0 means feature is not selected Pbest and gbest is given in the equation below (15.4 and 15.5).

$$Pbest_i(t+1) = (\,x_i(t+1),\ if\ F(x_i(t+1)) < F(pbest_i(t)) \quad (15.4)$$

$$Pbest_i(t),\ otherwise$$

$$gbest(t+1) = (\,pbest_i(t+1),\ if\ F(pbest_i(t+1)) < F(gbest(t)) \quad (15.5)$$

$$gbest(t)\ ,\ otherwise$$

The flowchart for particle swarm optimization is given below:
Feature selected using particle swarm optimization is

- Major axis length
- Minor axis length
- Eccentricity
- Orientation
- Circularity
- Equivalent diameter
- Solidity
- Extent
- Min Feret diameter
- Min Feret angle

Next, we will produce the result of classification accuracy for different cases. Different classifiers like SVM, KNN, logistic regression, discriminant analysis, etc. are used. In all, the cases we can see that the classification accuracy is higher for the case of selected features instead of all features.

The image source taken for the project is described below [17].

15.4 Results

The classification accuracy is shown in the table below. Different types of classifiers are used and in all the cases the classification accuracy for selected feature is much higher than the classification accuracy with all features. So, from the table, we can understand the necessity for feature selection [18–22].

15.5 Future Scope

In this project we have taken 100 MRI images as input (Table 15.1) and all the images are contrast enhanced, clustered and edge detected. Then feature is extracted and selected using GA and PSO. Finally, classifiers are used to check the classification accuracy (Table 15.2). Maximum classification accuracy was 80% for ensemble and tree classifier. Other optimization algorithm, like ant colony optimization, Cuckoo search algorithm, etc., will be used.

In the future, we shall try to increase the classification accuracy. More images will be taken as input. We shall design a new type of classifier that can increase the classification accuracy. We shall try to implement the whole as a software where the MRI image will be taken as input and the system will automatically detect the nature of tumor which will reduce our dependency on doctors.

Table 15.1 List of open source datasets for MRI of brain.

Sl. no.	Dataset	No. of images	Source	Link
1.	Benign Tumor MRI	50	The Whole Brain Atlas-Harvard Medical School	https://www.med.harvard.edu/aanlib/
2.	Malignant Tumor MRI	50	The Whole Brain Atlas-Harvard Medical School	https://www.med.harvard.edu/aanlib/

Table 15.2 The classification accuracy for different classifiers with all features and with selected features.

Classifier name	Classification accuracy with all features	Classification accuracy with selected features
Fine Tree	71.4%	78.6%
Medium Tree	71.4%	78.6%
Coarse Tree	74.3%	80%
Linear Discriminant	64.3%	70%
Logistic Regression	62.9%	68.6%
Gaussian Naïve Bayes	51.4%	55.7%
Kernal Naïve Bayes	61.4%	62.9%
Linear SVM	64.3%	72.9%
Cubic SVM	62.9%	67.1%
Coarse Gaussian SVM	62.9%	68.6%
Fine KNN	61.4%	62.9%
Medium KNN	62.9%	68.6%
Cubic KNN	57.1%	64.3%
Weighted KNN	65.7%	70%
Cosine KNN	54.3%	62.9%
Bagged Trees	74.3%	78.6%
Subspace Discriminant	67.1%	72.9%

15.6 Conclusion

This project is related to feature extraction and selection of brain tumor MRI images. The image was contrast-enhanced, then k-means clustered and finally edged using canny edge detection algorithm. Then using region props command we have extracted all the features which includes area, major axis length, minor axis length, solidity, extant, min Feret diameter, max Feret diameter, and many more. Eighteen features among them are taken to be selected through genetic algorithm and particle swarm optimization algorithm. Finally, nine features among them are selected. Then, we have used classifiers like linear svm, coarse tree, eighted knn, and ensemble to check the classification accuracy. The classification accuracy with all features and selected features are detected with the help of classifier learner application. With selected features, the classification accuracy was higher with respect to all features. Also, we have understood the importance of feature selection in this project. The classification accuracy was 80%.

References

1. Ricard, D., Idbaih, A., Ducray, F., Lahutte, M., Hoang-Xuan, K., Delattre, J.Y., Primary brain tumours in adults. *Lancet*, 379, 9830, 1984–1996, 2012.
2. Davies, E. and Clarke, C., Early symptoms of brain tumor. *J. Neurol. Neurosurg. Psychiatry*, 75, 8, 1200–1207, 2004.
3. Herholz, K., Langen, K.-J., Schiepers, C., Mountz, J.M., Brain tumors. *Semin. Nucl. Med.*, 42, 6, 356–370, 2012.
4. Rogowska, J., Overview and fundamentals of medical image segmentation, in: *Handbook of Medical imaging*, Academic Press, Applied Physics Laboratory, Johns Hopkins University, Laurel, Maryland 2000. https://www.sciencedirect.com/book/9780120777907/handbook-of-medical-imaging
5. Simmons, A., Tofts, P.S., Barker, G.J., Addrige, S.R., Source of intensity non-uniformity in spin echo images at 1.5T. *Magn. Reson. Med.*, 32, 121–128, 1994.
6. McVeigh, E.R., Bronskil, M.J., Henkelman, R.M., Phase and sensitivity of receivers coil in magnetic resonance imaging. *Med. Phys.*, 13, 806–814, 1986.
7. Gonzalez, R.C. and Woods, R.E., *Digital image processing*, Second Edition, McGrawHill Education Pvt Ltd, India, 2017.
8. Nordin, P. and Banzhaf, W., Programmatic compression of images and sound, in: *Proceedings of the 1st Annual Conference on Genetic Programming*, J.R. Koza, *et al.* (Eds.), Morgan Kaufmann, San Francisco, CA, 1996.
9. Yin, P.-Y., A fast scheme for optimal thresholding using genetic algorithms. *Signal Process.*, 72, 85–95, 1999.
10. Singh, S. and Gill, N., Analysis and study of K-means clustering algorithm. *I.J.E.R.T.*, 2, 7, 2546–2551, 2013.
11. Poernama, I.A., Soesanti, I., Wahyunggoro, O., Feature extraction and feature selection meathods in classification of brain MRI images: A review. *I.B.I.T.e.C.*, 23-24 October 2019, Special Region of Yogyakarta, Indonesia, IEEE, pp. 58–63, 2019.
12. Melanie, M., *An introduction to genetic algorithm, fifth printing*, A Bradford Book The MIT Press, USA, 1999. https://www.boente.eti.br/fuzzy/ebook-fuzzy-mitchell.pdf

13. Too, J. and Abdullah, R.A., A new and fast rival genetic algorithm for feature selection. *J. Supercomput.*, 77, 2844–2874, 2020.
14. Shorbagy, E. and Hassanien, E.A., Particle swarm optimization from theory to application. *I.J.R.S.D.A.*, 5, 2, 1–24, 2018.
15. Too, J., Abdullah, A.R., Mohd Saad, N., A new co-evolution binary particle swarm optimization with multiple inertia weight strategy for feature selection. *Informatics*, 6, 2, 21, 2019.
16. Too, J., Abdullah, A., Mohd Saad, N., Tee, W., EMG feature selection and classification using a pbest-guide binary particle swarm optimization. *Computation*, 7, 1, 12, 2019.
17. Vidoni, E.D., The whole brain atlas. *J. Neurol. Phys. Ther.*, 36, 2, 108, June 2012, https://journals.lww.com/jnpt/fulltext/2012/06000/The_Whole_Brain_Atlas__www_med_harvard_edu_aanlib_.10.aspx
18. WWW.Mathworks.com.
19. Patel, H. and Prajapati, P., Study and analysis of decision tree based classification algorithm. *IJCSE*, 6, 10, 74–78, 2018.
20. Ladicky, L. and Torr, H.P., *Proceedings of the 28th International Conference on Machine Learning, ICML 2011*, Bellevue, Washington, USA, June 28 - July 2, 2011, 2011.
21. Gou, J., Du, L., Zhang, Y., Xiong, T., A new distance weighted K-nearest neighbor classifier. *J. Inf. Comput. Sci.*, 9, 6, 1429–1436, 2012.
22. Kotsiantis, S., Tsekouras, E.G., Pintels, P.E., Conference: Advances in informatics. *10th Panhellenic Conference on Informatics, PCI 2005*, Volos, Greece, November 11-13, 2005, 2005, Proceedings.

16

Student's Self-Esteem on the Self-Learning Module in Mathematics 6

Ariel Gulla Villar[1] and Biswadip Basu Mallik[2]*

[1]*Department of Education, Schools Division of Calbayog City, Calbayog 5 District. Northwest Samar State University, Brgy. Aguit-itan, Calbayog City, Samar, Philippines*
[2]*Department of Basic Science and Humanities, Institute of Engineering & Management, Salt Lake Electronics Complex, Kolkata, West Bengal, India*

Abstract

The current study on the student's self-esteem on the self-learning module in mathematics 6 utilized a descriptive correlation design. The average difference between the academic performance before the use of the self-learning mode and after the use is significantly meaningful in the sense that there is no teacher facilitation of instruction during the conduct of the study due to COVID 19 cases. Then base on the result, the self-learning module is effective to the students. There is a significant relationship between the result before and after the use of the self-learning module. Meaning, the self-learning module is applicable to all students in any field in mathematics provided it undergo quality assurance. All the statements are strongly agreed except on the four statements that only agreed especially on daring themselves in any complicated tasks in mathematics self-learning module, the activities in self-learning module in mathematics is easy and understandable, and the statement that able to express thoughts, self-learning module help them cope enthusiasm especially problem solving, and mathematical skills in self-learning module. The null hypothesis failed to reject. Hence, there is no significant relationship on the academic performance of grade 6 students in mathematics on the utilization of the Self-Learning Module and the response on the statements of the student's self-esteem on the self-learning module in mathematics 6. The academic performance cannot affect the student's self-esteem on self-learning module in Mathematics. The current

**Corresponding author*: basumallik.biswadip@gmail.com

Sharmistha Ghosh, M. Niranjanamurthy, Krishanu Deyasi, Biswadip Basu Mallik, and Santanu Das (eds.) Mathematics and Computer Science Volume 1, (317–332) © 2023 Scrivener Publishing LLC

researchers recommend the self-study matrix on the utilization of the self-learning module in Mathematics 6.

***Keywords*:** Self-learning module, student's self-esteem, alternative delivery mode, new learning mode during pandemic, learning module in Mathematics 6

16.1 Introduction

The self-learning module is one of the modality utilized by the Department of Education Philippines, for use as alternative mode due to COVID 19 pandemic. The said modality has undergone expert consultation and was found to be one way of the pursuit to continuing education as said by Secretary Leonor M. Briones that "education cannot wait" in consideration of the needs in education in this pandemic times. As one of the modality, students are expected to answer the module without facilitation of the teacher for every details and parts. As we all know, mathematics is one of the most crucial subject in the academic milieu and is expected to have logical and rational thinking capability to understand the subject. Module on the other way around is somehow posing a problem by the current researcher on his curiosity whether the module is enjoyed, learned, and appreciated by the students. It was then found out by the researcher, that there is a need to conduct the study on the student's self-esteem on the self-learning module in Mathematics 6.

According to Secretary Leonor M. Briones, the self-learning module is delivered in printed format to schools that are located in coastal areas, far-flung provinces, and communities without access to the internet or electricity [1].

According to Magsambol [2], the self-learning module will be distributed to the parents before the opening of the classes, representatives for students whose parents are not available to get the pick-up points in their barangays.

The purpose of the study is to give light to other researchers that the self-learning module utilized by the Department of Education is effective and appreciated by the learners, as well as to provide the necessary study guide on how to utilize the self-learning module in Mathematics 6.

16.1.1 Research Questions

The aimed of the study is to investigate the effectiveness of the self-learning module in Mathematics 6 with the end view of providing the self-study matrix on the utilization of the self-learning module in Mathematics 6 through the following the guided questions specifically stated:

1. What is the academic performance of grade 6 students in mathematics before and after the utilization of the Self-Learning Module?
2. What is the response on the statements of the student's self-esteem on the self-learning module in Mathematics 6?
3. Is there a significant relationship on the academic performance of grade 6 students in mathematics on the utilization of the self-learning module and the response on the statements of the student's self-esteem on the self-learning module in Mathematics 6?
4. What is the proposed self-study matrix on the utilization of the self-learning module in Mathematics 6?

16.1.2 Scope and Limitation

The scope of the research is within the school year 2020 to 2021 with 13 students of grade 6 in Saputan Elementary School, Calbayog 5 District. The current researcher would like to limit the study on the academic performance of grade 6 students in mathematics before and after the utilization of the self-learning module. This is also limited in the response on the statements of the student's self-esteem on the self-learning module in mathematics 6 and its significant relationship on the academic performance of grade 6 students in mathematics on the utilization of the self-learning module and the response on the statements of the student's self-esteem on the self-learning module in mathematics 6 with the end view of providing the proposed self-study matrix on the utilization of the self-learning module in Mathematics 6.

16.1.3 Significance of the Study

The current researcher may find the result of this study giving its significance to the following;

Students. The students will be beneficial on the Module and the matrix of the study during the pandemic time. This is also applicable during face-face class instruction.

Teachers. The teachers nay find ease in providing instructions to the students and can give technical assistance to the colleagues.

School Heads. The researcher may give the result to the school heads on the view that this may help them analyze the current situation especially on the midst of COVID 19 in the Philippines. This find significant because this study may be used as guide in providing technical assistance in the curriculum and instructions.

Researchers. This may help the future researcher with the hope of giving them light that there is self-esteem to the answering of modules especially in Mathematics 6. The result of this study can be used for the conduct of further investigation.

16.2 Methodology

16.2.1 Research Design

The current study on the student's self-esteem on the self-learning module in mathematics 6 utilized a descriptive correlation design. This design was used because there is a need for proving its significant relationship to the described academic performance of the students. The description about academic performance is essential on the art of the current research, hence, the study is about utilization of the self-learning module in Mathematics 6.

16.2.2 Respondents of the Study

The respondents of the study are the grade 6 students of Saputan Elementary School, Calbayog 5 District, Schools Division of Calbayog City, Eastern Visayas Region, Department of Education, Philippines. The respondents are represented by 13 students, by which 6 are males and another 6 are females.

16.2.3 Sampling Procedure

The current researcher utilized the complete enumeration for all grade 6 students. Universal sampling is the most appropriate way to get the responses based on the availability of the participants.

16.2.4 Locale of the Study

The setting where the study was conducted is in the whole District of Calbayog 5, Schools Division of Calbayog City, Eastern Visayas Region,

Department of Education, Philippines. The schools represented by the District in the Elementary level are; San Policarpo Central Elementary school, Gadgaran Integrated School, Saputan Elementary School, Gabay Elementary School, Jimautan Elementary School, Looc Elementary and Roxas Elementary School. These were the elementary schools were the 83 students are represented and randomly selected.

16.2.5 Data Collection

The researcher personally facilitated and explained to every learner/respondent using questionnaire guide on the student's self-esteem on the self-learning module in Mathematics 6. The researcher with the permission letter on the conduct of the study were addressed to the parents and to the grade 6 learners of the Calbayog 5 Districts. After the approval of the parents to permit the researcher, the class adviser instructed the respondents on the conduct of the study in order to avoid prejudice. The researcher had further explained the purpose of the study and will discuss the manner of answering the questionnaires and for the actual administration of the test. After retrieving the necessary data, the researcher was the one to tally the responses and treated the data with the right statistical tool and from the result the researcher was able to answer the result.

16.2.6 Instrument of the Study

The instrument used in the study is composed of two parts the first part is the profile of the respondents where you can find the name which is optional and the academic rating before and after the utilization of the self-learning module. The second part are the statements where the researcher is expected to respond using the Likert Scale checklist which represented by (5) means "strongly agree," (4) means "agree," (3) means "undecided," (2) means "disagree," and (1) means "strongly disagree."

16.2.7 Validation of Instrument

The instrument was validated by the three (3) experts of the field and undergone rigorous changes before the utilization to the respondents. The instrument was also carefully analyzed by the head validator for assuring the quality of the output.

16.3 Results and Discussion

The mean difference between the academic performance before the use of the self-learning mode and after the use is 4.39 as shown in Table 16.1, the result is significantly meaningful in the sense that there is no teacher facilitation of instruction during the conduct of the study due to COVID-19 cases.

In Table 16.2 the result shows that there is a significant relationship between the result before and after the use of the self-learning module. The null hypothesis is rejected, hence, 0.856** is very high significant at 0.01 level of marginal error in relation to the result before and after the use of the self-learning module.

Table 16.1 Academic performance of the students before and after the use of self-learning module.

$$Mean = \frac{Total\ number\ of\ Responses}{Number\ of\ Population}$$

Before the use of the self-learning mode	Result	After the use of the self-learning mode	Result	Difference
Mean	84.51	Mean	88.9	4.39

Table 16.2 Significant relationship between the academic performance of the students before and after the use of self-learning module.

t test for correlation		Before	After
BEFORE **Academic performance**	Pearson Correlation	1	.856**
	Sig. (2-tailed)		.000
	N	200	200
AFTER **Academic performance**	Pearson Correlation	.856**	1

*Correlation is significant at the 0.05 level (2-tailed).
**Correlation is significant at the 0.01 level (2-tailed).

Table 16.3 The response on the statements of the student's self-esteem on the self-learning module in Mathematics 6.

$$Mean = \frac{Total\ number\ of\ Responses}{Number\ of\ Population}$$

Students' self-esteem on the self-learning module in Mathematics 6	N	Mean	Remark
1. I can learn mathematics through self learning module.	83	4.66	SA
2. I feel my learning interest is high.	83	4.51	SA
3. I dare myself in any complicated tasks in mathematics self learning module.	83	4.49	A
4. I feel drive and excitement on every self-learning module.	83	4.51	SA
5. I feel self-learning module in mathematics challenging.	83	4.63	SA
6. I understand every activities presented in the self-learning module particularly in mathematics.	83	4.66	SA
7. Self-learning module in mathematics can help me in my studies.	83	4.51	SA
8. Self-learning module help me cope my enthusiasm especially problem solving.	83	4.49	A
9. I like to broaden my interest through challenging activities presented in the self-learning module.	83	4.51	SA
10. My calculation strategies are sensible and dynamic using self-learning module in mathematics.	83	4.63	SA
11. I don't easily give up when I face difficulty in answering self-learning module in mathematics.	83	4.69	SA
12. It is difficult for me to do what I want in the self-learning module in mathematics without the help of others.	83	4.51	SA

(*Continued*)

Table 16.3 The response on the statements of the student's self-esteem on the self-learning module in Mathematics 6. (*Continued*)

Students' self-esteem on the self-learning module in Mathematics 6	N	Mean	Remark
13. Activities in self-learning module in mathematics is easy and understandable.	83	4.49	A
14. I blame myself when I make mistake in any calculations with self-learning module in mathematics.	83	4.51	SA
15. I am open for criticisms given to me regarding my answers on Self Learning Module in Mathematics.	83	4.63	SA
16. I believe other reactions on my answer toward my answers on self-learning module in mathematics.	83	4.66	SA
17. I avoid any situations in answering self-learning module in mathematics while others are observing me	83	4.51	SA
18. I am able to express my thoughts and mathematical skills in self-learning module.	83	4.49	A
19. I not getting nervous and confused on answering the self-learning module.	83	4.53	SA
20. I can manage mathematics subject through self-learning module as medium of my self-learning ways.	83	4.66	SA

Scale	Description	Code
4.51–5.00	Strongly Agree	(SA)
3.51–4.50	Agree	(A)
2.51–3.50	Moderately Agree	(MA)
1.51–2.0	Moderately Disagree	(MD)
1.00–1.50	Strongly Disagree	(SD)

Table 16.4 Significant relationship on the academic performance of grade 6 students in mathematics on the utilization of the self-learning module and the response on the statements of the student's self-esteem on the self-learning module in mathematics 6.

$$t = \frac{x_1 - x_2}{\sqrt{\frac{(n_1-1)(sd_1)^2 + (n_2-1)(sd_2)^2}{n_1+n_2-2}\frac{1}{n_1}+\frac{1}{n_2}}}$$

t test for correlation		Academic performance BEFORE	Academic performance AFTER
1. I can learn Mathematics through Self Learning Module.	Pearson Correlation	0.102	0.059
	Sig.(2-tailed)	0.361	0.599
	N	83	83
2. I feel my learning interest is high.	Pearson Correlation	0.135	0.176
	Sig. (2-tailed)	0.224	0.111
	N	83	83
3. I dare myself in any complicated tasks in Mathematics Self Learning Module.	Pearson Correlation	−0.005	−0.079
	Sig. (2-tailed)	0.964	0.48
	N	83	83
4. I feel drive and excitement on every Self Learning Module.	Pearson Correlation	0.135	0.176
	Sig. (2-tailed)	0.224	0.111
	N	83	83
5. I feel Self Learning Module in Mathematics challenging.	Pearson Correlation	−0.099	−0.114
	Sig. (2-tailed)	0.376	0.306
	N	83	83

(*Continued*)

Table 16.4 Significant relationship on the academic performance of grade 6 students in mathematics on the utilization of the self-learning module and the response on the statements of the student's self-esteem on the self-learning module in mathematics 6. (*Continued*)

t test for correlation		**Academic performance BEFORE**	**Academic performance AFTER**
6. I understand every activities presented in the Self Learning Module particularly in Mathematics.	Pearson Correlation	0.102	0.059
	Sig. (2-tailed)	0.361	0.599
	N	83	83
7. Self-Learning Module in Mathematics can help me in my studies.	Pearson Correlation	0.135	0.176
	Sig. (2-tailed)	0.224	0.111
	N	83	83
8. Self-learning Module help me cope my enthusiasm especially problem solving.	Pearson Correlation	-0.005	-0.079
	Sig. (2-tailed)	0.964	0.48
	N	83	83
9. I like to broaden my interest through challenging activities presented in the Self Learning Module.	Pearson Correlation	0.135	0.176
	Sig. (2-tailed)	0.224	0.111
	N	83	83
10. My calculation strategies are sensible and dynamic using Self Learning Module in Mathematics.	Pearson Correlation	−0.099	−0.114
	Sig. (2-tailed)	0.376	0.306
	N	83	83

(*Continued*)

Table 16.4 Significant relationship on the academic performance of grade 6 students in mathematics on the utilization of the self-learning module and the response on the statements of the student's self-esteem on the self-learning module in mathematics 6. (*Continued*)

t test for correlation		**Academic performance BEFORE**	**Academic performance AFTER**
11. I don't easily give up when I face difficulty in answering Self Learning Module in Mathematics.	Pearson Correlation	0.046	0.008
	Sig. (2-tailed)	0.68	0.945
	N	83	83
12. It is difficult for me to do what I want in the Self Learning Module in Mathematics without the help of others.	Pearson Correlation	0.135	0.176
	Sig. (2-tailed)	0.224	0.111
	N	83	83
13. Activities in Self Learning Module in Mathematics is easy and understandable.	Pearson Correlation	−0.039	−0.091
	Sig. (2-tailed)	0.724	0.413
	N	83	83
14. I blame myself when I make mistake in any calculations with Self Learning Module in Mathematics.	Pearson Correlation	0.135	0.176
	Sig. (2-tailed)	0.224	0.111
	N	83	83
15. I am open for criticisms given to me regarding my answers on Self Learning Module in Mathematics.	Pearson Correlation	-0.099	-0.114
	Sig. (2-tailed)	0.376	0.306
	N	83	83

(*Continued*)

Table 16.4 Significant relationship on the academic performance of grade 6 students in mathematics on the utilization of the self-learning module and the response on the statements of the student's self-esteem on the self-learning module in mathematics 6. (*Continued*)

t test for correlation		Academic performance BEFORE	Academic performance AFTER
16. I believe other reactions on my answer towards my answers on Self Learning Module in Mathematics.	Pearson Correlation	0.102	0.059
	Sig. (2-tailed)	0.361	0.599
	N	83	83
17. I avoid any situations in answering Self Learning Module in Mathematics while others are observing me	Pearson Correlation	0.135	0.176
	Sig. (2-tailed)	0.224	0.111
	N	83	83
18. I am able to express my thoughts and mathematical skills in Self Learning Module.	Pearson Correlation	−0.005	−0.079
	Sig. (2-tailed)	0.964	0.48
	N	83	83
19. I not getting nervous and confused on answering the Self Learning Module.	Pearson Correlation	0.115	0.145
	Sig. (2-tailed)	0.3	0.19
	N	83	83
20. I can manage Mathematics subject through Self Learning Module as medium of my self-learning ways.	Pearson Correlation	−0.136	−0.129
	Sig. (2-tailed)	0.219	0.246
	N	83	83

The result in Table 16.3 shows that all the statements are strongly agreed except on the four statements that only agreed especially on daring themselves in any complicated tasks in mathematics self-learning module, the activities in self-learning module in mathematics is easy and understandable, and the statement that able to express thoughts, self-learning module help them cope enthusiasm especially problem solving, and mathematical skills in self-learning module.

Table 16.4 shows that there is no significant relationship on the academic performance of grade 6 students in mathematics on the utilization of the self-learning module and the response on the statements of the student's self-esteem on the self-learning module in mathematics 6. The null hypothesis failed to reject.

16.4 Conclusion

The following conclusions are sought based on the findings of the study.

1. The average difference between the academic performance before the use of the self-learning mode and after the use is significantly meaningful in the sense that there is no teacher facilitation of instruction during the conduct of the study due to COVID-19 cases. Then base on the result, the self-learning module is effective to the students.
2. There is a significant relationship between the result before and after the use of the self-learning module. Meaning, the self-learning module is applicable to all students in any field in mathematics provided it undergo quality assurance.
3. All the statements are strongly agreed except on the four statements that only agreed especially on daring themselves in any complicated tasks in mathematics self-learning module, the activities in self-learning module in mathematics is easy and understandable, and the statement that able to express thoughts, self-learning module help them cope enthusiasm especially problem solving, and mathematical skills in self-learning module.
4. The null hypothesis failed to reject. Hence, there is no significant relationship on the academic performance of grade 6

students in mathematics on the utilization of the self-learning module and the response on the statements of the student's self-esteem on the self-learning module in mathematics 6. The academic performance cannot affect the student's self-esteem on self-learning module in mathematics.

16.5 Recommendation

The following recommendations are sought based on the findings of the study:

The current researchers recommend the self-study matrix given on Table 16.5 on the utilization of the self-learning module in Mathematics 6.

Table 16.5 Self-study matrix on the utilization of the self-learning module in Mathematics 6.

Activity	Remarks		Difficulty	
	Done	Undone	Easy	Hard
1. Read the instructions				
2. Read the key concept				
3. Understand the problem				
4. Follow the instructions in the example given.				
5. Answer the exercises.				
6. Solve the problems based on the given example				
7. Check your answer				
8. If the problem is difficult, please go back to no. 1				

References

1. Department of Education, Republic of The Philippines. DepEd prepares Self-Learning Modules for education's new normal, 2020, July 1. https://www.deped.gov.ph/2020/07/02/deped-prepares-self-learning-modules-for-educations-new-normal/
2. Magsambol, B., *FAST FACTS: DepEd's modular learning.* Rappler, 2020, September 22. https://www.rappler.com/newsbreak/iq/things-to-know-deped-modular-learning/

17

Effects on Porous Nanofluid due to Internal Heat Generation and Homogeneous Chemical Reaction

Hiranmoy Mondal[1] and Sharmistha Ghosh[2]**

[1]Department of Applied Mathematics, Maulana Abul Kalam Azad University of Technology, West Bengal, India
[2]Department of Basic Science & Humanities, Institute of Engineering & Management, Kolkata, India

Abstract

A detailed numerical exploration of MHD porous nanofluid flow with blended impact of thermal dissipation along with mass diffusion is reported in this work. The equation of energy accounts for absorption or generation of heat, while the mass diffusion equation takes care of the chemical reaction of the species diffusing with fluid. The concerned equations are transformed into a dimensionless form with the introduction of suitable similarity transformation parameters. The transformed equations, which represent nonlinear ODEs, coupled with suitable conditions at the boundary, are then solved using Runge-Kutta-Fehlberg scheme with shooting technique. A thorough graphical investigation of the effect of flow parameters on temperature, velocity, and concentration distribution is exhibited. The computed results are also validated through comparison with those available in literature.

***Keywords*:** Porous medium, nanofluid, mixed convection, heat generation, chemical reaction

**Corresponding author*: sharmisthag71@gmail.com

Sharmistha Ghosh, M. Niranjanamurthy, Krishanu Deyasi, Biswadip Basu Mallik, and Santanu Das (eds.) Mathematics and Computer Science Volume 1, (333–352) © 2023 Scrivener Publishing LLC

Nomenclature

k	thermal conductivity of the fluid
K	permeability of the porous medium
ν	kinematic viscosity of the fluid
μ	dynamic viscosity
ρ	fluid density
f	dimensionless stream function
θ	dimensionless temperature
ϕ	dimensionless concentration
Da	Darcy number
Le	Lewis number
Pr	Prandtl number
Re_L	Reynolds number
Gr	Grashof number for temperature
Gr^*	Grashof number for concentration
D_T	thermophoretic diffusion coefficient
D_B	Brownian diffusion coefficient
λ	buoyancy parameter owing to temperature
λ^*	buoyancy parameter owing to concentration
N	ratio of buoyancy forces
Nt	thermophoresis parameter
Nb	Brownian motion parameter
Q	heat generation or absorption parameter
α	material parameter of the fluid
β_T	volumetric coefficient of thermal expansion
β_C	volumetric coefficient of concentration
Δ	chemical reaction parameter

17.1 Introduction

It is acclaimed that nanofluids represent a type of fluids that consists of nanoparticles suspended in a base fluid. Generally, these particles are metals or metal oxides, which are known to enhance the convection, as well as conduction properties. This, in turn, allows better heat transfer out of the coolants. The main advantages of nanofluids are that they have enhanced spreading and wetting properties on solid surface and are more stable with admissible viscosity. In this era of energy awareness, dearth of profuse resources of clean energy and the extensive usage of battery operated devices, such as laptops and cell phones, have initiated the necessity for smart technological handling of energy sources. Furthermore, nanofluids are being investigated for medical applications, such as cancer therapy, as well as for numerous engineering applications and industries, including electronics industry, heating, ventilation, air conditioning, transportation, and many more. It is highly desired if maximum possible thermal properties can be achieved with the least possible concentration of the nanoparticles.

RamReddy *et al.* [1] have investigated the magnetic effects along with viscous dissipation with uniform nanoparticle concentration and wall temperature. Rehman *et al.* [2] have studied couple stress fluid consisting of nanoparticles that flows in porous medium. An exploration on the challenges, as well as applications of nanofluids as coolant in automobile radiator, has been reported by Bhogare and Kothawale [3]. Kuppalapalle *et al.* [4] have performed a numerical study on the impact of varying viscosity on flow and transfer of heat through nanoparticles. Dhal [5] has examined unsteady MHD nanofluid in porous medium along with suction. Takabi and Salehi [6] have investigated the augmented heat transfer in hybrid nanofluid flow. Sheikholeslami *et al.* [7] have examined MHD nanoparticle flow with account of Brownian motion and thermophoresis effects. Uddin *et al.* [8] have studied thermosolutal nanofluid flow in porous medium with slip conditions. Makinde *et al.* [9] have analyzed heat transfer effects in Berman flow of nanofluids accompanied by viscous dissipation, Navier slip, convective cooling. The flow in the boundary layer across a stretching sheet was investigated by Khan and Sanjayanand [10]. Afzal [11] analyzed thermal, as well as momentum boundary layers along an axisymmetric or two-dimensional nonlinear stretching sheet in stationary fluid. Nandeppanava *et al.* [12] examined the effect of heat transfer due to nonuniform source of heat across a nonlinear stretching sheet. Mondal *et al.* [13–16] have numerically studied unsteady stagnation-point MHD flow in different nanofluids. De *et al.* [17] analyzed the MHD nanofluids with nonisothermal wedge with stretching surface.

The MHD flow across a surface submerged in saturated porous media is experienced in various engineering problems, such as ceramic processing, design of nuclear reactors, drilling of crude oil, geothermal energy conversion, usage of fibrous material for thermal insulation of buildings, heat exchangers, catalytic reactors, transfer of heat due to agricultural product storage that produce heat due to metabolism, storage of nuclear wastes, petroleum reservoirs, and so on. Rehena *et al.* [18] studied heat transfer in nanofluids due to buoyant forces within a closed chamber. A similar study was conducted by Hwang *et al.* [19] within a rectangular cavity. Maghrebi *et al.* [20] investigated heat transfer in nanofluids within a porous channel due to forced convection. The impact of nonequilibrium particles on forced convective nanofluid flow past a porous channel has been analyzed by Armaghani *et al.* [21, 22]. Nazari *et al.* [23] have presented new models with heat flux splitting and thermal nonequilibrium.

Muthucumaraswamy *et al.* [24, 25] analyzed the different effects along a vertical plate. Seddeek [26] examined the influence of variable viscosity with chemical reaction on hydromagnetic flow with transfer of mass and heat. The impact of chemical reaction, as well as mass and heat transfer across a wedge with heat source and injection or suction, was established by Kandasamy *et al.* [27, 28]. The chemically reactive viscous flow in a magnetic field was also investigated by the authors in the references [29–33].

In this article, the authors investigate the effects of porous nanofluid flow with internal heat generation and chemical reaction. The governing equations coupled with suitable conditions at the boundary are solved numerically by applying Runge-Kutta-Fehlberg scheme along with shooting method. The impact of different physical parameters on temperature, velocity, and concentration fields are scrutinized.

17.2 Mathematical Formulations

Two-dimensional, laminar, steady, incompressible, viscous nanofluid flow across a semi-infinite porous vertical plate is considered here. The physical model for the flow configuration is displayed in Figure 17.1. Here, we take the x-axis in the vertical direction along the plate. The surface is preserved at a constant mass flux m_w and heat flux q_w. Further, we place a source of heat or sink inside the flow. It is presumed that concentration of the diffusing species is extremely low as compared to that of other chemical species away from the surface, and is considered to be infinitesimal. In this flow model, all thermal and physical properties except density are considered to be constant and the Boussinesq approximation is applied. Under these presumptions, the governing equations are as follows (Patil *et al.* [34]):

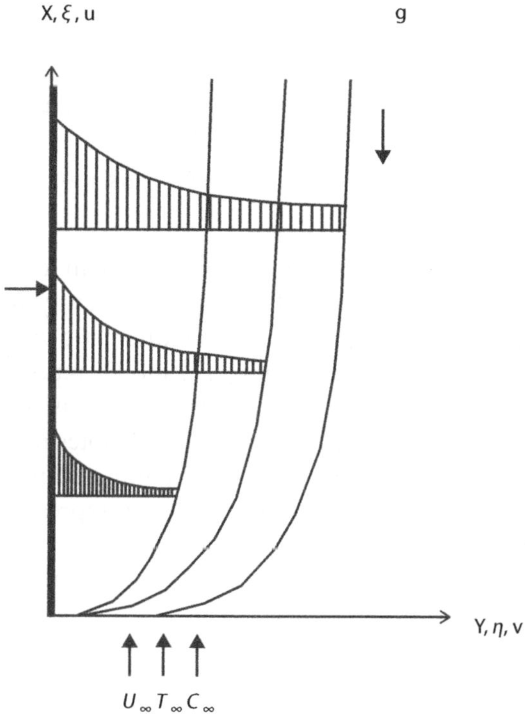

Figure 17.1 Physical flow model.

$$\frac{\partial \bar{u}}{\partial x}+\frac{\partial \bar{v}}{\partial y}=0, \tag{17.1}$$

$$\bar{u}\frac{\partial \bar{u}}{\partial x}+\bar{v}\frac{\partial \bar{u}}{\partial y}=\nu\frac{\partial^2 \bar{u}}{\partial y^2}+g[\beta_T(T-T_\infty)+\beta_c(C-C_\infty)]-\frac{\nu}{K}(\bar{u}-U_\infty), \tag{17.2}$$

$$\bar{u}\frac{\partial T}{\partial x}+\bar{v}\frac{\partial T}{\partial y}=\frac{k}{\rho C_p}\frac{\partial^2 T}{\partial y^2}+\frac{Q_0}{\rho C_p}(T-T_\infty)+\tau\left[D_B\frac{\partial C}{\partial y}\frac{\partial T}{\partial y}+\frac{D_T}{T_\infty}\left(\frac{\partial T}{\partial y}\right)^2\right], \tag{17.3}$$

$$\bar{u}\frac{\partial C}{\partial x}+\bar{v}\frac{\partial C}{\partial y}=D_B\frac{\partial^2 C}{\partial y^2}+\frac{D_T}{T_\infty}\frac{\partial^2 T}{\partial y^2}-k_1(C-C_\infty)^n. \tag{17.4}$$

The adjoining boundary conditions are:

$$\bar{u}=0,\quad -k\frac{\partial T}{\partial y}=q_w,\quad -D\frac{\partial C}{\partial y}=m_w\quad at\ y=0,$$

$$\bar{u}\to U_\infty,\quad T\to T_\infty,\quad C\to C_\infty\quad as\ y\to\infty. \tag{17.5}$$

Here, $\bar{u}$ and $\bar{v}$ are the components of velocity in x and y directions, respectively, U_∞ is the velocity in the x-direction away from the wall, g denotes acceleration due to gravity, v is the kinematic viscosity, T is temperature, C is the species concentration, T_∞, and C_∞ represent these quantities away from the wall, C_p stands for specific heat at constant pressure, Q_0 represents coefficient of heat generation, D denotes mass diffusivity, k_1 represents the first order rate of chemical reaction, n stands for the chemical reaction exponent, q_w is the heat flux, while m_w represents the mass flux per unit area of the plate.

The nondimensional variables introduced to obtain a similarity transformation are the following:

$$\bar{\xi}=\frac{x}{L},\quad \bar{\eta}=\left(\frac{U_\infty}{vx}\right)^{\frac{1}{2}}y,\ \psi(x,y)=(vU_\infty x)^{\frac{1}{2}}f(\bar{\xi},\bar{\eta}),$$

$$(T-T_\infty)=\left(\frac{q_w}{k}\right)\left(\frac{vx}{U_\infty}\right)^{\frac{1}{2}}\theta(\bar{\xi},\bar{\eta}),$$

$$(C-C_\infty)=\left(\frac{m_w}{D}\right)\left(\frac{vx}{U_\infty}\right)^{\frac{1}{2}}\phi(\bar{\xi},\bar{\eta}),$$

$$\bar{u}=U_\infty\frac{\partial f}{\partial\bar{\eta}},\bar{v}=-\frac{(vU_\infty x)^{1/2}}{2x}(f(\bar{\xi},\bar{\eta})+2\bar{\xi}f_{\bar{\xi}}-\bar{\eta}f_{\bar{\eta}}),$$

$$f_{\bar{\eta}}(\bar{\xi},\bar{\eta})=f'(\bar{\xi},\bar{\eta})=F(\bar{\xi},\bar{\eta}),f''(\bar{\xi},\bar{\eta})=F_{\bar{\eta}}(\bar{\xi},\bar{\eta}),$$

$$\text{and}\quad f'''(\bar{\xi},\bar{\eta})=F_{\bar{\eta}\bar{\eta}}(\bar{\xi},\bar{\eta}). \tag{17.6}$$

Here, $\bar{\xi}$ and $\bar{\eta}$ denote the transformed variables and L stands for the characteristic length. The stream function ψ is introduced in a way such that $\bar{u}=\frac{\partial\psi}{\partial y}$ and $\bar{v}=-\frac{\partial\psi}{\partial x}$ and thus the following derivatives are obtained as follows:

$$\frac{\partial \bar{u}}{\partial x}=\frac{U_\infty}{L}\frac{\partial f'}{\partial \bar{\xi}}-\frac{U_\infty}{2}\frac{\bar{\eta}}{x}f'',\quad \frac{\partial \bar{u}}{\partial y}=U_\infty\left(\frac{U_\infty}{\nu x}\right)^{\frac{1}{2}},\text{and }\frac{\partial^2 \bar{u}}{\partial y^2}=U_\infty\frac{U_\infty}{\nu x}f''' \tag{17.7}$$

$$\frac{\partial T}{\partial x}=\frac{q_w}{k}\left(\frac{\nu x}{U_\infty}\right)^{\frac{1}{2}}\left[\frac{1}{L}\frac{\partial \theta}{\partial \bar{\xi}}-\frac{\bar{\eta}}{2x}\theta'+\frac{\theta}{2x}\right],$$

$$\frac{\partial T}{\partial y}=\frac{q_w}{k}\theta',\quad \text{and}\quad \frac{\partial^2 T}{\partial y^2}=\frac{q_w}{k}\left(\frac{U_\infty}{\nu x}\right)^{\frac{1}{2}}\theta'' \tag{17.8}$$

$$\frac{\partial C}{\partial x}=\frac{m_w}{D_B}\left(\frac{\nu x}{U_\infty}\right)^{\frac{1}{2}}\left[\frac{\phi}{2x}+\frac{1}{L}\frac{\partial \phi}{\partial \bar{\xi}}-\frac{\bar{\eta}}{2x}\phi'\right],$$

$$\frac{\partial C}{\partial y}=\frac{m_w}{D_B}\phi',\quad \text{and}\quad \frac{\partial^2 C}{\partial y^2}=\frac{m_w}{D_B}\left(\frac{U_\infty}{\nu x}\right)^{\frac{1}{2}}\phi''. \tag{17.9}$$

Substituting equations (17.6) to (17.9) in the equations (17.1) to (17.4), we now get the following transformed equations:

$$F_{\bar{\eta}\bar{\eta}}+\frac{f}{2(1+\alpha)}F_{\bar{\eta}}-\frac{\bar{\xi}}{DaRe_L}(F-1)+\frac{\lambda}{(1+\alpha)}\bar{\xi}^{-\frac{3}{2}}(\theta+N\phi)$$
$$=\frac{\bar{\xi}}{1+\alpha}[FF_{\bar{\xi}}-f_{\bar{\xi}}F_{\bar{\eta}}], \tag{17.10}$$

$$\theta_{\bar{\eta}\bar{\eta}}+\frac{Pr}{2}f\theta_{\bar{\eta}}+Pr\left(Q\bar{\xi}-\frac{1}{2}F\right)\theta+Pr\left(Nb\theta_{\bar{\eta}}\phi_{\bar{\eta}}+Nt\theta_{\bar{\eta}}^2\right)$$
$$=Pr\bar{\xi}\left[F\theta_{\bar{\xi}}-f_{\bar{\xi}}\theta_{\bar{\eta}}\right], \tag{17.11}$$

$$\phi_{\bar{\eta}\bar{\eta}} + \frac{Le}{2} f\phi_{\bar{\eta}} - \frac{Le}{2} F\phi - Le\Delta\bar{\xi}^{\frac{n+1}{2}}\phi^n + \frac{Nt}{Nb}\theta_{\bar{\eta}\bar{\eta}}$$
$$= Le\xi[F\phi_{\bar{\xi}} - f_{\bar{\xi}}\phi_{\bar{\eta}}]. \tag{17.12}$$

Now, equations (17.10) to (17.12) convert to the following equations:

$$f''' + \frac{f}{2(1+\alpha)} f'' - \frac{\bar{\xi}}{DaRe_L}(f'-1) + \frac{\lambda}{(1+\alpha)}\bar{\xi}^{\frac{3}{2}}(\theta + N\phi)$$
$$= \frac{\bar{\xi}}{1+\alpha}\left[f' \frac{\partial f'}{\partial \bar{\xi}} - \frac{\partial f}{\partial \bar{\xi}} f'' \right], \tag{17.13}$$

$$\theta'' + \frac{Pr}{2} f\theta' + Pr\left(Q\bar{\xi} - \frac{1}{2} f' \right)\theta + Pr(Nb\theta'\phi' + Nt\theta'^2)$$
$$= Pr\bar{\xi}\left[f' \frac{\partial \theta}{\partial \bar{\xi}} - \frac{\partial f}{\partial \bar{\xi}}\theta' \right], \tag{17.14}$$

$$\phi'' + \frac{Le}{2} f\phi' - \frac{Le}{2} f'\phi - Le\Delta\bar{\xi}^{\frac{n+1}{2}}\phi^n + \frac{Nt}{Nb}\theta''$$
$$= Le\bar{\xi}\left[f' \frac{\partial \phi}{\partial \bar{\xi}} - \frac{\partial f}{\partial \bar{\xi}}\phi' \right]. \tag{17.15}$$

Here

$$\alpha = \frac{1}{\nu},\quad Da = \frac{K}{L^2},\quad Re_L = \frac{U_\infty L}{\nu},\quad Q = \frac{Q_0 L}{\rho C_p U_\infty},\quad Gr = \frac{g\beta_T q_w L^4}{k\nu^2 Re_L^{\frac{1}{2}}},$$

$$Gr^* = \frac{g\beta_C m_w L^4}{D_B \nu^2 Re_L^{\frac{1}{2}}},\quad \lambda = \frac{Gr}{Re_L^2},\quad \lambda^* = \frac{Gr^*}{Re_L^2},\quad N = \frac{\lambda^*}{\lambda},\ Pr = \frac{\rho\nu Cp}{k},$$

$$\Delta = k_1\left(\frac{m_w}{D}\right)^{n-1}(Re_L)^{\frac{n-1}{2}}\left(\frac{L^n}{U_\infty}\right),\quad Le = \frac{\nu}{D_B},$$

$$Nb = \frac{(\rho c)_p D_B (C_w - C_\infty)}{(\rho c)_f \nu},\quad Nt = \frac{(\rho c)_p D_T (T_w - T_\infty)}{(\rho c)_f \nu T_\infty} \tag{17.16}$$

stand for the different standard parameters of the fluid.

Using the boundary conditions (17.5) together with the equations (17.8) and (17.9), we get the dimensionless conditions at the boundary as follows:

$$F=0.0,\quad \theta'=-1.0,\quad \phi'=-1.0 \quad at\ \bar{\eta}=0,$$
$$f'=F\rightarrow 1.0,\quad \theta\rightarrow 0.0,\quad \phi\rightarrow 0.0 \quad as\ \bar{\eta}\rightarrow\infty. \tag{17.17}$$

The physical parameters that play a significant role in flow problems with mass and heat transfer are the Nusselt and Sherwood numbers, as well as coefficient of skin friction. These parameters characterize the temperature, concentration and velocity fields and may be obtained from the following relations:

$$Re_L^{-\frac{1}{2}}Nu_x=\frac{\bar{\xi}^{\frac{1}{2}}}{\theta(\bar{\xi},0)}, \tag{17.18}$$

$$Re_L^{-\frac{1}{2}}Sh_x=\frac{\bar{\xi}^{\frac{1}{2}}}{\phi(\bar{\xi},0)}, \tag{17.19}$$

$$Re_L^{\frac{1}{2}}C_f=2(1+\alpha)\bar{\xi}^{-\frac{1}{2}}f''(\bar{\xi},0). \tag{17.20}$$

17.3 Method of Local Nonsimilarity

In this section, we put forth the local nonsimilarity technique that is deployed to simplify the equations (17.13) to (17.15). If we consider the equations up to second level of truncation, the results are fairly accurate and are comparable to the solutions obtained using other methods known in the literature. For the purpose, we propose the functions:

$$A=\frac{\partial f}{\partial\xi},\quad B=\frac{\partial\theta}{\partial\xi},\quad C\frac{\partial\phi}{\partial\xi} \tag{17.21}$$

Now, we differentiate the equations (17.13) to (17.15) with respect to $\bar{\xi}$ to obtain the following:

$$A''' - \frac{1}{DaRe_L}(\bar{\xi}A' + f' - 1) + \frac{1}{2(1+\alpha)}(Af'' + fA'')$$
$$+ \frac{\lambda}{1+\alpha}\bar{\xi}^{\frac{1}{2}}\left(\bar{\xi}(B+NC) + \frac{3}{2}(\theta + N\phi)\right)$$
$$= \frac{1}{1+\alpha}(f'A' - Af'') + \frac{\bar{\xi}}{1+\alpha}(A'^2 - AA'') \tag{17.22}$$

$$B'' + \frac{Pr}{2}(fB' + \theta' A) + Pr\left[\left(Q - \frac{1}{2}A'\right)\theta + \left(Q\bar{\xi} - \frac{1}{2}f'\right)R\right]$$
$$+ Pr(Nb(\theta' C' + \phi' B') + 2Nt\theta' B')$$
$$= Pr((f'B - A\theta') + \bar{\xi}(BA' - AB')), \tag{17.23}$$

$$C'' + \frac{Le}{2}(A\phi' + fC' - A'\phi - f'C)$$
$$- Le\Delta\left(\phi^n \frac{n+1}{2}\bar{\xi}^{\frac{n-1}{2}} + \bar{\xi}^{\frac{n+1}{2}}C^n\right) + \frac{Nt}{Nb}B''$$
$$= Le((f'C - G\phi') + \bar{\xi}(CA' - AC')). \tag{17.24}$$

Next, as we differentiate the boundary conditions (17.17) with respect to $\bar{\xi}$, we obtain

$$\begin{aligned} &A'(\bar{\xi},\bar{\eta}) = 0, \quad B(\bar{\xi},\bar{\eta}) = 0, \quad C(\bar{\xi},\bar{\eta}) = 0 \quad at \quad \bar{\eta} = 0, \\ &A'(\bar{\xi},\bar{\eta}) \to 0, \quad B(\bar{\xi},\bar{\eta}) \to 0, \quad C(\bar{\xi},\bar{\eta}) \to 0 \quad as \quad \bar{\eta} \to \infty. \end{aligned} \tag{17.25}$$

17.4 Results and Discussions

The transformed equations that comprise a set of nonlinear ODEs, together with the conditions at the boundary, are then numerically solved employing the Runge-Kutta-Fehlberg scheme combined with shooting method [35]. The impact of chemical reaction on mixed convective nanofluid flow past a stretching sheet in porous media with the occurrence of internal heat generation is reported here through the study of various parameters of interest. The results are displayed graphically in Figures 17.2 to 17.11 and analyzed. All these computations have been performed at the parameter values $\alpha = 1.0$, $\lambda = 2.0$, $\Delta = 0.5$, $Da = 1.0$, $Re_L = 5.0$, $n = 1.0$, $\xi = 1.0$, $Q = 0.5$, $Pr = 0.7$. The results obtained in this study have also been compared in Table 17.1 with that available in literature.

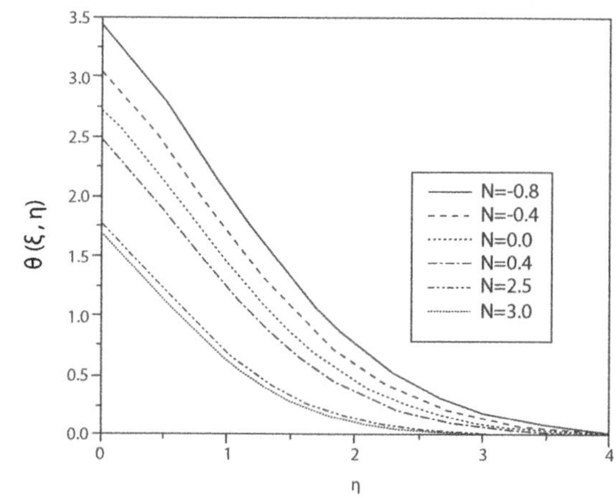

Figure 17.2 Temperature distribution for various values of N.

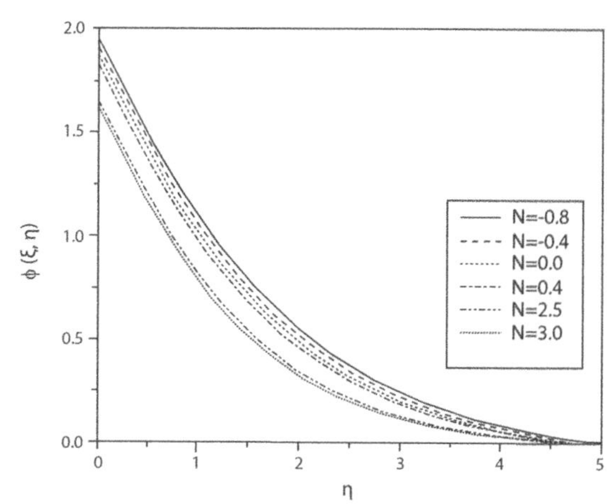

Figure 17.3 Concentration profiles for various values of N.

The temperature profiles for different values of the buoyancy ratio parameter, N are plotted in Figure 17.2 for $Nb = 0.2$, $Nt = 0.2$ and $Le = 0.22$. It may be noticed that temperature of the fluid at the wall decreases with rising values of N. A similar behavior may be observed in Figure 17.3 for the concentration profile. As $\eta \to \infty$, both the temperature as well as concentration profiles diminish to the free stream value zero. The velocity profiles for various values of N are displayed in Figure 17.4. It may be noted that for $N > 0$ when the flow is driven by buoyancy, significant overshoot is noticed in velocity profiles in the vicinity of the wall. Further, for buoyancy assisting flow, ($\lambda > 0$), the velocity increases near the wall with increment

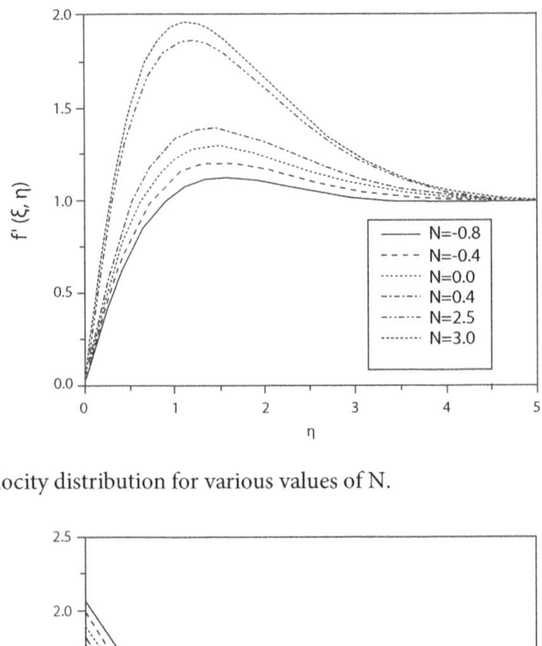

Figure 17.4 Velocity distribution for various values of N.

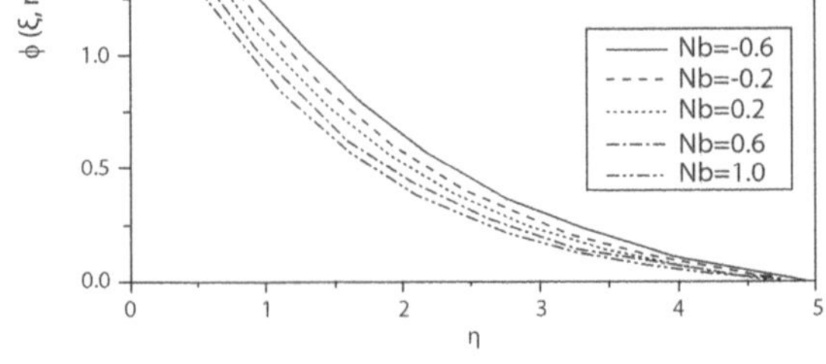

Figure 17.5 Concentration profiles for various values of Nb.

in N. Physically, this may be accounted to the fact that pressure gradient accelerates in the boundary layer. In fact, the parameter N has a substantial impact on the velocity field since N occurs explicitly in velocity. Also, the conjoint impact of assisting buoyant force caused by thermal as well as concentration gradients act as a favourable pressure gradient and escalate the acceleration of the fluid.

The effect of Brownian motion parameter Nb on concentration, temperature, and velocity is shown in Figures 17.5, 17.6, and 17.7 for $N = -0.4$, $Nt = 0.2$ and $Le = 0.22$. It may be noted that the concentration profiles

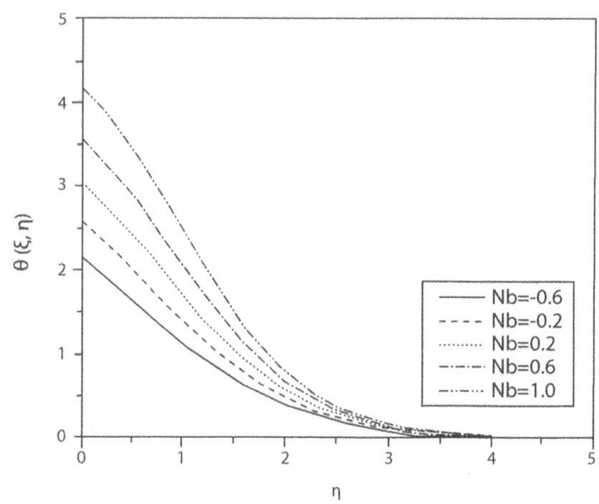

Figure 17.6 Temperature distribution for different values of Nb.

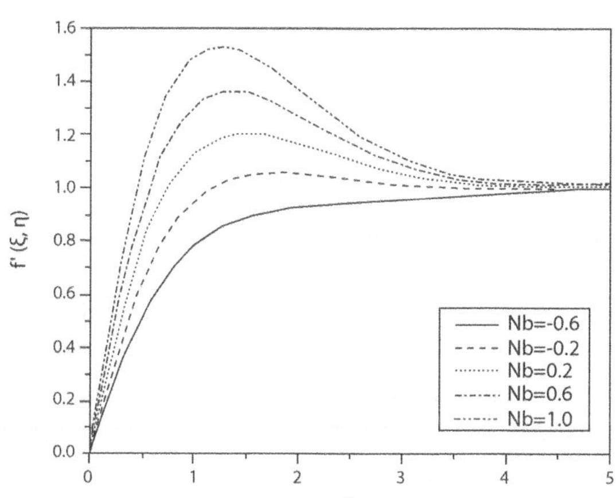

Figure 17.7 Velocity profiles for various values of Nb.

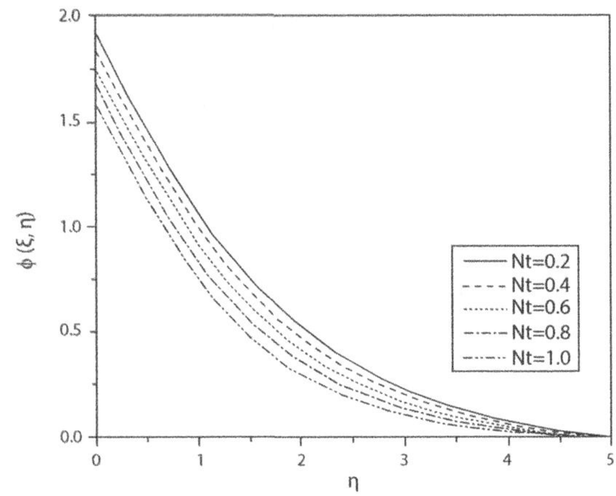

Figure 17.8 Concentration profiles for various values of Nt.

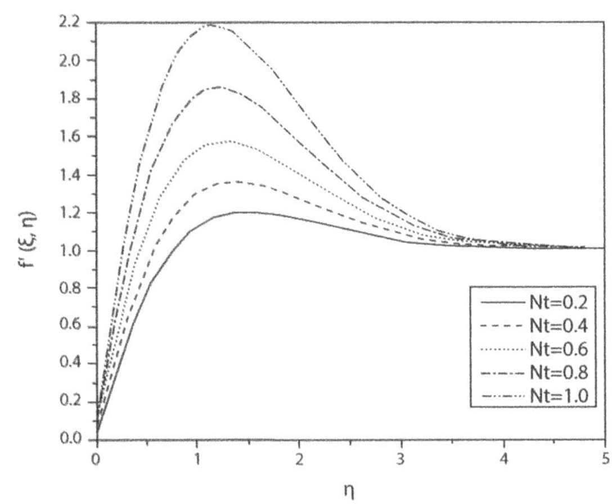

Figure 17.9 Velocity distribution for various values of Nt.

(Figure 17.5) show a decreasing trend with increment in the Brownian motion parameter *Nb*. However, Figures 17.6 and 17.7 demonstrate that in the boundary layer, temperature, as well as velocity, rises with increment in this parameter. Brownian motion of the nanoparticles may actually increase thermal conduction by either of the two mechanisms, namely, direct heat transport by nanoparticles or indirect contribution caused by microconvection of fluid in the vicinity of nanoparticles. Thus, the boundary layer is warmed up due to Brownian motion, and this, in turn, aggravates deposition of particles away from the fluid. This justifies the reduction in magnitude of concentration as displayed in Figure 17.5.

The effect of thermophoresis parameter *Nt* on concentration, velocity, and temperature profiles is exhibited in Figures 17.8, 17.9, and 17.10 for $N = -0.4$, $Nb = 0.2$, and $Le = 0.22$. Increment in both velocity and temperature profiles may be observed with increasing values of *Nt*. On the other hand, concentration diminishes with increase in *Nt*. It may be noted that

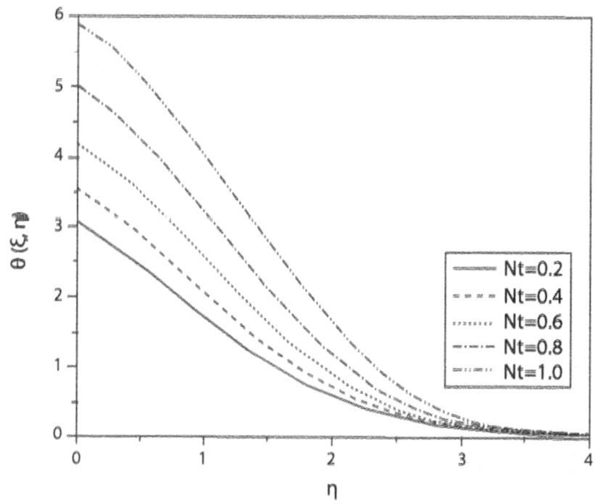

Figure 17.10 Temperature profiles for various values of Nt.

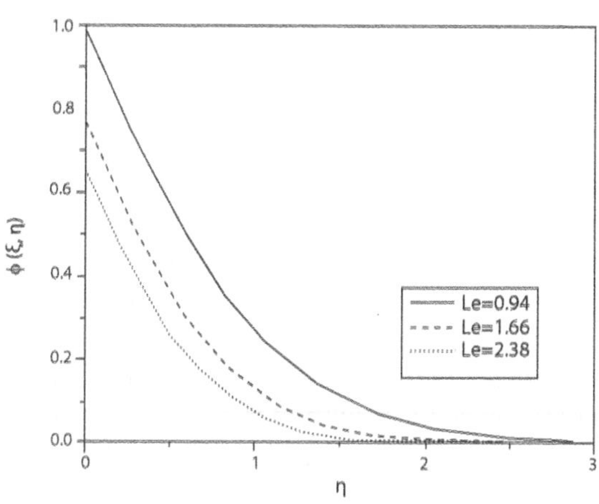

Figure 17.11 Concentration profiles for various values of Le.

Table 17.1 Comparison of Nusselt number for various values of *Pr*.

Pr	Lee *et al.* [36]	Patil [37]	Chang and Lee [38]	Current results
0.10	0.263412	0.263413	0.26391401	0.2638801
0.70	0.483801	0.483823	0.48377611	0.4839401
7.00	0.869703	0.869715	0.87002602	0.8689002

increment in Nt leads to a rise in the difference of temperature between the ambient fluid and the wall.

The impact of Lewis number Le on concentration profile may be noted from Figure 17.11 for $N = -0.4$, $Nt = 0.2$ and $Nb = 0.2$. One may observe that concentration profile diminishes with increment in Lewis number. The decrement in concentration is actually accompanied by reduction in concentration boundary layer thickness.

The Nusselt number as obtained in this study for various values of Pr has been presented in Table 17.1. It may be clearly observed that the current results are in excellent agreement with that of Lee *et al.* [36], Patil [37], Chang and Lee [38].

17.5 Concluding Remarks

The mixed convective nanofluid flow in a porous medium in presence of absorption or generation of heat along with homogeneous chemical reaction of nth order is investigated in this work. A comparison of the current results with earlier work available in literature in special cases shows notable agreement. The impact of different pertinent parameters is studied in details. The conclusions acquired from this analysis are as follows:

1. Increment in the value of N, the buoyancy ratio parameter, enhances the velocity profile while it diminishes both the thermal and concentration boundary layers.
2. The Brownian motion, as well as thermophoresis parameters, enhance the velocity and thermal boundary layer thickness, while reducing the thickness of the concentration profile.
3. Concentration reduces with increment in Lewis number.
4. The skin-friction coefficient and the Nusselt number are affected significantly due to the thermophoresis and Brownian motion parameters.

References

1. RamReddy, Ch., Murthy, P.V.S.N., Chamkha, A.J., Rashad, A.M., Influence of viscous dissipation on free convection in a non-darcy porous medium saturated with nanofluid in the presence of magnetic field. *Open Transp. Phenom. J.*, 5, 20–29, **2013**.
2. Rahman, A., Nadeem, S., Malik, M.Y., Stagnation flow of couple stress nanofluid over an exponentially stretching sheet through a porous medium. *J. Power Technol.*, 93, 2, 122–132, **2013**.
3. Bhogare, R.A. and Kothawale, B.S., A review on applications and challenges of Nanofluids as coolant in automobile radiator. *Int. J. Sci. Res. Publ.*, 3, 8, 1–11, **2013**.
4. Kuppalapalle, V., Vinayaka, P.K., Chiu-On, N.G., The effect of variable viscosity on the flow and heat transfer of a viscous Ag- water and Cu- water nanofluids. *J. Hydrodynam.*, 25, 1, 1–9, **2013**.
5. Dhal, R.K., Unsteady MHD free convection flow in a nanofluid through porous medium with suction and heat source. *Int. J. Sci. Innov. Discoveries*, 4, 1, 45–55, **2014**.
6. Takabi, B. and Salehi, S., Augmentation of the heat transfer performance of a sinusoidal corrugated enclosure by employing hybrid nanofluid. *Adv. Mech. Eng.*, 2014, 16, **2014**, doi.org/10.1155/2014/147059, Article ID 147059.
7. Sheikholeslami, M., Gorji-Bandpy, M., Ganji, D.D., Rana, P., Soleimani, S., Magnetohydrodynamic free convection of Al_2O_3-water nanofluid considering Thermophoresis and Brownian motion effects. *Comput. Fluids*, 94, 147–160, **2014**.
8. Uddin, M.J., Beg, O.A., Ismail, A.I.Md., Mathematical modelling of radiative hydromagnetic thermosolutal nanofluid convection slip flow in saturated porous media. *Math. Probl. Eng.*, 2014, Article ID 179172, 11pp, **2014**, doi.org/10.1155/2014/179172, Article ID 179172.
9. Makinde, O.D., Khamis, S., Tshehla, M.S., Franks, O., Analysis of heat transfer in Berman flow of nanofluids with Navier slip, viscous dissipation and convective cooling. *Adv. Math. Phys.*, 2014, 13, **2014**, doi.org/10.1155/2014/809367, Article ID 809367.
10. Khan, S.K. and Sanjayanand, E., Viscoelastic boundary layer flow and heat transfer over an exponential stretching sheet. *Int. J. Heat Mass Transf.*, 48, 8, 1534–1542, **2005**.
11. Afzal, N., Momentum and thermal boundary layers over a two-dimensional or axisymmetric non-linear stretching surface in a stationary fluid. *Int. J. Heat Mass Transf.*, 53, 1–3, 540–547, **2010**.
12. Nandeppanava, M.M., Vajravelu, K., Abel, M.S., Ng, C.-O., Heat transfer over a nonlinearly stretching sheet with non-uniform heat source and variable wall temperature. *Int. J. Heat Mass Transf.*, 54, 23–24, 4960–4965, **2011**.

13. Mondal, H. and Bharti, S., Spectral quasi-linearization for MHD nanofluid stagnation boundary layer flow due to a stretching/shrinking surface. *J. Appl. Comput. Mech.*, 6, 4, 1058–1068, **2020**.
14. Mondal, H. and Sibanda, P., Spectral quasi-linearization method for entropy generation using the Cattaneo-Christov heat flux model. *Int. J. Comput. Methods*, 17, 05, 1940002, **2020**.
15. Mondal, H., Das, S., Kundu, P.K., Influence of an inclined stretching cylinder over MHD mixed convective nanofluid flow due to chemical reaction and viscous dissipation. *Heat Transf.*, 49, 4, 2183–2193, **2020**.
16. Mondal, H., Mishra, S., Kundu, P.K., Sibanda, P., Entropy generation of variable viscosity and thermal radiation on magneto nanofluid flow with dusty fluid. *J. Appl. Comput. Mech.*, 6, 1, 171–182, **2020**.
17. De, P., Mondal, H., Bera, U.K., Influence of nanofluids on magnetohydrodynamic heat and mass transfer over a non-isothermal wedge with variable viscosity and thermal radiation. *J. Nanofluids*, 3, 391–398, **2014**.
18. Rehena, N., Alim, M.A., Chamkha, A.L., Buoyancy-driven heat transfer of water Al_2O_3 nanofluid in a closed chamber: Effects of solid volume fraction, Prandtl number and aspect ratio. *Int. J. Heat Mass Transf.*, 55, 7355–7365, **2012**.
19. Hwang, K.S., Lee, J.H., Jang, S.P., Buoyancy-driven heat transfer of water-based Al_2O_3 nanofluids in a rectangular cavity. *Int. J. Heat Mass Transf.*, 50, 4003–4010, **2007**.
20. Maghrebi, M.J., Nazari, M., Armaghani, T., Forced convection heat transfer of nanofluids in a porous channel. *Transp. Porous Media*, 93, 3, 401–413, **2012**.
21. Armaghani, T., Maghrebi, M.J., Chamkha, A.J., Nazari, M., Effects of particle migration on nanofluid forced convection heat transfer in a local thermal non-equilibrium porous channel. *J. Nanofluids*, 3, 1, 51–59, **2014**.
22. Armaghani, T., Chamkha, A.J., Maghrebi, M., Nazari, M., Numerical analysis of a nanofluid forced convection in a porous channel: A new heat flux model in LTNE condition. *J. Porous Media*, 17, 637–646, **2014**.
23. Nazari, M., Maghrebi, M.J., Armaghani T., T., Chamkha, A.J., New models for heat flux splitting at the boundary of a porous medium: Three energy equations for nanofluid flow under thermal non-equilibrium condition. *Can. J. Phys.*, 92, 11, 1312–1319, **2014**, DOI: 10.1139/cjp-2013-0451.
24. Muthucumaraswamy, R. and Ganeshan, P., On impulsive motion of a vertical plate with heat flux and diffusion of chemically reactive species. *Forsch. Ingenieurwes.*, 66, 17–23, **2000**.
25. Muthucumaraswamy, R. and Ganeshan, P., First order chemical reaction on flow past an impulsively started vertical plate with uniform heat and mass flux. *Acta Mech.*, 147, 45–57, **2001**.
26. Seddeek, M.A., Finite element method for the effects of chemical reaction, variable viscosity, thermophoresis and heat generation/absorption on a

boundary layer hydromagnetic flow with heat and mass transfer over a heat surface. *Acta Mech.*, 177, 1–18, **2005**.
27. Kandasamy, R., Perisamy, K., Sivagnana Prabhu, K.K., Effects of chemical reaction, heat and mass transfer along a wedge with heat source and concentration in the presence of suction or injection. *Int. J. Heat Mass Transf.*, 48, 1388–1394, **2005**.
28. Kandasamy, R., Perisamy, K., Sivagnana Prabhu, K.K., Chemical reaction, heat and mass transfer on MHD flow over a vertical stretching surface with heat source and thermal stratification effects. *Int. J. Heat Mass Transf.*, 48, 4557–4561, **2005**.
29. Raptis, A. and Perdikis, C., Viscous flow over a non-linearly stretching sheet in the presence of chemical reaction and magnetic field. *Int. J. Non-Linear Mech.*, 42, 527–529, **2006**.
30. Rudraswamy, N.G. and Gireesha, B.J., Influence of chemical reaction and thermal radiation on MHD boundary layer flow and heat transfer of a nanofluid over an exponentially stretching sheet. *J. Appl. Math. Phys.*, 2, 24–32, **2014**.
31. Hammed, A.O., Sibanda, P., Mondal, H., On MHD mixed convective stagnation point flow of an Eyring-Powell nanofluid over a stretching cylinder with thermal slip conditions. *J. Cent. South Univ.*, 26, 1172–1183, **2019**.
32. Goqo, S.P., Oloniiju, S.D., Mondal, H., Sibanda, P., Motsa, S.S., Entropy generation on MHD radiative viscous nanofluid flow over a porous wedge using bivariate spectral quasi-linearization method. *Case Stud. Therm. Eng.*, 12, 774–788, **2018**.
33. Mondal, H., Ghosh, S., Roy, P.K., Chatterjee, S., Effects of ion-slip and Hall currents on magnetohydrodynamic nanofluid flow with thermal diffusion using spectral quasi-linearization method. *J. Nanofluids*, 10, 4, 608–615, **2021**.
34. Patil, P.M., Chamkha, A.J., Roy, S., Effects of chemical reaction on mixed convection flow of a polar fluid through a porous medium in presence of internal heat generation. *Meccanica*, 47, 483–499, **2012**.
35. Kuznetsov, A.V. and Nield, D.A., Natural convective boundary-layer flow of a nanofluid past a vertical plate. *Int. J. Therm. Sci.*, 49, 243–247, **2010**.
36. Lee, S.L., Chen, T.S., Armaly, B.F., Mixed convection along vertical cylinders and needles with uniform surface heat flux. *ASME J. Heat Transf.*, 109, 711–716, **1987**.
37. Patil, P.M., Chemical reaction effects on free convective flow of a polar fluid from a vertical plate with uniform heat and mass fluxes. *IOSR J. Math.*, 6, 5, 66–85, **2013**.
38. Chang, C.L. and Lee, Z.Y., Free convection on a vertical plate with uniform and constant heat flux in a thermally stratified micropolar fluid. *Mech. Res. Commun.*, 35, 421–427, **2008**.

18

Numerical Solution of Partial Differential Equations: Finite Difference Method

Roushan Kumar[1], Rakhi Tiwari[2*] and Rashmi Prasad[1]

[1]*Department of Mathematics, Central University of South Bihar, Gaya, Bihar, India*

[2]*Department of Mathematics, Nitishwar College, a Constituent Unit of Babasaheb Bhimrao Ambedkar Bihar University, Muzaffarpur, Bihar, India*

Abstract

This book chapter provides a broad description of the finite difference methods for parabolic differential equations (heat equation). Section 18.1 covers an overview of second-order partial differential equation via: classification, initial, and boundary conditions. Section 18.2 discusses the finite difference method, in which we provide the discretization of the domain and finite difference approximation of heat equation and some primary definitions (consistency, convergence, stability). Section 18.3 describes the general explicit method and its characteristics, e.g., truncation error, stability (non-Neumann), and some well-known explicit methods. Section 18.4 presents general two-level implicit methods and some characteristics, e.g., truncation error, stability (non-Neumann), an explicit method. Lastly, we provide the numerical example solved by a different numerical method for verification and validation of the computed solution; after that, concluding remarks are presented.

Keywords: Partial differential equations, finite difference method

18.1 Introduction

Partial differential equations (PDE) arise in the mathematical modeling of many problems in science and engineering, e.g., in waves and

**Corresponding author*: rakhitiwari.rs.apm12@itbhu.ac.in

Sharmistha Ghosh, M. Niranjanamurthy, Krishanu Deyasi, Biswadip Basu Mallik, and Santanu Das (eds.) Mathematics and Computer Science Volume 1, (353–372) © 2023 Scrivener Publishing LLC

electromagnetics, fluid mechanics, heat propagation in solid, elasticity, quantum mechanics, electromagnetic theory, potential theory. Partial differential equations of second order govern various critical physical phenomena. The general linear partial differential equation of the second order in two independent variables x and y, is

$$a\frac{\partial^2 u}{\partial x^2}+b\frac{\partial^2 u}{\partial x\partial y}+c\frac{\partial^2 u}{\partial y^2}+g\left(\frac{\partial u}{\partial x},\frac{\partial u}{\partial y},u,x,y\right)=0. \tag{18.1}$$

Such a partial differential equation is said to be

i. Linear, if a, b, and c are functions of x and y, and g is a linear function of $x, y, \frac{\partial u}{\partial x}$, and $\frac{\partial u}{\partial y}$.
ii. Semilinear, if a, b and c are functions of independent variables x, and y.
iii. Quasilinear, if a, b and c are functions of $x, y, \frac{\partial u}{\partial x}$, and $\frac{\partial u}{\partial y}$.

Otherwise, it is nonlinear.

The most general second-order PDE in two independent variables x and y has the form:

$$a(x,y)\frac{\partial^2 u}{\partial x^2}+b(x,y)\frac{\partial^2 u}{\partial x\partial y}+c(x,y)\frac{\partial^2 u}{\partial y^2}+d(x,y)\frac{\partial u}{\partial x}$$
$$+e(x,y)\frac{\partial u}{\partial y}+f(x,y)u+g(x,y)=0 \tag{18.2}$$

If the free term (that contains no unknown function u) $g(x, y)$ of PDE is identically zero, equation 18.2 is said to be homogenous otherwise nonhomogeneous.

The function $u(x, y)$ is called solution of equation 18.2 if it has continuous derivatives of all orders and satisfies equation 18.2. It represents an integral surface $u(x, y)$ in the plane (x, y, u). If there exists curve on integral surface across, which the partial derivatives $\frac{\partial^2 u}{\partial x^2}$, $\frac{\partial^2 u}{\partial x\partial y}$ and $\frac{\partial^2 u}{\partial y^2}$ are discontinuous or indeterminate called characteristics.

The classification of PDE is governed from the following:
Equation 18.2 is recognized as

(i) Elliptic, if $b^2 - 4ac < 0$,
(ii) Parabolic, if $b^2 - 4ac = 0$,
(iii) Hyperbolic, if $b^2 - 4ac > 0$.

The well-known examples of three types of PDE are as follows:

1. Heat equation: $\frac{\partial^2 u}{\partial x^2} = \frac{\partial u}{\partial t}$ (parabolic type)
2. Wave equation: $\frac{\partial^2 u}{\partial x^2} = \frac{\partial^2 u}{\partial t^2}$ (hyperbolic type)
3. Elliptic equation: $\frac{\partial^2 u}{\partial x^2} + \frac{\partial^2 u}{\partial y^2} = 0$ (elliptic type)

The parabolic and hyperbolic types of PDE have either initial value problem or initial-boundary value problem. Elliptic PDE is always a boundary value problem. The boundary conditions are of the following types:

i. Dirichlet condition: the solution is prescribed along the boundary.
ii. Neumann condition: the derivative of solution is prescribed along the boundary.
iii. Mixed condition: the solution and its derivative are prescribed along the boundary.

Few classes of PDE can be solved by analytical methods, which are also complicated by requiring the use of advanced mathematical techniques. In most cases, it is easier to develop approximate solutions by numerical methods. Of all the numerical methods available for partial differential equations, the method of finite differences is most commonly used. In this method, their finite difference approximations replace the derivatives appearing in the equation and the boundary conditions. Then the given equation is changed to a system of linear equations, solved by iterative procedures. This process is slow but produces good results in many boundary value problems. An added advantage of this method is that electronic computers can carry the computation. We assume throughout our discussion that a mathematical problem is well-posed.

18.2 Finite Difference Method

The finite difference method (FDM) is famous as a fundamental tool for obtaining the solutions of linear and non-linear partial differential equations in the purview of various initial and boundary conditions. Its mathematical background is obtained in various books [1, 2], as well as some research articles [3]. At the present scenario, this technique/method is gaining popularity due to the increasing usage of computers for determining the solutions of partial differential equations in FDM. Additionally, FDM is found to be simple for solving PDE compared to the other mathematical tools of solving PDE due to the derivation of discretization [4–7]. This chapter presents brief numerical solution of partial differential equation.

18.2.1 Finite Difference Approximations to Derivatives

Let us consider a rectangular region R in the x, y plane. Divide this region into a rectangular network of sides $\Delta x = h$ and $\Delta y = k$ as shown in Figure 18.2.

$$\frac{\partial u}{\partial x}=\frac{u(x+h,y)-u(x,y)}{h}+O(h), \tag{18.3}$$

$$\frac{\partial u}{\partial x}=\frac{u(x,y)-u(x-h,y)}{h}+O(h), \tag{18.4}$$

$$\frac{\partial u}{\partial x}=\frac{u(x,+hy)-u(x-h,y)}{h}+O(h^2), \tag{18.5}$$

Following the similar pattern

$$\frac{\partial^2 u}{\partial x^2}=\frac{u(x+h,y)-2u(x,y)+u(x-h,y)}{h}+O\left((h^2)\right). \tag{18.6}$$

18.2.2 Discretization of Domain

In the finite difference method, we divide on the region into the mesh of line as follows as shown in Figures 18.1 and 18.2:

Figure 18.1 Mesh in one-dimension.

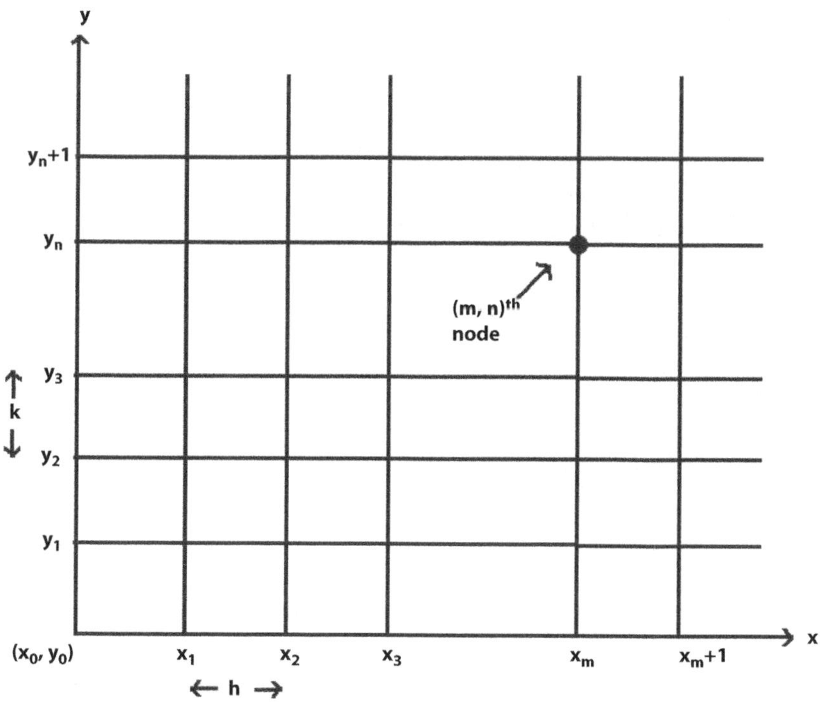

Figure 18.2 Mesh in two-dimension.

a. One-dimensional case

$$x_m = x_0 + mh \; ; m = 0,1,2... \tag{18.7}$$

b. Two-dimensional case

$$x_m = x_0 + mh \; ; m = 0,1,2... \tag{18.8}$$

$$y_m = y_0 + nk \; ; n = 0,1,2... \tag{18.9}$$

where h and k are mesh size in x and y direction, respectively.

Nodes: The point of intersection of the mesh lines are known as nodes.

Note: If an initial value problem is considered then we have $t_k = tk$. Since, in initial value problem $t_0 = 0$, where k is the step length in the t direction.

18.2.3 Difference Scheme of Partial Differential Equation

The partial derivatives in the differential equation are substituted by appropriate finite difference. Then differential equation is converted into the differential equation at each node. The given boundary or initial data is used the difference equation at the nodes near or on the boundary. The solution of this system of equation gives the numerical solution of the given initial boundary value problem.

The difference method must satisfy three basic requirements:

a. Consistency
b. Convergency
c. Stability

Consistency
Consider one dimensional heat equation

$$u_t = u_{xx} \tag{18.10}$$

Take numerical solution at (x_m, t_n) as u_m^n and exact solution as $u(x_m, t_n)$.

Approximating $k\frac{\partial}{\partial t} = \Delta_t$ and $h^2\frac{\partial^2}{\partial x^2} = \delta_x^2$, then equation 18.10 is

$$\frac{1}{k}\Delta u_m^n = \frac{1}{h^2}\delta_x^2 u_m^n,$$

or

$$u_m^{n+1} = u_m^n + \beta(u_{m+1}^n - 2u_m^n + u_{m-1}^n). \tag{18.11}$$

where $\beta = \frac{k}{h^2}$.

Now truncation error is defined as

$$T.E. = u(x_m, t_{n+1}) - u(x_m, t_n) - \beta[u(x_{m+1}, t_n) - 2u(x_m, t_n) + u(x_{m-1}, t_n)].$$

Applying Taylor series expansion and using the fact $\frac{\partial}{\partial t} = \frac{\partial^2}{\partial x^2}$, we find

$$T.E. = \frac{k^2}{2}\frac{\partial^2 u}{\partial t^2} - \frac{kh^2}{12}\frac{\partial^4 u}{\partial x^4} + O(k^3 + kh^4). \qquad (18.12)$$

The order of truncation method is defined by $\frac{1}{k}(T.E.)$ and from equation 18.12 it is of $O(k + h^2)$.

Since from equation 18.10, we have $u_{tt} = u_{xxxx}$.

Therefore equation 18.11 becomes

$$T.E. = \frac{kh^2}{2}(6\beta - 1)\frac{\partial^4 u}{\partial x^4} + O(k^3 + kh^4). \qquad (18.13)$$

For $\beta = \frac{1}{6}$, order of TE is $O(k^2 + h^4)$.

When $h \to 0, k \to 0$ and $T.E. \to 0$.

Thus, difference equation is equivalent to differential equation and is said to be consistent with the differential equation. Consistency means

Difference equation ≡ Differential equation

Consistency does not generate that the solution of the difference equation approximates the solution of the given differential equation.

Convergency

When the solution of the differential equation is determined by approaching $h \to 0, k \to 0$. Then the difference equation becomes convergent.

Numerical solution → Exact solution as h → 0, k → 0.

The numerical solution contains error like round-off error. We can write

$$\varepsilon_m^n = u(x_m, t_n) - u_m^n, \qquad (18.14)$$

where ε_m^n is error occurd at $(m, n)^{th}$ node. Since ε_m^n also satisfies equation 18.11 i.e.

$$\varepsilon_m^{n+1} = \varepsilon_m^n + \beta(\varepsilon_{m+1}^n - 2\varepsilon_m^n + \varepsilon_{m-1}^n). \qquad (18.15)$$

Stability

If the error occurred at one stage and does not grow in next step when the computation is continued, we say that difference scheme 18.12 is stable. Since the error at $t = 0$ propagate according to equation 18.11. To check stability of difference method, we use Von Neumann stability Method. For the linear difference scheme with constant coefficient, we consider

$$\varepsilon_m^n = A\xi^n e^{i\alpha h} \tag{18.16}$$

Error at initial level does not grow it is necessary and sufficient that

$$|\xi| \le 1. \tag{18.17}$$

Substituting equations 18.16 into 18.15, yields

$$\begin{aligned}\xi - 1 &= \beta(e^{i\alpha h} - 2 + e^{-i\alpha h}),\\ &= 2\beta(\cos\alpha h - 1),\\ &= -4\beta\sin^2\frac{\alpha h}{2}.\end{aligned} \tag{18.18}$$

From 18.17, we have

$$-1 \le 1 - 4\beta\sin^2\frac{\alpha h}{2} \le 1.$$

Since the above inequality satisfies for all value of α. Therefore, $\beta \le \frac{1}{2}$.

Note: The equation 18.12 is known as Schmidt explicit difference scheme.

18.3 Multilevel Explicit Difference Schemes

The general three-level explicit difference scheme for equation 18.10 can be written as

$$F(\nabla_t)u_m^{n+1} = \beta\delta_x^2 u_m^n, \tag{18.19}$$

Where $F(\nabla_t) = (1 - \alpha\nabla_t)^{-1}(\nabla_t + \tau\nabla_t^2)$,

Then $(1 - \alpha\nabla_t)^{-1}(\nabla_t + \tau\nabla_t^2)u_m^{n+1} = \beta\delta_x^2 u_m^n$,

Or $$(\nabla_t + \tau\nabla_t^2)u_m^{n+1} = \beta(1-\alpha\nabla_t)\delta_x^2 u_m^n. \qquad (18.20)$$

Now the truncation error (TE) at $(m, n)^{th}$ level can be written

$$T_m^n = (\nabla_t + \tau\nabla_t^2)u(x_m, t_{n+1}) - \beta(1-\alpha\nabla_t)\delta_x^2 u(x_m, t_n).$$

Now expending the dependent variable at $(m, n + 1)th$ level in terms of $(m, n)th$ level using Taylore series

$$\delta_x^2 u(x_m, t_n) = u(x_{m+1}, t_n) - 2u(x_m, t_n) + u(x_{m-1}, t_n).$$
$$= h^2\frac{\partial^2 u}{\partial x^2} + \frac{h^4}{12}\frac{\partial^4 u}{\partial x^4} + \frac{h^6}{360}\frac{\partial^6 u}{\partial x^6} + \cdots \qquad (18.21)$$

$$\nabla_t u(x_m, t_n) = u(x_m, t_n) - u(x_m, t_{n-1}).$$
$$= k\frac{\partial u}{\partial t} - \frac{k^2}{2}\frac{\partial^2 u}{\partial t^2} + \frac{k^3}{6}\frac{\partial^{18} u}{\partial t^{18}} + \cdots \qquad (\text{using } u_t = u_{xx})$$
$$= k\frac{\partial^2 u}{\partial x^2} - \frac{k^2}{2}\frac{\partial^4 u}{\partial x^4} + \frac{k^6}{6}\frac{\partial^6 u}{\partial x^6} + \cdots \qquad (18.22)$$

Using equations 18.21 and 18.22, yields

$$\nabla_t \delta_x^2 u(x_m, t_n) = \left(k\frac{\partial^2}{\partial x^2} - \frac{k^2}{2}\frac{\partial^4}{\partial x^4} + \frac{k^6}{360}\frac{\partial^6}{\partial x^6} + \cdots\right)$$
$$\left(h^2\frac{\partial^2 u}{\partial x^2} + \frac{h^4}{12}\frac{\partial^4 u}{\partial x^4} + \frac{k^3}{6}\frac{\partial^6 u}{\partial x^6} + \cdots\right).$$

$$= kh^2\frac{\partial^4 u}{\partial x^4} + \left(kh^4 - \frac{k^2h^2}{2}\right)\frac{\partial^6 u}{\partial x^6} + \cdots$$

Using the above equation,

$$(1-\alpha\nabla_t)u(x_m,t_n)=h^2\frac{\partial^2 u}{\partial x^2}+\left(\frac{h^2}{12}-\alpha^2\right)\frac{\partial^4 u}{\partial x^4}$$
$$+\left(\frac{h^6}{360}-\frac{\alpha kh^4}{12}+\frac{\alpha k^2 h}{2}\right)\frac{\partial^6 u}{\partial x^6}+\cdots$$

With similar argument

$$(\nabla_t+\tau\nabla_t^2)u(x_m,t_n)=k\frac{\partial^2 u}{\partial x^2}+\left(\frac{k^2}{2}+\tau k^2\right)\frac{\partial^4 u}{\partial x^4}+\frac{k^3}{6}\frac{\partial^6 u}{\partial x^6}+\cdots$$

Stability

The error equation of equation 18.20 is

$$(\nabla_t+\tau\nabla_t^2)\varepsilon_m^{n+1}-\beta(1-\alpha\nabla_t)\delta_x^2\varepsilon_m^n=0.$$

Now $$(\nabla_t+\tau\nabla_t^2)\varepsilon_m^{n+1}=(\varepsilon_m^{n+1}-\varepsilon_m^n)+\tau(\varepsilon_m^{n+1}-2\varepsilon_m^n+\varepsilon_m^{n-1}),$$

$$\varepsilon_m^{n+1}-\varepsilon_m^n+\tau(\varepsilon_m^{n+1}-2\varepsilon_m^n+\varepsilon_m^{n-1})-\beta[\delta_x^2\varepsilon_m^n-\alpha\delta_x^2(\varepsilon_m^n-\varepsilon_m^{n-1})]=0,$$
$$\Rightarrow(1+\tau)\varepsilon_m^{n+1}+\tau\varepsilon_m^{n+1}-(1+2\tau)\varepsilon_m^n-\beta[(1-\alpha)\varepsilon_x^2\varepsilon_m^n-\alpha\delta_x^2\varepsilon_m^{n-1}]=0. \quad (18.23)$$

Substituting $\varepsilon_m^n=A\xi^n e^{i\alpha h}$ in equation 18.3, yields

$$\xi(\xi-1)+\tau(\xi-1)^2+4\beta\sin^2\left(\frac{\theta h}{2}\right)[(1-\alpha)\xi+\alpha]+2\beta\sin^2\left(\frac{\theta h}{2}\right)=0. \quad (18.24)$$

This is quadratic characteristic equation of the scheme.

Instead of applying Newman Criteria of stability on each not on the above, let us use Routh Hurwitz criteria in which $|\xi|\le 1$ is transformed into $z\le 0$ plane by transformation,

$$\xi=\frac{1+Z}{1-Z} \quad (18.25)$$

Substituting equation 18.25 in equation 18.24, we have

$$Z^{2}\left[(1+2\tau)-2\beta(1-2\alpha)\sin^{2}\left(\frac{\theta h}{2}\right)\right]+Z\left[1-4\alpha\beta\sin^{2}\left(\frac{\theta h}{2}\right)\right]$$
$$+2\lambda\sin^{2}\left(\frac{\theta h}{2}\right)=0. \qquad (18.26)$$

Therefore, from Hurwitz Criteria, we have

(i) $(1+2\tau)-2\beta(1-2\alpha)\sin^{2}\left(\frac{\theta h}{2}\right)\geq 0.$

(ii) $1-4\alpha\beta\sin^{2}\left(\frac{\theta h}{2}\right)\geq 0.$

(iii) $\sin^{2}\left(\frac{\theta h}{2}\right)\geq 0.$

As $0\leq\sin^{2}\left(\frac{\theta h}{2}\right)\leq 1.$

Since the above inequality will satisfied for all θ if it reduces to

(i) $(1+2\tau)-2\beta(1-2\alpha)\geq 0$

(ii) $1-4\alpha\beta\geq 0$

Now, Dufort-Frankel 3-level explicit scheme is obtained for $\tau=\beta-\frac{1}{2}$. which is

$$\left[\nabla_t+\left(\beta-\frac{1}{2}\right)\nabla_t^2\right]u_m^{n+1}=\beta\,\delta_x^2\,u_m^n,$$

Or $$(1+2\beta)u_m^{n+1}=(1-2\beta)u_m^{n-1}+2\beta(u_{m+1}^n+u_{m-1}^n). \qquad (18.27)$$

Dufort-Frankel method is unconditionally stable (stable for all β) and order of truncation error is $O\left(k^2+h^2+\left(\frac{k}{h}\right)^2\right)$.

For $\alpha=0,\tau=-\frac{1}{2}$, gives Richardson formula

i.e. $$\left(\nabla_t-\frac{1}{2}\nabla_t^2\right)u_m^{n+1}=\beta\,\delta_x^2 u_m^n,$$

$$\Rightarrow \quad u_m^{n+1}=u_m^{n-1}+2\lambda\delta_x^2\,u_m^n. \qquad (18.28)$$

Stability criteria gives $-\beta \geq 0$ or $\beta < 0$, which is not possible, i.e., Richardson Scheme (18-level explicit) is unstable.

18.4 Two-Level Implicit Scheme

The general two-level implicit scheme for heat equation is

$$(1+\tau\delta_x^2)\nabla_t u_m^{n+1} = \beta(1-\alpha\nabla_t)\delta_x^2 u_m^{n+1},$$

$$\Rightarrow \qquad (1+\tau\delta_x^2)\nabla_t u_m^{n+1} = \beta[\delta_x^2 u_m^{n+1} - \alpha\delta_x^2(u_m^{n+1} - u_m^n)],$$

$$\text{or} \qquad [1+(\tau+\alpha\beta-\beta)\delta_x^2]u_m^{n+1} = [1+(\tau+\alpha\beta)\delta_x^2]u_m^n. \qquad (18.29)$$

The error equation will be

$$[1+(\tau+\alpha\beta-\beta)\delta_x^2]\epsilon_m^{n+1} - [1+(\tau+\alpha\beta)\delta_x^2]\epsilon_m^n = 0. \qquad (18.30)$$

Let $\varepsilon_m^n = A\xi^n e^{i\theta mh}$

The characteristic equation is

$$\xi\left[1-4(\tau+\alpha\beta-\beta)\sin^2\left(\frac{\theta h}{2}\right)\right]-\left[1+(\tau+\alpha\beta)\sin^2\left(\frac{\theta h}{2}\right)\right]=0. \qquad (18.31)$$

Transform $\xi = \dfrac{1+Z}{1-Z}$,

$$\Rightarrow \quad Z\left[1-4\tau\sin^2\left(\frac{\theta h}{2}\right)+2\beta(1-2\alpha)\sin^2\left(\frac{\theta h}{2}\right)\right]+2\beta\sin^2\left(\frac{\theta h}{2}\right)=0.$$

$$(18.32)$$

Stability criteria

$$1 - 4\tau + 2\beta(1 - 2\alpha) > 0.$$

Since, $0 \leq \sin^2\left(\dfrac{\theta h}{2}\right) \leq 1$ which is true for $\beta > 0$ when $\tau \leq \dfrac{1}{4}$ and $\alpha \leq \dfrac{1}{2}$. It means that two-level implicit scheme is unconditionally stable.

If $\tau<\frac{1}{4}$ and $\beta>\frac{1}{2}$, the scheme is conditionally stable for $0<\beta\leq\frac{1-4\tau}{2(2\alpha-1)}$.

Truncation error

The truncation error at $(m, n)^{th}$ grid is given by

$$T_m^n=[1+(\tau+\alpha\beta-\beta)\delta_x^2]u(x_m,t_{n+1})-[1+(\tau+\alpha\beta)\delta_x^2]u(x_m,t_n).$$

As calculated previously, we find

$$k^{-1}T_m^n=\left[k\left(\alpha-\frac{1}{2}\right)+h^2\left(\tau-\frac{1}{12}\right)\right]\frac{\partial^4 u}{\partial x^4}+[k^2(\ldots)+h^4(\ldots)+\ldots].$$

So, order of approximation is

(i) $((k+h^2)$ for $\alpha\neq\frac{1}{2}$, $\tau\neq\frac{1}{12}$.

(ii) (k^2+h^4) for $\alpha=\frac{1}{2}$, $\tau=\frac{1}{12}$.

As unconditional stability criteria hold good for $\alpha<\frac{1}{2}$. So, scheme is of $O(k+h^2)$ for $\alpha<\frac{1}{2}$ and $\frac{1}{2}<\tau<\frac{1}{4}$.

Some unconditionally stable two-level implicit scheme.

(i) $\alpha=\frac{1}{2}$, $\tau=0$, Laasonon-scheme is obtained i.e.,

$\nabla_t u_m^{n+1}=\beta\delta_x^2 u_m^{n+1}$.

(ii) $\alpha=\frac{1}{2}$, $\tau=0$, Crank-Nikolson scheme is obtained i.e.,

$\nabla_t u_m^{n+1}=\frac{\beta}{2}\delta_x^2 u(_m^{n+1}+u_m^n)$.

(iii) $\tau=\frac{1}{12}$, $\alpha=\frac{1}{2}$, crandit scheme is obtained.

Example [4]

Solve $\frac{\partial u}{\partial t}=\frac{\partial^2 u}{\partial x^2}$,

Under the boundary condition,

$$u(x,0)=\sin\pi x,\ 0<x<1$$

$$u(0,t)=u(1,t)=0.$$

Solution. Using the Method

(i) The Schmidt method
(ii) The Laasenon method
(iii) Crack-Nicolson method
(iv) Dufort-Nicolson method

The nodal points are shown in Figure 18.3.

We take $h = \frac{1}{3}, k = \frac{1}{36}$

Now, since, $\beta = \frac{k}{h^2} = \frac{1}{4}$.

From initial condition, $u(x, 0) = \sin \pi x$,

Gives $u_0^0 = 0, u_1^0 = \sin\frac{\pi}{3} = \frac{\sqrt{3}}{2}, u_2^0 = \sin\frac{2\pi}{3} = \frac{\sqrt{3}}{2}, u_3^0 = 0.$

And boundary condition $u_0^n = 0 = u_3^n$.

(i) For Schmidt Method

$$u_m^{n+1} = \frac{1}{4}(u_{m+1}^n + 2u_m^n + u_{m-1}^n).$$

Put $n = 0$;

$$u_m^1 = \frac{1}{4}(u_{m+1}^0 + 2u_m^0 + u_{m-1}^0).$$

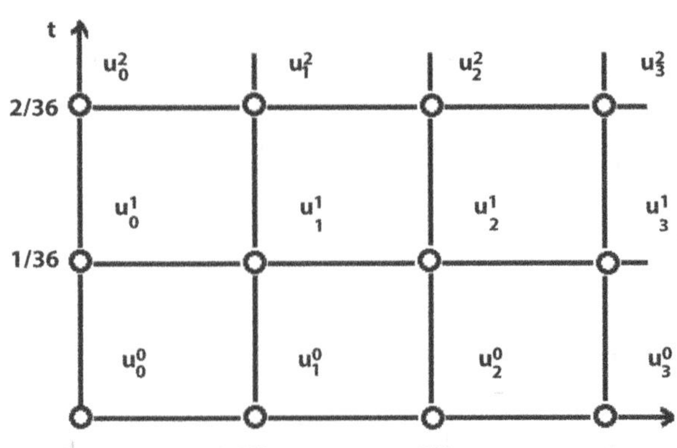

Figure 18.3 Representation of nodal points.

Substitute $m = 1, 2$;

$$u_1^1 = \frac{1}{4}(u_2^0 + 2u_1^0 + u_0^0) = \frac{1}{4}\left(\frac{\sqrt{3}}{2} + \frac{2\sqrt{3}}{2} + 0\right) = 0.6495,$$

$$u_2^1 = \frac{1}{4}(u_3^0 + 2u_2^0 + u_1^0) = \frac{1}{4}\left(0 + \frac{2\sqrt{3}}{2} + \frac{\sqrt{3}}{2}\right) = 0.6495.$$

Put $n = 1$;

$$u_m^2 = \frac{1}{4}(u_{m+1}^1 + 2u_m^1 + u_{m-1}^1).$$

Put $m = 1, 2$;

$$u_1^2 = \frac{1}{4}(u_2^1 + 2u_2^1 + u_0^1) = \frac{1}{4}(0.6495 + 1.299 + 0) = 0.4871,$$

$$u_2^2 = \frac{1}{4}(u_2^1 + 2u_2^1 + u_0^1) = \frac{1}{4}(0 + 1.299 + 0.6495) = 0.4871.$$

(ii) Laasenon method

For $\beta = \frac{1}{4}$, The scheme is $-\frac{1}{4}u_{m-1}^{n+1} + \frac{3}{2}u_m^{n+1} - \frac{1}{4}u_{m+1}^{n+1} = u_m^n$.

Put $n = 0$

$$-\frac{1}{4}u_{m-1}^{n+1} + \frac{3}{2}u_m^{n+1} - \frac{1}{4}u_{m+1}^{n+1} = u_m^n.$$

Put $m = 1$, gives

$$-\frac{1}{4}u_0^0 + \frac{3}{2}u_1^1 - \frac{1}{4}u_2^1 = u_1^0,$$

$$\frac{3}{2}u_1^1 - \frac{1}{4}u_2^1 = \frac{\sqrt{3}}{2}. \quad (18.32)$$

Put $m = 2$, gives

$$
\begin{aligned}
-\frac{1}{4}u_1^1 + \frac{3}{2}u_2^1 - \frac{1}{4}u_3^1 &= u_2^0, \\
-\frac{1}{4}u_1^1 + \frac{3}{2}u_2^1 = u_2^0 + \frac{1}{4}u_3^1 &= \frac{\sqrt{3}}{2}. \\
\frac{3}{2}u_1^1 - \frac{1}{4}u_2^1 &= \frac{\sqrt{3}}{2}, \\
-\frac{1}{4}u_1^1 + \frac{3}{2}u_2^1 &= \frac{\sqrt{3}}{2}.
\end{aligned} \tag{18.33}
$$

Solving 18.32 and 18.33, we get

$$u_1^1 = 0.6928 \quad \text{and} \quad u_2^1 = 0.6928.$$

Similarly, for $n = 1$, we have

$$
\begin{aligned}
\frac{3}{2}u_1^2 - \frac{1}{4}u_2^2 &= 0.6928, \\
-\frac{1}{4}u_1^2 + \frac{3}{2}u_2^2 &= 0.6928.
\end{aligned}
$$

After solving, we get

$$u_1^2 = 0.5542 \quad \text{and} \quad u_2^2 = 0.5542.$$

(iii) Using Crank-Nicolson Method
for $\beta = \frac{1}{4}$,

$$-\frac{1}{8}u_{m-1}^{n+1} + \frac{5}{4}u_m^{n+1} - \frac{1}{8}u_{m+1}^{n+1} = \frac{1}{8}u_{m-1}^n + \frac{3}{4}u_m^n + \frac{1}{8}u_{m+1}^n.$$

for $n = 0$ and $m = 1$;

$$-\frac{1}{8}u_0^1+\frac{5}{4}u_1^1-\frac{1}{8}u_2^1=\frac{1}{8}u_0^0+\frac{3}{4}u_1^0+\frac{1}{8}u_2^0,$$

or
$$\frac{5}{4}u_1^1-\frac{1}{8}u_2^1=0+\frac{3\sqrt{3}}{8}+\frac{\sqrt{3}}{16},$$

or
$$\frac{5}{4}u_1^1-\frac{1}{8}u_2^1=\frac{7\sqrt{3}}{16}. \tag{18.34}$$

for $n = 0$ and $m = 2$

$$-\frac{1}{8}u_1^1+\frac{5}{4}u_2^1-\frac{1}{8}u_3^1=\frac{1}{8}u_1^0+\frac{3}{4}u_2^0+\frac{1}{8}u_3^0,$$

or
$$-\frac{1}{8}u_1^1+\frac{5}{4}u_2^1=\frac{\sqrt{3}}{16}+\frac{3\sqrt{3}}{8},$$

or
$$\frac{5}{4}u_1^1-\frac{1}{8}u_2^1=\frac{7\sqrt{3}}{16}. \tag{18.35}$$

Solving equations 18.34 and 18.35, we have

$$u_1^1=0.6736 \quad \text{and} \quad u_2^1=0.6736.$$

Similarly, for $n = 0$ and $m = 1, 2$;

$$\frac{5}{4}u_1^2-\frac{1}{8}u_2^2=0.5894,$$

$$-\frac{1}{8}u_1^2+\frac{5}{4}u_2^2=0.5894.$$

Solving $u_1^2=0.5239$ and $u_2^2=0.5239$.

(iv) Dufort-Frankel scheme
The difference scheme is

$$u_m^{n+1}=\frac{1}{3}[u_m^{n-1}+u_{m+1}^n+u_{m-1}^n],$$

This is a three-level explicit method; therefore, we first find the solution of first level by above three methods.

We take the solution at first level by Schmidt method as

$$u_1^1 = 0.6495 \quad \text{and} \quad u_2^1 = 0.6495$$

Now for the solution of second level, we use
for $n = 1$ and $m = 1$

$$u_1^2 = \frac{1}{3}[u_1^0 + u_2^1 + u_0^1] = \frac{1}{3}\left[\frac{\sqrt{3}}{2} + 0.6495 + 0\right] = 0.5052,$$

$$u_2^2 = \frac{1}{3}[u_2^3 + u_3^1 + u_1^1] = \frac{1}{3}\left[\frac{\sqrt{3}}{2} + 0 + 0.6495\right] = 0.5052.$$

The exact solution of this problem is

$$u(x,t) = e^{-\pi^2 t} \sin \pi x$$

Hence,

$$u_1^1 = u\left(\frac{1}{3}, \frac{1}{36}\right) = e^{-\frac{\pi^2}{36}} \sin\frac{\pi}{3} = 0.6583,$$

$$u_2^1 = u\left(\frac{2}{3}, \frac{1}{36}\right) = e^{-\frac{\pi^2}{36}} \sin\frac{2\pi}{3} = 0.6583,$$

$$u_1^2 = u\left(\frac{1}{3}, \frac{1}{3}\right) = e^{-\frac{\pi^2}{36}} \sin\frac{\pi}{3} = 0.5005,$$

$$u_2^2 = u\left(\frac{2}{3}, \frac{1}{3}\right) = e^{-\frac{\pi^2}{36}} \sin\frac{2\pi}{3} = 0.5005.$$

Table 18.1 expresses the list of solutions obtained from various schemes for Ex-4.

Table 18.1 List of solutions for different methods.

	Schmidt	Laasenon	Crank Nicolson	Dufort Frankel	Exact
u_1^1	0.6495	0.6928	0.6736	0.6495	0.6583
u_2^1	0.6495	0.6928	0.6736	0.6495	0.6583
u_1^2	0.4871	0.5542	0.5894	0.5032	0.5005
u_2^2	0.4871	0.5542	0.5894	0.5032	0.5005

18.5 Conclusion

In this chapter, we have attempted to supply to the reader some basic numerical methods for the parabolic partial differential equation. Many important methods have been excluded due to the limitation of space. Many important topics such as the multilevel implicit method and cubic spline finite difference method are found in an excellent textbook by Chawla [8, 9] studied the L-stable difference scheme. The von-Neumann method for stability of finite difference scheme is studied in O'Brien *et al.* [10]. Iyenger [6] treated the problem on a cylindrical heat conduction equation. ADI method can be found in the study of Jain [5].

References

1. Forsythe, G.E. and Wasow, W.R., *Finite-difference methods partial differential equations*, Wiley, New York, 1960.
2. Smith, G.D., *Numerical solution of partial differential equations: Finite difference methods*, Clarendon Press, Oxford, 1978.
3. Richardson, L.F., On the approximate arithmetical solution by finite differences of physical problems involving differential equations, with an application to the stresses in a masonry dam. *Proc. Trans. R. Soc. London, Ser. A, Containing Papers of a Mathematical or Physical Character*, 818, 18185–18186, 1910, https://doi.org/10.1098/rspa.1910.0020.
4. Jain, M.K., Iyengar, S.R.K., Jain, R.K., *Computational methods for partial differential equations*, New Age International (P) Limited, 202AD, New Delhi, 2002.
5. Jain, M.K., *Numerical solution of differential equations*, New Age International (P) Limited, New Delhi, 2014.

6. Iyengar, S.R.K. and Mittal, R.C., High accuracy difference schemes for the cylindrical heat conduction equation. *IMA J. Appl. Math.*, 22, 1821–18180, 1978, https://doi.org/10.10918/imamat/22.18.1821.
7. Chapra, S.C. and Raymond, P.C., *Numerical methods for engineers*, Tata McGraw-Hill, New Delhi, 2000.
8. Chawla, M.M., Al-Zanaidi, M.A., Evans, D.J., Generalized trapezoidal formulas for parabolic equations. *Int. J. Comput. Math.*, 70, 429–4418, 1999, https://doi.org/10.1080/00207169908804765.
9. Chawla, M.M., Al-Zanaidi, M.A., Al- Shammeri, A.Z., High-accuracy finite-difference schemes for the diffusion equation. *Neural Parallel Sci. Comput.*, 6, 5218–5185, 1998.
10. O'Brien, G.G., Hyman, M.A., Kaplan, S., A study of the numerical solution of partial differential equations. *J. Math. Phys.*, 29, 2218–251, 1950, https://doi.org/10.1002/sapm19502912218.

19

Godel Code Enciphering for QKD Protocol Using DNA Mapping

Partha Sarathi Goswami[1*] **and Tamal Chakraborty**[2]

[1]*CST, Behala Government Polytechnic, Kolkata, West Bengal, India*
[2]*Computer Sc., Mrinalini Datta Mahavidyapith, Kolkata, West Bengal, India*

Abstract
Security is today's foremost concern over the transmission channel. Quantum computing is a very potent tool in the research of a secure communication channel. Godel code encrypts a number that is allocated to each alphabets of the message, thereby obtaining a sequence of natural numbers that signifies a sequence of symbols. The mapping between the plain text to Godel code is done by a DNA sequence. The idea of using quantum computing and DNA code set encoded with Godel's numbers to perform the quantum key distribution scheme is that each qubit pattern has its own unique properties and so, using a scheme that applies both scores over a scheme that uses any one alone. In this paper, a plain text is first transformed to its DNA equivalent sequence that is further encoded using Godel code to obtain a first level of encoded message. This message is then encrypted using a key that is obtained by a quantum key distribution protocol to guard against the threats in the transmission channel.

***Keywords*:** Cryptography, quantum computing, information security, quantum key distribution (QKD)

**Corresponding author*: goswamipsg@rediffmail.com

Sharmistha Ghosh, M. Niranjanamurthy, Krishanu Deyasi, Biswadip Basu Mallik, and Santanu Das (eds.) Mathematics and Computer Science Volume 1, (373–396) © 2023 Scrivener Publishing LLC

19.1 Introduction

Of late, secure data communication has assumed a very substantial importance in modern research. Biologically inspired encipherment schemes have been adopted to improve the safety of the network medium. Genetic information is encoded as an arrangement of nucleotide bases Adenine (A), Thymine (T), Cytosine (C) and Guanine (G).

This paper applies a unidirectional doorway mapping based on quantum axioms. It facilitates the authentic participants with a manageable problem while making the adversaries face a computationally infeasible solution. A qubit or a quantum bit exists in two-quantum polarization states viz upright and straight polarizations. A qubit has a special property of being in a superposition of several quantum states. The algorithm translates the input into a string of nucleotides and subsequently applies a suitable quantum encoding for the quantum session key exchange.

QKD permits the communicating parties to create a secret key using the principles of atomic physics, thus increasing the underlying safety of the quantum transmission network. In the procedure, the key is undisclosed among the users, hence, the session key is never revealed to anyone. Also, with termination of each session, the previous key is substituted with a new one so that no information associated with the data and session key can be hacked by anyone in the network. The use of DNA sequences encoded with Godel code helps in discarding some of the security threats like the man-in-the-middle, intercept resend attack, etc. Moreover, with increase in error level to a certain value, the photon deflection angle also increases, thereby easing the clients to detect the attackers in the quantum channel.

In Section 19.2, a literature survey is undertaken. Sections 19.3 briefly covers the concepts of DNA code set, whereas Section 19.4 briefs the Godel coding technique. Section 19.5 glimpses the key exchange protocol. In Section 19.6, encipherment and decipherment of the input message using our proposed Quantum Key Distribution Protocol is detailed along with illustration. Section 19.7 discusses the experimental set-up along with Section 19.8 showing the calculation of the detection probability and dark count. Section 19.9 deliberates on the safety measures of the methodology where comparison of the QBER tolerance of different protocols are shown graphically. Section 19.10 gives the denouement.

19.2 Related Work

The laws of quantum physics [1] are employed in the QKD framework [2, 3] to assure the confidentiality of data transmission. Bennett and Brassard (1984) [4] were the first to incorporate the concept of quantum cryptography based on the Uncertainty Principle of Heisenberg and named it as BB84. B92, a more elementary form of BB84 was given later by Bennett (1992) [5]. Combining BB84 and B92, a modified version of Quantum Key Exchange (QKE) methodology was proposed by Ching-Nung Yang and Chen-Chin Kuo (2002) [6]. The idea was to enhance the efficiency of QKE by 42.9% and obtain a complexity of $O(n^{2.86})$. Houshmand and Hosseini (2011) [7] compared their procedure with BB84 using the rules of Quantum Key Distribution (QKD) given by Cabello [8]. Devitt *et al.* (2011) [9] gave an idea of a high-performance quantum computing system where a large cluster lattice was used for multi-user quantum computing. They used a three-dimensional network topology for increasing a large topological group mainframe. To communicate securely between clients Odeh *et al.* (2013) [10] gave a novel model of QKD to be shared among three parties. It has a trusted center to facilitate clients to share vital secret information by eliminating duplicate rounds of checking and verification of quantum bases. Incapacitating the flaws in BB84 and B92, A. Aldhaheri *et al.* (2014) [11] put forward a concept in which, over the quantum channel the session key was exchanged. Also, by substituting nonce and arbitrary basis the privacy and users' authentication was preserved. Xu *et al.* (2016) [12] proposed two protocols that were based on an interference of a single-photon in quantum computing. The first protocol can be easily realized whereas the second protocol was a flexible loss tolerant one. Also, both data and user privacy were analyzed by them. Gueddana and Lakshminarayanan (2018) [13] proposed a view of viability of circuit implementations and theoretical modeling by comparing BB84 QKD) with the optimized version of Quantum Dense Coding. Rusca, Davide *et al.* (2018) [14] proposed a key exchange protocol that employed a minimum number of states which uses a simple detention arrangement. Security was ensured by presenting a complete study contrary to coherent attacks. Huawang *et al.* (2019) [15] projected a quantum secret sharing arrangement by means of the notion of orbital angular momentum. The sender arbitrarily generates a single particle in angular position. Private keys are encrypted by the receiver

Table 19.1 DNA to binary to decimal conversion code set.

DNA	Code	Decimal
A	00	0
C	01	1
G	10	2
T	11	3

using quantum Fourier transforms. To gain the shared key the sender then employs the single-particle measurements. Price *et al.* (2020) [16] gave a quantum key exchange protocol that was designed for interpretation of false users that were connected to the two parties to control the link and deny the service. By decreasing the security in BB84 to quantum-safe cipher, key was generated by them from two-photon pulses.

19.3 The DNA Code Set

The process of encryption begins with an input message "$\mathcal{M}$". "$\mathcal{M}$" is first encoded by a bioinformatic scheme [17] to a DNA triplet code [28]. The receiver upon accepting the nucleotide sequence decodes it back to the original text, according to Table 19.1 below.

19.4 Godel Code

Godel code is a method of assigning to every character and a predefined formula of some formal language with an exclusive number that is called its Godel's numbers or Godel code. A Godel code is construed as an encryption in which a number is allocated to every character of a mathematical notation thereby obtaining a sequence of natural numbers that signifies a sequence of symbols. These sequences of natural numbers are again signified by a single natural number, enabling their use in formal theories of arithmetic [18].

Godel code uses prime factorization method. Let there be a sequence of positive integers ($x_1, x_2, x_3, \ldots x_n$), then the Godel encoding, EG of ($x_1, x_2, x_3, \ldots x_n$) is given by

$E_G(x_1, x_2, \ldots x_n) = 2^{x_1} * 3^{x_2} * 5^{x_3} * \ldots\ldots * p_n^{x_n}$, where the numbers 2, 3, 5,. …, p_n are prime numbers.

Godel code works by a two-step process. In the first step, it replaces each symbol by its corresponding symbol number where a sequence of symbols converts to a sequence of numbers. And in the second step by means of the powers of primes coding, encodes sequences of formulas representing a unique single number as its code.

Now by fundamental theorem of arithmetic, any number that can be exclusively factorized into prime factors, can also be recovered to the original sequence from any given sequence of number of symbols that is to be encoded.

Let the text be (p+q) and the match for each symbol (, p, +, q,) be 4, 6, 7, 8 and 12 respectively.

So, the sequence of numbers is 4, 6, 7, 8, and 12. But these numbers are only a sequence of numbers and not a number itself. We have to make this sequence into a number without losing any information, so that it is possible to obtain the number back to its original text.

Godel's encoding method uses prime numbers. The text has five numbers and to start with the first five prime numbers 2, 3, 5, 7, and 11 are considered.

Now, multiplying the first prime, 2, by itself 4 times we get 16. Similarly, the other numbers are 729, 78125, 5764801, 3138428376721. So, each bit becomes 16, 729, 78125, 5764801, and 3138428376721, respectively.

Multiplying all these together, we have 16486713209345820741011250000.

Thus, the matching Godel code for the text is (p+q) = 16486713209345820741011250000

To decrypt the code first divide by the first prime number, 2 until a remainder is obtained, as is shown below:

16486713209345820741011250000 ÷ 2 = 8243356604672910370505625000
8243356604672910370505625000÷2=4121678302336455185252812500
4121678302336455185252812500÷2=2060839151168227592626406250
2060839151168227592626406250÷2=1030419575584113796313203125
1030419575584113796313203125÷2=515209787792056898156601562,

with remainder 1.

Thus, 2s were multiplied by each other 4 times. So, the first number in the sequence is 4. Repeating the same process for the other numbers we have 3 multiplied by itself 6 times, 5 by 7 times, 7 by 8 times, and 11 by

12 times. Thus, the input number sequence was 4, 6, 7, 8, 12 and thereby the matching text was (p+q).

19.5 Key Exchange Protocol

Let "$\mathcal{M}$" be the sender's message and K_A be the Key. Let K_B be the Key of the Receiver and the XOR operation is denoted by $\oplus$. Then the scheme is shown in Figure 19.1 as under. The protocol facilitates both the sender and receiver to securely communicate among themselves devoid of sharing the secret keys.

19.6 Encoding and Decoding of the Plain Text—The QKD Protocol

Encryption and decryption of the proposed Quantum Key Distribution Protocol is explained under different subsections as follows.

Sender sends a message "$\mathcal{M}$" after encrypting it with his key K_A. On the receiver end the encrypted message $\mathcal{M} \oplus K_A$ is received and resend to the sender for authentication by encrypting it with his key K_B. Sender validates $\mathcal{M} \oplus K_A \oplus K_B$ and sends it again to the Receiver encrypting it with his

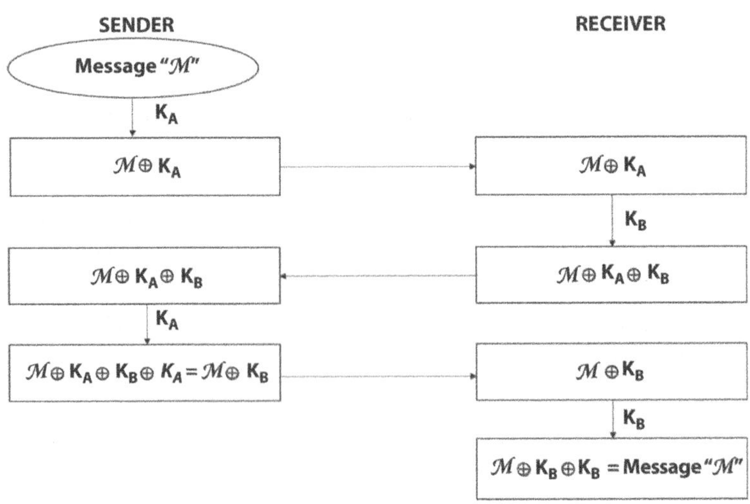

Figure 19.1 Scheme for key exchange protocol.

key K_A. Receiver accepts $\mathcal{M} \oplus K_B$ and decrypts the encoded message by his key K_B to get "$\mathcal{M}$".

19.6.1 Plain Text to Encoded Text and Vice-Versa

The following are the steps to convert a simple plain text to its DNA-coded equivalent and vice-versa.

Step 1: Transform the plain text to its DNA equivalent code.
Step 2: Convert this DNA equivalent code to its decimal equivalent code using Table (19.1).
Step 3: Obtain the Godel code for the corresponding decimal code as per the method discussed below.

Let the encoded text after transformation be MAY and the mapping for each symbol be M to 4, A to 6, and Y to 7. So, the sequence of the numbers is 4, 6, and 7. These numbers are a sequence of numbers and not a number itself.

As Godel's encoding method uses prime numbers, the input text MAY has three numbers and the first three prime numbers 2, 3, and 5 is considered. Multiplying the first prime 2 by itself 4 times we get 16. Similarly, the other numbers are 729 and 78125.

So, each bit becomes 16, 729 and 78125 respectively. Multiplying all these together we have 91,12,50,000. Thus, the matching Godel Code for the text is - MAY = 911250000
Step 4: Obtain the binary equivalent code of the Godel Code and let it be "$\mathcal{M}$".
Step 5: The sender carries out steps 1 to 4 to generate the enciphered Godel code. The following are the steps performed by the receiver for the deciphering process. In between, the encryption-decryption of the message the key exchange is done using the quantum protocol that is described in subsection 19.6.2.
Step 6: On the receiver side, the message "$\mathcal{M}$" in binary form is converted to its equivalent decimal number. This decimal number is the Godel code.
Step 7: Decrypt the Godel Code by the method as discussed below.

To decrypt the code, divide the number by the first prime number, i.e., by 2 until a remainder is obtained.

$911250000 \div 2 = 455625000$
$455625000 \div 2 = 227812500$
$227812500 \div 2 = 113906250$

113906250 ÷ 2 = 56953125
56953125 ÷ 2 = 28476562, with remainder 1.

The prime number 2 were divided by each other 4 times before a remainder is obtained. Thus, 2s were multiplied by each other 4 times. So, the first number in the sequence is 4. Repeating the same methodology for the other succeeding prime numbers it is observed that 3 was multiplied by itself 6 times and 5 by 7 times, respectively. Thus, the input number sequence was 4, 6, and 7 and in that way the matching text obtained is MAY.

Hence, we can say that the method of encrypting generates a unique number that retains the originality of the input text.

Step 8: Convert the decrypted code so obtained to its DNA equivalent code as given in Table 19.1.

Step 9: Convert the DNA equivalent code to its plain text equivalent to get the original input text.

19.6.2 The Proposed Message Passing Scheme

Below are the states of a Qubit under consideration.

$$|00\rangle = \begin{pmatrix} 1 & 1 \\ 0 & 0 \end{pmatrix} = |\Psi_A\rangle$$

$$|01\rangle = \begin{pmatrix} 1 & 0 \\ 0 & 1 \end{pmatrix} = |\Psi_C\rangle$$

$$|10\rangle = \begin{pmatrix} 0 & 1 \\ 1 & 0 \end{pmatrix} = |\Psi_G\rangle$$

$$|11\rangle = \begin{pmatrix} 0 & 0 \\ 1 & 1 \end{pmatrix} = |\Psi_T\rangle$$

Let "$\mathcal{M}$" consist of "n" binary digits then the enciphered quantum states may possibly be defined as

$$\mathcal{M} = |m_1\rangle \otimes |m_2\rangle \otimes |m_3\rangle \otimes ..... \otimes |m_k\rangle$$

where $\{m_k \mid m_k = \{00, 01, 10 \text{ or } 11\}, k = 1, 2, 3, \ldots, n\}$ and $\oplus$ = tensor product.

Since $2n$ qubits are encoded by a n-bit message, the states of the kth qubit then undergoes a rotation by angle ϕ_k such that $\phi_k = \frac{\pi}{2^{(k-1)}}$ for any constant natural number "n", where "n" is arbitrarily selected from the set of positive integers. Then the angular-rotation matrix can be expressed as

$$\hat{\rho}(\phi_k) = \begin{pmatrix} \cos\frac{\phi_k}{2} & -\sin\frac{\phi_k}{2} \\ \sin\frac{\phi_k}{2} & \cos\frac{\phi_k}{2} \end{pmatrix} \tag{19.1}$$

The angle $\frac{\phi_k}{2}$ is the encipherment key. Similarly, the decipherment is achieved with inverse rotation by the angle $\left(-\frac{\phi_k}{2}\right)$.

Given the fact that every qubit has a different angular rotation, the encryption resembles a one-time pad [19, 20]. The distinguishing point being the key which remains undisclosed. Let λ_S and λ_R be the secret key of the sender and receiver, respectively. Then, for every session

$$\lambda = \left\{ \frac{\phi_k}{2} \middle| 0 \le \frac{\phi_k}{2} < \pi, k = 1, 2, 3, \ldots, n \right\}$$

19.6.3 Illustration

Let plain text be MAY. Therefore, by DNA conversion table (also shown in Table 19.2)

$$\text{MAY} = \text{ACA ACT GAC}$$

The encryption scheme using Godel code is shown in Table 19.3.

Let this message $\mathcal{M} = 111001011100101101101001000 11$ be depicted as $\mathcal{M} = |\Psi_A\rangle$. The process is started by the sender.

Both sender as well as receiver now generate their individual session key.

Let $\lambda_S = \frac{\phi_S}{2}, \lambda_R = \frac{\phi_R}{2}$

Table 19.2 DNA conversion table.

Text	DNA code
M	ACA
A	ACT
Y	GAC

Table 19.3 Encryption using Godel code.

Text	DNA code	Decimal code (x)	Corresponding prime (y)	Using Godel code (x^y)
M	A	0	2	1
	C	1	3	3
	A	0	5	1
A	A	0	7	1
	C	1	11	11
	T	3	13	2197
Y	G	2	17	289
	A	0	19	1
	C	1	23	23
Product =				481914147
Binary Equivalent =11100101110010110110100100011				

Sender encrypts $\mathcal{M}$ with λ_s, (encryption key). Thus,

$$\underset{\lambda_S}{\varepsilon}[\mathcal{M}]:\hat{\rho}\left(\frac{\phi_S}{2}\right)|\Psi_A\rangle=\underset{\lambda_S}{\varepsilon}[\mathcal{M}]:\hat{\rho}\left(\frac{\phi_S}{2}\right)|00\rangle$$

$$=\begin{pmatrix} cos\frac{\phi_S}{2} & -sin\frac{\phi_S}{2} \\ sin\frac{\phi_S}{2} & cos\frac{\phi_S}{2} \end{pmatrix}\begin{pmatrix} 1 & 1 \\ 0 & 0 \end{pmatrix}$$

$$=\begin{pmatrix} cos\frac{\phi_S}{2} & cos\frac{\phi_S}{2} \\ sin\frac{\phi_S}{2} & sin\frac{\phi_S}{2} \end{pmatrix}$$

$$=cos\frac{\phi_S}{2}\begin{pmatrix} 1 & 1 \\ 0 & 0 \end{pmatrix}+sin\frac{\phi_S}{2}\begin{pmatrix} 0 & 0 \\ 1 & 1 \end{pmatrix}$$

$$=cos\frac{\phi_S}{2}|\Psi_A\rangle+sin\frac{\phi_S}{2}|\Psi_T\rangle=|\Psi'\rangle\,(say) \tag{19.2}$$

Here, the superposition state $\underset{\lambda_S}{\varepsilon}$ represents encipherment with λ_S. Sender then sends $|\Psi'\rangle$ as in equation (19.2) that is accepted by the recipient and by λ_R he encrypts it.

$$\underset{\lambda_R}{\varepsilon}\left[\underset{\lambda_S}{\varepsilon}[\mathcal{M}]\right]: \hat{\rho}\left(\frac{\phi_R}{2}\right)|\Psi'\rangle$$

$$= \begin{pmatrix} cos\frac{\phi_R}{2} & -sin\frac{\phi_R}{2} \\ sin\frac{\phi_R}{2} & cos\frac{\phi_R}{2} \end{pmatrix} \begin{pmatrix} cos\frac{\phi_S}{2} & -sin\frac{\phi_S}{2} \\ sin\frac{\phi_S}{2} & cos\frac{\phi_S}{2} \end{pmatrix} \begin{pmatrix} 1 & 1 \\ 0 & 0 \end{pmatrix}$$

$$= \begin{pmatrix} cos\frac{(\phi_S+\phi_R)}{2} & -sin\frac{(\phi_S+\phi_R)}{2} \\ sin\frac{(\phi_S+\phi_R)}{2} & cos\frac{(\phi_S+\phi_R)}{2} \end{pmatrix} \begin{pmatrix} 1 & 1 \\ 0 & 0 \end{pmatrix}$$

$$= \begin{pmatrix} cos\frac{(\phi_S+\phi_R)}{2} & cos\frac{(\phi_S+\phi_R)}{2} \\ sin\frac{(\phi_S+\phi_R)}{2} & sin\frac{(\phi_S+\phi_R)}{2} \end{pmatrix}$$

$$= cos\frac{(\phi_S+\phi_R)}{2}\begin{pmatrix} 1 & 1 \\ 0 & 0 \end{pmatrix} + sin\frac{(\phi_S+\phi_R)}{2}\begin{pmatrix} 0 & 0 \\ 1 & 1 \end{pmatrix}$$

$$= cos\frac{(\phi_S+\phi_R)}{2}|00\rangle + sin\frac{(\phi_S+\phi_R)}{2}|11\rangle$$

$$= cos\frac{(\phi_S+\phi_R)}{2}|\Psi_A\rangle + sin\frac{(\phi_S+\phi_R)}{2}|\Psi_T\rangle = |\Psi''\rangle\,(say) \tag{19.3}$$

$|\Psi''\rangle$ is sent back after rotating it in the opposite direction by an angle $\left(-\frac{\phi_S}{2}\right)$, decryption is done by the sender.

$$\mathcal{D}_{\lambda_S}\left[{}^{\varepsilon}\lambda_R\left[{}^{\varepsilon}\lambda_s[\mathcal{M}]\right]\right] = {}^{\varepsilon}\lambda_R[\mathcal{M}] : \hat{\rho}\left(-\frac{\phi_S}{2}\right)|\Psi''\rangle$$

$$= \begin{pmatrix} \cos\left(-\frac{\phi_S}{2}\right) & -\sin\left(-\frac{\phi_S}{2}\right) \\ \sin\left(-\frac{\phi_S}{2}\right) & \cos\left(-\frac{\phi_S}{2}\right) \end{pmatrix} \begin{pmatrix} \cos\frac{(\phi_S+\phi_R)}{2} & \cos\frac{(\phi_S+\phi_R)}{2} \\ \sin\frac{(\phi_S+\phi_R)}{2} & \sin\frac{(\phi_S+\phi_R)}{2} \end{pmatrix}$$

$$= \begin{pmatrix} \cos\frac{\phi_R}{2} & \cos\frac{\phi_R}{2} \\ \sin\frac{\phi_R}{2} & \sin\frac{\phi_R}{2} \end{pmatrix}$$

$$= \cos\frac{\phi_R}{2}\begin{pmatrix} 1 & 1 \\ 0 & 0 \end{pmatrix} + \sin\frac{\phi_R}{2}\begin{pmatrix} 0 & 0 \\ 1 & 1 \end{pmatrix}$$

$$= \cos\frac{\phi_R}{2}|\Psi_A\rangle + \sin\frac{\phi_R}{2}|\Psi_T\rangle = |\Psi'''\rangle\ (say) \tag{19.4}$$

Here, $\mathcal{D}_{\lambda_S}$ represents decipherment with λ_S. Recipient accepts it and by inversely rotating it with the angle $\left(-\frac{\phi_R}{2}\right)$, he decrypts to get $\mathcal{M}$.

$$\mathcal{D}_{\lambda_R}\left[\mathcal{E}_{\lambda_S}[\mathcal{M}]:\hat{\rho}\left(-\frac{\phi_R}{2}\right)|\Psi'''\rangle\right]$$

$$=\begin{pmatrix}\cos\left(-\frac{\phi_R}{2}\right) & -\sin\left(-\frac{\phi_R}{2}\right)\\ \sin\left(-\frac{\phi_R}{2}\right) & \cos\left(-\frac{\phi_R}{2}\right)\end{pmatrix}\begin{pmatrix}\cos\frac{\phi_R}{2} & \cos\frac{\phi_R}{2}\\ \sin\frac{\phi_R}{2} & \sin\frac{\phi_R}{2}\end{pmatrix}$$

$$=\begin{pmatrix}\cos 0 & \cos 0\\ \sin 0 & \sin 0\end{pmatrix}$$

$$=\begin{pmatrix}1 & 1\\ 0 & 0\end{pmatrix}$$

$$=|00\rangle$$

$$=|\Psi_A\rangle=\mathcal{M} \tag{19.5}$$

Now, $\mathcal{M}$=11100101110010110110100100011. Its corresponding decimal equivalent is 481914147. The decryption scheme using Godel code is shown in Table 19.4.

Table 19.4 Decryption using Godel code.

Decimal No.	**For Prime No. = 2**	**No. of Time (s) without Remainder**
481914147	240957073.5	**0**
Decimal No.	**For Prime No. = 3**	**No. of Time (s) without Remainder**
481914147	160638049	**1**
160638049	53546016.33	

(Continued)

Table 19.4 Decryption using Godel code. (*Continued*)

Decimal No.	**For Prime No. = 5**	**No. of Time (s) without Remainder**
481914147	96382829.4	0
Decimal No.	**For Prime No. = 7**	**No. of Time (s) without Remainder**
481914147	68844878.14	0
Decimal No.	**For Prime No. = 11**	**No. of Time (s) without Remainder**
481914147	43810377	1
43810377	3982761.545	
Decimal No.	**For Prime No. = 13**	**No. of Time (s) without Remainder**
481914147	37070319	3
37070319	2851563	
2851563	219351	
219351	16873.15385	
Decimal No.	**For Prime No. = 17**	**No. of Time (s) without Remainder**
481914147	28347891	2
28347891	1667523	
1667523	98089.58824	
Decimal No.	**For Prime No. = 19**	**No. of Time (s) without Remainder**
481914147	25363902.47	0
Decimal No.	**For Prime No. = 23**	**No. of Time (s) without Remainder**
481914147	20952789	1
20952789	910990.8261	

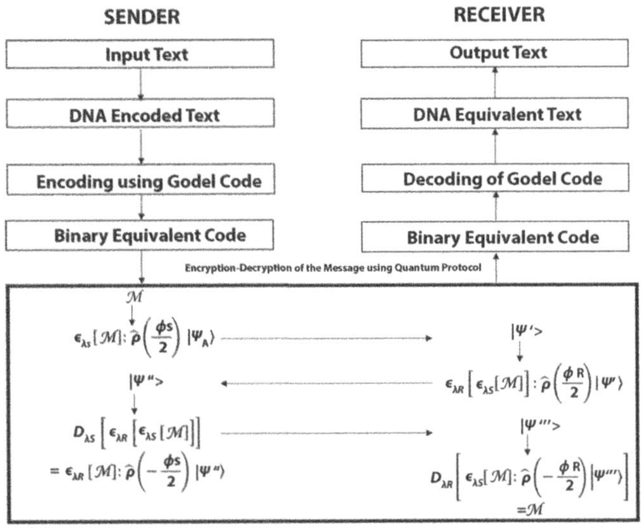

Figure 19.2 Overview of our proposed quantum key distribution protocol.

So, the DNA code is = ACA ACT GAC
And its corresponding text = MAY i.e., ACAACTGAC = MAY
The overall idea of the protocol is summarized in Figure 19.2 also.

19.7 Experimental Setup

A couple of beams of photons, namely the ideal and the signal beams, respectively, are generated by a QKD laser source (•) as shown in Figure 19.3. The two beams are parted by a Photon Beam Splitter η_{BS}. On detection of the idle beam trigger, sender encrypts a qubit into a signal photon beam with the help of the photon detector η_D. Here the polarization controller η_C modulates the signal beam to encrypt the qubit. As the sender

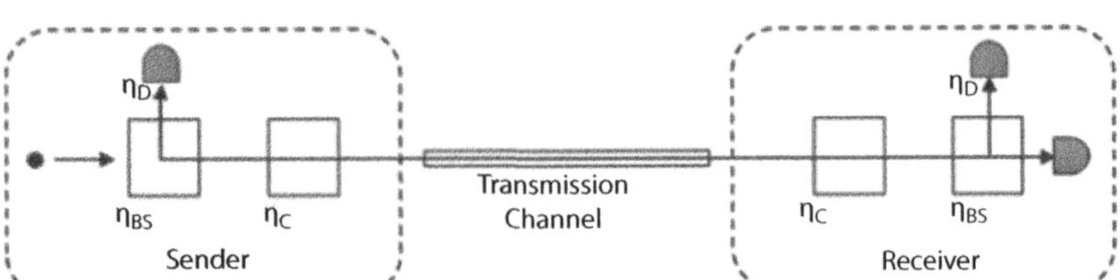

Figure 19.3 Experimental setup for triggering of photons in the QKD.

needs the information of the number of photons in the triggering mode, at the sender's end, a single photon number resolving detector is enough. On the contrary, the recipient chooses a basis by means of the polarization controller η_C to find the measurements using two threshold photon detectors, η_D that are adapt at sensing clicks instead of adding up the photon numbers.

19.8 Detection Probability and Dark Counts

The probability of "n" number of photons that are present in any pulse when the photon beams from the QKD laser beam source has a mean number of photons per pulse, ρ, follows Poisson distribution as

$$P(n,\rho)=\frac{\rho^n}{n!}e^{-\rho}$$

Now, if P_T is the true detection probability from a non-photon number resolving detector, i.e., the detection of the photon only without considering the noise, then we have

$$P_T = 1 - e^{-\rho\sigma}$$

where ρ = mean photon number to be measured, and σ = detection efficiency of the single-photon detector that is standardized at the mean spectral frequency, γ_f.

$$\therefore \rho = -\frac{1}{\sigma T(\gamma_f)} log_e(1-P_T)$$

where $T(\gamma_f)$ = transmittance at γ_f of any further optical elements that are introduced amongst the QKD laser beam source and the single-photon detector.

Deprived of illumination, photon detectors may incorrectly report detection owing to a property known as dark count. Chances of occurrences of dark count, P_D is acquired by tracking the number of detections in unit time without any lighting photon beam. For simulating the measurement of P_D, the triggering and the signal beams are attached, respectively, to a frequency counter and an event timer. Also, the detector is jammed by a shutter to discard unwanted beams. Now, if T_{click} = total number of clicks/unit time in the detector and N = total number of gates/unit time from the triggering beam, then we have

$$P_D = \frac{T_{click}}{N}$$

The false detections as a result of the dark count, usually of thermal origin, can be significantly minimized with a cooled form of detector.

19.9 Security Analysis of Our Algorithm

Our proposed algorithm also withstands different security threats that are discussed as follows:

1. To prevent attackers from copying the message, quantum superposition is utilized as the communication channel. Replication of any quantum state ($p_{00}|00\rangle + p_{01}|01\rangle + p_{10}|10\rangle + p_{11}|11\rangle$, *where* p_{ij} = *random variables and* $\sum_{i,j}|p_{ij}|^2 = 1$) are not feasible. So, it is infeasible to copy the encrypted message to be send in superposition state precisely.
2. In the given environment [27] a hacker may act as the receiver to sender or as a sender to the receiver (man-in-the-middle attack [26]). The algorithm handles the man-in-the-middle attack by inserting dual safety layers of nucleotide encoding and angular qubit rotation, respectively.
3. In the individual particle attack [23], obtainment of facts from the ongoing communication is attempted using the properties of quantum entanglement. But due to the inherent randomness of the algorithm, this type of threat is mitigated.

The fact that the proposed protocol is a two-way communication protocol in which multiple transmission of the same photon is permissible makes it susceptible to Trojan horse attacks (invisible photon attacks). To secure the protocol from such attacks, the sender and receiver may add devices such as wavelength quantum filter and photon number splitters (PNS) to sense the presence of an attacker. The wavelength quantum filter is deployed to allow only the wavelengths close to the operating wavelength, thereby discarding the invisible photons sent by the attacker [24]. The PNS equips the discovery of delay photons by splitting each sample photon and identifying the occurrence of unreasonably high rate of multiple photons (if any) in the quantum signal [25]. It may be noted that the

instance of Trojan horse attack as described above is not due to a limitation of the protocol itself, but rather a manifestation of certain inadequacies of single photon detectors. Moreover, there are infinitely many possibilities of generating ciphertexts by altering the encryption angles. Hence, Trojan-Horse attacks tend to become unsuccessful even on external interferences.
4. Intercept-resend attack may be applied by the Attacker, but this proposed method secures it from that. Quantum Bit Error Rate better known as QBER represents the relation between the false discoveries and the complete number of discoveries in any transmission network.

To find the QBER acceptance in the given methodology, assume that the intruder attempts to interrupt a qubit tuple in logic zero state before encryption, then the qubit tuple that is being received is –

$$|\Psi'\rangle = cos\frac{\phi}{2}|00\rangle + sin\frac{\phi}{2}|11\rangle \tag{19.6}$$

where the angle $\frac{\phi}{2}$ is unidentified by the intruder.

$cos\frac{\phi}{2}$ and $sin\frac{\phi}{2}$ both represent the chances of occurrence of the qubit.

Suppose the intruder attempts to crack the information in this qubit, he will notice $|00\rangle$ and $|11\rangle$ only, the possibility being $\left|cos\frac{\phi}{2}\right|^2$ and $\left|sin\frac{\phi}{2}\right|^2$ respectively.

If the hacker attempts to crack a qubit in logic one state prior to enciphering, the qubit state obtained is

$$|\Psi'\rangle = sin\frac{\phi}{2}|00\rangle + cos\frac{\phi}{2}|11\rangle \tag{19.7}$$

Hence, the chances of cracking the information by the intruder will be at all times

$$\frac{1}{2}\left|cos\frac{\phi}{2}\right|^2 + \frac{1}{2}\left|sin\frac{\phi}{2}\right|^2 = \frac{1}{2} \tag{19.8}$$

for $|00\rangle$ and $|11\rangle$.

Therefore, the intruder at all times will observer $|00\rangle$or $|11\rangle$with probability $\frac{1}{2}$, as is shown in Figure 19.4.

Alternatively, we can say that any effort to intercept and resend the message hikes the QBER tolerance by 50%. So, our system can easily detect the

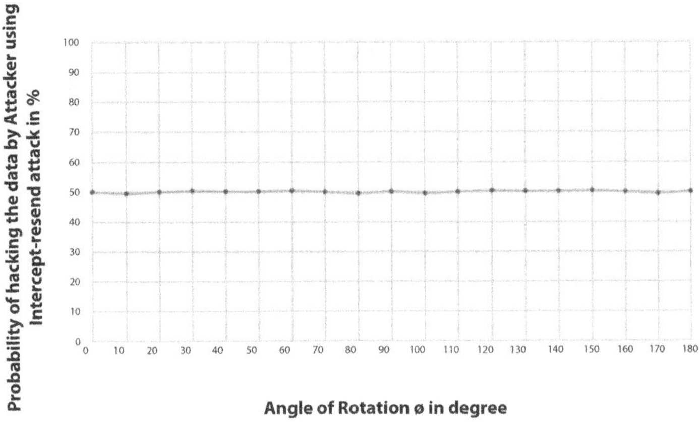

Figure 19.4 Probability of hacking the data by attacker using intercept-resend attack.

error caused by the attacker and the noise in the transmission network provided the QBER is less than 50%. Hence, by the proposed method, communication confidentiality is maintained as without any error, it is impossible for the attacker to recommunicate the captured data to the receiver.

Figure 19.5 shows the QBER tolerance of other protocols in quantum computing. In case of BB84 the QBER tolerance is 25%. In QSS6S or Quantum Secret Sharing with 6 states [22], it is 14.1% while in QSSES or

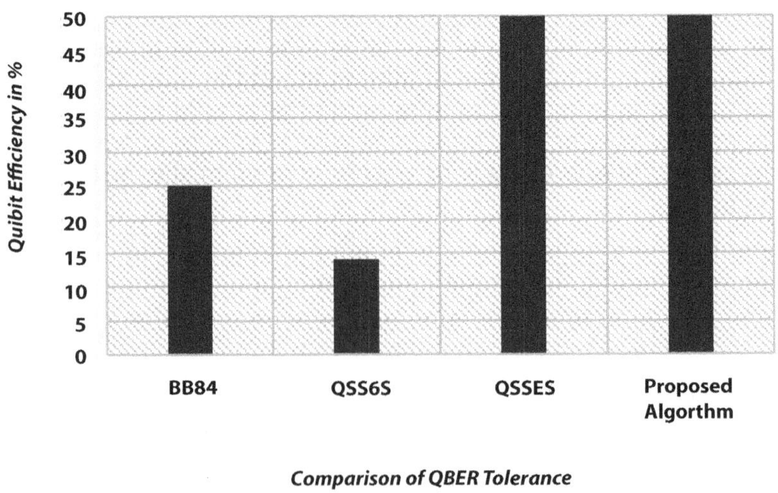

Figure 19.5 QBER tolerance in various algorithms like BB84, QSS6S, QSSES, and proposed algorithm.

Quantum Secret Sharing with Entanglement Swapping [21] bit error rate is 50%.

19.10 Conclusion

The article gives a glimpse of QKD algorithm in the light of bioinformatic encoding and using Godel code for secured communication between the clients. The concept of DNA-encoded encryption added with Godel code is concatenated with the axioms of quantum physics to lessen the security challenges of the proposed procedure. Security analysis of the proposed algorithm in the framework of known attacks (like the attacks due to man-in-the-middle and intercept-resend) are also discussed. Probability of detection and dark counts have also been computed. The implementation of superposition and rotation of the quantum bit at the time of key distribution unlocks the possibility for more investigative survey in this domain.

References

1. Nielsen, M.A. and Chuang, I.L., *Quantum computation and quantum information*, Cambridge University Press, Cambridge, London, 2000.
2. Bouwmeester., D., Ekert, A., Zeilinger., A., *The physics of quantum information*, Springer, New York, 2000.
3. Jun, L., Yi-Min., L., Yan, X., Zhan-Jun, Z., Revisiting controlled quantum secure direct communication using a non-symmetric quantum channel with quantum superdense coding. *Commun. Theor. Phys.*, 49, 887–890, 2008.
4. Bennett, C.H. and Brassard, G., Quantum cryptography: Public key distribution and coin tossing. *Proceedings of IEEE International Conference on Computers Systems and Signal Processing*, Bangalore, India, December 1984, pp. 175–179.
5. Bennett, C.H., Quantum cryptography using any two non-orthogonal states. *Phys. Rev. Lett.*, 68, 3121–3124, 1992.
6. Yang, C.-N. and Kuo, C.-C., Enhanced quantum key distribution protocols using BB84 and B92. *Proceedings of the 2002 International Computer Symposium*, pp. 951–959, 2002.
7. Houshmand, M. and Hosseini-Khayat, S., An entanglement based quantum key distribution protocol. *Information Security and cryptology (ISCISC), 8th International ISC Conference*, IEEE, pp. 45–48, 2011.
8. Adán, C., Quantum key distribution in the Holevo limit, *Phys. Rev. Lett.*, 85, 5635–5638, 2000.
9. Devitt, S.J., Munro, W.J., Nemoto, K., High performance quantum computing. *Quantum Inf. Technol.*, 8, 49–55, 2011.

10. Odeh, A., Elleithy, K. *et al.*, Quantum key distribution by using public key algorithm (RSA). *Third International Conference on Innovative Computing Technology (INTECH), IEEE*, August 2013, United Kingdom, London.
11. Aldhaheri, A., Elleithy, K., Alshammari, M., Ghunaim, H., A novel secure quantum key distribution algorithm, *IEEE Long Island Systems, Applications and Technology (LISAT) Conference 2014*, pp. 1–4, 2014.
12. Xu, S., Sun, Y., Lin, S., Quantum private query based on single-photon interference. *Quantum Inf. Process.*, 15, 3301–3310, 2016, https://doi.org/10.1007/s11128-016-1326-8.
13. Gueddana, A. and Lakshminarayanan, V., Physical feasibility of QKD based on probabilistic quantum circuits. *IET Inf. Secur.*, 12, 6, 521–526, November 2018.
14. Rusca, D. *et al.*, Security proof for a simplified Bennett-Brassard 1984 quantum-key-distribution protocol. *Phys. Rev. A*, 98, 5, 052336, 2018.
15. Huawang, Q., Tso, R., Dai, Y., Quantum secret sharing by using fourier transform on orbital angular momentum. *IET Inf. Secur.*, 13, 2, 104–108, March 2019.
16. Price, A.B., Rarity, J.G., Erven, C., A quantum key distribution protocol for rapid denial of service detection. *EPJ Quantum Technol.*, 7, 8, 1–20, 2020, https://doi.org/10.1140/epjqt/s40507-020-00084-6.
17. Britannica, The Editors of Encyclopaedia, genetic code. *Encyclopedia Britannica*, 2019.
18. Gödel, K., Über formal unentscheidbare Sätze der Principia Mathematica und verwandter Systeme I. *Mon. Math. Phys.*, 38, 173–198, 1931, archived from the original (PDF) on 2018-04-11, retrieved 2013-12-07.
19. Goswami, P.S. and Chakraborty, T., Design of a quantum one-way trapdoor function, in: *Emerging Technology in Modelling and Graphics. Advances in Intelligent Systems and Computing*, vol. 937, J. Mandal and D. Bhattacharya (Eds.), Springer, Singapore, 2020, https://doi.org/10.1007/978-981-13-7403-6_48.
20. Goswami, P.S., Chakraborty, T., Chattopadhyay, A., Knapsack encoding for secured quantum key distribution protocol. *Mod. Phys. Lett. A*, 35, 36, 2050295 (16 pages), 2020, https://doi.org/10.1142/S0217732320502958.
21. Lin, S., Gao, F., Qin, S-j, Wen, Q-y, Zhu, F-c, Quantum secret sharing between multiparty and multiparty with entanglement swapping. *J. China Univ. Posts Telecommun.*, 15, 4, 63–68, 2008.
22. Venkatramulu, S. and Veena, S., Secure communication using two party authenticated quantum key distribution protocols. *Int. J. Comput. Sci. Netw. Secur.*, 10, 8, 233–238, 2010.
23. Kanamori, Y., Yoo, S.M., Al-Shurman, M., A Quantum no-key protocol for secure communication. *ACM Southeast Conference (ACMSE-05)*, Atlanta, March 2005, pp. 92–93.
24. Cai, Q.Y., Eavesdropping on the two-way quantum communication protocols with invisible photons. *Phys. Lett. A*, 351, 1-2, 23C25, 2006.

25. Deng, F.G., Li, X.H., Zhou, H.Y., Improving the security of multi party quantum secret sharing against Trojan horse attack. *Phys. Rev. A*, 72, 4, 044302, 2005.
26. Svozil, K., Feasibility of the interlock protocol against man-in-the-middle attacks on quantum cryptography. *Int. J. Quantum Inf.*, 3, 4, 649–654, 2005.
27. Yang, L., Wu, L.-A., Liu, S., Quantum three-pass cryptography protocol. *Quantum Opt. Comput. Commun. SPIE*, 4917, 106–111, 2002.
28. Goswami, P.S., Chakraborty, T., Chattopadhyay, A., A secured quantum key exchange algorithm using fermat numbers and DNA encoding. *2021 Fourth International Conference on Electrical, Computer and Communication Technologies (ICECCT)*, pp. 1–8, 2021, doi: 10.1109/ICECCT52121.2021.9616749.

20

Predictive Analysis of Stock Prices Through Scikit-Learn: Machine Learning in Python

Vikash Kumar Mishra[1]*, Richa Binyala[2], Pratibha Sharma[2] and Simran Singh[2]

[1]SCSE, Galgotias University, Uttar Pradesh, India
[2]School of Business and Management, Christ (Deemed to be University), Delhi NCR, India

Abstract
Scikit-learn, a tool for developing machine learning algorithms, is a standard library of python. Through Scikit-learn, a trained model for predictive analysis can be developed. Such models aim to provide accurate predictions. Stock predictions are based on changes and patterns identified in the historical dataset. Following the trends and patterns of the historical changes of stocks, machine learning algorithms can be developed for achieving accurate outcomes. An effective model is developed, which enhance the working pattern or performance of the machine that further helps to draw a precise analysis of stocks.

***Keywords*:** Predictive analysis, Scikit-learn, machine learning, stock market

20.1 Introduction

Scikit-learn is one of the libraries in python that serves as a robust tool for the branch of artificial intelligence and machine learning. Machine learning focuses on the use of data and algorithms to imitate the way humans learn, which gradually improves its accuracy. One of the major functions of machine learning is predictive analysis, meaning the system uses the data to predict what will happen in the future [1]. Accurate predictions can be

**Corresponding author*: mail2dr.vikash@gmail.com

Sharmistha Ghosh, M. Niranjanamurthy, Krishanu Deyasi, Biswadip Basu Mallik, and Santanu Das (eds.) Mathematics and Computer Science Volume 1, (397–404) © 2023 Scrivener Publishing LLC

achieved through machine learning and scikit-learns provides a means to do it [2–4]. Problems encountered in machine learning can be approached in two different ways, supervised and unsupervised models. Supervised learning is a structured way of teaching or training machines to work effectively [5, 6]. On the other hand, under unsupervised learning, the software itself analyzes the unlabeled dataset through observations to obtain and identify patterns and relationships in the unlabeled dataset by creating clusters in it [7].

Scikit-learn, an open sources machine learning library under python, inculcates several featured algorithms such as classification, regression, and clustering. Clustering is an automatic grouping of similar data into a dataset, a feature useful in the predictive analysis of stock prices. Stock prices are an industry with a continuous cycle of fluctuations. The game of risk and return serve as a livelihood for many. There are several tools and techniques identified and established to study the historical patterns of stock fluctuations based on which future predictions can be determined. Scikit-learn are one such tool that can be used to predict stock price future fluctuations. It provides us with the feature of clustering that is useful in such an area of predictive analysis. Following the trends and patterns of the historical changes of stocks, machine learning algorithms can be developed for achieving accurate outcomes. This study explores the predictive analysis of stock prices through Scikit-learn.

20.2 Study Area and Dataset

For this study, a sample of historical data of Equity and Small & Medium Enterprises of Kotak Mahindra bank have been taken from the National Stock exchange of India (India, n.d.) [8].

There is no specificity while choosing historical data for predictive analysis. To ease the understanding of results obtained the study focuses on 1-week prices fluctuations starting from November 22, 2021, to November 29, 2021. This type of dataset is also used and analyzed by investors and companies. The analysis of the dataset is important to determine the risk-return factor. Prices of the stock fluctuate due to various factors and extensive study and observation could assist in determining the structure and patterns. Once a relationship is established in the historical dataset a prediction can be made. The format of the dataset is downloaded in CSV format directly from NSE India [9].

The graphical representation of stock price fluctuations has been taken from the money control website and shown in Figure 20.1. This assists in the interpretation of results and deriving insights.

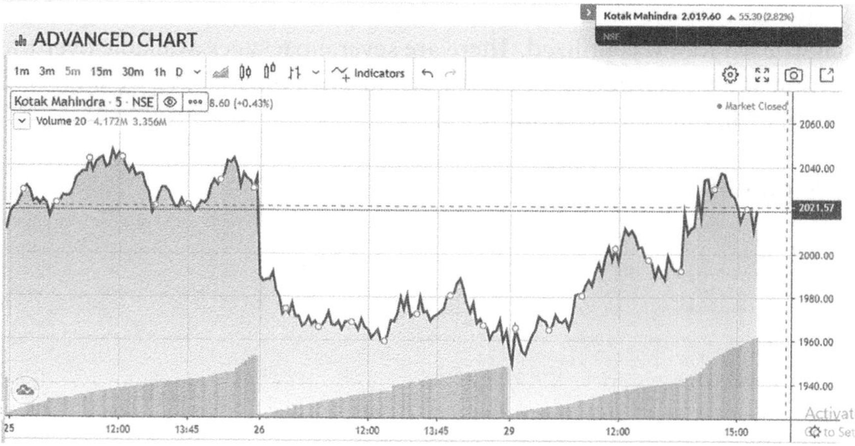

Figure 20.1 Visualization of stock market fluctuation of Kotak Mahindra (Moneycontrol.com, n.d.).

20.3 Methodology

The application of machine learning in the stock prices prediction is achieved via a tool Scikit-learn. The flowchart chart shown in Figure 20.2 depicts the process undertaken within the methodology for the predictive analysis [10].

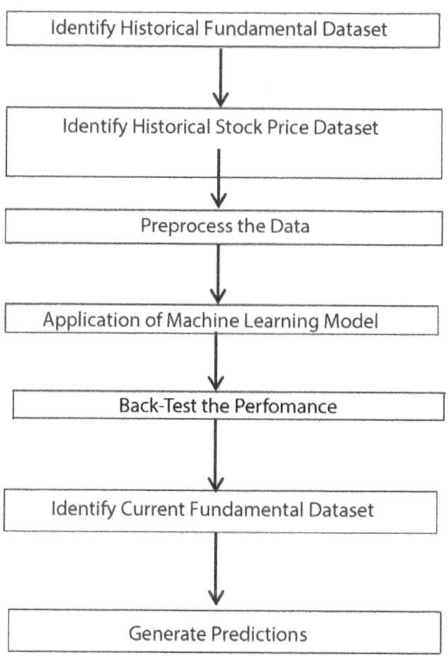

Figure 20.2 Workflow to use machine learning for stock price predictions (using Python and Scikit-Learn to Make Stock Predictions | PythonRepo, n.d.) [11].

In order to create a supervised machine learning model in python, scikit-learn tool was utilized. There are several packages available over the python and for the purpose of this study in stock predictions, we imported the following mentioned libraries:

Syntax of the program

```
import numpy as np
import pandas as pd
import tensorflow as tf
import datetime
import dataframe


import matplotlib.pyplot as plt

from matplotlib.pylab import rcParams
rcParams['figure.figsize']=20,10
from keras.models import Sequential
from keras.layers import LSTM,Dropout,Dense
from sklearn.preprocessing import MinMaxScaler

from sklearn.tree import DecisionTreeRegressor
from sklearn.linear_model import LinearRegression
from sklearn.model_selection import train_test_split
plt.style.use('bmh')

#read Dataset
kot=pd.read_csv("C:\\Users\\Simran Singh\\Desktop\\Final Kot\\Kstock.csv")
kot.head()
print(kot)
#no. of Trading days
kot.shape

#close price data
plt.figure(figsize=(16,8))
plt.title('Kotak')
plt.xlabel('Days')
plt.ylabel('Close Price')
plt.plot(kot['Closing Price '])
plt.show()
```

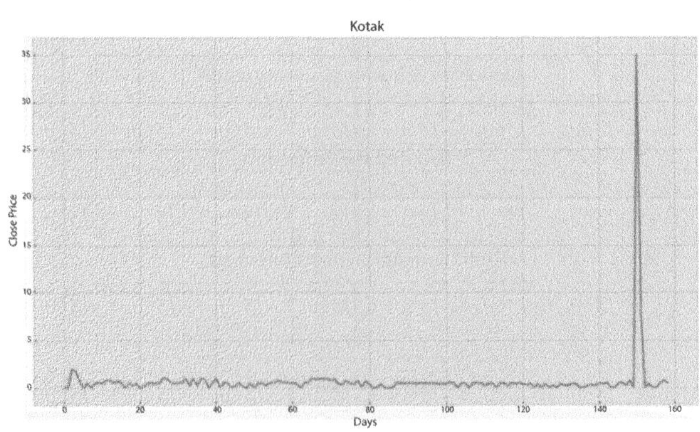

Figure 20.3 Closing price of Kotak bank.

Following flowchart, the first step was to identify the historical dataset, which contained opening and closing price of the stock. Further for preprocessing the imported data, DecisionTreeRegressive from sklearn, the first model selection was made. Before moving on to the predictions, a command to select 10 random future days has to be written [12, 13]. Figure 20.3 shows the command function of the closing price of Kotak Mahindra Bank during 159 days. As visible in the above graph, a sudden peak only confirms the instability that goes along in the stock market and why their predictions are complex [14, 15].

20.4 Results

The model prepared in Figure 20.4a for generating stock predictions was successful. The diagram represents Kotak Stock Price Predictions for first and last 4 days in the duration of 159 days. A total number of rows and columns can be viewed at the end of the program outcome, which is 159 rows and 12 columns, also shown in Figure 20.4b.

Syntax of the program

```
import numpy as np
import pandas as pd
import tensorflow as tf
import datetime
import dataframe


import matplotlib.pyplot as plt

from matplotlib.pylab import rcParams
rcParams['figure.figsize']=20,10
from keras.models import Sequential
from keras.layers import LSTM,Dropout,Dense
from sklearn.preprocessing import MinMaxScaler

from sklearn.tree import DecisionTreeRegressor
from sklearn.linear_model import LinearRegression
from sklearn.model_selection import train_test_split
plt.style.use('bmh')

#read Dataset
kot=pd.read_csv("C:\\Users\\Simran Singh\\Desktop\\Final Kot\\Kstock.csv")
kot.head()

future_days=10
kot['Prediction']=kot[['Closing Price ']].shift(-future_days)
kot.head(4)
print(kot)

#create the feature dataset 'x' and convert it to a numpy array and remove the last 'x' rows/days
X=np.array(kot.drop(['Prediction'],1))[:-future_days]
print(X)

#create the target data set (y) and convert it to a numpy array and get all of the target except the last 'x' rows/days
y=np.array(kot['Prediction'])[:-future_days]
print(y)
```

Figure 20.4a Syntax performed.

```
IDLE Shell 3.9.6
File Edit Shell Debug Options Window Help
Python 3.9.6 (tags/v3.9.6:db3ff76, Jun 28 2021, 15:26:21) [MSC v.1929 64 bit (AMD64)] on win32
Type "help", "copyright", "credits" or "license()" for more information.
>>>
================ RESTART: C:/Users/Simran Singh/Desktop/final4k.py ================
          Date  Open Price  ...  no of trades  Prediction
0   29-11-2021        0.50  ...             7        0.50
1   26-11-2021        0.50  ...             3        0.75
2   26-11-2021        1.00  ...             5        0.79
3   26-11-2021        0.75  ...            10        0.50
4   25-11-2021        0.50  ...            70        0.80
..         ...         ...  ...           ...         ...
154 06-09-2021        0.50  ...             6         NaN
155 06-08-2021        0.01  ...             1         NaN
156 06-07-2021        0.48  ...             1         NaN
157 06-03-2021        0.97  ...             2         NaN
158 06-01-2021        0.50  ...             6         NaN

[159 rows x 12 columns]
>>> |
```

Figure 20.4b Stock price predictions.

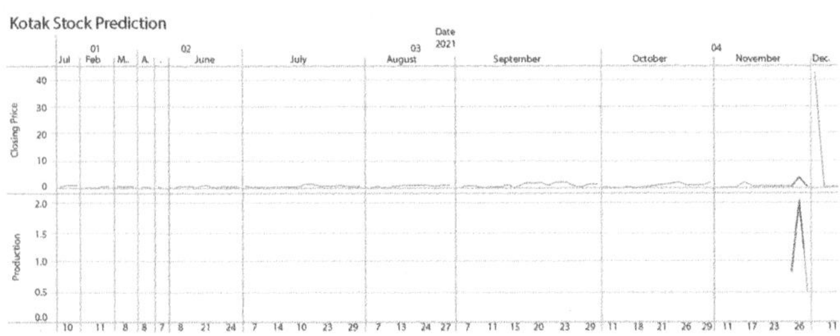

Figure 20.5 Final prediction.

Based on the last 159 days' stock data of Kotak, the model predicted a sudden growth in the month of November, which will gradually fall thereon. This prediction is also depicted through Figure 20.5.

20.5 Conclusion

Through this study, it can be observed that using machine learning and scikit-learn tools and techniques, model generation, and selection are possible. The model created for stock predictions succeeded in returning the stock predictions based on the historical data for 159 days. For the generation of any model a supervised algorithm is needed. It cannot be solved using unsupervised machine learning. The data of Kotak Mahindra Bank selected for the purpose of this study, returned the outcome of the sudden rise in its stock prices and fall at the end of the month of November.

Another thing confirmed over the study was instability of stock market, but it only increases the risk return factor for the investors.

References

1. Brownlee, J., A gentle introduction to scikit-learn, Machine Learning Mastery, 2014, April 15, https://machinelearningmastery.com/a-gentle-introduction-to-scikit-learn-a-python-machine-learning-library/.
2. Fact vs. fiction: How does machine learning actually work?, Expert.Ai, 2021, May 18, https://www.expert.ai/blog/how-does-machine-learning-work/.
3. How does machine learning work?, Dummies. (n.d.). Retrieved November 30, 2021, from https://www.dummies.com/programming/big-data/data-science/how-|does-machine-learning-work/.
4. How is Python used in finance?, Python applications in finance. Retrieved November 29, 2021, from https://www.softkraft.co/how-is-python-used-in-finance/.
5. India, N. S. E., Kotak Mahindra bank limited future derivatives & quotes options, F&O Analysis, NSE India. Retrieved November 29, 2021, from https://www.nseindia.com/get-quotes/derivatives?symbol=KOTAKBANK.
6. Introduction to machine learning in Python with scikit-learn (video series), Data School, 2015, July 18, https://www.dataschool.io/machine-learning-with-scikit-learn/.
7. Machine learning, explained, MIT Sloan. Retrieved November 30, 2021, from https://mitsloan.mit.edu/ideas-made-to-matter/machine-learning-explained.
8. Moneycontrol.com., Kotak Mahindra Share Price, Kotak Mahindra Stock Price, Kotak Mahindra Bank Ltd. Stock Price, Share Price, Live BSE/NSE, Kotak Mahindra Bank Ltd. Bids Offers. Buy/Sell, (n.d.).
9. Kotak Mahindra Bank Ltd., News & tips, & F&O Quotes, NSE/BSE Forecast News and Live Quotes. Retrieved November 29, 2021, from https://www.moneycontrol.com/india/stockpricequote/banks-private-sector/kotakmahindrabank/KMB.
10. Nifty & Sensex Charts & News, NSE India. Retrieved November 29, 2021, from https://www.nseindia.com/market-data/live-equity-market.
11. Pedregosa, F., Varoquaux, G., Gramfort, A., Michel, V., Thirion, B., Grisel, O., Blondel, M., Prettenhofer, P., Weiss, R., Dubourg, V., Vanderplas, J., Passos, A., Cournapeau, D., Brucher, M., Perrot, M., Duchesnay, E., Louppe, G., Scikit-learn: Machine learning in Python. *J. Mach. Learn. Res.*, 12, 2012.
12. Stock price prediction—Machine learning project in Python—DataFlair. Retrieved November 30, 2021, from https://data-flair.training/blogs/stock-price-prediction-machine-learning-project-in-python/.

13. Using Python and scikit-learn to make stock predictions, PythonRepo. (n.d.). Retrieved November 29, 2021, from https://pythonrepo.com/repo/robertmartin8-MachineLearningStocks-python-deep-learning.
14. What are the types of machine learning? | by Hunter Heidenreich | Towards Data Science. (n.d.). Retrieved November 30, 2021, from https://towardsdatascience.com/what-are-the-types-of-machine-learning-e2b9e5d1756f
15. What is machine learning | Types of machine learning, Simplilearn.Com. Retrieved November 30, 2021, from https://www.simplilearn.com/tutorials/machine-learning-tutorial/what-is-machine-learning.

21

Pose Estimation Using Machine Learning and Feature Extraction

J. Palanimeera* and K. Ponmozhi

Department of Computer Apllications, Kalasalingam Academy of Research and Education, Krishnankoil, Tamil Nadu, India

Abstract

The recognition rate, robustness, and operational efficiency of machine vision-based yoga action detection of human postures is all typically low. The subtlety of the heritage and the diverse yoga asanas, blockages, and self-closures make contributions to this. This paper proposes a function extraction approach that mixes directional gradient of depth function (D-GoD) and local distinction of depth function (L-DoD) and employs a unique method that consists of eight close by factors round a picture element for joint evaluation for you to calculate the distinction among the picture elements for you to resolve this problem. This paper presents a method for reliably recognizing diverse Yoga asana using machine learning methods. A dataset comprising seven yoga asanas (Pranamasana, Dhanurasana, Dandasana, Gomukhasana, Garudasana, Padmavrikshasana, and Padmasan) was constructed using ten people (five men and five women) and a standard RGB webcam. The random-forest classifier is then professional within the use of the real-time statistics set. The random-forest classifier then educated the use of the real-time information set, and the picture elements on diverse areas of the human frame intensity image categorized the use of a random forest two-way vote casting procedure. Finally, the center of gravity (CoG) of each part is established, and a suitable joint for detaching the human skeleton is selected. The experimental results show that by evaluating our approach, using the provided data set, resilience, and accuracy are considerably improved, as is competitive operating efficiency. This study also tested the system in real time on a different sample of five persons

**Corresponding author*: jpalanimeera@mail.com

Sharmistha Ghosh, M. Niranjanamurthy, Krishanu Deyasi, Biswadip Basu Mallik, and Santanu Das (eds.) Mathematics and Computer Science Volume 1, (405–424) © 2023 Scrivener Publishing LLC

(two men and three women) and discovered that it was 82.92% accurate. The experiment findings provide a qualitative assessment of the process, as well as a comparison to the present state of the art.

Keywords: Machine vision, yoga acts, recognizing, estimated in real time, center of gravity

21.1 Introduction

Human activity recognition is a well-known computer vision issue that has encountered a variety of challenges throughout the years [1]. It refers to the difficulty of determining key points and the posture of a human structure using sensor data. Activity recognition is useful in biometrics, video surveillance, human–computer interaction, assisted living, sports arbitration, in-home health monitoring, and other fields [2]. Monitoring and recognising an individual's behaviours can be used to assess and forecast their health [3]. Action recognition is a way of labelling video frames with actions such as leaping, playing, punching, strolling, and running, allowing the system to recognise a wide range of human gestures effectively and automatically [4]. In recent years, many techniques to Human activity recognition have been employed, including wearable sensor-based, video-based, and wireless sensor network-based Human activity identification [5]. However, because to their high recognition rate and simplicity of use, video-based Human activity identification techniques are gaining prominence [6].

Yoga is an ancient Indian science that has been practiced for thousands of years. It is the remover of suffering and destroyer of anguish, according to the Bhagavad Gita. Yoga has recently gained popularity across the world owing to its physical, mental, and spiritual advantages. The United Nations General Assembly recognized the 21st of Jun as the 'International-Day of Yoga' in 2014 [7]. Yoga has gained significant attention in the medical community over the last decade, and multiple studies have been presented for a variety of medical applications, including cardiac rehabilitation [7], positive structure image intervention [8], psychiatric diseases, and so on. Yoga can totally treat several disorders without the need of drugs [9].

In motion analysis, identifying poses is a tough task. The main purpose is to derive postural features from image sequences of diverse elements of the humans, such as real-world location are 3-dimensional spaces or joints angle. The postural measurements stated above can be utilised to reconstruct human structure motion in three dimensions. Human posture reputation algorithms primarily based totally on device imaginative and prescient are presently divided into types: those who utilize well-known RGB pictures and people that appoint intensity images. The most important distinction is that picture elements in an RGB image collect colour information about the item, but picture elements in a depth image record the object's distance from the camera. In human posture recognition based on RGB pictures, Observable homes which include 'HOG' function [10] and 'contour-function' [11] are typically employed. These methods, on the other hand, are frequently impacted through the outside environment, and are mainly at risk of light, ensuing in low identification accuracy. Furthermore, significant differ in human structure variation; these algorithms are only suitable for a small number of sites and people. When compared to ordinary sensors, the rate of human posture-detection has grown notably in current years with the improvement of intensity sensors, considerably Microsoft's Kinect, which affords each coloration and intensity information (RGB-D). The main reason for this is because depth shots have a number of advantages than RGB photos. For starters, depth images are resistant to colour and light fluctuations. Furthermore, the RGB image includes less information than the 3D depth image. Human posture reputation technologies encompass model-primarily based totally techniques and function-learning. In model-primarily occupying totally human posture-detection, the human frame was split within severe element, that are then included right within a framework, and the human posture is derived through upend the Microsoft`s-kinematics or addressing gain challenges. A unique articulated posture model was proposed based on image morphology [13]. To accomplish human-posture recognition, function learning seeks into extract superior functions from intensity pictures by assessing every one picture element and utilising a number of machine learning algorithms [12, 14]. Two methods for estimating human structure positions were proposed. One of the methods for categorising each picture element in the depth image is to utilise a random forest. Another method may be used to estimate the position of a human joint. To train random forest classifiers, both techniques rely on a huge variety of artificial and actual human intensity photos. To suggest graph cut optimization, Shotton's technique [15] was applied SVM and super picture element [16] were used to develop a new human posture estimation technique. Deep learning methods also implemented to overcome the target's pose-estimation [17], and the convolution neural network (CNN) is employed to analyze enormous data sets.

The benefit of model-based human pose identification is that it does not require the creation of a vast data collection; instead, it simply builds a few models. Because the model matched, it had a greater recognition rate for the posture. However, there are certain drawbacks to this strategy. Complex human structure models, for example, are challenging to create due to the wide range of human postures in real-life scenarios. The fundamental advantage of feature method is that it does not need the creation of complicated human joint model; therefore it's not bound by it and can be used in a variety of settings. However, there are certain drawbacks to this strategy. On the one hand, it must provide a massive data set that may be used in a variety of settings. Many feature extraction approaches, on the other hand, are quite sophisticated and cannot match real-time needs. As a result, this research proposes a human posture detection approach based on depth image multi feature fusion. The bodily components were first encoded using numbers in deep image, and then a data's was created.

The depth feature's positional gradient and near variation are then obtained and utilised to train a random-forest classifier. Finally, the (CoG) establishers, and any joints are dominated out. The close fluctuation of depth feature and positional gradient of depth feature can fulfil the real-time condition because they have a lower processing cost than other approaches. Furthermore, combining approaches improve the rate of human posture recognition. The words near variation of depth feature and positional gradient of depth feature are interchangeable.

The remain of this paper is organised as follows: Section 21.2 follow-up work Section 21.3 delves into each aspect of posture detection and the algorithms that go with it, as well as how the joints are positioned in the yoga asana image, and creates a random forest classifier. The experimental findings are examined in Section 21.4. The final portion is Section 21.7.

21.2 Related Work

Many researches on movement reputation were posted within side the literature, with a focal point on classical and CNN aspects [18]. In nowadays and technology, CNN plays admirably within side the area of system studying. Ullah *et al.* [19], for example, provided a stepped forward and green CNN-primarily based totally deep auto-encoder (DAE) method for Human Action Recognition (HAR). The advised method takes enter statistics from real-global non-static surveillance environments and applies a pre-skilled CNN version to extract deep features. Optimized differential algebraic equations (DAE) are hired to observe movement fluctuations in surveillance sequences, and quadratic SVM is used for human movement categorization. In the closing stage, the schooling version`s parameters are up to date via way of means of along with the iterative fine-tuning version. The proposed method is examined on publicly to be had datasets which include HMDB51, YouTube movement dataset, UCF50, and UCF101. In contrast to current approaches, the consequences display that the brand new approach works extra successfully in phrases of time and computational cost. Yang *et al.* [20] brought an uneven unidirectional deep three-D-CNN technique for human movement reputation. Micro nets are used on this approach to enhance the function studying talents of an uneven three-D convolution. During the pre-processing phase, multi-supply stepped forward enter is added, which mixes wished RGB and waft structure characteristics. The advised structure is tested the use of the human movement datasets UCF-one hundred and one and HMDB-51. The received consequences display an development in overall performance whilst as compared to preceding three-D-CNN algorithms. Wang *et al.* [21] defined a way for aggregating structure-stage statistics for human movement identity primarily based totally on temporal pooling. For structure-stage representation, the temporal convolutional technique is advanced for the extraction of dynamic data whilst additionally keeping the strong fee of version parameters. To display the approaches overall performance in contrast to current approaches, it changed into examined on 3 datasets: UCF101, Hollywood2, and HMDB51. Wu *et al.* [22] proposed the MPCANet technique for human movement categorization, that's primarily based totally on deep studying. Tensor interplay is used on this method to enhance reputation rates. Projection dictionaries, a projection encoder layer, and a pooling layer make up this.

21.3 Proposed Work

Yoga posture recognition using multi-feature function in depth images:

- Once removing the human target from the original depth picture, the various regions of the segmented structure may be easily identified with a unique code.
- The 'Directional-Gradient of Depth' and 'Local-Difference of Depth' function are then extracted and utilised to train a random forest classifier using multiple decision trees.
- The test samples' physiological components are categorised using the classifier.
- At last, the region of the human parts inside aspect the human body picture was determined.

21.3.1 Yoga Posture Identification

21.3.1.1 Deep Extraction of a Normal Image

This study commonly focuses on specific areas known as ROIs in image processing (regions of interest). These sections usually provide a plethora of feature information. As a result, the target region must be separated from the background in order to identify and assess the target. This may be used for feature extraction and human structure identification. The human structure is segmented using depth background difference because the genuine scenario in this post is fixed. The intensity fee is quantized right into a greyscale area spanning from zero to 255, with a decrease quantity indicating extra intensity. As a result, the 3-d photograph can be proven as a 2D greyscale photograph, with every picture element fee having a wonderful which means from that of a widespread RGB photograph. Leg records of the human frame cannot be analysed because the digital cam focuses Dhanurasana from the pinnacle of the head. D_near and D_far regulate the depth range, which may be represented asD_near, D_far. To reduce noise and minimise depth data drift, Gaussian filtering was first executed to the unique intensity data. The foreground goal is then recovered the usage of the brink T after subtracting the unique intensity photo from the heritage photo, as visible below:

$$D(x,y)=|I(x,y)-B(x,y)|,$$

$$T(x,y)=\begin{cases} 1, & D(x,y)\geq T, \\ 0, & D(x,y)<T, \end{cases} \tag{21.1}$$

where B (x, y) represents the heritage photograph, I (x, y) represents the authentic image, and T (x, y) represents the b-image.

The intensity photograph of the matching location is then retrieved, as visible below:

$$S(x,y)=\begin{cases} I(x,y), & D(x,y)\geq T, \\ 0, & D(x,y)<T, \end{cases} \tag{21.2}$$

where *S* (*x*, *y*) represents the depth ratio area and S (*x*, *y*) $\subseteq [D_{near}, D_{far}]$.

21.3.1.2 Human Joints Identification

This study generated a real-time data collection that comprised yoga asanas including Pranamasana, Dhanurasana, Dandasana, Gomukhasana, Garudasana, Padmavrikshasana, and Padmasan because there is no standard human position depth photo library. The random forest learning algorithm is a supervised learning approach, which means that the data samples must be labelled and reflect a predefined category. The tagging technique splits the human frame into sixteen joints, with the relaxation serving because of the backdrop; the approximate area of every human frame aspect within side the intensity photograph is evaluated, and the location is then tagged with an appropriate colours. In Figure 21.1, the legitimate locations within side the rectangle of the top region are all marked in Dark brown.

Table 21.1 displays the tagging results. This article separates the human structure joints into the following sections: head, neck, right-shoulder, left-shoulder, right-elbow, left-elbow, right-wrist, left-foot, right-foot, left-wrist, right-hip, left-hip, right-knee, left-knee, left-hand, and right-hand.

21.3.1.3 Extraction of L-DoD Features

The functions of 16 joints have to be retrieved from an intensity picture of a human frame that has been manually annotated. The neighbor-hood distinction function is used as a function representation in this research of the pixel, which is capable of replicating the pixel's community information. It uses the distinction between pixels and the various 8 community components to represent the pixel's characteristics. Figure 21.2 shows the location of the eight community pixels.

The local difference of depth functionality may be expressed as follows:

$$T_{i,j}(I,p)=\left|d_s(p_i)-d_s\left(p_j\right)\right|, \tag{21.3}$$

where $((i, j) \in \{1, 2, 3, 4, 5, 6, 7, 8\})$, $i \neq j$, and $ds\ (pi)$ is the depth value of pi. Assuming $\theta = \{\theta1, \theta2, \theta3,\ldots, \theta28,\}$ $T\theta k\ (s, p)$ is restore by $Ti,j\ (s, p)$ and $k \in \{1, 2, 3,\ldots, 28\}$.

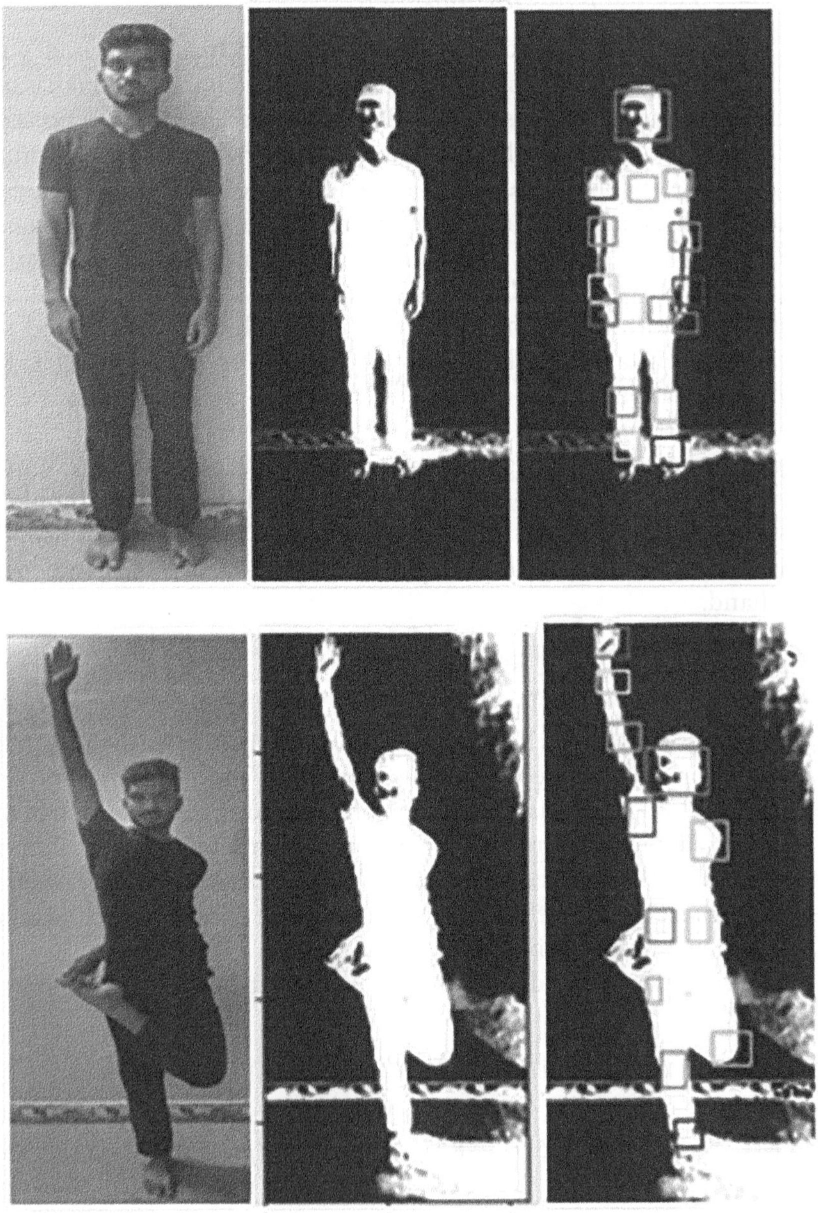

Figure 21.1 Determining which human joints should be high-lighted in a rectangular form.

Table 21.1 Tagging human joints.

S. no	Human joints	Identifying colors	S. no	Human joints	Identifying colors
1	Head	Red	9	Left hand	Sky blue
2	Right shoulder	Light pink	10	Neck	Light green
3	Left-shoulder	Dark brown	11	Right hip	Dark green
4	Right-elbow	Gray	12	Left hip	Sky blue
5	Left-elbow	Orange	13	Right knee	Light brown
6	Right-wrist	Yellow	14	Left knee	Lavender
7	Left-wrist	Dark blue	15	Right foot	Block
8	Right-hand	Purple	16	Left foot	Dark yellow

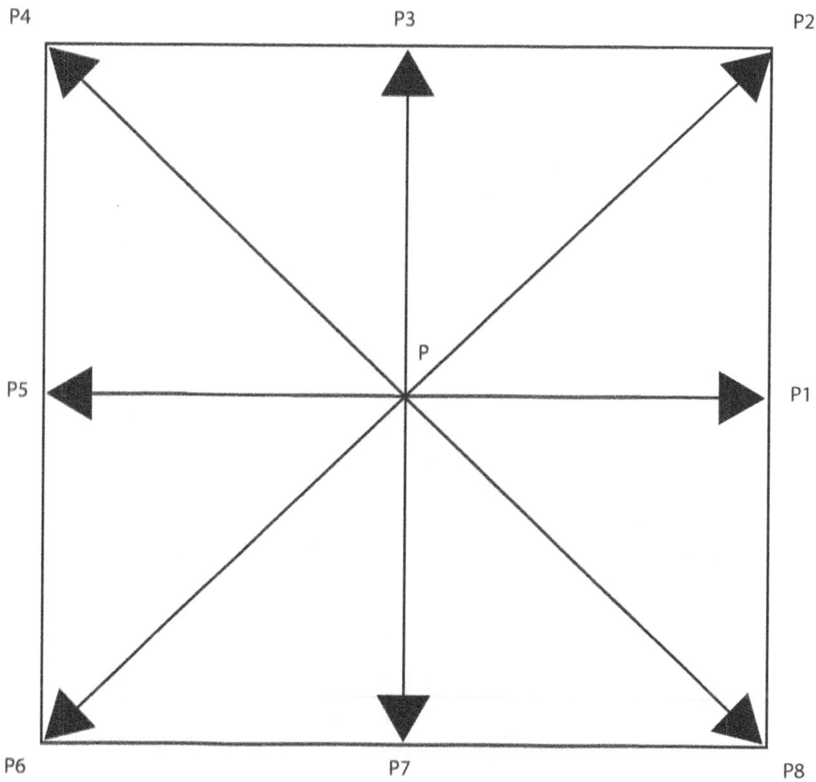

Figure 21.2 Locations of the eight neighborhood picture elements.

A point's function-vector may be written as

$$T_\theta(I,p)=\{T_{\theta_1}(I,p),T_{\theta_2}(I,p),\ldots,T_{\theta_{28}}(I,p)\}. \tag{21.4}$$

Characteristics of the same type of picture element are typically identical according to the L-DoD feature; however, features of distinct picture element types range substantially. As a result, this feature has a high degree of division for various human structure parts. Figure 21.3 shows a split depth image and an enlarged view of the left elbow.

As seen in the pictures, picture elements p_6 and p_7 are within side the frame area; however, picture element p_4 was outside the structure region and has a range of 0. As a result, $T_{6,7}$ (S, p) is less and $T_{4,7}$ (S, p) is greater. A grow view of the proper decrease arm is proven in Formula (21.4). $T_{6,7}$ (S, p) has a higher value, while $T_{4,7}$ (S, p) has a lower value. As a result, these two values may identify the human structure's left and right lower arms.

Figure 21.3 Split depth picture and the magnified image of the left elbow.

This feature has a very low computational cost. Subtraction is the only operation used in Formula (21.3). It also possesses spatial movement and spins invariance and may be used into human posture alterations.

21.3.1.4 Extraction of D-GoD Features

As a result, the extracted features represent the object's distance from the depth camera.

The angle relationship between the plane where the picture element is located and the plane where the depth camera is positioned can purely be determined by calculating the arctangent of the gradient, i.e., the D-GoD feature, which can be calculated:

$$\begin{aligned} \mathrm{DGoD}_{S(x,y)} &= \tan^{-1}\frac{dy}{dx}, \\ dy &= S(x,y+i) - S(x,y-i), \\ dx &= S(x+i,y) - S(x-i,y), \end{aligned} \tag{21.5}$$

where i = 1, 2, 3. 3 D-GoD function were chosen, and they are depicted as $G_{\theta 1}$(S, x, y), $G_{\theta 2}$(S, x, y), and $G_{\theta 3}$ (S, x, y).

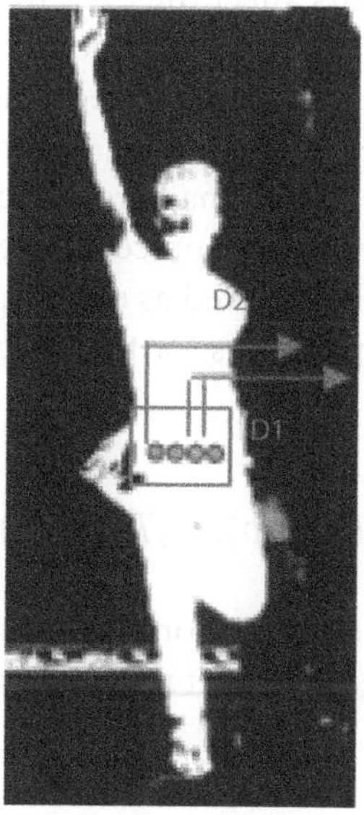

Figure 21.4 Depicts the D-GoD feature diagram.

The directional gradients have a range of {0° to 360°}. The direction gradients are comparable in size whenever the picture element coordinates was all at the identical plane; otherwise, the direction-gradients were considerably different. Figure 21.4 depicts the DGoD feature diagram. From left to right, the green dots are p_1, p_2, p_3, and p_4. Formula (21.5) shows that the range of D_1 was little and the value of D_2 was high, indicating (p_2 and p_3) are in the identical plane, but p_1 and p_4 are in separate planes.

21.3.2 The Random Forest Classifier's Design

21.3.2.1 Construction of a Random Forest Model

At the moment, decision tree was considered one among the maxima appreciably used inductive inference methods. Its created guidelines were neat and straight forward. Because decision tree can classify image factors intensive photos unexpectedly and effectively, they will be extensively hired in goal detection and recognition. A single selection tree, on the alternative hand, would possibly effortlessly cause over becoming, ensuing in the wrong categorization.

The random tree is made from numerous decision tree, and every selection tree is skilled with numerous information units and parameters, which now no longer best reduces over becoming; however, additionally improves class accuracy seeing that its output is voted on through several decision tree. Many elements affect the class impact of random woodland classifiers, inclusive of the scale of the schooling set D, the size comparable pattern function vector N, the variety of decision tree K, the most intensity of every tree d, the proper vector size n, and the close circumstance for an increase of every tree. The human frame turned separated within 18 unique components within side the preceding sections, after which (LDoD and DGoD) traits had been retrieved as entering to the random woodland classifier. All preparatory paintings for the advent of the classifier version have been completed.

$$S=\left\{T_{\theta_1}(I,p),T_{\theta_2}(I,p),\ldots,T_{\theta_{28}}(I,p),\mathrm{DGoD}_{S(x,y)}\right\}. \tag{21.6}$$

Each decision tree in a random- forest is trained using the ID3 decision tree method. The training sample set may be written as

$$Q=\{P,C\}, \tag{21.7}$$

where P = $p_1, p_2, p_3,..., p_i,..., p_n$ is a collection of training picture elements and C = $c_1, c_2, c_3,...,c_{16}$ Is a group of classes to which a image detail belongs, i.e. sixteen sections of the human structure.

The parameter written

$$\varphi = (\theta, \tau_1, \tau_2), \tag{21.8}$$

where θ is the attribute parameter and the thresholds are τ_1 and τ_1.

To begin, the extraction approach employs putting back, and the training set D_i, that is the equal length as D, is extracted along-with D to offer K subsets. Then, a tree node is generated, and if it comes to a close condition, the process is closed along with the existing node turns within a leaf node. Alternatively, n capabilities are retrieved from the N-dimensional function set extracted with a fixed-scale and no again extraction technique. The 1-dimensional function is determined by the function attribute's metric, and the current node is divided into the left subset $D_{1\varphi}$ and the right subset $D_r\varphi$:

$$\begin{cases} p_i \in D_1(\varphi), & T_\theta \le \tau_1, G_\theta \le \tau_2, \\ p_i \in D_r(\varphi), & T_\theta > \tau_1, G_\theta > \tau_2. \end{cases} \tag{21.9}$$

The details are used to choose a decision-tree's attribute of splitting, which may be computed as follows:

$$\varphi_* = \arg\max \text{Gain}(D,\varphi),$$
$$\text{Gain}(D,\varphi) = \text{Ent}(D) - \sum_{v\in\{l,r\}} \frac{|D^v(\varphi)|}{|D|} \text{Ent}(D^v(\varphi)), \tag{21.10}$$

where Ent(D) is the entropy of information.

21.3.2.2 *Random Forest Two-Way Voting*

The sample is assessed and voted on by each decision tree in the typical random-forest classification. Every tree has the same ability to make decisions. In this research, the random-forest two-way voting method was used with various award privileges. "In-bag and out-bag data" are separated in the data set. When utilized to generate a random-forest, the data subset is referred to as In-bag data. A data subset that is not in the bag is referred to as out-bag. A tree's decision right is determined by the results of testing out-bag data. The tree was voted when the outcomes are true. The weight of a decision tree increases as the number of vote's increases. The following are the basic phases in the two-way voting algorithm.

1. Make a total of K decision trees. For each tree, in-bag data and out-bag data are created.
2. Conduct a performance review. To put it another way, a tree is judged based on a particular quantity of out-bag data. If the categorization result of the decision tree is correct, and the tree is correct voted.
3. To the decision tree, assign the total number of votes as follows: All decision trees' weights should be weighed and normalized.
4. Use the sorted random forest to input the test sample. Techniques and the class end result done multiply the load used to decide the very last class end result, that is displayed as follows:

$$R(x) = \sum_{i=1}^{K} T_i r_i(x), \tag{21.11}$$

where (R (x)) are the at last identification result, T_i is the i-th decision tree's weight coefficient, and $T_1 + T_2 + T_3 + \dots\dots\dots + T_k = 1$. The assessment outcome of i-th decision tree is r_i (x).

21.3.3 Joint Positioning in Humans

The final phase in human posture identification is determining the human structure joints. The random forest classifier was used to categorize 16 joints of the human structure picture in the preceding chapters, but the joint location was not identified. The gravity centers of the 16 structure components in this article will be used to identify the joints. The p + q moment m_{pq} and middle moment at the picture element I (x, y) may be determined using formulas (21.12) and (21.13), respectively, for the depth picture of size (M *N).

$$m_{pq} = \sum_{y=1}^{N}\sum_{x=1}^{M} x^p y^q \cdot I(x,y), \tag{21.12}$$

$$\mu_{pq} = \sum_{y=1}^{N}\sum_{x=1}^{M} (x - x_c)^p \cdot (y - y_c)^q \cdot I(x,y), \tag{21.13}$$

The center of gravity (COG), x_c y_c, may be computed using formula (21.14) and (21.15). The joint was computed using the COG of the upper arm and the lower arm, as given in formula (21.16).

$$x_c = \frac{m_{10}}{m_{00}},$$
$$y_c = \frac{m_{01}}{m_{00}}, \tag{21.14}$$

$$x_c = \frac{\sum_{y=1}^{N}\sum_{x=1}^{M} x \cdot I(x,y)}{\sum_{y=1}^{N}\sum_{x=1}^{M} I(x,y)},$$
$$y_c = \frac{\sum_{y=1}^{N}\sum_{x=1}^{M} y \cdot I(x,y)}{\sum_{y=1}^{N}\sum_{x=1}^{M} I(x,y)}, \tag{21.15}$$

$$x_{\text{elbow}} = \frac{\left(\sum_{y=1}^{N_{\text{up}}}\sum_{x=1}^{M_{\text{up}}} x \cdot I(x,y) / \sum_{y=1}^{N_{\text{up}}}\sum_{x=1}^{M_{\text{up}}} I(x,y)\right) + \left(\sum_{y=1}^{N_{\text{low}}}\sum_{x=1}^{M_{\text{low}}} x \cdot I(x,y) / \sum_{y=1}^{N_{\text{low}}}\sum_{x=1}^{M_{\text{low}}} I(x,y)\right)}{2} + \Delta x,$$
$$y_{\text{elbow}} = \frac{\left(\sum_{y=1}^{N_{\text{up}}}\sum_{x=1}^{M_{\text{up}}} y \cdot I(x,y) / \sum_{y=1}^{N_{\text{up}}}\sum_{x=1}^{M_{\text{up}}} I(x,y)\right) + \left(\sum_{y=1}^{N_{\text{low}}}\sum_{x=1}^{M_{\text{low}}} y \cdot I(x,y) / \sum_{y=1}^{N_{\text{low}}}\sum_{x=1}^{M_{\text{low}}} I(x,y)\right)}{2} + \Delta y, \tag{21.16}$$

where $M_{up} * N_{up}$ denotes the area of the upper arm, $M_{low} * M_{low}$ denotes the area of the (lower arm), and (x and y) denote the offsets.

21.4 Outcome and Discussion

This study, 500 intensity pictures are used to educate the classifier model, and 70 pix are given to check it, with ten wonderful people's stances included. Colab is used to program the algorithm. Figure 21.5 depicts the results of human structure part identification and human joints placement in ten yoga poses. The segmented depth pictures are in the first column, the random forest classifier outputs are in the second column, the COG of each component is in the third column, and the skeletons built of the joints are in the last column. The random-forest category can accurately categories most of the picture elements in the human structure picture, such as the structure and the head, as shown in Figure 21.5.

At the junction of the two portions, incorrect categorization is common. Fortunately, the joints are almost perfectly aligned, allowing for the

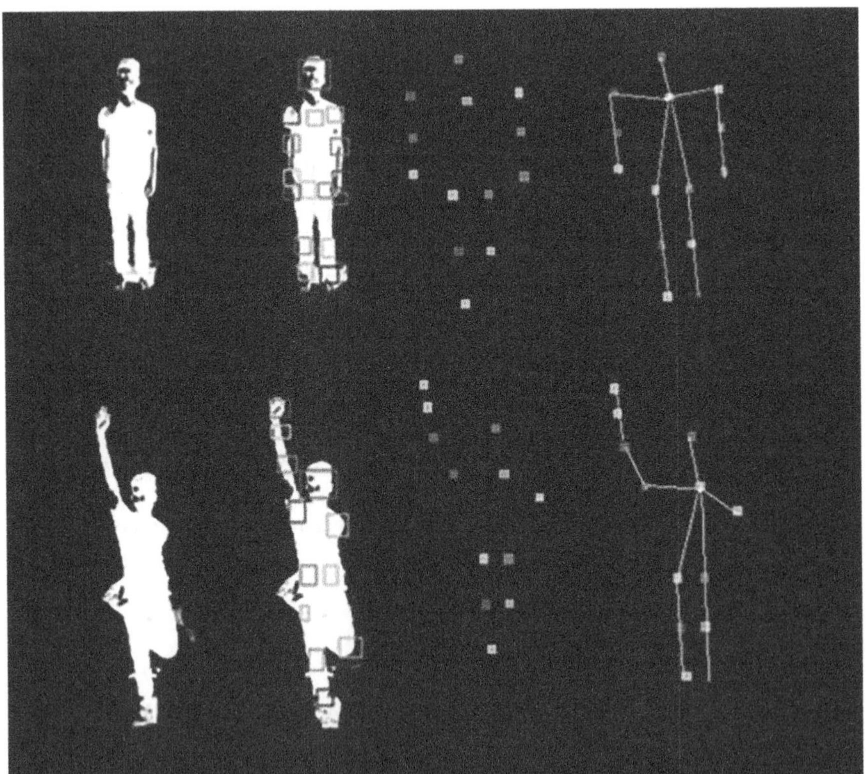

Figure 21.5 Recognition of yoga poses and joint point location in various postures.

creation of a decent human skeleton. Experiments on different parameters of classifiers are compared. The wide variety of decision-tree K, the most intensity of value d, and the minimal wide variety of samples in leaf-nodes N-node can all effect the classifier's effectiveness while constructing a random wooded area techniques. The experiment begins by training five example photos to identify the best classifier parameters.

Figure 21.6 shows how different parameters affect the algorithm's classification accuracy and training time. The recognition rates of several algorithms are compared. As shown in Figure 21.7, this article contrasts the identity price of every element with a single function, L-DoD, and the popularity price with an aggregate of features, D-GoD and L-DoD. The random-forest method with multifeature fusion has a much higher recognition rate, reaching over 82.92%. The left and right arms have the lowest identification rates among these 16 sections, owing to the complicated motions of the upper-limbs. Furthermore, the number of samples gathered grows, so does the recognition rate.

In this study, the two-way vote casting mechanism and the same old vote casting mechanism of random forests are compared. The category accuracy of the random-forest two-way vote casting tool is an awful lot more than the standard one-way vote casting mechanism.

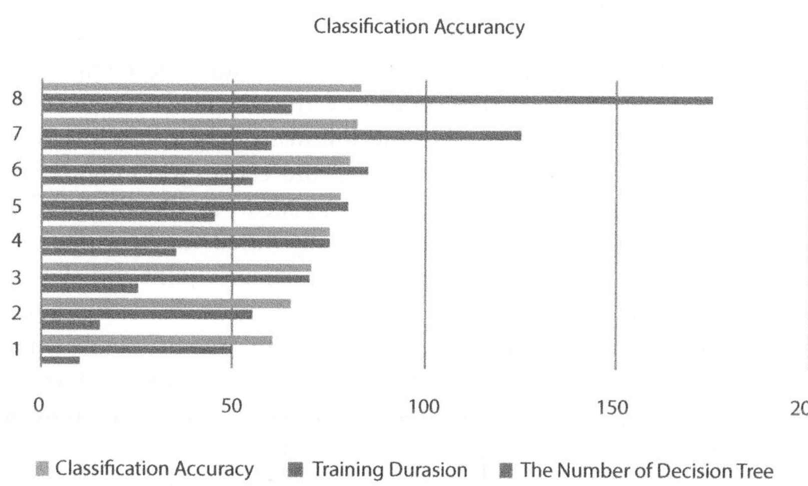

Figure 21.6 Classification accuracy and training time.

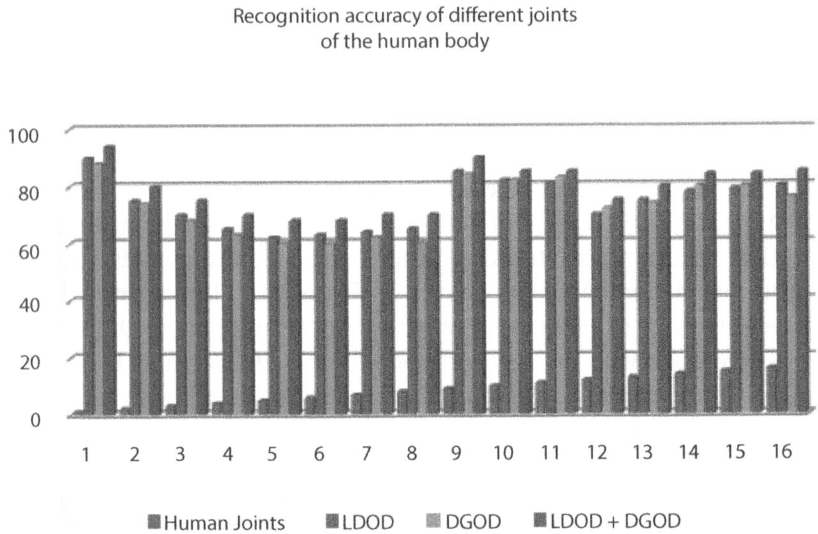

Figure 21.7 Identification charge of every component with a single feature, L-Dod, and the popularity charge with a mixture of function, D-God and L-Dod.

21.5 Conclusion

This study presents a human posture identity device primarily based totally on the combination of L-DoD and D-GoD traits on this paper. In order to recognize yoga poses, first create our personal pattern information set, which incorporates intensity snap shots coded with a specific code. The L-DoD and D-GoD traits are then extracted from the pattern. The aforementioned traits are smooth to compute. As a result, the computation time is extensively lowered. These traits are then applied to educate the random-forest classifier within side the following stage. Random-forest way vote casting device is used to discover and classes awesome regions of the human shape as a way to increase class accuracy. Finally, the COG of diverse shape additives is decided primarily based totally at the categorization results, making an allowance for particular joint and skeleton reconstruction. The random-forest classifier has an extra class accuracy and resilience, in line with the findings of the experiments. Furthermore, while as compared to preceding approaches, our technique is much less pricey to compute and fulfils real-time requirements. However, no device is faultless in terms of identifying yoga poses, for this reason it is crucial to have observe the subsequent factors. In destiny, the usage of many asana datasets and locating 18 joints attitude values.

References

1. Gao, Z., Zhang, H., Liu, A.A. *et al.*, Human action recognition on depth dataset. *Neural Comput. Appl.*, 27, 2047–2054, 2016.
2. Weinland, D., Ronfard, R., Boyer, E., A survey of vision based methods for action representation, segmentation and recognition. *Comput. Vis. Image Underst.*, 115, 224–241, 2011.
3. Suto, J., Comparison of offline and real-time human activity recognition results using machine learning techniques. *Neural Comput. Appl.*, 32, 20, 2018.
4. Zhang, P., Lan, C., Xing, J., Zeng, W., Xue, J., Zheng, N., View adaptive neural networks for high performance skeleton-based human action recognition. *IEEE Trans. Pattern Anal. Mach. Intell.*, 41, 8, 1963–1978, 2019.
5. Sharif, A., Khan, M.A., Javed, K., Gulfam, H., Iqbal, T., Saba, T. *et al.*, Intelligent human action recognition: A framework of optimal features selection based on Euclidean distance and strong correlation. *J. Control Eng. Appl. Inform.*, 21, 3–11, 2019.
6. Kamel, A., Sheng, B., Yang, P., Li, P., Shen, R., Feng, D.D., Deep convolutional neural networks for human action recognition using depth maps and postures. *IEEE Trans. Syst. Man Cybern.: Syst.*, 49, 9, 1806–1819, 2018.
7. Guddeti, R.R., Dang, G., Williams, M.A., Alla, V.M., Role of yoga in cardiac disease and rehabilitation. *J. Cardiopulm. Rehabil. Prev.*, 39, 3, 146–152, 2018, https://doi.org/10.1097/hcr.0000000000000372.
8. Halliwell, E., Dawson, K., Burkey, S., A randomized experimental evaluation of a yoga-based structure image intervention. *Struct. Image*, 28, 119–127, 2019, https://doi.org/10.1016/j.structureim.2018.12. 005.
9. Sathyanarayanan, G., Vengadavaradan, A., Bharadwaj, B., Role of yoga and mindfulness in severe mental illnesses: A narrative review. *Int. J. Yoga*, 12, 3–28, 2019, https://doi.org/10.4103/ijoy. IJOY_65_1.
10. Konečný, J. and Hagara, M., One-shot-learning gesture recognition using HOG-HOF features. *J. Mach. Learn. Res.*, 15, 1, 2513–2532, 2017.
11. Rasche, C., Rapid contour detection for image classification. *IET Image Process.*, 12, 4, 532–538, 2018.
12. Supančič, J.S., Rogez, G., Yang, Y., Shotton, J., Ramanan, D., Depth-based hand pose estimation: Methods, data, and challenges. *Int. J. Comput. Vis.*, 126, 11, 1180–1198, 2018.
13. Pishchulin, L., Jain, A., Andriluka, M., Thormahlen, T., Schiele, B., Articulated people detection and pose estimation: Reshaping the future. *IEEE Conference on Computer Vision and Pattern Recognition*, Providence, Rhode Island, June 2012, pp. 3178–3185.
14. Shotton, J., Sharp, T., Kipman, A. *et al.*, Real-time human pose recognition in parts from single depth images. *Commun. ACM*, 56, 1, 116–124, 2013.
15. Hernández-Vela, A., Zlateva, N., Marinov, A. *et al.*, Graph cuts optimization for multi-limb human segmentation in depth maps. *Conference on Computer*

Vision and Pattern Recognition, Providence, Rhode Island, June 2012, pp. 726–732.

16. Kim, H., Lee, S., Lee, D., Choi, S., Ju, J., Myung, H., Real-time human pose estimation and gesture recognition from depth images using superpicture elements and SVM classifier. *Sensors*, 15, 6, 12410–12427, 2015.
17. Oliveira, G.L., Valada, A., Bollen, C., Burgard, W., Brox, T., Deep learning for human part discovery in images. *IEEE International Conference on Robotics and Automation (ICRA)*, Stockholm, Sweden, pp. 1634–1641, May 2016.
18. Sharif, M., Akram, T., Raza, M., Saba, T., Rehman, A., Hand-crafted and deep convolutional neural network features fusion and selection strategy: An application to intelligent human action recognition. *Appl. Soft Comput.*, 87, 105986, 2020.
19. Ullah, A., Muhammad, K., Haq, I.U., Baik, S.W., Action recognition using optimized deep autoencoder and CNN for surveillance data streams of non-stationary environments. *Future Gener Comput. Syst.*, 96, 386–397, 2019.
20. Yang, H., Yuan, C., Li, B., Du, Y., Xing, J., Hu, W. *et al.*, Asymmetric 3d convolutional neural networks for action recognition. *Pattern Recognit.*, 85, 1–12, 2019.
21. Wang, P., Liu, L., Shen, C., Shen, H.T., Order-aware convolutional pooling for video based action recognition. *Pattern Recognit.*, 91, 357–365, 2019.
22. Wu, J., Qiu, S., Zeng, R., Kong, Y., Senhadji, L., Shu, H., Multilinear principal component analysis network for tensor object classification. *IEEE Access*, 5, 3322–3331, 2019.

22
E-Commerce Data Analytics Using Web Scraping

Vikash Kumar Mishra[1], Bosco Paul Alapatt[2*], Aaditya Aggarwal[2] and Divya Khemani[2]

[1]*Galgotias University, Gautam Buddha Nagar, Uttar Pradesh, India*
[2]*Christ (Deemed to be University), Delhi NCR Ghaziabad, India*

Abstract
Some companies, like Twitter and others, provide an application programming interface (API) to fetch the information. If the API is not available, we will have to search other websites to get the data in a structured format. The primary way to get data from a web page is through web scraping. The basic idea of web scraping is to pull data from a website and convert it into a format that can be used for analysis. In this paper, we will discuss the simple explanation of how we can use Beautiful Soup to scratch data into Python and later save the extracted data in an Excel spreadsheet and do the spreadsheet analysis later. We will pull data from the Flipkart website to know the cell phone name, cell phone price, cell phone rating, and cell phone specification.

***Keywords*:** Web scraping, data analysis, implementing web scraping, spreadsheet analysis, cell phone

22.1 Introduction

Data analysis is the method for finding solutions to the problems of querying and interpreting data. The analysis process involves discovering problems, resolving the accessibility of relevant data, determining the method that can help solve the problem, and producing the results.

**Corresponding author*: bosco.paul@christuniversity.in

Sharmistha Ghosh, M. Niranjanamurthy, Krishanu Deyasi, Biswadip Basu Mallik, and Santanu Das (eds.) Mathematics and Computer Science Volume 1, (425–434) © 2023 Scrivener Publishing LLC

For the analysis, the data must be broken down into different phases, such as B. Compiling, organizing, cleaning up, reanalyzing your specifications, applying models and algorithms, and then obtaining results [1].

Web scraping is also known as screen scraping, web harvesting or web data extraction. It is a technique to naturally extract large amounts of data from websites and store it in a file or database. The retrieved data is usually saved as a table or spreadsheet [2]. Data published by websites can only be viewed with a web browser. Most websites do not allow us to save or download this data. When we need the data, the only option is to manually copy and paste it [2]. Web scraping is the technique used to automate this process. Instead of manually copying data from websites, we can use web scraping software that does the same job in a fraction of the time.

Web scratching is divided into two parts, namely the crawler and the scraper. Crawler is an artificial intelligence algorithm that crawls the web for specific data needed by following the link on the internet. The scraper is a special tool created to get data from the website. The design of the scraper can vary depending on the complexity and scope of the project so that it can extract data quickly and accurately [3].

22.1.1 Uses of Web Scraping

Web scraping is used for obtaining data. In the contemporary world, the access to accurate data, having methods to analyze it and performing intelligent actions based on analysis can make an enormous difference in the success and growth of most businesses. Data collection and evaluation is crucial even for government authorities, nonprofit and educational institutions. The data collected can be applied in various industries for the purpose of price monitoring, market research, news monitoring, sentiment analysis, and E-mail marketing [3].

22.2 Research Objective

The paper illustrates the need to fetch information from various websites using the Python programming language with the help of two Python libraries: Requests and Beautiful Soup. Further, save the data in an Excel file to create a database. The database, which is created after obtaining all the information from various sources, is further analyzed through the analytical process of collecting, organizing, cleaning, reanalysis, applying models, and algorithms to finally determining the results.

22.3 Literature Review

The cases studied by us explains how the data can be examined through web scrapping.

Extracting the hotel prices can provide many companies daily price data to automate and optimize the process of comparing hotel prices. For the companies in travel industry, it is very important to use the entire World Wide Web as a database for quick access to accurate and well-structured data. If we take a look at the recently supported use cases for these requirements. Customer Famous Indian budget hotel chains had to update hotel prices twice a day. Customers needed price data for themselves and their competitors every day and were looking for a web data scraping service provider to help with data scraping. Consistent Data Hotel Price Scraping where the hotels have programmed the scrapers to search for files on shared servers at specific times and select them when they are accessible. The crawler can search for files both in the morning and in the afternoon. Except for the early stages of the implementation, the process was fully automated. Service interruptions from the support team are automatically updated to ensure a smooth crawl. By crawling inventory every 2 months, we can have a frequent access to updated data and where the hotels can be in a good position to expand their presence nationwide. The reduced processing time allows them to use the collected data more efficiently [4].

It is understood by the case that it is not difficult to examine and modify the data as per the needs and requirements of the demands. The data is based on various factors that help in understanding the extracted data in order to verify the customer analysis. Web scraping can also be used in technologies, such as market research. In addition, web scrapping technology is used on price comparison sites where it analyzes the price of an item or part from various websites. Various government and private guard dogs use web scraping to monitor malicious activity on the Internet. The Netucon Company operating in Ahmadabad offers its customers absolute solutions with the help of innovative and creative software development services. Netucon understands the concerns of customers and customers and produces the software. They produced a LinkedIn Connection Creator (LCC). This LCC was effective in eliminating CEOs, building B2C connections, lead generators, digital marketers, and bloggers posting their blogs on LinkedIn [5].

22.4 Feasibility and Application

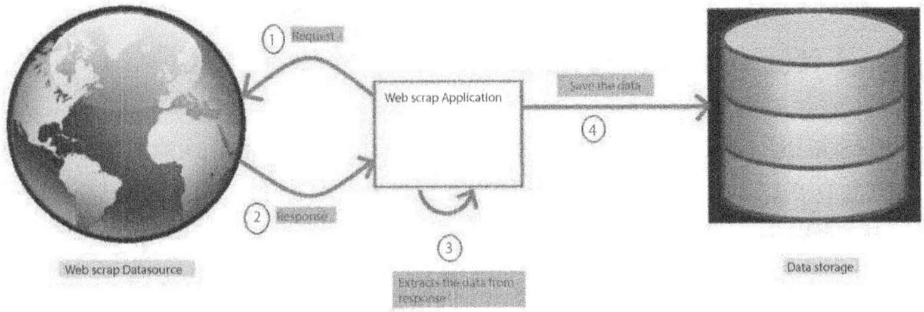

Figure 22.1 Source: web-parsing.com.

22.4.1 Web Scrapers Process

Web scrapers can fetch all data from a specific site or a specific data, depending on the needs of the user. For example, if a user wants to get an Amazon page for the TV that is available, but is only interested in a model specification for different TVs and does not want customer reviews. So Figure 22.1, when a web scraper needs to sort out a site in this case, it first checks the URLs provided. Then all of the HTML for those sites will be loaded. The scraper then fetches the required data from this HTML code and generates this data in the format specified by the user. Usually, it is an Excel table or a CSV file [3, 6].

22.5 Proposed Methodology

The methodology used for the project is to collect all the data related to cell phones, which is extracted from the online Flipkart website using the scripts written in Python. The data collected is further analyzed using spreadsheet analysis.

The data extracted includes:

1. Name of the mobile phones
2. Price of the mobile phones
3. Ratings of the mobile phones
4. Specifications of the mobile phones

22.5.1 Coding Phase

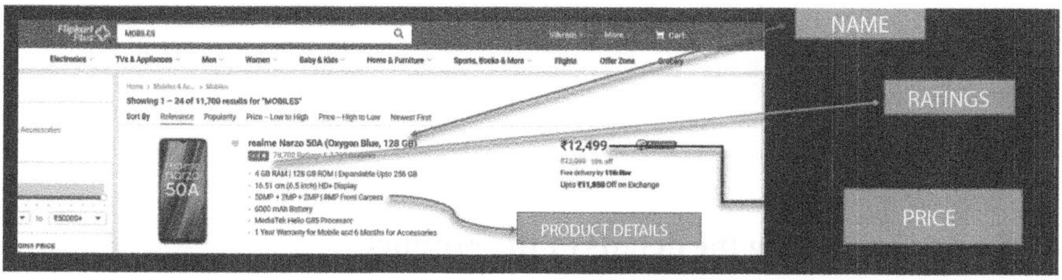

Figure 22.2 Flipkart webpage: from where the data needed to be extract.

The mobile data will be collected from the Flipkart official site as shown in Figure 22.2. All the information about the mobile phones, including name, price, reviews, and specifications, will be taken directly from the source.

Step 1: Install/ Import the necessary libraries

```
#import libraries
import bs4
from bs4 import BeautifulSoup as bs
import requests
import xlsxwriter
import pandas as pd
```

Step 2: Choose the data you need to extract

```
#enter url
link="https://www.flipkart.com/search?q=mobiles&as=on&as-show=on&otracker=AS_Que
```

Step 3: Send HTTP request to the URL of the page you want to scrape

```
#send request
page = requests.get(link)
```

We can print the content of the page using the below code. It will give the HTML content of the page and it's in string format [7–10].

```
page.content
```

Step 4: Extracting the Name of the Product

```
#get name of the product
name=soup.find('div',class_="_4rR01T")
print("PRODUCT NAME:",name.text)
```

Step 5: Extracting the Rating of the Product

```
#get ratings of the product
rating=soup.find('div',class_=" 3LWZlK")
print("PRODUCT RATINGS:",rating.text)
```

Step 6: Extracting the details of the product

```
#get other details and specifications of the product
specification=soup.find('div',class_="fMghEO")
print("PRODUCT DETAILS:\n",specification.text)
```

Step 7: Extracting the price of the product

```
#get price of the product
price=soup.find('div',class_=' 30jeq3 _1_WHN1')
print("PRICE:",price.text)
```

Output:

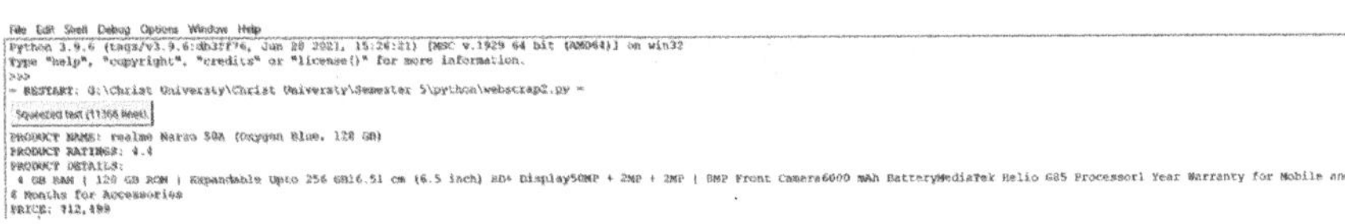

Figure 22.3 Scraped data for a particular mobile phone available on website.

We have successfully got the Name, Rating, Other Specifications, and Price of a mobile phone as shown in Figure 22.3. Till now we have just extracted the data of one mobile phone but now we need the data of all the mobile phones available on the page so we will be run a loop and store the data in the defined lists [7–10].

Step 8: Defining the lists to store the value of each feature

```
Names=[]
Ratings=[]
Specifications=[]
Prices=[]
```

Step 9: Make a loop and inspect the tag where all the information is present and using it to extract all the features together

```
for data in soup.findAll('div',class_='_3pLy-c row'):
    Name=data.find('div', attrs={'class':'_4rR01T'})
    Price=data.find('div', attrs={'class':'_30jeq3 _1_WHN1'})
    Rating=data.find('div', attrs={'class':'_3LWZlK'})
    Specification = data.find('div', attrs={'class':'fMghEO'})
    Names.append(Name.text)
    Prices.append(Price.text)
    Ratings.append(Rating.text)
    Specifications.append(Specification.text)

print(Names)
print(Prices)
print(Ratings)
print(Specifications)
```

After executing the above code, all the features are extracted and stored in the respective lists. [7–10].

Output:

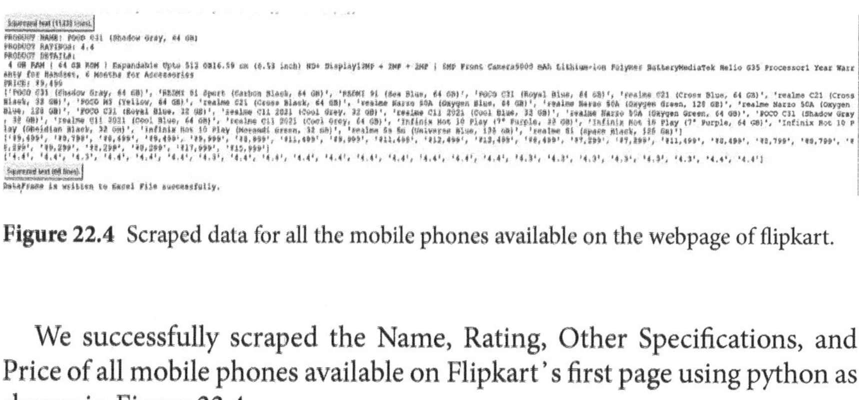

Figure 22.4 Scraped data for all the mobile phones available on the webpage of flipkart.

We successfully scraped the Name, Rating, Other Specifications, and Price of all mobile phones available on Flipkart's first page using python as shown in Figure 22.4.

Step 10: Store the data into the structured format in the Data Frame

```
df=pd.DataFrame({'PRODUCT NAME':Names,'PRICE':Prices,'RATINGS':Ratings,'Specification':Specifications})
```

Step 11: Extract the data in a spreadsheet

```
file_name = 'MobileData.xlsx'
df.to_excel(file_name)
print('DataFrame is written to Excel File successfully.')
```

Final output in the Spreadsheet

A	B	C	D	
	PRODUCT NAME	PRICE	RATINGS	Specification
0	POCO C31 (Shadow Gray, 64 GB)	₹9,499	4.4	4 GB RAM | 64 GB ROM | Expandable Upto 512 GB16.59 cm (6
1	REDMI 9i Sport (Carbon Black, 64 GB)	₹8,799	4.4	4 GB RAM | 64 GB ROM | Expandable Upto 512 GB16.59 cm (6
2	REDMI 9i (Sea Blue, 64 GB)	₹8,499	4.3	4 GB RAM | 64 GB ROM | Expandable Upto 512 GB16.59 cm (6
3	POCO C31 (Royal Blue, 64 GB)	₹9,499	4.4	4 GB RAM | 64 GB ROM | Expandable Upto 512 GB16.59 cm (6
4	realme C21 (Cross Blue, 64 GB)	₹9,999	4.4	4 GB RAM | 64 GB ROM | Expandable Upto 256 GB16.51 cm (6
5	realme C21 (Cross Black, 32 GB)	₹8,999	4.4	3 GB RAM | 32 GB ROM | Expandable Upto 256 GB16.51 cm (6
6	POCO M3 (Yellow, 64 GB)	₹11,499	4.3	6 GB RAM | 64 GB ROM | Expandable Upto 512 GB16.59 cm (6
7	realme C21 (Cross Black, 64 GB)	₹9,999	4.4	4 GB RAM | 64 GB ROM | Expandable Upto 256 GB16.51 cm (6
8	realme Narzo 50A (Oxygen Blue, 64 GB)	₹11,499	4.4	4 GB RAM | 64 GB ROM | Expandable Upto 256 GB16.51 cm (6
9	realme Narzo 50A (Oxygen Green, 128 GB)	₹12,499	4.4	4 GB RAM | 128 GB ROM | Expandable Upto 256 GB16.51 cm
10	realme Narzo 50A (Oxygen Blue, 128 GB)	₹12,499	4.4	4 GB RAM | 128 GB ROM | Expandable Upto 256 GB16.51 cm
11	POCO C31 (Royal Blue, 32 GB)	₹8,499	4.4	3 GB RAM | 32 GB ROM | Expandable Upto 512 GB16.59 cm (6
12	realme C11 2021 (Cool Grey, 32 GB)	₹7,299	4.4	2 GB RAM | 32 GB ROM | Expandable Upto 256 GB16.51 cm (6
13	realme C11 2021 (Cool Blue, 32 GB)	₹7,299	4.4	2 GB RAM | 32 GB ROM | Expandable Upto 256 GB16.51 cm (6
14	realme Narzo 50A (Oxygen Green, 64 GB)	₹11,499	4.4	4 GB RAM | 64 GB ROM | Expandable Upto 256 GB16.51 cm (6
15	POCO C31 (Shadow Gray, 32 GB)	₹8,499	4.4	3 GB RAM | 32 GB ROM | Expandable Upto 512 GB16.59 cm (6
16	realme C11 2021 (Cool Blue, 64 GB)	₹8,799	4.3	4 GB RAM | 64 GB ROM | Expandable Upto 256 GB16.51 cm (6
17	realme C11 2021 (Cool Grey, 64 GB)	₹8,799	4.3	4 GB RAM | 64 GB ROM | Expandable Upto 256 GB16.51 cm (6
18	Infinix Hot 10 Play (7° Purple, 32 GB)	₹8,299	4.3	3 GB RAM | 32 GB ROM | Expandable Upto 256 GB17.32 cm (6
19	Infinix Hot 10 Play (7° Purple, 64 GB)	₹9,299	4.3	4 GB RAM | 64 GB ROM | Expandable Upto 256 GB17.32 cm (6
20	Infinix Hot 10 Play (Obsidian Black, 32 GB)	₹8,299	4.3	3 GB RAM | 32 GB ROM | Expandable Upto 256 GB17.32 cm (6
21	Infinix Hot 10 Play (Morandi Green, 32 GB)	₹8,299	4.3	3 GB RAM | 32 GB ROM | Expandable Upto 256 GB17.32 cm (6
22	realme 8s 5G (Universe Blue, 128 GB)	₹17,999	4.4	6 GB RAM | 128 GB ROM | Expandable Upto 1 TB16.51 cm (6.
23	realme 8i (Space Black, 128 GB)	₹15,999	4.4	6 GB RAM | 128 GB ROM | Expandable Upto 256 GB16.76 cm

Figure 22.5 Extracted data on the Excel sheet in a structured form.

Hence, we are able to scrap the Mobile phone data from the Flipkart webpage on the spreadsheet as shown in Figure 22.5 [7–10].

22.5.2 Spreadsheet Analysis and Results

A	B	C	D	
	PRODUCT NAME	PRICE	RATINGS	Specification
22	realme 8s 5G (Universe Blue, 128 GB)	₹17,999	↑ 4.4	6 GB RAM | 128 GB ROM | Expandable Upto 1 TB
23	realme 8i (Space Black, 128 GB)	₹15,999	↑ 4.4	6 GB RAM | 128 GB ROM | Expandable Upto 256 G
9	realme Narzo 50A (Oxygen Green, 128 GB)	₹12,499	↑ 4.4	4 GB RAM | 128 GB ROM | Expandable Upto 256 G
10	realme Narzo 50A (Oxygen Blue, 128 GB)	₹12,499	↑ 4.4	4 GB RAM | 128 GB ROM | Expandable Upto 256 G
6	POCO M3 (Yellow, 64 GB)	₹11,499	↓ 4.3	6 GB RAM | 64 GB ROM | Expandable Upto 512 G
8	realme Narzo 50A (Oxygen Blue, 64 GB)	₹11,499	↑ 4.4	4 GB RAM | 64 GB ROM | Expandable Upto 256 G
14	realme Narzo 50A (Oxygen Green, 64 GB)	₹11,499	↑ 4.4	4 GB RAM | 64 GB ROM | Expandable Upto 256 G
4	realme C21 (Cross Blue, 64 GB)	₹9,999	↑ 4.4	4 GB RAM | 64 GB ROM | Expandable Upto 256 G
7	realme C21 (Cross Black, 64 GB)	₹9,999	↑ 4.4	4 GB RAM | 64 GB ROM | Expandable Upto 256 G
0	POCO C31 (Shadow Gray, 64 GB)	₹9,499	↑ 4.4	4 GB RAM | 64 GB ROM | Expandable Upto 512 G
3	POCO C31 (Royal Blue, 64 GB)	₹9,499	↑ 4.4	4 GB RAM | 64 GB ROM | Expandable Upto 512 G
19	Infinix Hot 10 Play (7° Purple, 64 GB)	₹9,299	↓ 4.3	4 GB RAM | 64 GB ROM | Expandable Upto 256 G
5	realme C21 (Cross Black, 32 GB)	₹8,999	↑ 4.4	3 GB RAM | 32 GB ROM | Expandable Upto 256 G
1	REDMI 9i Sport (Carbon Black, 64 GB)	₹8,799	↑ 4.4	4 GB RAM | 64 GB ROM | Expandable Upto 512 G
16	realme C11 2021 (Cool Blue, 64 GB)	₹8,799	↓ 4.3	4 GB RAM | 64 GB ROM | Expandable Upto 256 G
17	realme C11 2021 (Cool Grey, 64 GB)	₹8,799	↓ 4.3	4 GB RAM | 64 GB ROM | Expandable Upto 256 G
2	REDMI 9i (Sea Blue, 64 GB)	₹8,499	↓ 4.3	4 GB RAM | 64 GB ROM | Expandable Upto 512 G
11	POCO C31 (Royal Blue, 32 GB)	₹8,499	↑ 4.4	3 GB RAM | 32 GB ROM | Expandable Upto 512 G
15	POCO C31 (Shadow Gray, 32 GB)	₹8,499	↑ 4.4	3 GB RAM | 32 GB ROM | Expandable Upto 512 G
18	Infinix Hot 10 Play (7° Purple, 32 GB)	₹8,299	↓ 4.3	3 GB RAM | 32 GB ROM | Expandable Upto 256 G
20	Infinix Hot 10 Play (Obsidian Black, 32 GB)	₹8,299	↓ 4.3	3 GB RAM | 32 GB ROM | Expandable Upto 256 G
21	Infinix Hot 10 Play (Morandi Green, 32 GB)	₹8,299	↓ 4.3	3 GB RAM | 32 GB ROM | Expandable Upto 256 G
12	realme C11 2021 (Cool Grey, 32 GB)	₹7,299	↑ 4.4	2 GB RAM | 32 GB ROM | Expandable Upto 256 G
13	realme C11 2021 (Cool Blue, 32 GB)	₹7,299	↑ 4.4	2 GB RAM | 32 GB ROM | Expandable Upto 256 G

Figure 22.6 Spreadsheet analysis for the data.

1. The above data was scrapped from Flipkart webpage. It has been analyzed according to the price of the mobile phones and ratings as shown in Figure 22.6.
2. The **green** highlighted cell shows us the highest price of the mobile phones available on the Flipkart first page as shown in Figure 22.6.
3. The **yellow** highlighted cell shows us the average price of the mobile phones available on the Flipkart first page as shown in Figure 22.6.
4. The **orange** highlighted cell shows us the below average price of the mobile phones available on the Flipkart first page as shown in Figure 22.6.
5. The **red** highlighted cell shows us the least price of the mobile phones available on the Flipkart first page as shown in Figure 22.6.
6. On the other hand, ratings are denoted with green and red, up and down arrows respectively. The rating of 4.4 is denoted with **Green Upward arrow** and the rating of below 4.4 with **Red Downward arrow** as shown in Figure 22.6.

22.6 Conclusion

From this, we can conclude that the big data available on the Internet offer many fruitful opportunities for researchers and practitioners. For researchers, these opportunities include using this data to better understand many old and new social phenomena with more timeliness and greater precision [11]. The analysis process includes finding problems, solving data accessibility, deciding how to solve the problem of interest, and communicating the results. For analysis purposes, the data should be broken down into different phases, such as B. Specification processing, organization, adjustment, reanalysis, application of models and algorithms and subsequent dissemination of the results. Web scraping, also known as screen scraping, web data mining, web harvesting, etc., is a widely used technique for extracting large amounts of data from a website and storing it in a file or database. Data is usually presented as a table or in a spreadsheet.

References

1. Thomas, D. and Mathur, S., Data analysis by web scraping using Python. *2019 3rd International Conference on Electronics, Communication and Aerospace Technology (ICECA)*, pp. 450–454.

2. What is web scraping, https://www.webharvy.com/articles/what-is-web-scraping.html.
3. What is web scraping and how to use it?, p. 1, 09 Nov, 2021, https://www.geeksforgeeks.org/what-is-web-scraping-and-how-to-use-it/ Harikiran.
4. Success story - Web scraping hotel prices on daily basis, https://www.xbyte.io/web-scraping-hotel-prices-on-daily-basis.php 2012-2021.
5. Shreesha, M., Srikara, S.B., Manjesh, R., A novel approach for news extraction using web. *Int. J. Eng. Res. Technol.*, 6, 15, 1–2, 2018, https://www.ijert.org/a-novel-approach-for-news-extraction-using-web-scrapping.
6. What is web scraping, https://www.imperva.com/learn/application-security/web-scraping-attack/ 22/11/2021.
7. Scraping flipkart data using Python: Web scraping tutorial, October 17, 2021, https://www.analyticsvidhya.com/blog/2021/10/scraping-flipkart-data-using-python-web-scraping-tutorial.
8. Web scraping and its practical uses, October 25, 2021, https://youtu.be/aUECp_LS_1g.
9. Jim Shaped Coding, Web scraping with Python - Beautiful soup crash course, November 18, 2020, https://www.youtube.com/watch?v=XVv6mJpFOb0.
10. Edureka, Web scraping with Python, December 12, 2017, https://www.youtube.com/watch?v=mKxFfjNyj3c.
11. Krotov, J. A., Legality and ethics of web scraping. *Commun. Assoc. Inf. Syst.*, 47, 555–579, January 2020. Retrieved from https://www.researchgate.net/publication/352014123_Legality_and_Ethics_of_Web_Scraping

23

A New Language-Generating Mechanism of SNPSSP

Prithwineel Paul*, Soumadip Ghosh and Anjan Pal

Department of Computer Science and Engineering Institute of Engineering and Management, Kolkata, West Bengal, India

Abstract

P system is a well-known parallel and distributed computing model in natural computing. Spiking neural P system (SNPS) is a popular variant of P system, which was inspired by the structure and functioning of spiking neural networks (SNNs). P systems work as a language generating and accepting device. SNPS with structural plasticity (SNPSSP) is a relatively new variant of SNPS. In this paper, a new language generating mechanism for SNPSSP is introduced. More specifically, we associate one label with each of the rules present inside the neurons. In each neuron, different rules can have same label, and the same rules cannot have different labels. Also, the same rule in different neurons can be assigned with different label. We also consider that, in any step of the computation only, the same labeled rules are applied in all neurons. Now concatenation of the labels of the rules applied from initial configuration to halting configuration, a string over the labels is obtained. The collection of this newly generated string is called *control language* of the SNPSSP model. In this paper, we compare the family of control languages of the SNPS with structural plasticity with the family of regular, context-free, context-sensitive, and recursively enumerable languages.

***Keywords*:** Spiking neutral networks, P systems, structural plasticity, control language, Chomsky hierarchy

**Corresponding author*: prithwineel.paul@iemcal.com

Sharmistha Ghosh, M. Niranjanamurthy, Krishanu Deyasi, Biswadip Basu Mallik, and Santanu Das (eds.) Mathematics and Computer Science Volume 1, (435–454) © 2023 Scrivener Publishing LLC

23.1 Introduction

In the last few decades, computer scientists and mathematicians have initiated many studies in order to solved complex mathematical problems, which traditionally cannot be solved by modern computers. In this regard, some new computing models were introduced [14]. The working of biological cells, DNA molecules, Quantum states, etc. inspired scientists to construct these new models. The study of membrane computing was initiated by the inherent structural properties and reactions happening inside the biological cells. In membrane computing, the objects inside the cells can travel from one cell to another by some rewriting rules. In fact, the objects, as well as the communication mechanism inside the cells, are mathematically expressed using automata and formal language theory framework. Membrane computing models are popularly known as *P systems*. It was introduced by Gh. Păun and is divided into cell-like P systems, tissue-like P systems, and neural-like P systems [1, 2, 4, 5, 15, 17].

Spiking neural networks (SNNs) [18] are well known as third-generation neural networks. In SNNs, the information is encoded in the form of time. Another distinct feature of SNNs is that they have individual spikes. These features also make SNNs more relatable with the features of biological neurons. Spiking neural P system is a relatively new variant of third generation of neural networks, and it was introduced by Ionescu, Păun, and Yokomori. SNPS models were inspired by the properties of SNNs. In the last decade, SNPS models have drawn attention from the community of computer scientists, mathematicians, bioinformaticians etc due to its use in solving real-life problems. Subsequently, different variants of SNPS models have been introduced that are mostly inspired by the features of the biological neurons [17]. Some of the popular variant of SNPS models are SNPS with astrocytes [3], scheduled synapses [6], structural plasticity [8–11], anti-spikes [25], etc. Most of the SNPS models and its variants are Turing complete [17] and have been used to solve problems in real-life applications in different areas, such as pattern recognition [19, 20], fault diagnosis in power systems [16], computational biology [21], hardware implementation [22, 24], biochip design [23], etc.

Associating label language/Szilard language/control language with grammars in Chomsky hierarchy and its variants is a well-known topic of research in formal language theory [28–30, 33, 37, 38]. Researchers also have associated the idea Szilard languages and control languages with DNA and membrane computing models and investigated properties of these languages in references [26, 27, 31, 32, 34–36, 39]. This paper is an extension of the idea of associating control languages with SNPSSP model introduced in Cabarle *et al.* [7]. It is inspired by the idea of structural plasticity in neural networks where the synaptic connections between the neurons are dynamic, i.e., the synaptic connection between the neurons can be created as well as removed. In general, SNPS models have spiking and forgetting rules. In SNPSSP, forgetting rules do not exist. Instead, SNPSSP models [6–11] have plasticity rules. Neurons in these models can establish as well as remove synaptic connections with other neurons using plasticity rules. Computational power of SNPS with structural plasticity and its variants have been investigated by many researchers recently [6–11]. In this paper, we investigate what kind of languages can be generated by the labels associated with the rules present inside the neurons in SNPS with structural plasticity. More specifically, we use the restriction that at each computation step only same labeled rules are applied. Now, when a halting configuration is reached, then we obtain a string by concatenating the labels of the rules applied in each step of the computation starting from the initial configuration. The collection of all such strings is called as *Control language* of the SNPS with structural plasticity. In this paper, we compare the family of control languages associated with SNPSSP models with REG, CF, CS and RE.

The organization of the paper is as follows: Section 1, and 8 discuss the SNPSSP model, the structure and working of labelled SNPSSP model, the main results and conclusion respectively.

23.2 Spiking Neural P Systems With Structural Plasticity (SNPSSP)

A SNPSSP [7] with degree $m \geq 1$ has the following mathematical structure: $\Pi = (0, \sigma_1, \ldots, \sigma_m, syn, in, out)$ where

- $O = \{a\}$ represents the alphabet where a is called spike;
- $\sigma_i = (n_i, R_i)$ represents the neurons where n_i is the spike present in the neuron. R_i is a finite set and it denotes the rules present inside the neuron σ_i. The rules in SNPS are of two types i.e., spiking and consuming rules. However, the rules in SNPSSP are of two types, i.e., (1) spiking rule; (2) plasticity rule.
  - *Spiking rule:* $E_1/a^c \rightarrow a$ ($c \geq 1$ and E_1 is a regular expression over the alphabet $\{a\}$).

- *Plasticity rule:* $E_1/a^c \rightarrow \alpha k(i, N_j)$ where $c \geq 1$, $\alpha \in \{+,-,\pm,\mp\}$, $k \geq 1, 1 \leq j \leq |R_i|$ and $N_j \subseteq \{1,2, \ldots, m\}$

- $syn \subseteq \{1,2, \ldots, m\} \times \{1,2, \ldots, m\}$ where $(i, i) \notin syn$ for $1 \leq i \leq m$. It represents the synaptic connection between the neurons.
- Input and output neuron are represented by *in* and *out* respectively and *in, out* $\in \{1,2, \ldots, m\}$.

Most of the SNPS models investigated until now have input neurons and output neurons. The output of the SNPS is based on the spiking/not-spiking of the output neuron. However, in the SNPSSP model introduced in [7], the idea of input as well as output neurons has been ignored. Let us consider a neuron σ_{i_1}. There exist synaptic connections towards σ_{i_1} as well as going out of σ_{i_1}. Based on these features the notations $pres(i_1)$ and $pos(i_1)$ are introduced. $pres(i_1) = \{j|(i_1, j) \in syn\}$ (i.e., neurons, which have a directed synapse starting at i_1 and ending at j) and $pos(i_1) = \{j|(j, i_1) \in syn\}$ (i.e., the set of all neurons, which has a directed synapse to σ_{i_1}).

The spiking rule $E_1/a^c \rightarrow a$ in a neuron σ_k of a SNPS with structural plasticity model works in the similar manner. This rule is applicable only when the neuron σ_k has b spikes, i.e., $a^b \in L(E_1)$ and $b \geq c$. Whenever the spiking rule is applied, c spikes are consumed and $(b - c)$ spikes remain inside the neuron. Also one spike is sent to all the neurons present in the set $pres(k)$. If $E_1 = a^c$, then the spiking rule is written as $a^c \rightarrow a$.

Again if at time t, the neuron σ_{i_1} has $b(\geq c)$ spikes and $a^b \in L(E_1)$ and we also have the plasticity rule $E_1/a^c \rightarrow \alpha k(i, N)$ σ_{i_1}. So the plasticity rule can be applied. Furthermore, c spikesare consumed by this rule and along with it synaptic connection established or existing synaptic connections are removed with the neurons present in the set N. Furthermore, based on α some more operations such as synapse creation and deletion insuccessive steps and synapse deletion and creation in successive steps are also possible.

When only a synaptic connection is established between any two neurons, it is represented by $\alpha = +$. Note that synaptic connections can be established with the neurons which do not have any synaptic connection with σ_{i_1}. Also if $Ninuspres(i_1) = \varnothing$ (i.e., there exists synaptic connection between σ_i and the targeted neuron set). In this case, nothing can be done. Similarly, if $\alpha = -$ and $pres(i_1) = \varnothing$, i.e., no synaptic connection exist between σ_{i_1} and the targeted neuron set N. So no synaptic connection can be removed.

Again, let σ_{i_1} can establish synaptic connection with at most k neurons and $|N \setminus pres(i_1)| \leq k$. In that case, synapses are created between σ_i and σ_l, $l \in N_j \setminus pres(i_1)$ deterministically. Similarly, if σ_{i_1} can establish synaptic connections with more than k neurons, i.e., $|N \setminus pres(i_1)| > k$, then at first k neurons are selected non-deterministically and in the next step synapses are created between them.

If σ_{i_1} has synaptic connection with at most k neurons, i.e., $|pres(i_1)| \leq k$, then for $\alpha = -$, k synapses joining σ_{i_1} with other neurons are removed deterministically. Again, if σ_{i_1} has synaptic connection with more than k number of neurons, i.e., $|pres(i_1)| > k$, then k synapses joining σ_{i_1} and other neurons are removed non-deterministically.

Proceeding in the similar manner, if a plasticity rule with $\alpha = \pm$ is applied then at any time t, neuron σ_i establishes synaptic connection with the neurons with which it does not have any synaptic connection. Also, at time $(t + 1)$, the newly established synaptic connections are removed. Similarly, when $\alpha = \mp$, at first, synapses are removed (if they exist) and then new synapses are created. Note that during this process of neuron creation and deletion, as well as neuron deletion and creation σ_{i_1} remains open, and it can receive spike. However, any other spiking and plasticity rules are applied only at $(t + 2)$-th time.

Also, note that whenever any plasticity rule with $\alpha \in \{+, \pm, \mp\}$ is applied along with establishing synaptic connections, one spike is also sent to the neurons. More specifically, if σ_i establishes synaptic connection with the neurons $\{\sigma_1, \sigma_2, \ldots, \sigma_k\}$, then at the same time, one spike is also sent to each of these neurons.

It may happen that a neuron contains two rules with different regular expressions E_1' and E_2', i.e., $L(E_1') \cap L(E_2') \neq \varnothing$. Also both of them can be applied at the same time. In this situation, one of the rules is selected non-deterministically and applied. Again, at any given time, one neuron can apply at most one rule. Furthermore, if any rule is applicable, then it must be applied. It shows that the computation happening inside a spiking neural P system is locally sequential as well as globally parallel.

The configuration of any SNPSSP is $(n_1, n_2, \ldots, n_m)$ where n_i denotes the number of spikes in σ_{i_1}. Any computation in SNPSSP starts from $(n_1^0, n_2^0, \ldots, n_m^0)$ (initial configuration; n_1^0 represents the number of spikes in membrane). If the SNPSSP reaches a configuration $(n_1^k, n_2^k, \ldots, n_m^k)$, where no further computation takes place, this is called as *halting configuration.* The computation in a SNPSSP can halt, or it may continue without halting. However, many different techniques are introduced to associate natural numbers with the output of the SNPSSP [6–11]. The popular method of collecting the output is based on two subsequent spiking of a particular neuron. In this paper we follow the same technique. Initially, we assign a neuron as output neuron, i.e., σ_{out}. Now if two consecutive spiking of σ_{out} happens at time t_1 and t_2, then the number $(t_1 - t_2)$ is computed by the SNPSSP model Π. Furthermore, the set of natural numbers generated by the SNPSSP Π is represented by $N_2(\Pi)$.

SNPSSP models work as number generating, as well as accepting device. We discussed about the generating case above, and in this case, we ignore the input neurons. However, in accepting case the SNPSSP model has a specified input neuron and in this case output neurons do not exist. Furthermore, two spikes are introduced into SNPSSP model through σ_{in} at times t_1 and t_2. Now the input $(t_1 - t_2)$ is accepted by Π if it reaches a halting configuration after processing it. Again, $N_{acc}(\Pi)$ represents the set of all numbers accepted by the SNPSSP Π. The family of natural numbers generated and accepted by Π is denoted by N_2SNPSP and $N_{acc}SNPSP$ respectively.

23.3 Labeled SNPSSP (LSNPSSP)

Let us consider an SNPS with structural plasticity rules $\Pi = (\{a\}, \sigma_1, \sigma_2, \ldots, \sigma_m, syn, in, out)$ with m neurons. Every neuron $\sigma_i = (n_i, R_i)$ has two components (a) n_i number of spikes and (b) the set of rules R_i. Let $|R_i| = n$ and $R_i = \{r_{i1}, r_{i2}, \ldots, r_{in}\}$. Each rule in R_i is associated with a label, i.e., $l: R_i \rightarrow \Sigma \cup (\lambda)$ where Σ is a finite set containing the labels of the rules and λ is the empty label. Furthermore, any two rules inside a neuron can have same label. However same rule cannot have more than one label associated with it. Again, same spiking or plasticity rule present in two different neurons can have same label. Proceeding in this manner, we can label all the rules in SNPSSP. Also, let the rules of the neuron σ_i have labels $\{l_{i1}, l_{i2}, \ldots, l_{im}\} \cup \{\lambda\}$. So, $\Sigma = \bigcup_{i=1}^{n} l_{ij} \cup (\lambda)$. (i.e., the rules in the SNPS can have a label l_{ij} or λ). Now, let $R = R_1 \cup R_2 \cup \ldots R_m$. Now after application of any l_{ij} rule or λ rule, the number of spikes in different neurons will change. Let us assume that initially we have $n_1, n_2, \ldots, n_m$ number of spikes inside the neurons $\sigma_1, \sigma_2, \ldots, \sigma_m$. The configuration of the system can be represented as $(n_1, n_2, \ldots, n_m)$. When the rule $l_{ij}: a \rightarrow a$ is applied in σ_i and σ_i has synaptic connection with the neuron σ_k, then one spike is reduced from n_i and similarly n_k becomes $n_k + 1$ and no change of number of spikes happens in other neurons. Similarly, if we apply $l_{ij}: a \rightarrow \pm t(i, \{i_1, i_2, \ldots, i_t\})$, then synaptic connection is established between σ_i and $\{\sigma_{i_1}, \sigma_{i_2}, \ldots, \sigma_{i_t}\}$. Furthermore, one spike is sent from σ_i to $\sigma_{i_1}, \sigma_{i_2}, \ldots, \sigma_{i_t}$.

23.3.1 Working of LSNPSSP

In this section, we investigate the language generated by LSNPSP. Let the initial configuration of the labeled LSNPSP system Π is $C_0 = (t_0^1, t_0^2, \ldots, t_0^m)$ where t_0^i represents the number of spikes in the neuron i. In this paper, we assume that at computation step only same labeled rule will be applied in each of the neurons. More specifically, let l be a non-empty label which is applied at any time t in σ_i. If the l rule in σ_i is applied, then in each of the neurons only the l-labeled rule will be applied if it contains such labeled rule. Otherwise,at time t no rule is applied in that specific neuron. In [12], the authors used similar kind of restrictions. However, in [12], authors considered that if any rule with label l is applied in any neuron σ_i and at the same time σ_j does not contain any rule with label l and it contains λ (empty) labeled rules, then the λ rule is applicable.

Now let we have a halting computation of the form

$$C_0 \Rightarrow^{b_1} C_1 \Rightarrow^{b_2} C_2 \ldots \Rightarrow^{b_n} C_n$$

where C_0 is the initial configuration and C_n is the halting configuration, i.e., no rule is applicable to the spikes present inside the neurons. We say that $b_1 b_2 \ldots b_n \in \Sigma^*$ is the *control word* generated by the SNPSSP and the collection of these words is called as *control language*. Also, note that during the above computation when a b_i rule is applied, it means that only the b_i rule present in the neurons will be applied. Other rules are not applied during this period. If the b_i rule is not present in a neuron, then at that moment the contents of the neuron remains unchanged. The labels of the rules can be a nonempty, i.e., b_i as well as λ (empty). When a λ rule is applied in a neuron, it means that only the λ rule is going to be applied in all the neurons of the SNPS with structural plasticity Π. The control language generated by such system Π is denoted by $CLSNPSP_m(\Pi)$ when there exists no λ rule in the system, and m denotes the number of neurons. If Π has any λ rule, then the control language generated by Π is denoted as $CL_\lambda SNPSP_m(\Pi)$. The family of control languages generated by a SNPS with structural plasticity with m neurons is represented by $CLSNPSP_m$ (if there exist no λ rule in the neurons) and $CL_\lambda SNPSP_m$ (if neurons contain λ rule). m is replaced by $*$ whenever it is not bounded.

Example 1. The language $\{b^n c | n \geq 0\}$ is obtained as control language of SNPSP Π. The construct of Π is as follows (see Figure 23.1):

In Π, initially only the neuron σ_1 contains one spike. The neurons σ_2 and σ_3 do not contain any spike. The neuron σ_1 can spike using both b

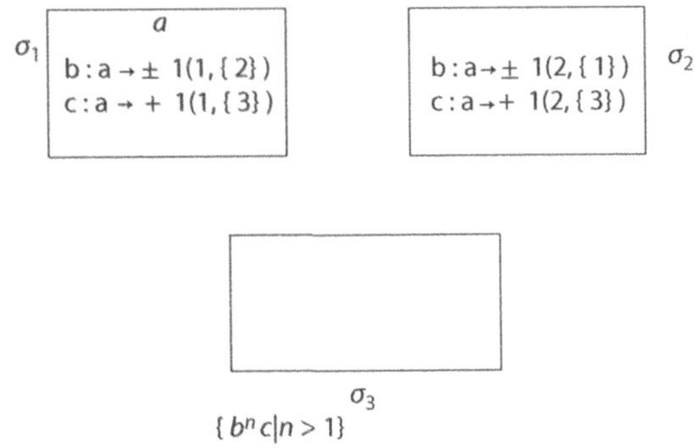

Figure 23.1 SNPSP Π generating {bn c|n ≥ 1} as control language.

and c labeled rules. If b labeled rule is applied at any time t, then at first one synapse is created with the neuron σ_2 and one spike is also sent to the neuron σ_2. At time $(t + 1)$, the synapse between the neuron σ_1 and σ_2 is removed. Note that at time $(t + 1)$, the neuron σ_2 contains one spike and both b and c labeled rule are applicable. If c labeled rule is applied, the computation stops. At time $(t + 2)$, after application of b rule one synapse is created between σ_2 and σ_1. Note that during this time also, the synapse created by application of the b rule in σ_1 is also removed. Since according to the definition in [7], during this period, the neuron σ_1 will not spike. However, it can receive spike. So at $(t + 2)$ -th time, σ_1 receives a spike from σ_2. Note that in this work, we consider that at any time only with the rules with same label is applicable and when we reach a halting configuration, the concatenation of the labels is performed. Hence, we have $CLSNPSP_3(\Pi) = \{b^n\ c|n \geq 1\}$.

23.4 Main Results

In [12], it was proved that $REG \subseteq CL_\lambda SNPS$, i.e., the λ rules was used. However we can prove that any regular language is represented as control language of a SNPS with plasticity rule without labeling any rule with λ.

Theorem 1. $REG \subseteq CLSNPSP_*$.
Proof. In Figure 23.2, we construct a SNPSSP by simulating the transition functions of a deterministic finite state automata. Let $L_1 \in REG$ and $M = (Q, \Sigma, \delta, i_0, F_M)$ be a DFA (deterministic finite state automata) such that $L(M_1) = L_1$. The states in M_1 are renamed in the following manner: if $q_i(1 \leq i \leq p)$ are the states, then $q_p = i_0$ and the transition functions are defined

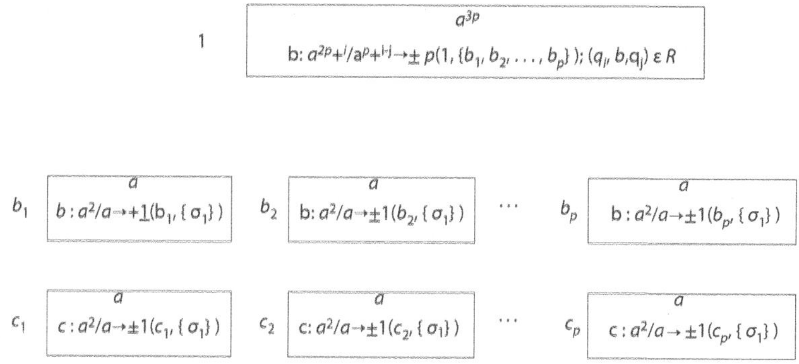

Figure 23.2 Simulation of the transition functions of a deterministic finite state automata using the rules in SNPSSP.

accordingly. Now we construct a SNPS with structural plasticity Π_1 such that $L(\Pi_1) = L(M_1)$. Let $\Pi_1 = (\{a\}, \sigma_1, \sigma_2, \ldots, \sigma_m, syn)$ where for each $b \in \Sigma$, there exist p neurons $\sigma_{b_1}, \sigma_{b_2}, \ldots, \sigma_{b_p}, (\sigma_{b_i} = (n_i, R_i))$ and each of these neurons contains one spike and one plasticity rule, i.e., $n_i = a$ and each $R_i(1 \leq i \leq p)$ contains the plasticity rule b: $a^2/a \rightarrow \pm1(b_i, \{1\}), 1 \leq i \leq p$.

The neuron σ_1 contains $3p$ number of spikes and the following plasticity rules: b: $a^{2p+i}/a^{p+i-j} \rightarrow \pm p(1, \{b_1, b_2, \ldots, b_p\})$ for each $(q_i, b, q_j) \in R$.

Initially, the σ_1 contains $3p$ spikes where p denotes the number of states in the FSA. Again, for each $(q_i, b, q_j) \in R$ in the FSA, we have a plasticity rule b labeled rule, which consumes a^{p+i-j} number of spikes and σ_1 is connected with the neurons $b_1, b_2, \ldots, b_p$. Note that in this case the transition function $(q_i, b, q_j) \in R$ in the FSA is simulated. For each b, we have the neurons $b_1, b_2, \ldots, b_p$ with one spike in each of them.

If at t-th time, we have $2p + i$ number of spikes in σ_1 and one spike in each of the corresponding neurons $t_1, t_2, \ldots, t_p$ where t represents the terminal alphabet. Since at each computation step only same labeled rules are applied and while simulating the transition function $(q_i, b, q_j) \in R$, we apply the b labeled rules present in each of these neurons. At $(t + 1)$ -th time, only $2p + i - (p + i - j)$ spikes are remaining in σ_1. At the same time σ_1 establishes synaptic connection with $b_1, b_2, \ldots, b_p$ and each of them receives one spike.

At time $(t + 2)$, the synapses connecting σ_1 with the neurons $b_1, b_2, \ldots, b_p$ are removed. Furthermore, at the same time, these neurons spike and send one spike to σ_1. So total p spikes are sent to σ_1 and it is done by creation of new synapses between $b_1, b_2, \ldots, b_p$ and σ_1.

At time $(t + 3)$, the neuron σ_1 has $2p + j$ number of spikes and it can apply the labeled plasticity rules based on which transition function will be simulated. Also at the same time, the synapses joining $b_1, b_2, \ldots, b_p$ and σ_1 are removed.

This process will continue until we reach a stage where we have to simulate the transition function $(q_i, t, q_0) \in R$. Since according to the construction above, we start with the state q_p (i.e., the initial state). Similarly, the final state is q_0. So when we simulate the transition function $(q_i, t, q_0) \in R$, we will apply the corresponding t rule in σ_1 and after receiving p spikes from $\sigma_{t_1}, \sigma_{t_2}, \ldots, \sigma_{t_p}$, it will have $2p$ number of spikes. Since $i > 0$, according to our assumption, no rule will be applicable now. So we have reached at the halting configuration.

Theorem 2. $(CF \setminus REG) \cap CLSNPSP_3 \neq \emptyset$.
Proof. We construct a snps with structural plasticity Π such that $CLSNPSP(\Pi) = \{b^n c^n | n \geq 1\}$ (see Figure 23.3).

It contains three neurons. Initially, there exist no synaptic connection between these neurons. Moreover, the neurons 1 and 2 contain one spike each. So if at t-th time the b rule in σ_1 is applied and synaptic connection is established with neurons 2 and 3 at time $(t + 1)$. Also, at time $(t + 1)$, σ_2 and σ_3 receives one spike each. At $(t + 2)$-th time, the synapses between σ_1, σ_3, and σ_2, σ_3 are removed.

Similarly, at t-th time, the b rule is applied in neuron 2. At $(t + 1)$-th time, neuron 2 and neurons 1,3 are connected by the synapses and one spike is sent to each neuron. However, at $(t + 2)$-th time, these synaptic connections are removed. At time $(t + 1)$, both σ_1 and σ_2 receive one spike from each other. However, during this time, the b rule cannot be applied, since the synaptic connection between the neurons still exists which are removed only at $(t + 2)$-th time. At $(t + 3)$-th time, again the b rule in σ_1 and σ_2 are applied.

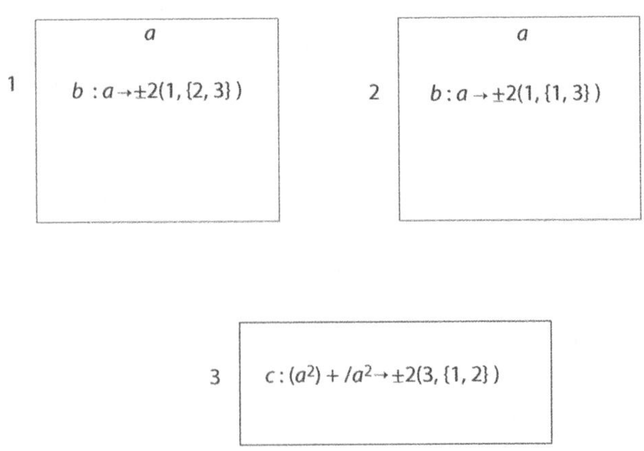

Figure 23.3 SNPSSP Π obtaining {bncn|n ≥ 1} as a control language.

Note that at $(t + 1)$-th time, the σ_3 has two spikes. If c rule is applied at time $(t + 2)$, then two spikes are consumed and one spike is received by neurons 1 and 2. During this period, the neurons 1 and 2 can receive spike but plasticity rules present in these neurons cannot be applied. So if c rule is applied, then neurons 1 and 2 will have 2 spikes and neuron 3 will have one no spike inside it. Hence a halting computation is obtained. In summary, after the b rule is applied, 2 spikes are received by σ_3. This process will continue until c rule is applied. After c rule is applied, the b rules cannot be applied. Furthermore, after application of the b rules, two spikes are received by σ_3 and c rule consumes two spikes. So the halting configuration is obtained only after application of b rule followed by equal number of c rule. It implies that $CLSNPSP_3(\Pi) = \{b^n c^n | n \geq 1\}$.

Theorem 3. $(CS \setminus CF) \cap CLSNPSP_4 \neq \emptyset$.

Proof. We construct one SNPS with structural plasticity Π such that $CLSNPSP_4(\Pi) = \{b^n w | w \in \{c, d\}^*, |w|_c = |w|_d\}$. See Figure 23.4.

The structure of Π is shown in Figure 23.4. Initially, only neurons 1 and 2 contain one spike. Other neurons do not contain any spike. Whenever b rule is applied in neurons 1 and 2, one spike is received by each of them and σ_3 and σ_4 receive one spike. The b rule in σ_1 and σ_2 are applicable any number of times and for each application of it 2 spikes are added in neurons 3 and 4. This process continues, until c rule or d rule is applied. Note that before application of the c rule or d rule, the neurons 3 and 4 have same number of spikes. Furthermore, after application of either c rule or d rule, two spikes from neuron 3 (respectively, neuron 4) are consumed. Also one

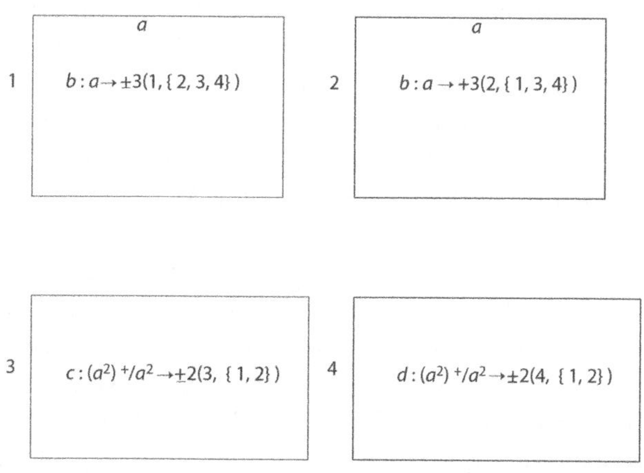

Figure 23.4 The structure of SNPSSP obtaining {bnw|w ∈ {c, d}*, |w|c = |w|d} as a control language.

spike is received by σ_1 and σ_2. So the neurons 1 and 2 now have two spikes and the plasticity rule, i.e., b rule cannot be applied.

Next only either c rule or d rule will be applied. Since neurons 3 and 4 contain same number of spikes, they can be applied only equal number of times.

So a halting configuration is derived only after application of the b rule followed by equal number of application of the c and d rule. Hence the control language generated by the SNPS with structural plasticity Π is $CLSNPSP_4(\Pi) = \{b^n w | n \geq 1, w \in \{c, d\}^*, |w|_c = |w|_d\}$.

Theorem 4. $RE = CL_\lambda SNPSP_*$.

Proof. In order to prove this result, we proceed in the similar manner as in the proof of Theorem 8 in [12]. However, the rules present in the neuron in [12] has delay associated with them. In this proof, we use plasticity rules to simulate the rules with delay. Furthermore, we use the spiking and plasticity rules to simulate the SUB and ADD unit.

From Church-Turing thesis, the inclusion $CL_\lambda SNPSP_* \subseteq RE$ follows directly. So $RE \subseteq CL_\lambda SNPSP_*$ remains to be proved.

Let us assume that $L(\subseteq \Sigma^* = \{b_1, b_2, \ldots, b_l\}^*) \in RE$. The encoding of each element of L is performed in the similar manner as in Chen *et al.* [13]. It is important to note that after processing an input i_0, the register machine M halts if and only if $e_1(w_1) = i_0$ where $w_1 \in L$ and encoding of w_1 is i_0. Moreover, we associate a special symbol \$ at the end of the string $w\$$ where \$ is not an element of the alphabet Σ. So, we construct a SNPSSP with labelled rules such that it will halt only after processing the strings $w\$$ where $w \in L$. Again, as in [12] we consider two subsystems of neurons for simulation of two register machines M_0 and M. (since SNPS can simulate any register machine [13]). So, to avoid the redundancy in the proof, we only discuss the overall appearance of the SNPSSP Π. The SNPSSP Π in Figure 23.5 has M_0 which represents the subsystem containing the neurons for simulation of $M_0 = (m_0', H_0', l_{0,0}, l_{h,0}, I_0')$ and M represents the subsystem for simulation of the register machine $M = (m', H', l_0, l_h, I')$ where $H_0' \cap H' = \emptyset$.

Initially, σ_X and σ_{13} have one spike. σ_X contains the rules $b_1: \pm 5(X, \{A1,2,3,4,5\})$; $b_j: a \rightarrow \pm 3(X, \{A_j, 2,3\})$, $j = 2, \ldots, p$. So whenever any b_j rule ($j = 2,3, \ldots, p$) is applied, at first a synapse is created between σ_X and σ_{Aj}($j = 2,3, \ldots, p$), σ_2, σ_3. Also, note that during this period, one spike is sent to these neurons. Furthermore, after application of the b_1 rule, synaptic connection is established as well as one spike is sent to neurons A_1, 2,3,4,5 from neuron X. However, in the next step these synaptic connections are removed. At the same time the rules λ: $a \rightarrow a$ in σ_2 and σ_3 are activated and both

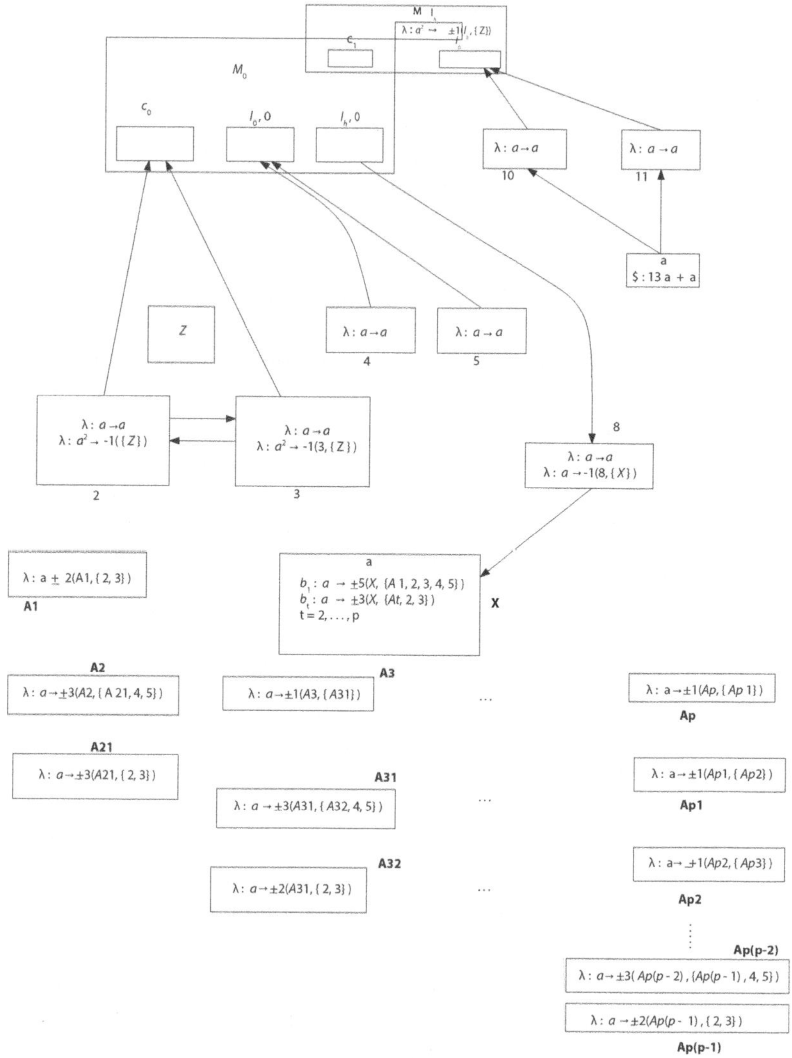

Figure 23.5 The structure of SNPSSP in the proof of Theorem 4.

of the neurons receive one spike from one another and also σ_{c0} receives two spikes from these neurons. This process is continued until one spike is received from the neurons $\sigma_{A1}, \sigma_{A21}, \ldots, \sigma_{Ap(p-1)}$.

Now, our aim is to store $2i$ spikes in σ_{c0} when b_i rule in σ_X is applied. It is done in the following manner: Let b_1 rule is applied at time t, then at time $(t + 1)$, the neurons σ_{A1}, σ_4, σ_5, σ_2 and σ_3 have one spike. The neuron σ_X

is now empty. At time $(t + 2)$, the synapses connecting σ_X and σ_4, σ_5, σ_2, σ_3 and σ_{A1} are removed. Also at the same time, σ_2 and σ_3 send one spike each to σ_{c0}. Since σ_3 and σ_2 receive one spike from one another and at time $(t + 2)$, receives one spike from σ_{A1}. So at $(t + 3)$-th time, the rule $\lambda: a^2 \rightarrow -1(2, \{Z\})$ and $\lambda: a^2 \rightarrow -1(3, \{Z\})$ are activated and two spikes are consumed.

Similarly, if at any time t, b_2 rule is applied, then at $(t + 1)$-th time, σ_{A_2}, σ_2 and σ_3 receive one spike. At $(t + 2)$-th time, σ_2 and σ_3 send one spike to each other and 2 spike to σ_{c0}. Also, σ_{A21}, 4 and 5 receive one spike from σ_{A_2}. So neuron σ_{A_2} becomes empty and $\sigma_{A_{21}}$ has one spike. Next at $(t + 3)$-th time, σ_{A21}, σ_2 and σ_3 spike. So the neuron σ_{c_0} receives another two spikes from σ_2 and σ_3. At the same time, after receiving the spike from $\sigma_{A_{21}}$, the neurons σ_2 and σ_3 have 2 spikes. So at $(t + 4)$-th time, rules with λ rule is applied and two spikes are consumed from both σ_2 and σ_3. Hence, when the b_2 rule in X is applied, 4 spikes are stored into c_0. Proceeding in the similar manner, when b_i rule is applied, we can store $2i$ spikes in σ_{c0}.

The neurons $\sigma_{A_2}, \sigma_{A_{31}}, \ldots, \sigma_{A_{p(p-2)}}$ contain the plasticity rules $\lambda: a \rightarrow \pm 3(A2, \{A21,4,5\})$, $\lambda: a \rightarrow \pm 3(A31, \{A32,4,5\})$, …$\lambda: a \rightarrow \pm 3(A_{p(p-2)}), \{A_{p(p-1)}, 4,5\})$. These rules send one spike to σ_4 and σ_5. All the spikes in σ_2 and σ_3 are consumed and at the same time σ_4 and σ_5 neurons send one spike to $\sigma_{l_{0,0}}$. So the neuron $a_{l_{0,0}}$ has received two spikes and the simulation of M_0 will start. Next, the subsystem for M_0 will start working and during this time the value contained in the neuron σ_{c_1} is multiplied with $l + 1$ and then added with $2i$.

After this process has halted, the neuron $\sigma_{l_{h,0}}$ is activated and σ_8 receives one spike from it. The neuron σ_8 contains two rules with label \$ and λ. If the λ rule is applied, σ_X receives one spike and then the computation is going to continue based on which b_j rule applied. However, after application of \$ labelled rule, the spike inside the σ_8 is consumed and at the same time the synapse connecting 8 and X is removed. Again, since at any step of computation we can apply only the rule with same label, the \$ rule in σ_{13} is also activated. Next, σ_{10} and σ_{11} receive one spike from σ_{13}. At the next step, σ_{10} and σ_{11} fires and send 2 spikes to σ_{l_0} in M. It will start the simulation of M. Whenever M halts for some input, σ_{l_h} receives two spikes. The neuron σ_{l_h} is associated with the label of the halt instruction of M. Next, the λ rule in σ_{l_h} is applied and all spikes are consumed. So the computation halts. During this process, we can see that the concatenation of the labels gives us a string of the form $w_1\$$ where $w_1 \in \Sigma^+$.

Since in this proof we are using the labeled plasticity rules. So we discuss the simulation of ADD and SUB unit in detail. For the simulation of the ADD instruction l_1: $(ADD(r), l_2)$, we have the corresponding neuron σ_{l_1}. It contains two λ rule. First λ rule consumes two spikes and creates two synapses joining l_1' and l_1'' with l_1. Also l_1' and l_1'' receive one spike each. Next, l_1' and l_1'' spikes and σ_r receives two spikes where σ_r represents the register r of the register machine and also send 2 spikes to σ_{l_2} associated with l_2-labeled instruction of register machine.

Next we discuss about the SUB unit. The simulation of the SUB unit of M_0 and M are described in Figures 23.6 and 23.7.

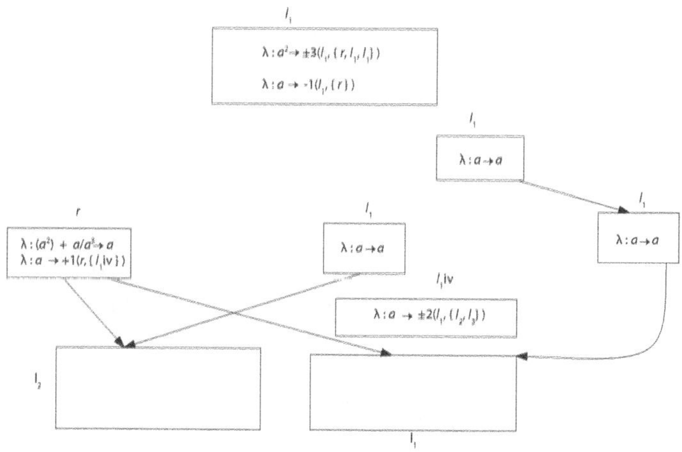

Figure 23.6 SUB unit.

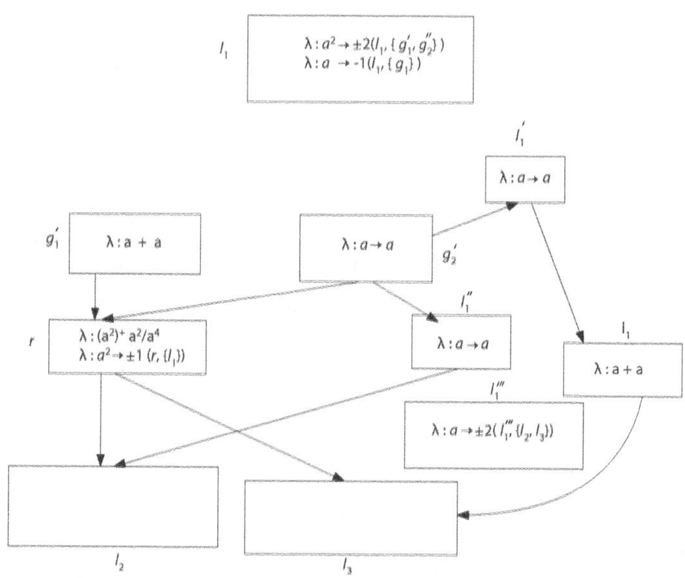

Figure 23.7 SUB unit.

The construction of simulation of the SUB unit is similar to the construction in [13]. However, in [13], neurons do not contain plasticity rules. Moreover, in order to differentiate the simulation of the SUB units of M_0 and M_1, we use two different simulations.

In Figure 23.6, if σ_{l_1} receives 2 spikes, the rule $\lambda: a^2 \rightarrow \pm 3(l_1, \{r, l_1', l_1''\})$ is applied. Next, synaptic connections are created between σ_{l_1} and $\sigma_r, \sigma_{l_1'}, \sigma_{l_1''}$.. The plasticity rule also sends one spike to these neurons. After receiving the spike from σ_{l_1}, the neurons $\sigma_{l_1'}$ and $\sigma_{l_1''}$ will spike immediately using $\lambda: a \rightarrow a$. If σ_r already contains 2 spikes, then after receiving the spike from σ_{l_1}, the rule $\lambda: (a^2)^+ a/a^3 \rightarrow a$ is applied and it sends one spike to the neuron σ_{l_2}. Again, $\sigma_{l_1'}$ sends one spike to σ_{l_2}. So at this moment, the neuron σ_{l_2} contains 2 spikes and σ_{l_3} contains one spike. Note that if σ_r does not contain any spike initially, i.e., it is empty, then after receiving one spike from σ_{l_1} the plasticity rule $\lambda: a \rightarrow \pm 1(r, \{l_i^{iv}\})$ is applied. It establishes synaptic connection between the neurons σ_r and $\sigma_{l_1^{iv}}$ and also sends one spike to $\sigma_{l_1^{iv}}$. Next, the neuron $\sigma_{l_1^{iv}}$ sends one spike to σ_{l_2} and σ_{l_3}. At the same time $\sigma_{l_1'''}$ also sends one spike to l_3. So σ_{l_3} has two spikes and σ_{l_2} has one spike. So in σ_{l_3}, the plasticity rule $\lambda: a^2 \rightarrow \pm 3(l_3, \{r, l_1', l_1''\}$ is applied. However, in σ_{l_2}, the rule $\lambda: a \rightarrow -1(l_2, \{r\})$ is applied and one spike is consumed.

In the simulation of the SUB unit of M (shown in Figure 23.7), neuron r contains the plasticity rules $\lambda: (a^2)^+ a^2/a^4 \rightarrow a$ and $\lambda: a^2 \rightarrow \pm 1(r, \{l_1'''\})$. These rules are applied based on the neuron r is non-empty or empty. Furthermore, these rules distinguishthe simulations of the SUB units of M and M_0.

23.5 Conclusion

In this paper, control languages of SNPSSP model have been investigated. Furthermore, we have investigated the comparisons of the family of control languages generated by SNPSSP models with the family of languages in Chomsky hierarchy. The study of control languages for SNPS can be further extended with different variants of SNPS models.

References

1. Păun, Gh., Computing with membranes. *J. Comput. Syst. Sci.*, 61, 108–143, 2000.
2. Ionescu, M., Păun, Gh., Yokomori, T., Spiking neural P systems. *Fundam. Inform.*, 71, 279–308, 2006.

3. Pan, L., Wang, J., Hoogeboom, H., Spiking neural P systems with astrocytes. *Neural Comput.*, 24, 805–825, 2012.
4. Su, Y., Wu, T., Xu, F., Paun, A., Spiking neural P systems with rules on synapses working in sum spikes consumption strategy. *Fundam. Inform.*, 156, 187–208, 2017.
5. Pan, L., Păun, Gh., Zhang, G., Neri, F., Spiking neural P systems with communication on request. *Int. J. Neural Syst.*, 27, Article ID 1750042, 2017.
6. Cabarle, F.G.C., Adorna, H.N., Jiang, M., Zeng, X., Spiking neural P systems with scheduled synapses. *IEEE Trans. Nanobiosci.*, 16, 792–801, 2017.
7. Cabarle, F.G.C., Adorna, H.N., Pérez-Jiménez, M.J., Song, T., Spiking neural P systems with structural plasticity. *Neural Comput. Appl.*, 26, 1905–1917, 2015.
8. de la Cruz, R.T.A., Cabarle, F.G.C., Macababayao, I.C.H. *et al.*, Homogeneous spiking neural P systems with structural plasticity. *J. Membr. Comput.*, 3, 10–21, 2021, https://doi.org/10.1007/s41965-020-00067-7.
9. Cabarle, F.G.C., Adorna, H.N., Pérez-Jiménez, M.J., Asynchronous spiking neural P systems with structural plasticity, in: *International Conference on Unconventional Computation and Natural Computation*, Springer, Berlin, pp. 132–143, 2015.
10. Cabarle, F.G.C., Adorna, H.N., Pérez-Jiménez, M.J., Sequential spiking neural P systems with structural plasticity based on max/min spike number. *Neural Comput. Appl.*, 27, 5, 1337–1347, 2016.
11. de la Cruz, R.T.A., Cabarle, F.G.C., Adorna, H.N., Generating context-free languages using spiking neural P systems with structural plasticity. *J. Membr. Comput.*, 1, 3, 161–177, 2019.
12. Ramanujan, A. and Krithivasan, K., Control languages associated with spiking neural P systems. *Rom. J. Inf. Sci. Technol.*, 15, 4, 301–318, 2012.
13. Chen, H., Freund, R., Ionescu, M., Păun, Gh., Pérez-Jiménez, M.J., On string languages generated by spiking neural P systems. *Fundam. Inform.*, 75, 141–162, 2007.
14. Adamatzky, A. (Ed.), *Handbook of unconventional computing*, WSPC Book Series in Unconventional Computing, volume 1: Theory, A. Adamatzky (Ed.), World Scientific, Singapore, 2021.
15. Martín-Vide, C., Pazos, J., Păun, Gh., Rodrigez-Paton, A., Tissue P Systems. *Theor. Comput. Sci.*, 296, 2, 295–326, 2003.
16. Zhang, G., Pérez-Jiménez, M.J., Gheorghe, M., *Real-life applications with membrane computing*, Emergence, Complexity and Computation Book Series (ECC), vol. 25, Springer, New York, NY, USA, 2017.
17. Păun, Gh., Rozenberg, G., Salomaa, A., *The Oxford Handbook of Membrane Computing*, Oxford University Press, London, UK, 2009.
18. Wolfgang, M., Networks of spiking neurons: The third generation of neural network models. *Neural Networks*, 10, 9, 1659–1671, 1997, doi: 10.1016/S0893-6080(97)00011-7.

19. Song, T., Pang, S., Hao, S., Rodríguez-Paton, A., Zheng, P., A parallel image skeletonizing method using spiking neural P systems with weights. *Neural Process. Lett.*, 50, 1485–1502, 2019.
20. Ma, T., Hao, S., Wang, X., Rodríguez-Patón, A., Wang, S., Song, T., Double layers self-organized spiking neural P systems with anti-spikes for fingerprint recognition. *IEEE Access*, 7, 177562–177570, 2017.
21. Chen, Z., Zhang, P., Wang, X., Shi, X., Wu, T., Zheng, P., A computational approach for nuclear export signals identification using spiking neural P systems. *Neural Comput. Appl.*, 29, 695–705, 2018.
22. Díaz, C., Sanchez, G., Duchen, G., Nakano, M., Perez, H., An efficient hardware implementation of a novel unary spiking neural network multiplier with variable dendritic delays. *Neurocomputing*, 189, 130–134, 2016.
23. Ishdorj, T.-O., Ochirbat, O., Naimannaran, C., A μ-fluidic biochip design for spiking neural P systems. *Int. J. Unconv. Comput.*, 15, 59–82, 2020.
24. Zhang, G., Shang, Z., Verlan, S., Martínez-del-Amor, M.A., Yuan, C., Valencia-Cabrera, L., Pérez-Jiménez, M.J., An overview of hardware implementation of membrane computing models. *ACM Comput. Surv.*, Article No.: 90, 1–38, 2020, https://doi.org/10.1145/3402456.
25. Pan, L. and Păun, Gh., Spiking neural P systems with anti-spikes. *Int. J. Comput. Commun. Control*, 4, 273–282, 2009.
26. Ramanujan, A. and Krithivasan, K., Control languages associated with spiking neural P systems. *Rom. J. Inf. Sci. Technol.*, 15, 4, 301–318, 2012.
27. Ramanujan, A. and Krithivasan, K., Control words of transition P systems, in: *BIC-TA 2012, Advances in Intelligent Systems and Computing*, vol. 201, Springer, Switzerland, 2013.
28. Penttonen, M., On derivation language corresponding to context-free grammars. *Acta Inform.*, 3, 285–291, 1974.
29. Penttonen, M., Szilard languages are *logn* tape recognizable. *Elektron. Inf. Verarb. Kybern.*, 13, 11, 595–602, 1977.
30. Păun, Gh., On Szilard's languages associated to a matrix grammar. *Inf. Process. Lett.*, 8, 2, 104–105, 1979.
31. Paul, P., On Szilard languages of labelled insertion grammars. *Fundam. Inform.*, 172, 53–72, 2020.
32. Paul, P. and Ray, K.S., Derivation languages and descriptional complexity measures of restricted flat splicing systems. *Theor. Comput. Sci.*, 16, 19–36, 2020.
33. Mäkinen, E., On context-free and Szilard languages. *BIT*, 24, 2, 164–170, 1984.
34. Mahalingam, K. and Paul, P., On Szilard languages of InsDel systems. *J. Autom. Lang. Comb.*, 25, 4, 321–348, 2020.
35. Mahalingam, K., Paul, P., Mäkinen, E., On derivation languages of a class of splicing systems. *Acta Cybern.*, 23, 1–13, 2018.

36. Mahalingam, K., Paul, P., Song, B., Pan, L., Subramanian, K.G., Derivation languages of splicing P systems, in: *BIC-TA 2017, Communications in Computer and Information Sciences*, vol. 791, pp. 487–501, 2017.
37. Duske, J., Parchmann, R., Specht, J., Szilard languages of IO-grammars. *Inf. Control*, 40, 3, 319–331, 1979.
38. Cojocaru, L. and Mäkinen, E., On some derivation mechanisms and the complexity of their Szilard languages. *Theor. Comput. Sci.*, 537, 87–96, 2014.
39. Ramanujan, A. and Krithivasan, K., Control languages associated with tissue P systems, in: *UCNC 2013, Lecture Notes in Computer Science*, vol. 7956, Springer, Switzerland, 2013.

24

Performance Analysis and Interpretation Using Data Visualization

Vikash Kumar Mishra[1], **Iyyappan, M.**[2*], **Muskan Soni**[2] and **Neha Jain**[2]

[1]*Indian Institute of Information Tecgnology, Allahabad, Devghat Jhalwa, Prayagraj, UP, India*

[2]*Christ (Deemed to be) University, Delhi NCR Ghaziabad, Uttar Pradesh, India*

Abstract

The matrix plot library (Matplotlib) is a unique feature in python that helps in the visualization of data via entering certain dataset and codes. It is a portable two-dimension of plot and images are mainly focused on visualizing scientific, technical, and financial data. These matrix plots are performing with the help of python programming and various user interface applications. Most familiar versions of joint photographic and supportable picture graphics are used for the picture visualization. These additional features include the various navigation processes, pages with the line, as well as images. The financial charts of open source website are used for tables and mathematical texts. The library is based on numerical python arrays, giving us visual access to massive quantities of data in readily consumable graphics. The problem statement here delves further into the functions of this feature, which will aid in a better understanding of Python's involvement in the data visualization.

***Keywords*:** Performance analysis, data visualization, Matplotlib in Python

24.1 Introduction

The matrix plotting library was developed on the basis of user request to create a sample visualization with a specific source code [1]. If histogram of any data required for the graph do not instantiate objects, call methods,

**Corresponding author*: iyyappan.m@christuniversity.in

Sharmistha Ghosh, M. Niranjanamurthy, Krishanu Deyasi, Biswadip Basu Mallik, and Santanu Das (eds.) Mathematics and Computer Science Volume 1, (455–462) © 2023 Scrivener Publishing LLC

or set properties, it should work fine [2]. The initial goals of this library modules are as follows: the plot should be a better quality, especially text (smoothing, rotation, etc.) PostScript output for integration into text documents [3, 4]. It can be embedded in a graphical user interface for application development and must be able to understand the code as well as easy to create plots [5]. The software is an open source, so it can be downloaded and used for free distribution [6]. This plot library can be used for various settings. Most of the users are very well acquainted with the source code of command line argument for interactively creating plots and images. This interface provides a pop-up window for viewing and editing data [7]. Imagine there is a person studying in grade 9. The student was struggling a lot with the academics as the person has scored relatively very less as compared with the last semester. So using Matplotlib, we can show what exactly the student is facing and how that problem can be solved by giving the exact status of his condition. Here, the role of the data visualization will take place in order to give a better insight [8].

24.2 Selecting Data Set

In this research, we have used the primary data. A simple dataset was created on the spot to explain how different types of graphs can be used to solve the problems [9]. The data is about the academics of a grade 9 student. The marks of two terms (Table 24.1) are used and how his day-to-day activities (Table 24.2) affected the students' marks is shown using visualization plots. Looking at the different tables, the student can easily understand where he is lacking in time. With the help of visualization, the student can improve his performance in the next grade.

Table 24.1 Dataset of term 1 and term 2 marks of the student.

Sl. no.	Subjects	Term 1	Term 2
1.	English	90	80
2.	Hindi	100	95
3.	Mathematics	63	56
4.	Science	70	61
5.	S.st	85	77

Table 24.2 Dataset of the student's activity in a day.

Sl. no.	Activities	Time (in hrs)
1.	School	6
2.	Eating	2
3.	Tuition	2
4.	Sleeping	8
5.	Sports Academy	2
6.	Miscellaneous	1
7.	Studying	3

24.3 Proposed Methodology

First of all, in order to learn the basics of Matplotlib, we will first learn how to create a simple graph (linear graph) (Figure 24.1).

Step 1: Import pyplot as plt
Step 2: Import style since we will be using gg plot
Step 3: Add dimensions in the form of x-axis and y-axis
Step 4: Add labels, legends and a name to the graph
Step 5: Add final syntax to show the figure (output).

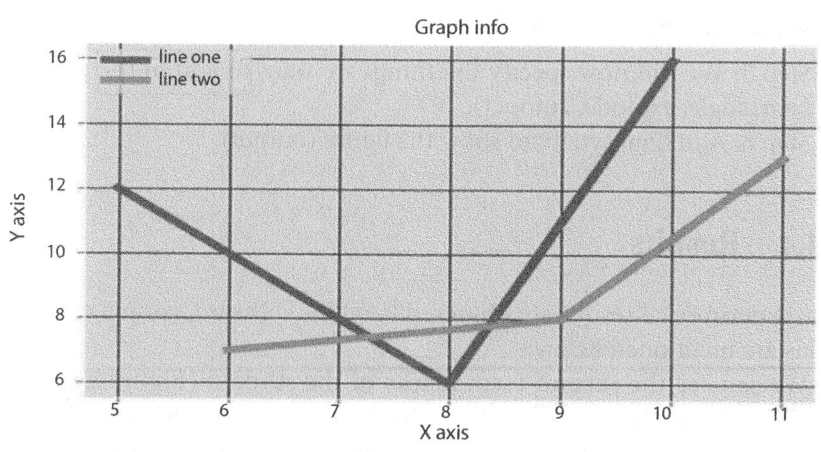

Figure 24.1 Linear graph example.

We can now start with how to work with our problem statement. In order to depict the score of the student in second term, we have created a bar chart (Figure 24.2):

Step 1: Import pyplot as plt.
Step 2: Add dimensions as of the dataset in the form of x and h.
Step 3: Specify the type of chart we have to create (in this case, it is bar graph).
Step 4: Add labels, legends and a name to the graph.
Step 5: Add final syntax to show the figure.

After that, we will be comparing the student's second term marks with that of the first term. For that we will be creating a group bar chart (Figure 24.3). It is a bit similar to a bar chart, we only have to add dimensions (x2, y2). We also have to import NumPy as np in Matplotlib. The width of the graph also has to be mentioned.

After we have visualized the first term and second term marks. To come up with the explanation as to why the student's grades might have dropped in the second term, we will now look into the activities of the student throughout the day. For that, we will be creating a pie chart (Figure 24.4) of all the activities performed by that particular student in 24 hours.

Step 1: Import pyplot as plt.
Step 2: We will add slices here instead of dimensions as it will help divide the entire pie chart according to the activities performed.
Step 3: Add the name of the activities performed (this will be our dataset).
Step 4: Add labels, and colors to each slice.
Step 5: We will now specify the things we want to add in the pie chart (startangle, explode, autopct).
Step 6: Add final syntax to show the figure (output).

24.4 Results

The experimental results are observed from the software; sample visualizations are mentioned below:

We can see the second term marks of the student (Figure 24.2) and figure out that he/she scored the highest in Hindi and the lowest in mathematics. This means that the student needs to work more on mathematics and science.

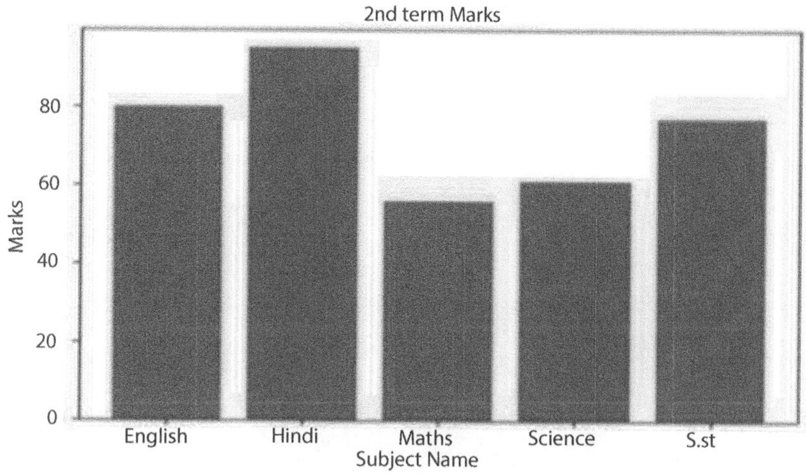

Figure 24.2 Bar graph of second term marks.

Upon comparing the marks of the student from term 1 to term 2 (Figure 24.3), we can figure out that the performance of the student has declined a bit in term 1 as compared to term 2, which is the reason the activities of the student was being analyzed.

Through this pie chart (Figure 24.4), we can see that only 12.5% of the student's time goes into studying. If the students increase their studying time by cutting down on one of the activities, the performance could improve in the next semester.

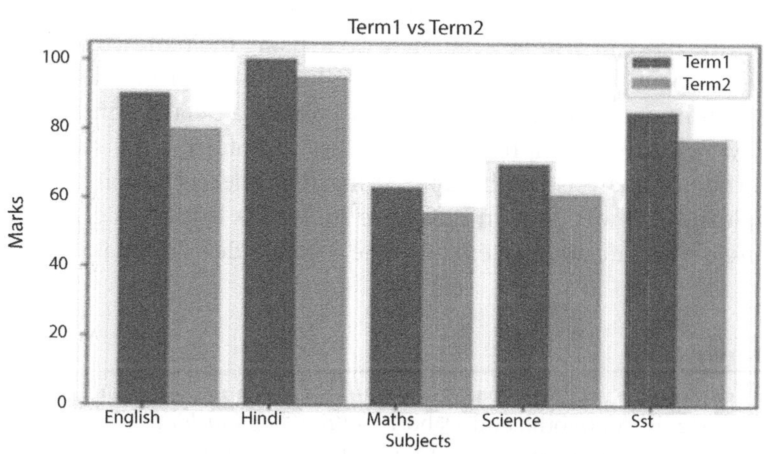

Figure 24.3 Double bar graph of first term and second term marks.

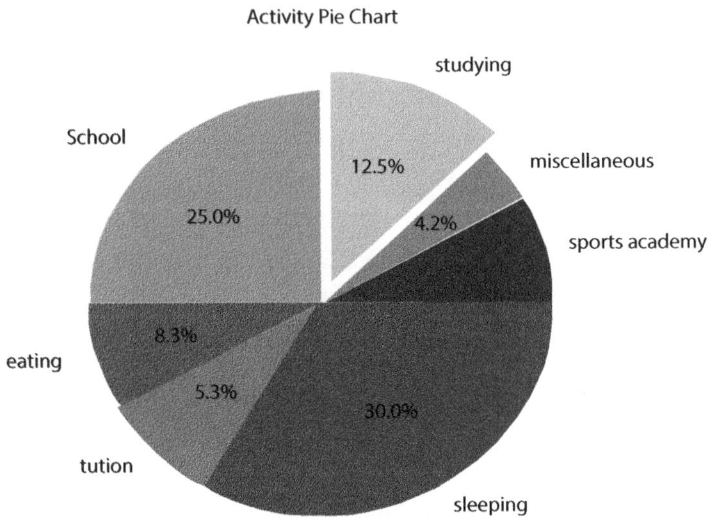

Figure 24.4 Pie chart of activities performed by the student.

24.5 Conclusion

In the current study, we have used imaginary primary data to showcase the solutions of the different problems using Matplotlib in python. The problem consists of the study cycle of a student who was facing difficulty in managing his day to day activities. Using Matplotlib, we have proposed a clear segregation of the activities that the student commences in his whole day [9]. With the help of the different charts used in the study, it becomes very easy for the student to prioritize the activities accordingly. The analysis of the day can also be accurately done using the plot, which becomes a helping hand for the student to solve his misery [10]. Matplotlib can be useful in data visualization and can provide a detailed analysis of the data, which can be helpful just like it, was helpful for the student [11]. Performance analysis can be done using the different visualization tools and plots like it was done in the above study. It is helpful in giving a clear picture of the data and can enhance the speed of decision making [12].

References

1. Bisong, E., Matplotlib and seaborn, in: *Building Machine Learning and Deep Learning Models on Google Cloud Platform*, Apress, Berkeley, CA, 2019, https://doi.org/10.1007/978-1-4842-4470-8_12.

2. Hunt, J., Graphing with Matplotlib pyplot, in: *Advanced Guide to Python 3 Programming*. Undergraduate Topics in Computer Science, Springer, Cham, 2019, https://doi.org/10.1007/978-3-030-25943-3_6.
3. Pajankar, A., Matplotlib, in: *Raspberry Pi Supercomputing and Scientific Programming*, Apress, Berkeley, CA, 2017, https://doi.org/10.1007/978-1-4842-2878-4_14.
4. Porcu, V., Matplotlib, in: *Python for Data Mining Quick Syntax Reference*, Apress, Berkeley, CA, 2018, https://doi.org/10.1007/978-1-4842-4113-4_10.
5. Nelli, F., Data visualization with Matplotlib, in: *Python Data Analytics*, Apress, Berkeley, CA, 2015, https://doi.org/10.1007/978-1-4842-0958-5_7.
6. Glover, J.C., Lazzarini, V., Timoney, J., Python for audio signal processing, in: *Linux Audio Conference 2011*, Maynooth, Ireland, May 6-8 2011, 2011. 7. Sial, A.H. *et al.*, *Int. J. Adv. Trends Comput. Sci. Eng.*, 10, 1, 277–281, January – February 2021.
7. Embarak, O., Data visualization, in: *Data Analysis and Visualization using Python*, Apress, Berkeley, CA, 2018, https://doi.org/10.1007/978-1-4842-4109-7_7.
8. Mehmood, H. *et al.*, Implementing big data lake for heterogeneous data sources. *2019 IEEE 35th International Conference on Data Engineering Workshops (ICDEW)*, pp. 37–44, 2019, doi: 10.1109/ICDEW.2019.00-37.
9. Ellson, J., Gansner, E., Koutsofios, L., North, S.C., Woodhull, G., Graphviz—Open-source graph drawing tools, in: *International Symposium on Graph Drawing*, 2001, September, Springer, Berlin, Heidelber, pp. 483–484.
10. Morra, G., Visualization, in: *Pythonic Geodynamics*. Lecture Notes in Earth System Sciences, Springer, Cham, 2018, https://doi.org/10.1007/978-3-319-55682-6_2.
11. Python matplotlib tutorial, Edureka, https://www.youtube.com/watch?v=yZTBMMd%20POww&t=1373s.
12. Matplotlib tutorialb(part1): Creating and customising our first plots, https://www.youtube.com/watch?v=UO98lJQ3QGI.

25

Dealing with Missing Values in a Relation Dataset Using the DROPNA Function in Python

Vikash Kumar Mishra[1*], Shoney Sebastian[2], Maria Iqbal[2] and Yashwin Anand[2]

[1]*School of Computing Science and Engineering, Galgotias University, Greater Noida, Uttar Pradesh, India*
[2]*Christ (Deemed to be) University Delhi-NCR, Ghaziabad, Uttar Pradesh, India*

Abstract

Python provides a rich data structure library called PANDAS, which provides fast and efficient data transformation and analysis. The word PANDAS is an abbreviation of Python Data Analysis Library. PANDAS facilitate optimized and dynamic data structure designs work with "relational" or "labeled" data. Python's approach is meant to provide a high-level, high-performance building block that can be used to do real-world analysis of data. PANDAS Library is allowing users to import data from different file formats, such as CSV, SQL, Microsoft Excel etc. [1]. It helps in data preparation, as well as in data modeling, for those projects, which aims data analysis for the extraction of information. Python's future will be built on this layer for statistical computing. In addition to discussing future areas of work and growth opportunities for statistics and data analytics applications built on Python, the study provides details about the language's design and features [2]. In this research paper, we intend to solve the problem of missing values in a dataset using the DROPNA function in Python using PANDAS library.

Keywords: Python, DROPNA, PANDAS, missing values

**Corresponding author*: vikash.mishra@galgotiasuniversity.edu.in

Sharmistha Ghosh, M. Niranjanamurthy, Krishanu Deyasi, Biswadip Basu Mallik, and Santanu Das (eds.) Mathematics and Computer Science Volume 1, (463–470) © 2023 Scrivener Publishing LLC

25.1 Introduction

PANDAS library is based on the NumPy library, and it offers a variety of operations and data structures for manipulating numerical data and time series. Developer Wes McKinney developed PANDAS in 2008. PANDAS is a fast, efficient, and customizable data frame object. Data objects are loaded in memory using tools for reading from and writing to various file formats. It enables the alignment and handling of missing data, reshaping, pivoting and indexing of large datasets, and it supports the merging and joining of large datasets that are high in quality [3]. The Python program PANDAS provides the ability to delete or add columns to a data structure. PANDAS package can be installed using pip function.

25.2 Background

When working with realistic datasets, missing data is a common issue. Knowing and analyzing the causes of missing values can help you see the steps you need to take to fix the problem. Python has a number of methods for analyzing and resolving the issue of unaccounted data. Here, we have created subsets of our dataset and implemented various functions of the PANDAS library on it. None and NaN are used interchangeably in PANDAS to indicate missing or null values [4]. PANDAS data frame gives many privileges and built-in function for different operations. Missing values in python can be dealt in following ways: (a) view and import the data: copy the file's path after downloading the dataset. Then, import the data from googleplaystore.csv into a variable using the PANDAS library: import PANDAS as pd.data = pd.read_csv('<path to googleplaystore.csv>') To confirm the data imported correctly, run: data.head() The aforementioned command should display the first few rows of data in tabular format. The presence of NaN values indicates that this dataset contains missing data. (b) Look for missing values: run the following command to see how many missing values are there per column: data.isnull().sum().

25.3 Study Area and Data Set

Google play store apps have been taken as a dataset for our study. In present scenario, the customer rating is playing an important role for the success of any application and good rating means customer's satisfaction, which motivate others to download highly rated applications. To get better ratings for their apps, developers utilize appealing app titles, demanding icons, and user-friendly interface. The data from Play Store has enormous potential to propel app-creation businesses to success [5]. Data for this analysis was selected using Kaggle. In the present study web scraped data set of 10,000 Play Store Applications have been considered with total of 10842 rows and 13 columns.

After examining the data, the app's "ratings" can be determined to be the most essential metric, as it depicts how well the app performs in comparison to other apps on the market. It also gives an indication of how well the company works to adopt user feedback. Ratings are based on a variety of variables. The link between these two will be examined in the following

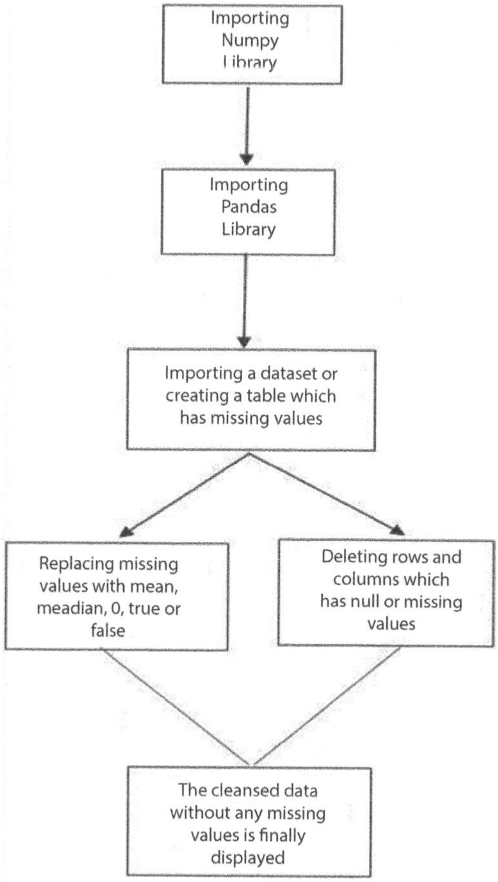

Figure 25.1 The flowchart of the proposed methodology.

section of this study. The dependent variable that we have chosen in number of installation and the independent variable is number of reviews and size of the app. Thus, the more the number of reviews, the more will be the number of installations and lower the size of the app, higher will be the installations. The dataset has a lot of potential for increasing corporate value and having a beneficial influence [6]. E-commerce and review sites are teeming with untapped data that has the potential to be transformed into useful insights that can aid in sound decision making. Here, we look at how data science and machine learning approaches can be used to gather. The proposed methodology is shown in Figure 25.1.

25.4 Methodology

Data gathered from the Google Play Store, one of the Internet's most popular destination. Missing data affects your algorithm and research in three ways: Missing values give the wrong impression about the data, resulting in ambiguity. Calculating an average for a column with half of the data unavailable or set to zero, for example, yields the incorrect metric. Some algorithms fail when data is unavailable [7–9]. With datasets containing Not a Number (NaN) values, some machine learning algorithms fail. A crucial factor is the pattern of missing data. In most cases, even if data from a dataset is missing at random, the information is still useful. However, if data is missing in a systematic way, the entire analysis will be skewed.

To identify the NULL values in a data frame, PANDAS' isnull() and notnull() methods are used. DataFrame.isnull() built-in function is used to detect the missing values whereas DataFrame.notnull() built-in function detects the non-missing value of an object.

```
from PANDAS import *
df = DataFrame([[10, 20, 30, 40], [7, 14, 21, 28], [5,]],
columns=['Apple', 'Orange', 'Banana', 'Pear'],
index=['Basket1', 'Basket2', 'Basket3'])
print("\n--------- DataFrame ---------\n")
print(df)
print("\n--------- Use of isnull() ---------\n")
print(df.isnull())
print("\n--------- Use of notnull() ---------\n")
print(df.notnull())
```

We have taken apple, orange, banana, and pear in three baskets as shown in Table 25.1. In baskets 1 and 2, there are numeric values of the number of fruits that are in the basket, whereas in basket 3 null values are there for orange, banana, and pear.

Here, in Table 25.2 we have used Isnull() function so that all the null values are shown as true and all the other values as false.

In Table 25.3, we have applied Notnull() function to see all the values which are not null as true and other values as false.

In Table 25.4, the dropna() function: The dropna() function makes an analysis of the relational table and finds the rows or the columns, which are having null entries, and it is going to drop such rows/columns.

Table 25.1 DataFrame.

	Apple	Orange	Banana	Pear
Basket1	10	20.0	30.0	40.0
Basket2	7	14.0	21.0	28.0
Basket3	5	NaN	NaN	NaN

Table 25.2 Use of Isnull() function on the DataFrame.

	Apple	Orange	Banana	Pear
Basket1	False	False	False	False
Basket2	False	False	False	False
Basket3	False	True	True	True

Table 25.3 Use of Notnull() function on the DataFrame.

	Apple	Orange	Banana	Pear
Basket1	True	True	True	True
Basket2	True	True	True	True
Basket3	True	False	False	False

Table 25.4 Use of DropNa() function on the DataFrame to find the rows and columns with missing values.

	First score	Second score	Third score	Fourth score
0	100.0	30.0	52	NaN
1	90.0	NaN	40	NaN
2	NaN	45.0	80	NaN
3	95.0	56.0	98	65.0

```
from PANDAS import *
from numpy import *
dict={'First_ Score':[100, 90, np.nan, 95],
'Second _Score':[30, np.nan, 45, 56],
'Third _Score':[52, 40, 80, 98],
'Fourth _Score':[np.nan, np.nan, np.nan, 65]}
df=DataFrame(dict)
print(df)
print(df.dropna())
```

25.5 Results

We have used the dropna() function in our problem statement because this function removes the rows with null values and only keeps those with numeric and float values. This helps in determining the right type of data in the dataset. The application of isnull() and notnull() functions is not recommended because isnull() function converts the null values into true and the not null values into false whereas notnull() function converts not null values into true and null values into false. This does not suggest anything about the data. So, this could be applicable and recommended for data cleansing approach for the preparation of data in data science.

25.6 Conclusion

It is believed that in coming years the role of Python libraries for data analysis will be at an apex. Since, the generation of data is going on at a rapid speed as 1.7 Mb of data is being generated by each person/second. Traditional statistical data analysis tools who may have previously chosen R, MATLAB, or another research environment may be replaced by Python modules as they are easy to use, efficient, and user-friendly. PANDAS and NumPy are well known and rich in built-in functions for the numerical data analysis [10–15].

25.7 Acknowledgment

The authors are thankful to Galgotias University Greater Noida and Christ University, Delhi NCR, Uttar Pradesh, India for providing sufficient infrastructure for the completion of this research paper.

References

1. Mckinney, W., PANDAS: A foundational Python library for data analysis and statistics. *Python High Performance Science Computer*, 2011.
2. Bokeh Development Team, Bokeh: Python library for interactive visualization, 2018. http://www.bokeh.pydata.org.
3. McKinney, W., PANDAS: A foundational Python library for data analysis and statistics, in: *Python High Performance Science Computer*, 2011.
4. Bryant, P.G. and Smith, M., *Practical data analysis: Case studies in business statistics*, Richard D. Irwin Publishing, Homewood, IL, 1995.
5. Fonnesbeck, C., Patil, A., Huard, D., PyMC: Bayesian stochastic modelling in Python. *J. Stat. Softw.*, 35, 4, 1, 2010.
6. McKinney, W., PANDAS: A python data analysis library, 2005. http://pandas. pydata. org.
7. Seabold, S. and Perktold, J., Statsmodels: Econometric and statistical modeling with python. In *Proceedings of the 9th Python in Science Conference* (Vol. 57, No. 61, pp. 10-25080), June 2010.
8. Hunter, J., Dale, D., Droettboom, M., Matplotlib: A Python 2D plotting library, 2008.
9. Geetha, J. and Harshit, N.G., Implementation and performance comparison of partitioning techniques in apache spark. In *2019 10th International Conference on Computing, Communication and Networking Technologies (ICCCNT)* (pp. 1–5). IEEE, July, 2019.
10. Stančin, I. and Jović, A., An overview and comparison of free Python libraries for data mining and big data analysis. In *2019 42nd International Convention on Information and Communication Technology, Electronics and Microelectronics (MIPRO)* (pp. 977–982). IEEE, May, 2019.

11. Leo, S. and Zanetti, G., Pydoop: A Python MapReduce and HDFS API for Hadoop. *Proceedings of the 19th ACM International Symposium on High Performance Distributed Computing*, pp. 819–825, 2010.
12. Stevens, J.-L.R., Elver, M., Bednar, J.A., An automated and reproducible workflow for running and analyzing neural simulations using Lancet and IPython Notebook. *Front. Neuroinform.*, 7, 44, 2013.
13. *What is Python? Executive summary. Python documentation*, Python Software Foundation, Retrieved 2011-08-1. https://www.python.org/doc/essays/blurb/
14. McKinney, W., *Python for data analysis: Data wrangling with PANDAS, NumPy, and IPython.* O'Reilly Media, Inc., 2012.
15. Paffenroth, R. and Kong, X., Python in data science research and education. In *Proceedings of the 14th Python in Science Conference* (Vol. 19, pp. 164–170), 2015.

26

A Dynamic Review of the Literature on Blockchain-Based Logistics Management

C. Devi Parameswari[1]* and M. Ilayaraja[2]

[1]Department of Computer Applications, Kalasalingam Academy of Research and Education, Krishnankoil, Tamil Nadu, India
[2]Department of Computer Science And Information Technology, Kalasalingam Academy of Research and Education, Krishnankoil, Tamil Nadu, India

Abstract
In recent years, both practitioners and academics have become interested in blockchain, a decentralized ledger system. Despite the fact that both researchers and professionals see the significance of blockchain innovation and its implementations, the most effective deployments of blockchain implementation remain in the field of cryptocurrencies. This study uses research and network analysis to results of studies to see how blockchain may help with logistics management. Blockchain platforms and digital contracts were also determined to be the more significant developing technology in this sector. This comprehensive study sheds light on a number of research disciplines, including the prospects of blockchain for privacy and security concerns in the food industry. The study also discovered that scientific research in this field is scarce. As a result, implementing blockchain in the real-world supply chain has a lot of new research opportunities.

***Keywords*:** Blockchain technology, smart contracts, supply chain management, visibility, transparency, literature review

26.1 Introduction

Increased openness of products, makers, and providers across the production process is vitally beneficial to consumers, government agencies, and nongovernmental organizations [1, 2]. As a result, integration of

**Corresponding author*: deviparameswari@klu.ac.in

Sharmistha Ghosh, M. Niranjanamurthy, Krishanu Deyasi, Biswadip Basu Mallik, and Santanu Das (eds.) Mathematics and Computer Science Volume 1, (471–484) © 2023 Scrivener Publishing LLC

environmental challenges is now significantly crucial for producers in effort to keep their brand's perfect reputation [3]. The likelihood of replica items being introduced into a production process is increased when there is an absence of traceability. In addition to the possibility of becoming accidentally implicated in social and environmental sustainability issues. For example, duplicating electrical parts can put firms' safety and revenues at risk, along with tarnish manufacturers' and distributors' reputations [4]. According to Machado *et al.* [5] manage and confidences in the logistics management are critical first steps in avoiding dangerous actions when integrating buyers and employees. Obtaining complete transparency and detecting counterfeit components, but in the opposite side, is highly difficult and costly [4, 5]. Chen *et al.* [2] suggested with the goal of creating a tightly regulated, open database for exchanging information about producers' descriptions and certification data could aid in overcoming the visibility and ecological issues that affect globalized global production chains [6]. Current academic and business efforts study the usage of blockchain innovation in production processes to overcome these transparency concerns [7, 8].

The blockchain innovation can be considered as a decentralized peer-to-peer network that is employed to process and verify data transactions. To make transactions practically immutable, it employs cryptographic methods, consensus techniques, and back-linked blocks [10]. Satoshi Nakamoto created the paradigm in 2008 when he officially released the popular Bitcoin policy document, launching the blockchain also with goal of upsetting the traditional finance system and eliminating the use of trusted third parties [11]. In 2013, Vitalik Buterin published the Ethereum policy statement, which expanded on the Bitcoin premise. The Ethereum protocol, in comparison to Bitcoin, goes much beyond simply using blockchain technology as a currency. Ethereum is a blockchain that includes a Turing-complete programming language [9]. Turing completeness is a measure of a computer language's computability that represents a mathematical idea. Complex elements, such as loops and conditions, are included in a Turing-complete language design, allowing general-purpose programmers to be written. So, using blockchain-based decentralized apps, Buterin [12] invented the phrase smart contract. Outside of the financial industry, decentralized apps are the foundation for blockchain-based use cases. In this context, blockchain technology's unchangeable, distributed, and protected properties were first offered as a solution to increase the visibility, safety, validity, and traceability of commodities in production processes [9, 12]. Meanwhile, enthusiasm for blockchain in manufacturing, logistics, and transportation, as well as smart contracts, is at an all-time high, contributing to the notion that production lines of any difficulty can be depicted on a blockchain [13].

26.2 Blockchain Concepts and Framework

A BC is a block chain, as its name suggests. Each block is made up of a head and a body and has the same structure. So that a chain is constructed, each block has a precise reference to the previous block (see Figure 26.1). Only the "Genesis block" is an exception. It is known as initial block of the network. Mining [9] is the method of join every exchange and save everything within the subsequent block. A block's header gives data including a timestamp and the block's version, and also, a previous block hash value. Every block has unique hash value in an effort to allow them easily identified by using algorithm. A chain of blocks grows longer and longer as an outcome of the reference to the previous block. Most participants can only handle for change; a block is deeply attached to this chain, which reduces the chance of manipulation. The most important feature of BC is that previous volumes cannot be updated without recreating the whole of chain of blocks [10]. Every block's most crucial component is its body. All transactions that were not included in the previous module will be pushed to this one. The amount of transactions per block is rigidly defined due to the fixed volume of each block. The body stores these activities, as well as all relevant details. In the Bitcoin example [11], this is information on the originator, destination, and Bitcoin money.

As mentioned earlier, blockchain's networks operate in a decentralized manner, not dependent on any centralized server or organization. A chain of nodes makes up a blockchain. Every client has a replica of the BC and can look over the fresh blocks. The more systems that transmit the

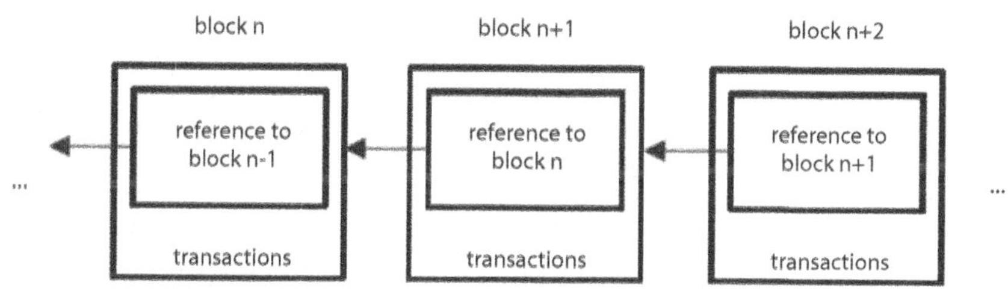

Figure 26.1 Blockchain diagram.

network, the more BC duplicates there still are, and the more secure the network obtains. This is accomplished using consensus processes, like as the proof of work (PoW) algorithm, which is applied by the Bitcoin BC [11].

A mathematical problem underpins the PoW algorithm. In the case of Bitcoin, the first valid solution will be issued with Bitcoin and with newly confirmed transactions. All other nodes may immediately verify the answer, and if it is correct, it can be appended to the BC, along with the transactions. The proof-of-stake (PoS) algorithm is an alternative to the PoW technique. In this situation, the miners will be responsible for a portion of their assets for the reliability of the block. This algorithm is significantly more energy efficient and scalable because no difficult mathematical problems need to be solved. This means the Internet is less expensive to run, but the network's security is weaker than using the PoW

Table 26.1 BCT properties.

BCT attribute	Explanation
Decentralized structure	Everyone has a replica of the similar object, which is distributed over the peer-to-peer network so that anyone can see it at any time and any place on the network.
Cryptography system	This makes it possible to follow data and transaction chains throughout time. Through a practical standpoint, each volume title contains a "hash value" that duplicates the previous volume's information and extends down from its earlier works to the chain's first volume. As a result, altering the previous record will not send the new data backwards, rendering the entire chain incorrect.
Consensus mechanism	To provide security level and authenticity, the consensus system ensured that operations are stored into blocks as per encryption criteria and sequential order. Distributed consensus approaches, such as majority, perception, or minimum votes, can be used to secure information.
Smart contract	This optional feature allows users to develop methods and policies that trigger exchanges across nodes with no need for external guidance.

algorithm [13]. BCT has four distinct properties that should be developed and defined (Table 26.1).

26.3 Study of the Literature

The primary goal of this article is to examine how blockchains have previously been employed in manufacturing operations. The literature also emphasizes on the use of blockchain innovation in logistics. We looked into two primary topics. Investigation of applications to a specific supply chain activity is one such area. Previous research, for example, looked into whether technological innovation could be applied for order placement [14], collaboration [15], shipping [16], and smart contracts [17]. Another area of interest is exploring ways in which blockchain characteristics can be applied to logistics operations. These characteristics include privacy, content sharing (or transparency), consistency, detectability, cost-effectiveness, reliability (or certainty), detectability, and faith. The privacy of blockchain innovation is ensured through the use of encoding techniques to protect data theft and tampering, as well as a consensus mechanism [17]. The blockchain's consensus process establishes confidence among its users. The transaction records (for example, supply chain legal and contracts documents) are sent to everyone production chain users participating in the transactions, demonstrating blockchain's shareability. The decentralization of the user system in the blockchain network and the community of blockchain participants certainly keep the blockchain infrastructure reliable—if one participant system occurs to corruption or departs the network of blockchain, the rest of the user nodes can take on to fulfill the blockchain operations [14].

26.3.1 Blockchain Technology and Supply Chain Trust

Many elements influence the use of blockchain innovation for data exchange in the logistics, but one of the major essential is the building of trust [18]. Each transaction is signed into a blockchain in a sealed, visible, and validated manner. As a result, once the transaction is approved by all parties involved, it cannot be changed or removed. Utilizing blockchain innovations can lead to greater security, detection, efficiency, and visibility in distribution management transactions that has the various attributes in terms of data integrity and security. Also on blockchain, records cannot be modified, thus all transactions and decisions made by supply chain members are recorded and documented. The correctness of these transactions is constantly visible to all supply chain members. Furthermore, when a company's operations have a controversial business effect, supply chain participants monitor them, consider them in their judgments, and modify their operations properly when concluding future contracts [19]. Smart contracts, which may be employed with blockchain technology, are another component that builds supply chain confidence. From policy returns to payment information, business transactions, product traceability, and patent protection, blockchain technologies and smart agreements could be utilized in a wide range of directions. As a reason, the range of blockchain-based smart contract technologies is steadily increasing [19]. Limited supply chains with many divisions of external providers can profit through smart contracts. Accepted contracts, software improvements depending on agreed changes, vent events, and patent filings can all be automated [20]. Modules in the blockchain run smart contracts remotely with variables when they are activated. A smart agreement is a program that is well and never changes. Before being uploaded, all shared ledger transactions must adhere to stringent protocols (database log). A digital contract is blockchain software that has been registered. Predefined transactions will take place after this file is published and verified on blockchain. Because the parameters of a digital contract are broadcast upon this blockchain, all parties may be comfortable that the deal has always been followed out as agreed, reducing trust issues [21].

Advantages of Using Blockchain in Supply Chain Management

- The entities associated with the production system form a mutually beneficial partnership.
- A transparent network is established so that anyone with an interest in the production process can audit it and gain access to the data.
- A traceability system is created so that products may be tracked throughout the supply chain's various processes.
- Fare trades are initiated between the people participating in the supply chain utilizing smart contracts.
- By incorporating blockchain technology into the agribusiness sector, we are able to establish an environment where the networking is transparent and the product's safety is routinely verified.

26.4 Challenges and Processes of Supply Chain Transparency

The production process currently comprises of a complex network of stakeholders from several industries that collaborate on duties and reach mutual agreements. The major supply chain difficulties are depicted in Figure 26.2 centralized systems, lack of transparency, scalability, and obstacles to emerging technology. Also, the excellent combination of fresh innovative technologies, traditional centralized supply chain systems are effectively struggling to meet some of the critical needs through work environments and reliable third parties [22]. The lack of trust in such independent databases leads to bad customer feedback and discontent. Furthermore, most distribution chains do not have reliable shared information, which is a fundamental visibility problem in a centralized network. Absence of visibility makes it difficult to innovate and be optimistic, as well as poor consumer feedback. Furthermore, as the product moves across multiple geographical zones, scalability becomes a huge issue.

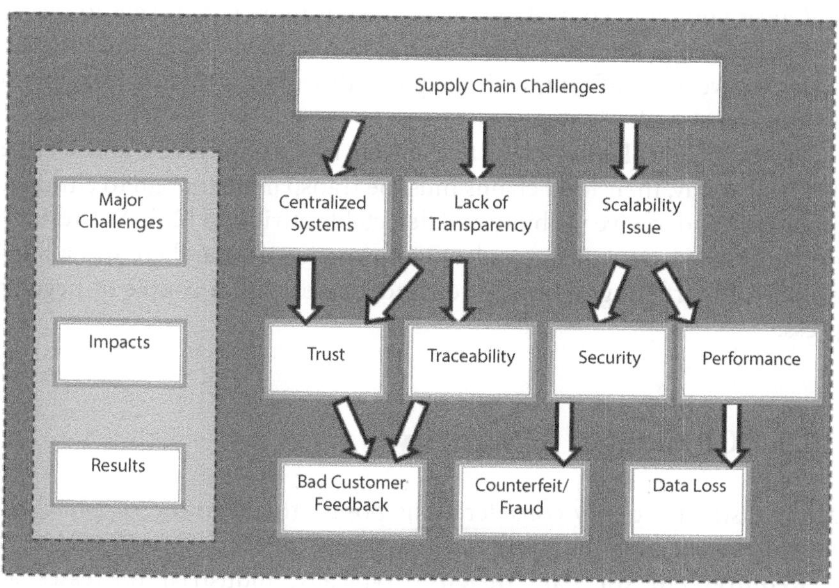

Figure 26.2 Challenges in the technical supply chain.

26.4.1 Motivation for Transparency in Data

Because of the increased demand for food chain visibility and detection, the supply chain has undergone substantial changes throughout time. These requirements are the driving force for the development of transparent systems. Apart from the aforementioned problems, the following are the motives for supply chain transparency:

Database Independent: The existing supply chain structure is made up of a number of centralized systems, which are owned by one or many distribution networks. These solutions rely largely on data handling platforms that are centralized, regularly varied, and individualized [23]. The databases used in the manufacturing process are diverse, ad hoc, and scattered.

Collaborative failure: The heterogeneity of the engaged stakeholders, diverse data types, and an ineffective communication across the relevant systems are the key supply chain problems. The way data is transferred among firms is determined by collaborative relationships, which are then projected to the underlying business processes. Modern organizations can use collaboration to improve their ties with their trading partners.

Loss of data: The widespread emergence of Internet of Things (IoT) has resulted in significant changes in global manufacturing [24]. IoT systems come in a variety of shapes and sizes, and they are often divided into many application components [25]. IoT technology boosts productivity and gives you more control, but it also means more data is sent to servers and peripherals. The risk of data destruction in today's distribution chain systems is very significant, with the possibility of information leakage and corruption at multiple stages.

Product Complexities: Due to the wide range of products and services available today, their food chains must be transparent enough to eliminate ambiguity and to give visibility and detectable attributes [26]. As a result of the centralized system's unregulated informational data, there was a number of counterfeiting, a number of trade losses, and a couple of negative press.

26.5 Challenges in Security

Many distinct security considerations pose a threat to the supply chain, potentially affecting the entire manufacturing process.

Requirement for data opacity: Typically, industrial processes are accompanied by a variety of private elements, such as adequate planning, recipes, manufacturing intelligence, and so on. One of the obvious problems in the supply chain domains is data privacy. As a result, all systems are vulnerable to data breaches, theft, leaks, unauthorized access, eavesdropping, and other threats. As a result, all of the parties in a supply chain must preserve data opacity.

Regulation of the Supply Chain: A thorough grasp of the intended commodities and their requirements is required to meet the supply chain transparency target on time. In addition, suppliers and processes must be mapped, and information gaps must be filled. Unfortunately, there is a gap of clarification in the existing literature, the transparency methods that depict a supply chain project's road map. It provides a practical approach to identify, understand and improve supply chain transparency in the global market. This is accomplished through the following steps: risk detection and visualization, to remove gaps in knowledge, transparent mechanisms are used to management and ultimately tracking. The following are methods of making good use of the supply chain under optimal conditions.

Self-identification: The first phase in establishing a production process is to identify the general components of the environment, both distributors and sub-suppliers.

Collect Details: The most delicate stage is gathering data regarding production methods, items, gaps, and other factors on the ground. Companies are progressively requiring much information from their providers these days. Throughout this process, getting appropriate information is necessary and has a direct impact on total information sharing.

Exposure to risk: At this point, the corporation has such a complete view of the production process and may make judgments. The decision depends on issues and challenges regulatory requirements and also the needs of internal and external customers. In furthermore, the organization should define how the data is shared.

26.6 Discussion: In Terms of Supply Chain Dynamics, Blockchain Technology and Supply Chain Integration

Companies first strive to prepare how they will deal with uncertainty and hazards in all of their activities. Furthermore, they are aware of fresh opportunities that they can exploit as a result of their intelligence. Companies that effectively arrange the operating cycles with optimum market stability face potential drawbacks and require an effective approach to maintain competitiveness throughout time. In conclusion, it is apparent that this is crucial information for a company. In that time, innovation of blockchain is characterized by the ability to more effectively massive data generating. This is one of the main significant properties of a company and save them in a very protection structure at low cost. Because of digital technology, interconnection has expanded significantly, encouraging collaboration and coordination in supply chains. Protecting this obligation from harmful attempts is a different challenge.

The plan that innovation of blockchain is benefit businesses prompted us to conduct this investigate in this context. We would like to see how new information technologies can affect transparency, flexibility, and trust, all of which are key to enabling a highly efficient supply chain process. We hope with the aim of this trend the measure of coordination and cooperation in supply chains will be improved. The more the production chain users trust one another, the more facts they may exchange, and the that much relevant data they give, the most transparent the production process will become, and the better integrated it will become. The supply chain's flexibility will result in improvement of increased cooperation.

The study's initial conclusion was that blockchain situation will make supply chain visibility. This conclusion is consistent with previous research [2–7]. Among the most pressing issues in production processes is transparency, which is essential for effective and stable interaction and communication. Visibility in distribution networks allows preemptive teamwork and coordination [26], and also accessibility of distribution of the material to clients [27], as per the research. As a result, stakeholder feedback is more effectively incorporated into the chain [28]. Companies can carry out more effective open-innovation efforts by involving more stakeholders in the process. Transparency can also help organizations improve their customer-involved product design processes. Companies can acquire a long-term competitive edge by improving product design processes [29]. As a result, businesses need to use innovation of blockchain to improve the visibility of the food production.

Second, we discovered that network of blockchain enhance the flexibility of food networks. Supply networks that are more flexible can better meet client expectations and deal with unpredictable and uncertain conditions. Blockchain technology will enable real-time data exchange throughout supply networks. Blockchain permits real-time data exchange throughout supply networks. These innovations are utilized to quickly combine all procurement operations since the corporate environment is more complex than ever. Sales forecasting, stock control, and identifying critical success factors [30] are all very effective and can help a company remain profitable.

The research's third outcome revealed that blockchain technology can be used to improve supply chain trust. Because effective collaboration that promotes sustainable performance demands a wide platform for details transfer [31] and trust in each other [32], interfirm trust is critical for supply chain success. Companies need to gain trust based on trust, especially in outsourcing situations like 3PL. Our findings corroborate comparable findings in the literature [33]. To improve collaboration and integration, businesses must create trust. As an outcome, blockchain technology helps businesses in both developing and sustaining trust.

26.7 Conclusion

Blockchain innovation is applied in a variety of businesses to manage supply chains. As a result, the main purpose of today's blockchain applications is to increase food chain visibility. Simplistic product complexity requirements, such as those in the food business, are already fully reflected on the blockchain. Food systems make up a significant portion of food management blockchain applications. There is no solution to supply chains dealing with complex areas aimed at increasing transparency and mapping of assembly processes, enabling efficient auditing of all assets and dynamic changes. The raw resources, intermediary components, the final products, and the transformation events all include very different properties, yet in complicated manufacturing supply chains, they all interact or merge at some point. Complete mapping of such supply chains for complex products is a significant problem for blockchain technology.

Acknowledgment

The first author expresses her gratitude to the management of Kalasalingam Academy of Research and Education for providing scholarship to carry out the research.

References

1. New, S., The transparent supply chain, *Harvard Bus. Rev.*, 1, 1–10, 2010.
2. Chen, S., Hang, Q., Zhou, Y.P., Impact of supply chain transparency on sustainability under NGO scrutiny. *Prod. Oper. Manage.*, 63, 9, 1–32, 2018.
3. Lemke, F. and Petersen, H., Managing reputational risks in supply chains, in: *Supply Chain Risk Management*, pp. 65–84, Springer, Singapore, 2018.
4. Collier, Z.A., Hassler, M.L., Lambert, J.H., DiMase, D., Linkov, I., Supply chains, in: *Cyber Resilience of Systems and Networks*, vol. 2, A. Kott, and I. Linkov, (Eds.), pp. 447–462, Springer International Publishing, Cham, 2019.
5. Machado, S.M., Paiva, E.L., da Silva, E.M., Counterfeiting: Addressing mitigation and resilience in supply chains. *Int. J. Phys. Distrib. Logist. Manage.*, 48, 2, 139–163, 2018.
6. Hinckeldeyn, J., Blockchain-technologie in der supply chain, in: *Essentials*, Springer, Wiesbaden, 2019.
7. Subramanian, N., Chaudhuri, A., Kayıkcı, Y., *Blockchain and supply chain logistics evolutionary case studies*, Springer, 2020.
8. Wu, H., Cao, J., Yang, Y., Tung, C.L., Jiang, S., Tang, B., Liu, Y., Wang, X., Deng, X.Y., Data management in supply chain using blockchain: Challenges and a case study. *International Conference on Computer Communication and Networks (ICCCN)*, vol. 28, pp. 1–8, 2019.
9. Gentemann, L., Blockchain in Deutschland – Einsatz, Potenziale, Herausforderungen, *Studienbericht*, 1–10, 2019.
10. Nakamoto, S., Bitcoin: A peer-to-peer electronic cash system, 2008. https://bitcoin.org/bitcoin.pdf
11. Buterin, V., Ethereum white paper: A next generation smart contract & decentralized application platform. *White Paper*, 3, 37, 2–1, 2013.
12. Lee, D.K.C. and Deng, R.H., ChinaTech, mobile security, and distributed ledger, in: *Handbook of Blockchain, Digital Finance, and Inclusion*, Academic Press, London, 2018.
13. Lu, Q. and Xu, X., Adaptable blockchain-based systems: A case study for product traceability. *IEEE Software*, 34, 6, 21–27, 2017.
14. Varriale, V., Cammarano, A., Michelino, F., Caputo, M., Sustainable supply chains with blockchain, IoT and RFID a simulation on order management. *Sustainability*, 13, 6372, 2021.
15. Li, Y., Wang, B., Yang, D., Research on supply chain coordination based on blockchain technology and customer random demand. *Discrete Dyn. Nat. Soc.*, 1–10, 2019.
16. Wang, M., Wu, Y., Chen, B., Evans, M., Blockchain and supply chain management: A new paradigm for supply chain integration and collaboration. *Oper. Supply Chain Manage.*, 14, 111–122, 2021.
17. Dolgui, A., Ivanov, D., Potryasaev, S., Sokolov, B., Ivanova, M., Werner, F., Blockchain-oriented dynamic modelling of smart contract design and execution in the supply chain. *Int. J. Prod. Res.*, 58, 2184–2199, 2020.

18. Batwa, A. and Norrman, A., Blockchain technology and trust in supply chain management: A literature review and research oper agenda. *Supply Chain Manage. Int. J.*, 14, 203–220, 2021.
19. de Giovanni, P., Blockchain and smart contracts in supply chain management: A game theoretic model. *Int. J. Prod. Econ.*, 228, 107855, 2020.
20. Vacca, A., Di Sorbo, A., Visaggio, C.A., Canfora, G., A systematic literature review of blockchain and smart contract development: Techniques, tools, and open challenges. *J. Syst. Software*, 174, 110891, 2021.
21. Ante, L., Smart contracts on the blockchain—A bibliometric analysis and review. *Telemat. Inform.*, 57, 101519, 2021.
22. Rejeb, A., Keogh, J.G., Treiblmaier, H., Leveraging the Internet of Things and blockchain technology in supply chain management. *Future Internet*, 11, 161, 2019.
23. Saberi, S., Kouhizadeh, M., Sarkis, J., Shen, L., Blockchain technology and its relationships to sustainable supply chain management. *Int. J. Prod. Res.*, 57, 2117–2135, 2019.
24. He, L., Xue, M., Gu, B., Internet-of-Things enabled supply chain planning and coordination with big data services: Certain theoretic implications. *J. Manage. Sci. Eng.*, 5, 1–22, 2020.
25. Aslam, J., Saleem, A., Khan, N.T., Kim, Y.B., Factors influencing blockchain adoption in supply chain management practices: A study based on the oil industry. *J. Innov. Knowl.*, 6, 2, 124–134, 2021.
26. Dutta, P., Choi, T.M., Somani, S., Butala, R., Blockchain technology in supply chain operations: Applications, challenges and research opportunities. *Transp. Res. Part E Logist. Transp. Rev.*, 142, 102067, 2020.
27. Modi, D. and Zhao, L., Social media analysis of consumer opinion on apparel supply chain transparency. *J. Fash. Mark. Manage. Int. J.*, 5, 33, 1–18, 2020.
28. Meidute-Kavaliauskiene, I., Çiğdem, S., Vasiliauskas, A.V., Yıldız, B., Green innovation in environmental complexity: The implication of open innovation. *J. Open Innov. Technol. Mark. Complex.*, 7, 107, 2021.
29. Goswami, M., Daultani, Y., De, A., Decision modeling and analysis in new product development considering supply chain uncertainties: A multi-functional expert based approach. *Expert Syst. Appl.*, 166, 114016, 2021.
30. Choudhary, A., De, A., Ahmed, K., Shankar, R., An integrated fuzzy intuitionistic sustainability assessment framework for manufacturing supply chain: A study of UK based firms. *Ann. Oper. Res.*, 1–44, 2021.
31. Goswami, M., De, A., Habibi, M.K.K., Daultani, Y., Examining freight performance of third-party logistics providers within the automotive industry in India: An environmental sustainability perspective. *Int. J. Prod. Res.*, 58, 7565–7592, 2020.
32. Aslam, J., Saleem, A., Khan, N.T., Kim, Y.B., Factors influencing blockchain adoption in supply chain management practices: A study based on the oil industry. *J. Innov. Knowl.*, 6, 2, 124–134, 2021.

33. Batwa, A. and Norrman, A., Blockchain technology and trust in supply chain management: A literature review and research agenda. *Oper. Supply Chain Manage. Int. J.*, 14, 203–220, 2021.

27

Prediction of Seasonal Ailments Using Big Data: A Case Study

K. Indhumathi[1]* and K. Sathesh Kumar[2]

[1]Department of Computer Application, Kalasalingam Academy of Research and Education, Kalasalingam University, Krishnan Kovil, Tamil Nadu, India
[2]Department of Computer Science and Information Technology, Kalasalingam University, Krishnan Kovil, Tamil Nadu, India

Abstract

Infection outbreaks are influenced by the state of the environment and climatic conditions. Depending on the temperature, some illnesses might influence how quickly they spread. Dengue fever, for example, spreads more during the monsoon and winter seasons due to the fact that these are wet seasons. As a result, damp conditions can encourage the generation of dengue mosquitoes. Atmospheric variations are an important factor in the onset of a new illness. It also creates new infections. Big Data is an effective component for forecasting seasonal infectious illnesses that are affected by climatic variations. The researcher addresses three deadly diseases in this paper: COVID-19, dengue fever, and flu. Along with how the weather pattern can influence the spread of these three outbreaks, researchers also forecast which season has an impact on which disease. In addition, numerous disease prediction methods will be discussed. Based on this analysis, it appears that the big data approach is best for predicting seasonal infectious illnesses. This research offers numerous important insights toward seasonal aliment prediction using big data.

***Keywords*:** Big data, ailments, atmospheric variations, seasonal infectious, disease prediction

**Corresponding author*: indhu16aug@gmail.com

Sharmistha Ghosh, M. Niranjanamurthy, Krishanu Deyasi, Biswadip Basu Mallik, and Santanu Das (eds.) Mathematics and Computer Science Volume 1, (485–492) © 2023 Scrivener Publishing LLC

27.1 Introduction

Big data analytics is a leading approach that may be applied to a variety of fields, including intelligent agriculture, weather forecasting, education, and smart traffic systems. Furthermore, it is an important component of healthcare analytics. Big data is utilized in the healthcare system to forecast the consequences of infestations and to identify preventive illnesses by retaining patient records. Heat, humid, and moisture all increase the population of infectious organisms in the atmosphere.

The World Health Organization discovered that variations in climate and precipitation can increase insect growth, which spreads illnesses [1]. The most serious sickness is coronavirus. As compared to other outbreaks, coronavirus has an extremely high death rate. A total of 99,30,284 persons in India were afflicted. In addition, 1,44,075 people have died. COVID-19 disease is a type of pulmonary disease. As a result, in the coolest climate, it has severe consequences.

The *Aedes albopictus* mosquitoes are the primary vectors of dengue disease transmission. The productivity increase of dengue mosquitoes is affected by climatic variables (hot and rainy) [2]. The break bone sickness parasites have been harmed by a variety of factors, including the environment, plants, and human-created environments, such as suburbanization [3].

When opposed to the chilly environment, however, the warmer temperature zone has a significant impact on dengue infections [4]. High temperatures in the air have a similar effect on malaria microorganisms [5]. Epidemiological distribution is also affected by geographical coordinates [6].

At the same period, the weather factors of rainfall and wind speed contribute to the spread of influenza infections. The flu sickness was reduced as a result of the temperature and moisture [7]. The researcher can anticipate future seasonal infections using big data analytics based on previous seasonal pandemics and atmospheric conditions. This data is extremely useful to health analysts. This allows us to take sensible precautions ahead of time during a disease.

27.2 Related Works

This review [8] examines the characteristics of COVID-19 and introduces a new multidimensional reference model to discover the epidemic. This research [1] provides a biological model for identifying the dengue transmission in the future based on RCP scenario.

The study [9] describes how the break bone fever is related to environmental changes. The paper [10] uses multiple regression models to analyze the associations between the dengue incidences and warmer and humidity environs.

The study [11] presents the forecasting model of dengue based on environmental factors. It includes machine learning algorithms (decision tree, regression analysis, and support vector machine, artificial neural network) to discover the DENV occurrence.

The studies [12, 13] provide the new knowledge about handling the coronavirus situations. In this study, expert system and massive inputs enhance that corona circumstances, as well as brings the study of stopping corona outcomes. In this paper, five machine learning models are used to find the pathogen's detection using supervised learning algorithms.

XGBoosting model is the highly reliable model to identify the virus families when compared to other ML procedures. It found 100% of severe acute respiratory syndrome and mosquito-borne flavivirus, 84.61% of flu viruses and 75% Zaire ebolavirus also hepatitis, as well as discover 88% to 89% of COVID-19.

This paper [14] compares the effects of coronavirus with other epidemics and determines how the ailments modify the entire society as well as the world economy. The research [7] presents a geographically weighted regression model for identifying the spread of influenza in Vellore.

This model produces a positive correlation between flu prevalence and meteorological conditions, such as precipitation and air velocity. Moreover, this generates a negative correlation between the influenza outcomes and the temperature and wet climate.

The review [15] examines the Pearson correlation coefficient framework utilized to identify the relationship between the vector-borne disease and climate change. At the same time, the maximum heat is positively correlated with malaria outbreaks and minimum temperature, storm; raindrops are negatively correlated with malaria parasites.

The research work [16] proposed a new hybrid machine learning approach to predict dengue occurrence. In this study, the patient medical dataset is used. Dataset is divided into 70% of training dataset and 30% of testing data. This model proposed a hybrid method to forecast the dengue prediction.

The study [17–19] proves that the environmental condition plays a primary factor for occurring mosquito-borne diseases. As well as it causes the major impact for forecasting the dengue spread. The analysis [15, 18] shown that the heat and precipitation was directly connected

Table 27.1 Comparison of various disease prediction algorithms.

Sl. no.	Reference no.	Algorithms used	Works carried out
1.	[8]	-	It provides a multidimensional reference model to identify the pandemics based on environs
2.	[12, 13]	Random forest, XGBoost, multilayer perceptron, support vector machines, and logistic regression	The XGBoost model gives more accuracy when compared to other methods for forecasting the pathogen's detection
3.	[14]	-	It analyses the causes of COVID-19 with other epidemics
4.	[1]	-	Produce a biological model for finding a break bone fever occurrences in future depending on RCP factors
5.	[11]	Decision tree, regression analysis, and Support vector machine, artificial neural network	It presents a forecasting model for predicting dengue occurrence based on meteorological variables
6.	[15]	-	Pearson correlation coefficient is very helpful to found that the associations among vector-borne diseases and weather conditions
7.	[7]	Geographically weighted regression model	It found that the occurrence of influenza prevalence in Vellore

(Continued)

Table 27.1 Comparison of various disease prediction algorithms. (*Continued*)

Sl. no.	Reference no.	Algorithms used	Works carried out
8.	[9]	-	It analysis how the climatic conditions are affected the productivity of dengue mosquitoes
9.	[10]	Multiple regression model	It determines the connection among dengue incidences and temperature and wet atmospheric zone
10.	[16]	Hybrid method (combining decision tree and Random forest)	To forecast the dengue occurrence using the hybrid model based on patient health records and climatic factors. it gives 79% of average accuracy
11.	[17–19]	-	To determine the climatic factor is the major factor for forecasting the occurrence of dengue
12.	[18, 20–24]	-	The study found that the heat and rain are directly connected to the outbreaks of dengue

to the spread of outbreaks. Table 27.1 descripts that the summary of the related works.

27.3 Conclusion

Weather condition has directly or indirectly influenced the pathogens, as well as human-made surroundings (urbanization and deforestation) plays an important role to produce insects that create pandemics. Through this unexpected meteorological modification creates a new epidemic such as

COVID-19. Till to date, the world have not unable to recover from coronavirus impacts. Lots and lots of people have died around the world. Still there is no medicine discovered for preventing from COVID-19.

Corona can increase their outbreaks in winter season. The dengue mosquitoes are maximizing their productivity in summer and winter. The outbursts of influenza are very high in hot weather. The utilization of big data is helpful to predict the future seasonal infectious diseases by using the combination of climate change data and disease data set. This information is very valuable to the people for preventing these types of epidemics.

References

1. Fan, J.C. and Liu, Q.Y., Potential impacts of climate change on dengue fever distribution using RCP scenarios in China. *Adv. Climate Change Res.*, 10, 1–8, 2019.
2. Bhatt, S., Gething, P.W., Brady, O.J., The global distribution and burden of dengue. *Nature*, 496, 7446, 504–507, 2013.
3. Messina, J.P., Brady, O.J., Pigott, D.M., The many projected futures of dengue. *Nat. Rev. Microbiol.*, 13, 4, 230–239, 2015.
4. Friedrich, M.J., Global temperature affects dengue. *JAMA*, 320, 3, 227, 2018.
5. Le, V.V.P., Kumar, P., Ruiz, M.O., Predicting the direct and indirect impacts of climate change on malaria in coastal Kenya. *PLoS One*, 14, 2, 1–18, 2019.
6. Xu, L., Stigem, L.C., Chan, K., Climate variation drives dengue dynamics. *Proc. Natl. Acad. Sci. U. S. A.*, 114, 1, 113–118, 2017.
7. Lopez, D., Gunasekaran, M., Murugan, S.B., Kaur, H., Spatial big data analytics of influenza epidemic in Vellore, India. *IEEE International Conference on Big Data*, 2014.
8. Jia, Q., Guo, Y., Wang, G., Barnes, J.S., Big data analytics in the fight against major public health incidents (including COVID-19): A conceptual framework. *Int. J. Environ. Res. Public Health*, 17, 6161, 2020.
9. Naish, S., Dale, P., Mackenzie, S.J., McBride, J., Mengersen, K., Tong, S., Climate change and dengue: A critical and systematic review of quantitative modelling approaches. *BMC Infect. Dis.*, 14, 167, 2014.
10. Colon-Gonzalez, F.J., Lake, I.R., Bentham, G., Climate variability and dengue fever in warm and humid Mexico. *Am. J. Trop. Med. Hyg.*, 84, 757–763, 2011.
11. Siriyasatien, P., Chadsuthi, S., Jampachaisri, K., Kesorn, K., Dengue epidemics prediction: A survey of the state-of-the-art based on data science processes. *IEEE Access*, 6, 53757–53795, 2018.
12. Pham, Q.V., Nguyen, D.C., Huynh-The, T., Hwang, W., Pathirana, N.P., Artificial intelligence (AI) and big data for coronavirus (COVID-19)

pandemic: A survey on the state-of-the-arts. *IEEE Trans. Artif. Intell.*, 8, 130820–130839, 2020.
13. Lai, C.C., Shih, T.P., Ko, W.C., Tang, H.J., Hsueh, P.R., Severe acute respiratory syndrome coronavirus 2 (SARS-CoV-2) and coronavirus disease-2019 (COVID-19): The epidemic and the challenges. *Int. J. Antimicrob. Agents*, 55, 3, 105924, 2020.
14. Gossling, S., Scott, D., Hall, C.M., Pandemics, tourism and global change: A rapid assessment of COVID-19. *J. Sustain. Tour.*, 29, 1, 1–20, 2020.
15. Lopez, D. and Sekaran, G., Climate change and disease dynamics - A big data perspective. *Int. J. Infect. Dis.*, 45, 1–477, 2016.
16. Sarma, D., Hossain, S., Mittra, T., Bhuiya, A.M., Saha, I., Chakma, R., Dengue prediction using machine learning algorithms. *2020 IEEE 8th R10 Humanitarian Technology Conference (R10-HTC), IEEE*, 2020.
17. Mathulamuthu, S.M., Asirvadam, V.S., Dass, S.C., Gill, B.S., Loshini, T., Predicting dengue incidences using cluster based regression on climate data. Control system, computing and engineering (ICCSCE). *2016 6th IEEE International*, pp. 245–250, 2016, https://doi.org/10.1109/ICCSCE.2016.7893579.
18. Soemsap, T., Wongthanavasu, S., Satimai, W., Forecasting number of dengue patients using cellular automata model. *Proc International Electrical Engineering Congress*, 2014, https://doi.org/10.1109/iEECON.2014.6925876.
19. Hii, Y.L., Zaki, R.A., Aghamohammadi, N., Rocklov, J., Research on climate and dengue in Malaysia: A systematic review. *Curr. Environ. Health Rep.*, 3, 1, 81–90, 2016, https://doi.org/10.1007/s40572-016-0078-z.
20. Andrick, B., Clark, B., Nygaard, K., Logar, A., Penaloza, M., Infectious disease and climate change: Detecting contributing factors and predicting future outbreaks. *Geoscience and Remote Sensing, IGARSS'97*, 1997, https://doi.org/10.1109/IGARSS.1997.609159.
21. Ibrahim, F., Faisal, T., Mohamad Salim, M.I., Taib, M.N., Non-invasive diagnosis of risk in dengue patients using bioelectrical impedance analysis and artificial neural network. *Med. Biol. Eng. Comput.*, 48, 11, 1141–1148, 2010, https:// doi.org/10.1007/s11517-010-0669-z.
22. Korstanje, M. and George, B., Media constructions of fear in the outbreak of an epidemic disease: The case of dengue fever in Argentina. *Int. J. Emerg. Serv.*, 5, 1, 95–104, 2016, https://doi.org/10.1108/IJES-01-2016-0001.
23. Burattini, M.N., Chen, M., Chow, A., Coutinho, F.A.B., Goh, K.T., Lopez, L.F., Ma, S., Massad, E., Modelling the control strategies against dengue in Singapore. *Epidemiol. Infect.*, 136, 3, 309–19, 2008, https://doi.org/10.1017/S0950 268807008667.
24. Mochammad, C.R., Achmad, B., Tri, H., Comparison of Montecarlo linear and dynamic polynomial regression in predicting dengue fever case. *Knowl. Creat. Intell. Comput.*, 2016, https://doi.org/10.1109/KCIC.2016.7883649.

28

Implementation of Tokenization in Natural Language Processing Using NLTK Module of Python

Vikash Kumar Mishra[1], Abhimanyu Dhyani[2*], Sushree Barik[3] and Tanish Gupta[3]

[1]Galgotias University, Gautam Buddha Nagar, Uttar Pradesh, India
[2]Indian Institute of Technology, Jodhpur, Rajsthan, India
[3]Christ (Deemed to be) University, Delhi NCR, India

Abstract
With the advancement of technologies, now it is possible to analyze the large amount of unstructured text circulated online with various tools and methods for understanding the changes as well to infer meaningful insights from the text data. In this work, the aim is to understand how Python can be used for text analytics by the help of various libraries available in it. The natural language processing (NLP) is being used to analyze and synthesize natural language and speech in Python.

***Keywords*:** Natural language processing (NLP), natural language toolkit (NLTK), tokenization

28.1 Introduction

According to one estimate, only a small fraction of today's data is structured. This includes everyday communication like speaking, tweeting, and sending messages through various platforms, such as WhatsApp, email, Facebook, Instagram, and text. For this data, the most common format is text, which is

**Corresponding author*: dhyani.2@iitj.ac.in

Sharmistha Ghosh, M. Niranjanamurthy, Krishanu Deyasi, Biswadip Basu Mallik, and Santanu Das (eds.) Mathematics and Computer Science Volume 1, (493–504) © 2023 Scrivener Publishing LLC

deeply unstructured. We must analyze the text data in order to gain meaningful insights. It can be done manually with one person and an excel spreadsheet but at large scale, this can be time-consuming, inefficient and inaccurate.

In order to be able to manage content in a data-driven way, text analysis aims to convert text into machine-readable data. We need to customize text mining based on our requirements. The machine learns and analyzes the unstructured data in order to identify patterns and trends.

It is then that text analytics helps make sense of these data after text analysis "preparates" the content. Natural language processing (NLP) is the process that helps us to convert unstructured to structured data by the help of existing metadata. Further, data visualization is used to covey our findings to the greater audience.

In text analytics, insights are transformed into quantitative data using data visualization tools. In text analysis, perspectives are gained through qualitative analysis.

With text analysis, tasks are performed such as finding the number of words in the article, the length of the sentence, and whether specific words appear in it or not. It also helps identify sentiment, locate entities in the article, and classify the blog or article.

It is possible, for instance, to analyze customer feedback in real time, such as the open-ended questions asked in Customer Satisfaction Score (CSAT) surveys, so that we can make improvements based on customer feedback We can utilize text analysis to discover the reasons for low CSAT scores, for example, in order to identify what is causing these results. The information we obtain allows us to make smart decisions that help increase customer satisfaction. Our customers may prefer to speak to a real person rather than a chatbot, or they may have been kept waiting too long for their first response.

28.2 Background

A natural language processing component of text mining allows a machine to "read" texts by performing linguistic analysis. Language processing is part of computer science and artificial intelligence, and it has to do with human language. For example, auto summarization, part-of-speech tagging, disambiguation, chunking, as well as natural language understanding and recognition, are some of the methodologies that it employs for understanding ambiguous sentences [1].

In the 1970s, information retrieval was started. It mostly concerns text retrieval. Web search engines are giant information retrieval systems.

The term data mining derives from three research areas: natural language processing, data mining, and information retrieval.

Research in the late 1990s began using text as data, giving rise to text mining. Data mining normally uses structured data, such as database tables. Data mining methods in the early days of text mining were mostly based on data mining and machine learning algorithms that did not utilize natural language processing (NLP) techniques, such as parsing, part-of-speech tagging, summary extraction, etc.

As a result, we can conclude that Natural Language Processing (NLP) is a much older field. It began in the 1950s with the purpose of allowing computers to understand human language [2].

28.3 Study Area and Data Set

The study area for the present research work is not specific. We have taken data from various sources using Kaggle platform, which is used in this study. Our dataset includes sentences and paragraphs, which are suitable for the present study.

28.4 Proposed Methodology

Proposed methodology for this study is something as follows:

The first step is the installation of Python software. Then, we need to install NLTK library, for building Python programs to work with human language data, which provides easy to use interface in Python.

After the installation of NLTK library, we need to open NLTK download manager as shown in Figure 28.1 below by typing the command

“import nltk
nltk. download()”

The NLTK downloader has various packages, which are used for performing different functions on the text based on our requirements like for tokenization, lexicon normalization, POS tagging, etc.

After installation of NLTK packages, we use it for tokenization as shown in Figure 28.2. It is the first step in text analytics. It is the process of breaking down a text paragraph into smaller chunks, such as words or sentence. The one particularly shown below is the sentence tokenization.

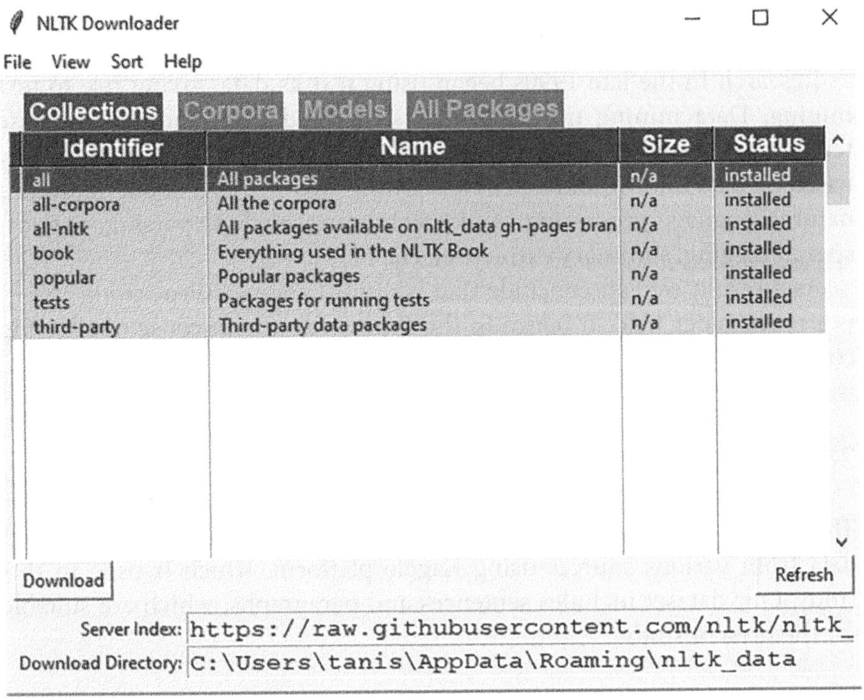

Figure 28.1 NLTK downloader.

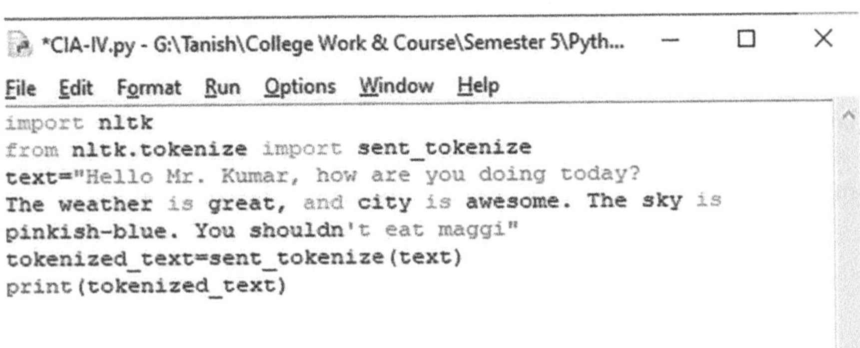

Figure 28.2 Sentence tokenization of paragraph.

After sentence tokenization, Figure 28.3 displays the code for work tokenization. It breaks paragraph or sentence into words.

Similarly, Figure 28.4 highlights the use of feature of word frequency distribution. Text analysis artificial intelligence service helps us determine which words or series of words reoccur by frequency. For example, doing a

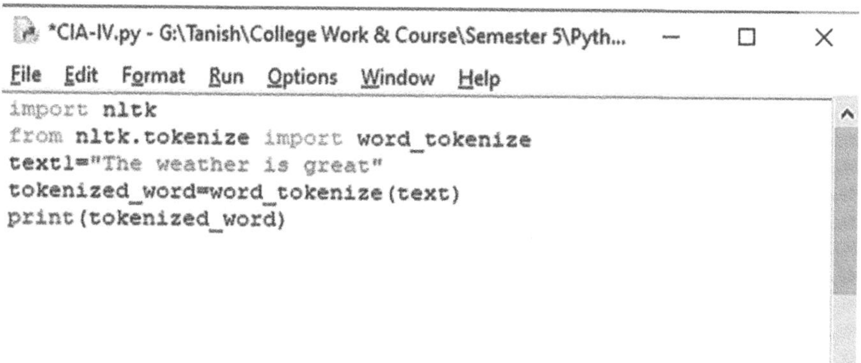

Figure 28.3 Word tokenization.

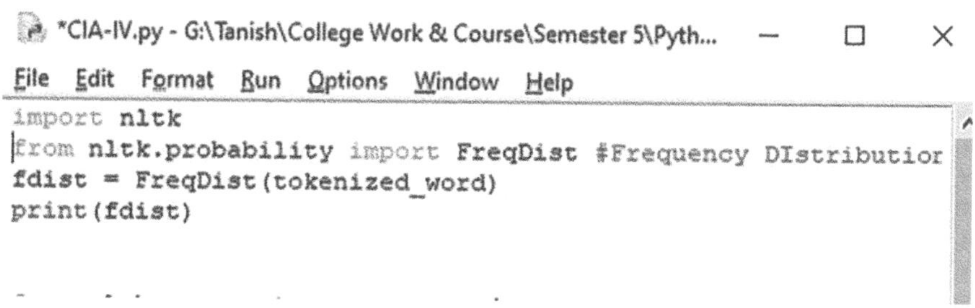

Figure 28.4 Word freqeuncy distribution.

find in Google docs, one can see that when you type in a word in a search box, it gives us the frequency of times the term is used in the text.

Next, we have discussed about removal of unnecessary words from sentence/paragraph in text analytics, which do not have any valuable output. These words are called stop words. Stop words are words that may not carry any valuable information, like articles ("the"), conjunctions ("and"), or propositions ("with"). "the" and "a" are the most common words in our dataset and does not tell us much about the data. It is showcased in Figure 28.5.

Secondly, stemming and lemmatization were performed. Stemming is a linguistic process that reduces a word to its root word or removes its derivative affix. The term connection, connected, and connecting word reduce to the word "connect," whereas lemmatization reduces words to their base word. It uses vocabulary and morphological analysis to transform root words.

Generally, stemmers work on an individual word without considering its context. For example, the word "good" has the lemma "better." Stemming misses this because it involves finding the word in a dictionary (Figure 28.6).

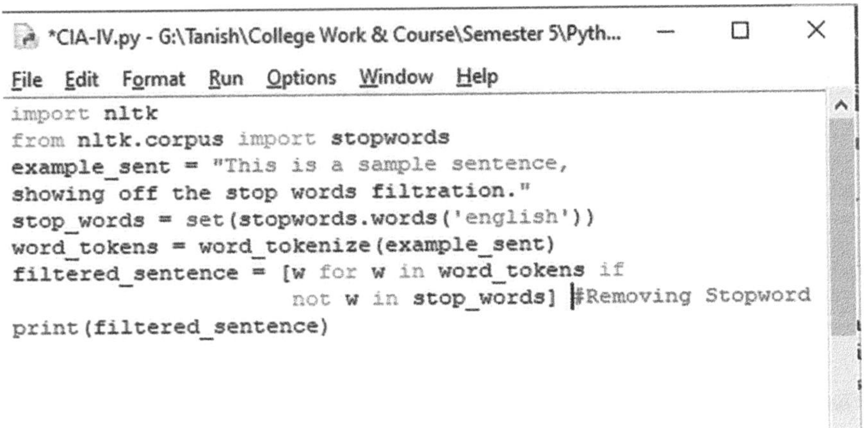

Figure 28.5 Stop word removal.

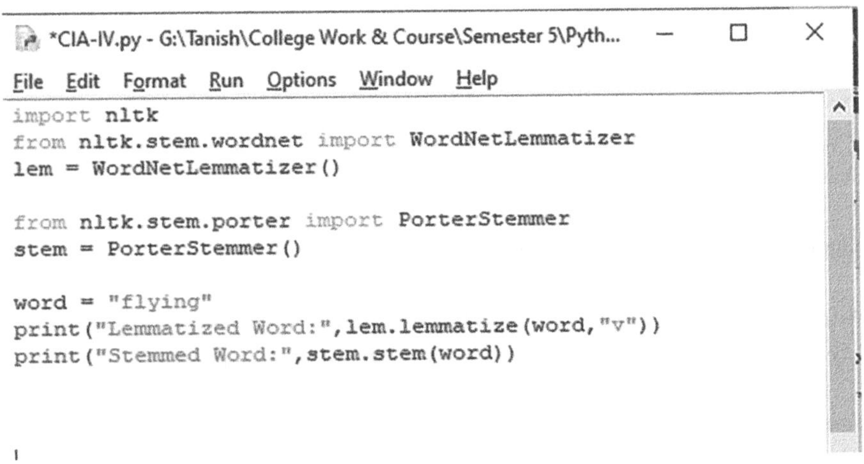

Figure 28.6 Lemmatization and stemming.

Its primary purpose is to identify a word's grammatical group, such as a noun, verb, preposition, adverb, conjunction, or adverbial phrase based on its contextPOS tagging assigns a corresponding tag to words that contain relationships within the sentence [3–10].

28.5 Result

The results of the abovementioned performed tests are shown below as an output of coding on Python.

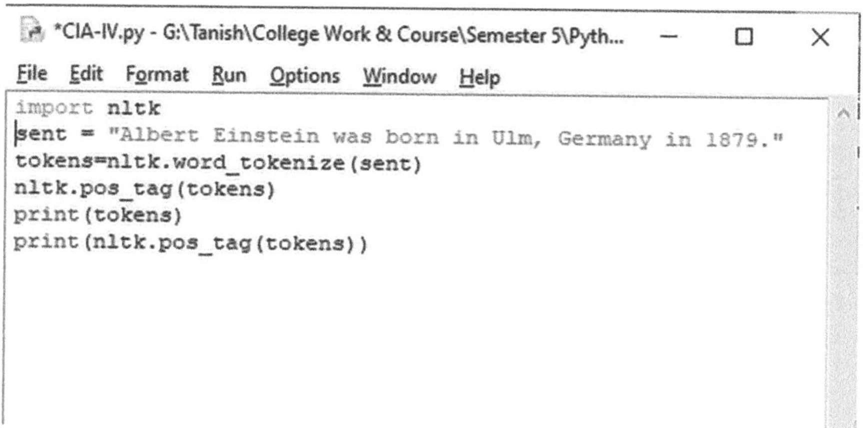


```
import nltk
sent = "Albert Einstein was born in Ulm, Germany in 1879."
tokens=nltk.word_tokenize(sent)
nltk.pos_tag(tokens)
print(tokens)
print(nltk.pos_tag(tokens))
```

Figure 28.7 Part of speech tagging.

In Figure 28.7, we can see the Part of speech tagging. In Figure 28.8, we can see the paragraph is broken into small sentences. Similarly, in Figure 28.9, the sentence is broken into words.

In Figure 28.10, we can see there are four samples and outcomes of the dataset.

In Figure 28.11, nonuseful words are removed, which will not etch us any output, and so on and so forth.

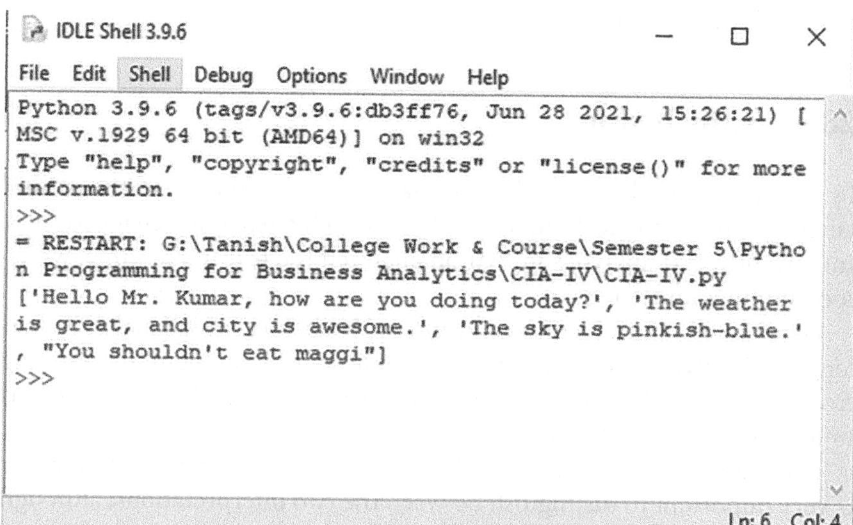


```
Python 3.9.6 (tags/v3.9.6:db3ff76, Jun 28 2021, 15:26:21) [
MSC v.1929 64 bit (AMD64)] on win32
Type "help", "copyright", "credits" or "license()" for more
information.
>>>
= RESTART: G:\Tanish\College Work & Course\Semester 5\Pytho
n Programming for Business Analytics\CIA-IV\CIA-IV.py
['Hello Mr. Kumar, how are you doing today?', 'The weather
is great, and city is awesome.', 'The sky is pinkish-blue.'
, "You shouldn't eat maggi"]
>>>
```

Figure 28.8 Sentence tokenization of paragraph.

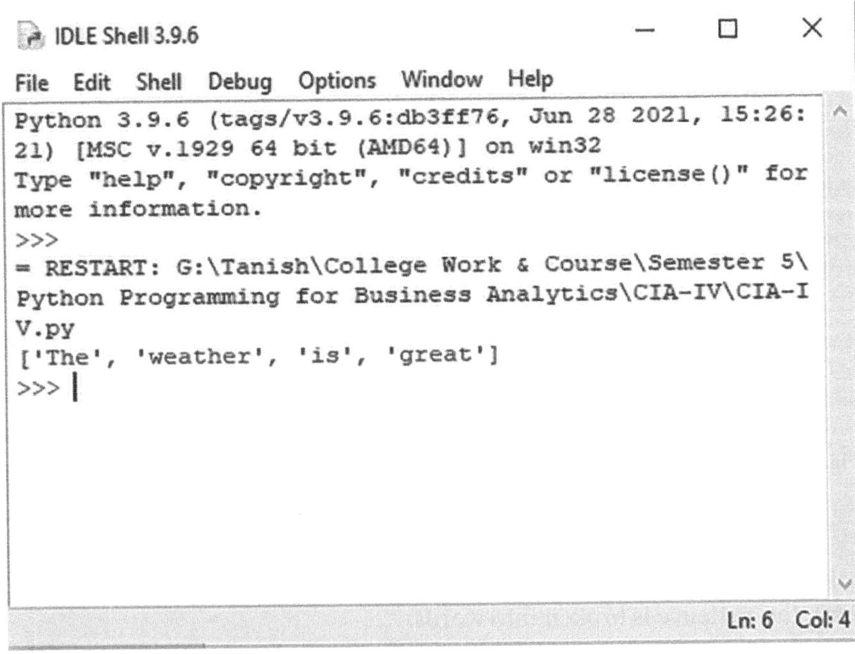

Figure 28.9 Word tokenization.

From all this, we can infer that performing test analysis on Python would be very easy for big data sets. It is consistent in terms of its results until any specific changes are made by us [11].

28.6 Conclusion

In the current study, we took small datasets and applied various natural language processing techniques on them, which fetched us reliable results. Importantly, text analytics in Python analyzed the text in real time, which took minutes and not days.

Data can be vague, inconsistent, and contradictory, which makes it challenging. Analyzing texts is further complicated by ambiguities resulting from differences in syntax, regional dialects, slang, and technical terms unique to a particular industry.

For example, an "apple" can refer to a fruit or a company, and context is very important to distinguish between the two interpretations, although we can try to mitigate it with the help of few commands.

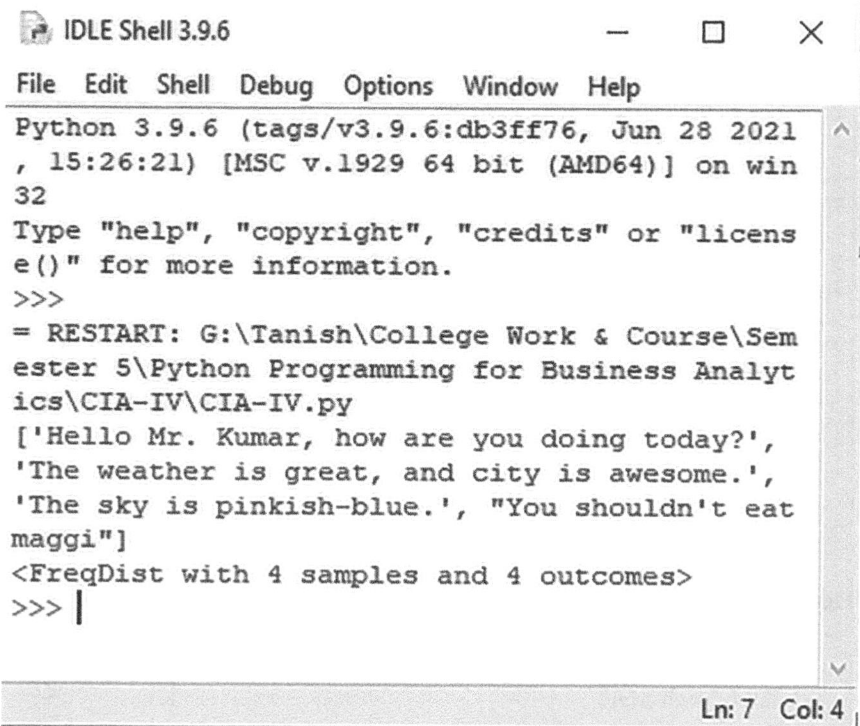

Figure 28.10 Word freqeuncy distribution.

Overall, we would say that it is an efficient and effective tool which can be used on large scale data to reduce human error and get the results of text analytics quickly, which would eventually help in fulfilling consumer needs [12–15]. In Figure 28.12, we can see Lemmatization and stemming. In Figure 28.13, we can see the Part of speech tagging.

28.7 Acknowledgment

The first and second authors are thankful to Kaggle Platform and Christ (deemed to be) University, Delhi NCR, India.

Conflicts of Interest/Competing Interests

No potential conflict of interest was reported by the authors.

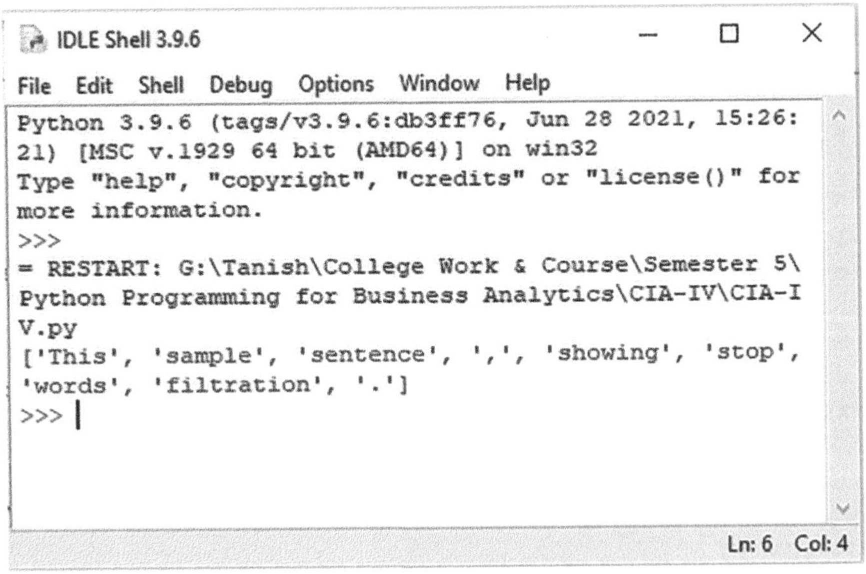

Figure 28.11 Stop word removal.

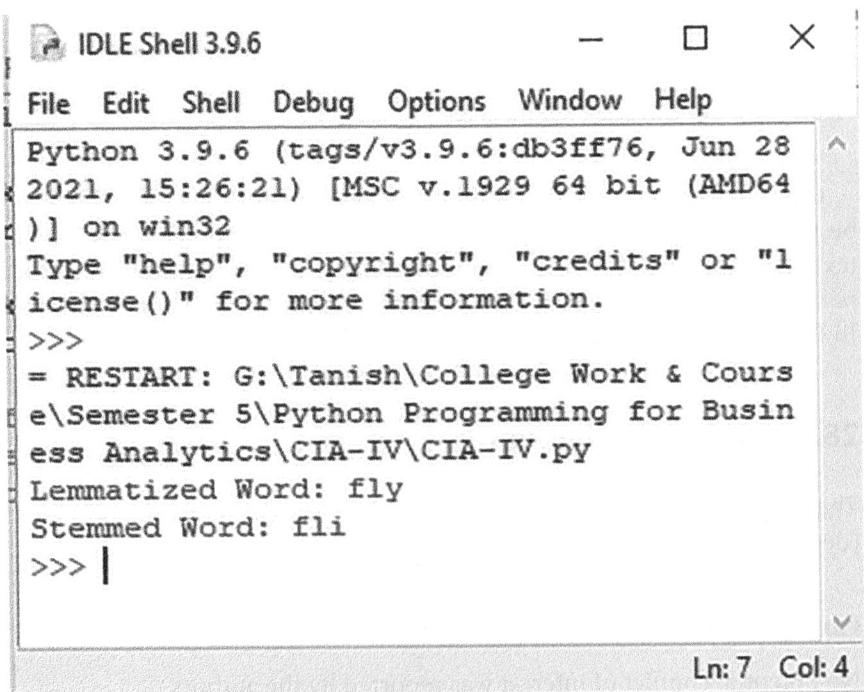

Figure 28.12 Lemmatization and stemming.

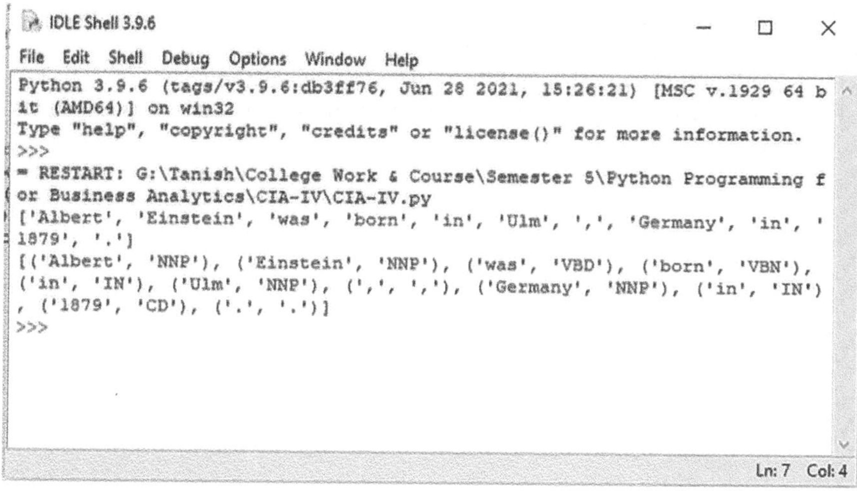

Figure 28.13 Part of speech tagging.

Availability of Data and Material

The authors show their gratitude to the https://www.kaggle.com/ and https://www.python.org/ for providing software and dataset for the research work.

References

1. https://towardsdatascience.com/getting-started-with-text-analysis-in-python-ca13590eb4f7.
2. https://www.datacamp.com/community/tutorials/text-analytics-beginners-nltk.
3. https://github.com/dipanjanS/text-analytics-with-python.
4. https://www.geeksforgeeks.org/text-analysis-in-python-3/.
5. https://www.analyticsvidhya.com/blog/2018/02/the-different-methods-deal-text-data-predictive-python/.
6. https://www.ibm.com/cloud/learn/text-mining.
7. https://www.kdnuggets.com/2020/05/text-mining-python-steps-examples.html.
8. https://link.springer.com/book/10.1007/978-1-4842-4354-1.
9. https://github.com/dipanjanS/text-analytics-with-python/tree/master/New-Second-Edition.
10. https://github.com/dipanjanS/text-analytics-with-python.
11. https://monkeylearn.com/text-analysis/.

12. https://searchbusinessanalytics.techtarget.com/definition/text-mining.
13. https://en.wikipedia.org/wiki/Text_mining#Text_analytics.
14. https://github.com/dipanjanS/text-analytics-with-python.
15. https://www.kdnuggets.com/2020/05/text-mining-python-steps-examples.html.

29

Application of Nanofluids in Heat Exchanger and its Computational Fluid Dynamics

M. Appadurai[1], E. Fantin Irudaya Raj[2*] and M. Chithambara Thanu[1]

[1]Department of Mechanical Engineering, Dr. Sivanthi Aditanar College of Engineering, Tamil Nadu, India

[2]Department of Electrical and Electronics Engineering, Dr. Sivanthi Aditanar College of Engineering, Tamil Nadu, India

Abstract

Nanofluids are colloids containing a base fluid (mostly water) and nanoparticles (polymers, metals, metal oxides, carbon nanotubes). It can be viewed as the future heat transfer fluid in a diverse array of plate heat exchangers. Due to the presence of dispersed nanoparticles with a high thermal conductivity, they are projected to perform better thermally than normal fluids. Evaluating the improvement of heat transfer resulting by the use of nanofluids has lately become a focus of attention for a large number of researchers. For the evaluation of heat transfer enhancement using nanofluids can be studied by using Computational fluid dynamics (CFD). It is a subset of fluid dynamics that uses data structures and numerical analysis to investigate and deal with problems involving fluid flows. It allows the implementation of fluid motion equations to a wide variety of complex situations, yielding quantitative as well as qualitative predictions. Computers are used to simulate the free-stream flow of a fluid and its interaction with boundary constraint surfaces. The present chapter discusses about basics of CFD and it also used to determine which Nanofluid is most efficient at transferring heat in a solar flat plate collector by taking into account a variety of physical factors such as density, specific heat, thermal conductivity, viscosity, and so on. Ansys-Fluent software package is utilized for this study. Working fluids like Al2O3 are designed to flow via a single flat

**Corresponding author*: fantinraj@gmail.com

Sharmistha Ghosh, M. Niranjanamurthy, Krishanu Deyasi, Biswadip Basu Mallik, and Santanu Das (eds.) Mathematics and Computer Science Volume 1, (505–524) © 2023 Scrivener Publishing LLC

plate collector. The simulations were run by changing the flow rate, a volumetric fraction of nanoparticles, and other parameters in the model to obtain the best power output, efficiency, useable heat, and heat losses, so that the same conditions may be applied to the actual system.

Keywords: Vanofluids, computational fluid dynamics, numerical analysis, heat exchanger, nanoparticles

29.1 Computational Fluid Dynamics

In the digital era, the usage of high-speed computers is rapidly increasing to execute many engineering and statistical tools. The computational fluid dynamics concept is popularized because of its higher productivity and better prediction of results [1]. The characteristics of several variable results are viewed in the contour plot from the computational fluid dynamics output. Basically, the accuracy of the computational fluid dynamics result depends on the proper boundary condition input and interpretation of attained results. The engineers should be concerned with the basic physics behind the application for the proper input of boundary conditions. The governing equation of the fluid flow phenomena is written in partial differential equation form and then it is converted into algebraic equations. The numerical iterations are performed to solve the algebraic problem to predict the several parameter values under the given boundary conditions [2].

29.1.1 Continuity Equation

The continuity equation for control volume is computed by means of Eulerian techniques. The control volume of the computational domain is considered as fixed and the fluid flows across the boundary of the given volume.

$$\frac{\partial \rho}{\partial t} + \nabla.(\rho V) = 0 \tag{29.1}$$

In the continuity equation, the density of the fluid is denoted as and fluid velocity is mentioned as V. The density change with respect to time is represented in the first portion of the equation and the rate of change of mass flux crossing the boundary is represented in the second portion of the equation.

Let x, y, z are considered as the parameters of the velocity vector and then the general continuity equation is rewritten in the Cartesian coordinate form as

$$\frac{\partial \rho}{\partial t}+\frac{\partial}{\partial x}(\rho u)+\frac{\partial}{\partial y}(\rho v)+\frac{\partial}{\partial z}(\rho w)=0 \tag{29.2}$$

For incompressible fluid the continuity equation is simplified as

$$\frac{\partial u}{\partial x}+\frac{\partial v}{\partial y}+\frac{\partial w}{\partial z}=0 \tag{29.3}$$

29.1.2 Momentum Equation

The equation of momentum is derived as follows for the mass of fluid passing through the fixed control volume based on Newton's second law. The rate of momentum increase for unit volume is given in the first term and the rate of momentum loss for unit volume due to convection is given in the second term of the momentum equation on the left-hand side. The body forces acting on unit volume is represented in the first part and the surface forces act on unit volume is denoted in the second part of the momentum equation right hand. Generally, the gravitational force is considered as the body force.

$$\frac{\partial}{\partial t}(\rho V)+\nabla.\rho VV=\rho f+\nabla.\Pi_{ij} \tag{29.4}$$

The above said momentum equation is applicable for non-continuum and continuum type fluid flow. In continuum flow, the stress at a region is directly proportional to the rate of strain. Generally, gas and most liquids are considered as continuum flow. The general deformation law is derived with these conditions and the stress tensor is presented as

$$\Pi_{ij}=-p\delta_{ij}+\mu\left(\frac{\partial u_i}{\partial x_j}+\frac{\partial u_j}{\partial x_i}\right)+\delta_{ij}\mu'\frac{\partial u_k}{\partial x_k} \qquad i,j,k=1,2,3 \tag{29.5}$$

u and x denote the velocity vector and position vector components, respectively. The coefficient of viscosity is denoted as μ and the second coefficient of viscosity is denoted as 'μ'. The two coefficients of viscosity are interlinked by the bulk viscosity coefficient, k = 2μ/3+μ'. For shock waves study itself, the k has reasonable value and the remaining factors the k is negligible. Substitute k=0 in the stress tensor, it is rewritten as

$$\Pi_{ij} = -p\delta_{ij} + \mu\left[\left(\frac{\partial u_i}{\partial x_j} + \frac{\partial u_j}{\partial x_i}\right) + \frac{2}{3}\delta_{ij}\frac{\partial u_k}{\partial x_k}\right] \quad i,j,k = 1,2,3 \quad (29.6)$$

Thus, the stress tensor equation is represented in two terms, i.e., Π_ij = - pδ_ij+τ_ij where the viscous shear stress is denoted as τ.

$$\tau_{ij} = \mu\left[\left(\frac{\partial u_i}{\partial x_j} + \frac{\partial u_j}{\partial x_i}\right) + \frac{2}{3}\delta_{ij}\frac{\partial u_k}{\partial x_k}\right] \quad i,j,k = 1,2,3 \quad (29.7)$$

Substituting the consequent equations on the momentum equation, finally, it is modified as

$$\rho\frac{DV}{Dt} = \rho f - \nabla p + \frac{\partial}{\partial x_j}\left(\mu\left(\frac{\partial u_i}{\partial x_j} + \frac{\partial u_j}{\partial x_i}\right) - \frac{2}{3}\delta_{ij}\mu\frac{\partial u_k}{\partial x_k}\right) \quad (29.8)$$

The above equation is called the Navier-stokes equation.

For the fluid flow is incompressible, the viscosity is not varied and so the equation is simplified from the general equation as

$$\rho\frac{DV}{Dt} = \rho f - \nabla p + \mu\nabla^2 V \quad (29.9)$$

29.1.3 Energy Equation

The energy equation is derived based on the thermodynamics first law for the fixed control volume. The computational domain energy equation is represented as

$$\frac{\partial E_i}{\partial t}+\nabla.E_iV=\frac{\partial Q}{\partial t}-\nabla.q+\rho f.V+\nabla.(\Pi_{ij}.V) \tag{29.10}$$

The total energy of the system per unit volume is denoted as Et,

$$Et=\rho\left(e+\frac{V^2}{2}+\text{potential energy}+\cdots\right) \tag{29.11}$$

This energy equation contains the rate of total energy enhancement, loss of total energy by convection, heat generated due to internal heat generation, and loss of heat due to conduction per unit volume.

The general energy equation in the Cartesian form is represented as

$$\begin{aligned}\frac{\partial E_i}{\partial t}-\frac{\partial Q}{\partial t}-\rho(f_xu+f_yv+f_zw)+\frac{\partial}{\partial x}(E_iu+pu-u\tau_{xx}-v\tau_{xy}-w\tau_{xz}+q_x)\\+\frac{\partial E_i}{\partial t}(E_iv+pv-u\tau_{xy}-v\tau_{yy}-w\tau_{yz}+q_y)\\+\frac{\partial}{\partial z}(E_iw+pw-u\tau_{xz}-v\tau_{yz}-w\tau_{zz}+q_z)=0\end{aligned} \tag{29.12}$$

The substitution of continuity equation in the general energy equation, the equation is modified as

$$\rho\frac{DE_i/\rho}{Dt}=\rho\frac{De}{Dt}+\rho\frac{DV^2/2}{Dt} \tag{29.13}$$

Consider the thermal conductivity coefficient is constant and the fluid flow type is incompressible, the energy equation is simplified into

$$\rho\frac{De}{Dt}=\frac{\partial Q}{\partial t}+k\nabla^2T+\varphi \tag{29.14}$$

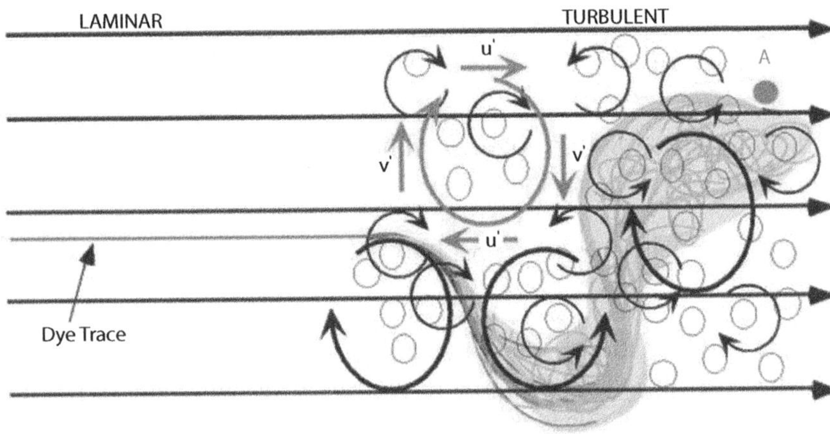

Figure 29.1 Laminar and turbulent flow.

29.1.4 Equations for Turbulent Flows

The fluid flow type is one of the important parameters in the computational fluid dynamics. The behavior of the fluid flow and heat transfer is completely different for the laminar and turbulent type fluid flow. The nature of the fluid flow generally depends on the fluid inertia and viscous force. In general, the Reynolds number is used to find the nature of fluid flow. The ratio of inertia force to viscous force of the fluid flow is said to be Reynolds number. The critical Reynolds number for transition state of the fluid flow is varied for different applications such as flow over flat plate, flow over sphere, internal flow, flow across bank of tubes, ground and surface flow. Figure 29.1 shows the laminar and turbulent boundary condition. The apparent turbulent flow equation is given as

$$\rho\frac{D\overline{u_1}}{Dt}=-\frac{\partial\overline{p}}{\partial x_i}+\frac{\partial\overline{(\tau_{ij})}_{lam}}{\partial x_j}+\frac{\partial\overline{(\tau_{ij})}_{turb}}{\partial x_j} \tag{29.15}$$

29.2 Nanofluids

In the modern compact appliances, the higher end technology is implemented for effective performance of equipment at all boundary conditions. Nanofluids are one of the good working fluids used in electronic devices, such as microsized devices, electronics cooling, and cameras. The higher temperature applications, such as solar flat plate collector, solar air collectors, space heating, internal combustion engine cooling, drilling fluids, medical fields, flue gas recovery unit, flue gas recovery, thermal energy storage, microwave tubes, and all major heat transfer applications [3]. Nanofluids are manufactured by the suspension of metal nanosized particles in conventional water. The higher thermal conductivity of the metals in nanosize form enhances the nanofluids thermal conductivity and diffusivity of the working fluid. The main benefits of the nanofluids utilization are larger heat diffusion due to the Brownian motion of the nanoparticles, higher heat transfer coefficient because of its higher thermal conductivity and final adjustable fluid thermal properties by changing its volume fraction [4].

However, there is a development in the heat transfer characteristics on one side, also there are limitations of having reasonable pressure drop based on the nanoparticles volume fraction. To overcome the pressure drop of the working fluid, the pump capacity must be increased. The production cost of nanosized particles and minimum specific heat capacities are the next level of limitations. The optimum point for the nanoparticles volume fraction is calculated by the researchers for every nanofluid in the global research. In this optimal volume fraction of nanofluids, the working fluid has superior thermal properties and reasonable pressure drop.

29.2.1 Viscosity

The resistance force offered by the fluid to change its shape is called viscosity. Several theories are proposed to calculate the viscosity of the nanofluid by considering the fraction of nanoparticles suspended into the base coolant. Of these, the Brinkman model is familiar, and it is extensively used in the numerical simulation of nanofluids studies. The Brinkman model to calculate the viscosity of nanofluids is given as

$$\mu_{nf} = \frac{\mu_f}{(1-\varphi)^{2.5}} \tag{29.16}$$

29.2.2 Density

The mass of the fluid for its unit volume is called density. The equation to find density of nanofluids is estimated by Vajjha [5], which is used in the fluid flow simulation to predict the flow behavior. If the nanosize metal particles density is indexed with suffix s and the base fluid is indexed with suffix bf, then the density of nanofluids is given by

$$\rho_{nf} = (1 - \varphi)\rho_{bf} + \varphi\rho_{s} \tag{29.17}$$

29.2.3 Heat Capacity

The multiplication of density with specific heat (C) is called heat capacity. The heat capacity of the nanofluid is calculated by

$$(\rho C_p)_{nf} = (1 - \varphi)(\rho C_p)_{bf} + \varphi(\rho C_p)_s \tag{29.18}$$

29.2.4 Thermal Conductivity

The general thermal conductivity of the nanofluid is derived from the volume fraction (φ) of the base fluid and nanofluids. The conductive heat transfer is calculated with the use of thermal conductivity property. The heat is transferred from the higher temperature region to lower temperature region due to the collision of nearby atoms or molecules is said to be conductive heat transfer. Normally, the thermal conductivity of base fluid is less but the thermal conductivity of metals is usually high. The net thermal conductivity of the nanofluids is calculated by

$$K_{nf} = \frac{K_p + 2K_{bf} - 2(K_{bf} - K_p)\varphi}{K_p + 2K_{bf} + 2(K_{bf} - K_p)\varphi} K_{bf} + 5\times10^4 \beta\varphi\rho_{bf}\sqrt{\frac{KT}{\rho_p d_p}} f(T,\varphi) \tag{29.19}$$

29.3 Preparation of Nanofluids

The nanofluid has superior thermal characteristics and is widely used in several applications. To attain superior thermal behavior, the preparation of nanofluid is the main parameter to be concentrated. The nanofluids can be obtained by suspension of nanosized particles, nanofibers, nanotubes and nanosheets with the base fluids. Normally, nanofluids have two phases of materials, i.e., solid phase and base liquid phase substances since agglomeration is the main problem arising from nanofluid utilization.

29.3.1 One-Step Method

The one-step nanofluid preparation method [6] uses the vapor digest system to disperse the nanoparticles throughout the base fluid. The methodology of one step preparation method is shown in Figure 29.2. The base fluid may be water, oil, refrigerant, any solvent or coolant. The bulk quantity of nanofluids can be prepared by this method. The dispersion of nanoparticles throughout the base fluid is uniform. The separate preparation of nanoparticles in dry form and suspension of nanoparticles with base fluid is eliminated in this process. The agglomeration of nanosize elements in one region is reduced, and so, the stability of nanofluids is high. The synthesis and processing cost is very high. The nanoparticles are produced and dispersed into the base fluid by chemical method in one step of processing. For example, the silver nanofluids is prepared by Korantin method, which coordinated to

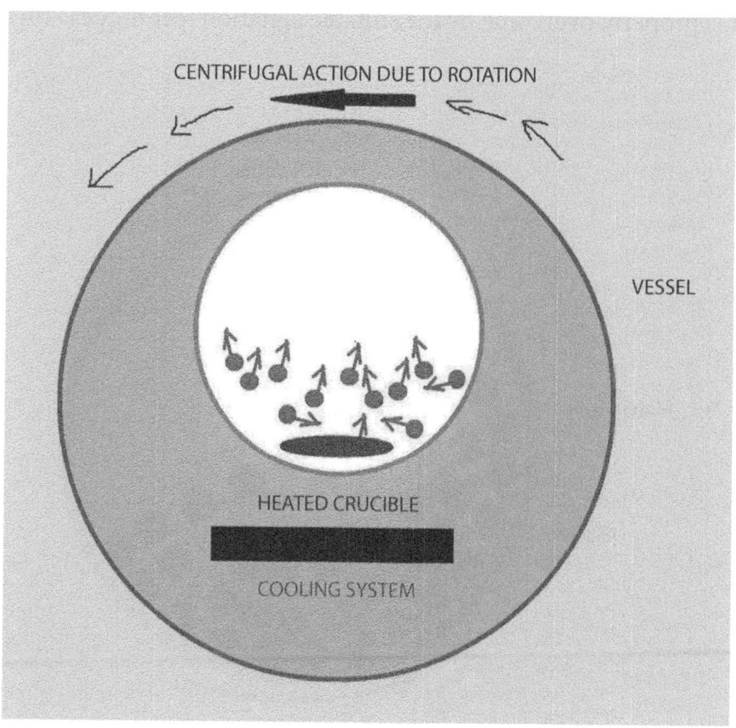

Figure 29.2 One step nanofluid preparation technique.

the silver particle surfaces via two oxygen atoms forming a dense layer around the particles.

29.3.2 Two-Step Method

This is the simple and most popular method for the preparation of nanofluids in lower quantum and larger commercial scale. First of all, the dry microsized particles are converted into nanosized particles though ball milling (Figure 29.3). Not only the nanoparticles but also other form of nanoelements is used in this preparation method. Then, the nanopowder is mixed with base fluids. The second step is the proper dispersion of nanoparticles by several methods, such as ultrasonic, magnetic force agitation [7]. Figure 29.4 illustrates the alumina nanofluid at different volume fraction with water. The suspension of nanoparticles into the base fluid is based on the period of agitation and the equipment capacity. The surfactant can also add to improve the stability of metal elements in the nanofluids for longer time. The surfactants addition is preferred especially for high-temperature application. The agglomeration is the main demerits, and it is properly overcome by the suitable agitation techniques. This is the

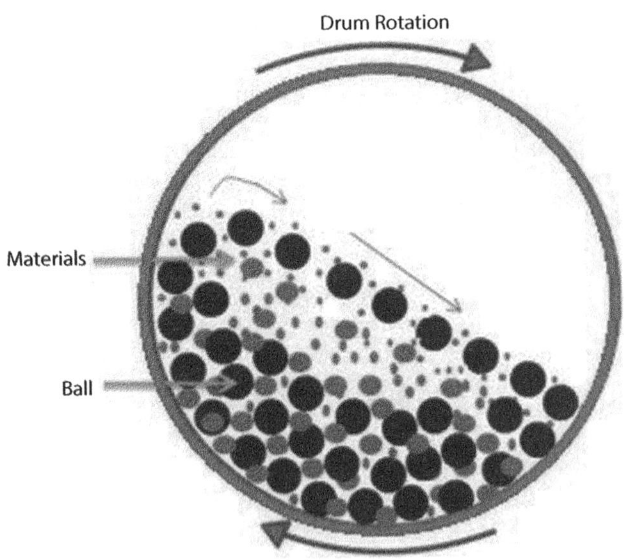

Figure 29.3 Ball milling.

Figure 29.4 Nanofluids.

most economic method of nanofluid preparation in the small scale and commercial quantity scale.

29.3.3 Nanofluids Implementation in Heat Exchanger

The heat exchanger is the device that transfers the thermal energy from hot fluid to cold fluid without mixing. The heat exchangers are widely implemented in major industries such as oil and gas companies, automotive sectors, aircrafts, and dairy product companies. The heat exchanger is used in different pressure and temperature boundary conditions, the fluid flow may be single phase or multiphase condition. The performance of the heat exchanger is improved by the optimization of the design variables and further the heat exchanger characteristics is enhanced by using nanofluids. The utilization of nanofluids in the heat exchanger increases its capacity and thermal energy transfer rate. Thus, the economic factor of the heat exchanger is improved. The compact size of the heat exchanger further reduces the pressure drop of the fluid flow across the heat exchanger. The pump work is reduced due to cutting the length of the exchanger tube due to the compact design.

Radiator is one of the best examples for heat exchanger. The radiator picture is shown in Figure 29.5. In the automotive vehicle, the engine is cooled by the conventional water to maintain the moderation temperature around the engine components. Of the total amount of heat generated, only a fraction of heat is converted as useful work, remaining waste heat is transferred through exhaust gas, cooling circuit, and lubricating oil. The heat absorbed by cooling water from the cylinder wall is often cooled

Figure 29.5 Radiator.

with the help of radiators. Improper cooling leads to increase of temperature on the engine components. At higher temperature, the viscosity of the lubricating oil decreases, and engine components expands which cause the engine to break down. An efficient cooling system is required for maintaining the engine in running condition. The glycol additives are added in the coolant water to increase the boiling of the liquid water coolant. These glycol additives do not change the thermal characteristic of the coolant. The total mixture of thermal conductivity and specific heat is low.

For this nanosized metal particles are added into the coolant to enhance the convective heat transfer coefficient of the coolant. The metal has higher thermal conductivity. When the metal particles are suspended into the coolant, the thermal conductivity of the mixture is increased. Further, the thermal diffusivity of coolant increases. This two-phase liquid is often experimented by Pak and Cho *et al.* [8], they identified that 45% of convective heat transfer coefficient enhancement in addition to 0.0134 volume fraction alumina nanoparticles. The experimentation is repeated for several volume fraction of alumina particles and found that the heat transfer coefficient is improved to 75% on 0.0278 volume fraction of nanoalumina particles.

Ali *et al.* [9] experimented with the alumina nanosized particles for various volume fractions ranging from 0.01 to 0.02. The nanoparticle mixed coolant is tested for the different engine loads and mass flow rate of coolant. The Nusselt number is calculated for the various coolant flow rate and found that the 9.51% enhancement of Nusselt number, 14.79% improvement of heat transfer rate, and 14.72% increased convective heat transfer coefficient in addition 0.01 fraction.

The behavior of ethylene glycol and alumina nanopowder mixed water coolant is analyzed by Sharma *et al.* [10] in the automotive radiator. From the analysis result it is found that the nanofluid heat transfer rate is higher in all loading conditions. Due to the enhanced thermal performance of the alumina nanofluid, the radiator size can be minimized. This is one the interesting benefits of the nana fluid utilization and the radiator size of 18% can be reduced. This leads to the reduction of radiator weight. This coolant behavior is tested on the high engine speed and maximum load. The behavior of the nanofluid coolant is better than conventional water coolant on thermal behavior and possible to work on higher temperatures due to the addition of ethylene glycol. Thus, the weight of the radiator can be reduced, and the capacity of the radiator is enhanced. Also, metal particles in the coolant support the coolant warmup on cold start conditions because of its higher thermal diffusivity.

29.4 Use of Computational Fluid Dynamics for Nanofluids

In the recent decades, the large number of developments is attained in the engineering fields with the support of the high-speed computers. The development of fast computers enhances the capability of the system to perform the large size computational analysis. Due to the latest high-speed computer, the computational time to perform the numerical analysis is rapidly decreased and the feasibility to do the computational analysis for a larger count of mesh models is increased [11]. The product development time for new product design is rapidly decreased due to these numerical computations. The computational analysis implemented in various engineering applications involving fluid flow analysis, multiphysics problems, heat, and mass transfer. The solution of the real time problems is previewed in the numerical results without entering an experimental stage. The parameters of the design model are improved with the help of numerical simulation [12].

Several engineering applications such as flat plate collectors, condensers, and heat exchangers utilized the heat transfer analysis in the computational domain. On the fluid flow with heat transfer problems, the convective heat transfer rate is enhanced with the help of fine size nanometal particles. Generally, metals have higher thermal conductivity compared to the conventional working fluid water. The thermal conductivity of the water is enhanced by suspension of the fine nanosize metal particles into the working fluid water. This new generation type coolants play a key role in the advanced engineering fields to make compact engineering models and effective heat transfer devices. The nanofluids are not only implemented in the heat exchangers but also in fuel cells, higher temperature water cooling systems, chemical fields, pharmaceutical applications and surfactants.

This advanced computational analysis is adopted to predict the heat transfer behavior of nanofluids on different volume fraction suspension and flow velocity. The hydraulic and thermal boundary theory phenomenon is also studied using numerical analysis. It is a powerful technique to solve several commercial applications. The remarkable enhancement of thermal characteristics with the suspension of nanofluids is simulated for several engineering applications.

29.5 CFD Approach to Solve Heat Exchanger

The solar flat plate collector utilized the thermal radiation from the sun to useful hot fluid. The diagram of solar flat plate collector is shown in Figure 29.6. This application is one of the green technologies, which supply hot water to domestic and industrial needs. The dependency of high-grade electrical energy to produce low energy grade hot water for day-to-day needs is rapidly reduced by this simple green solar flat plate collectors. This solar collector has the heat absorber plate to harvest the solar radiation as thermal energy source. The riser tube is fitted under the absorber plate has the water as working fluid, which transfer the thermal energy from absorber plate to the required applications. The top and bottom runner tube connect all riser tubes as a single inlet and outlet pipe respectively. Generally, the absorber plate, riser, and runner tubes are made up of copper material.

In this numerical analysis, one identical riser tube welded under the copper absorber plate is simulated. The model of the riser tube is shown in Figure 29.7. The characteristic of the chosen riser tube is similar to all of the remaining riser tubes of the solar flat plate collector. The fluid flow is considered as a fully developed steady state condition. The heat losses

Figure 29.6 Flat plate collector.

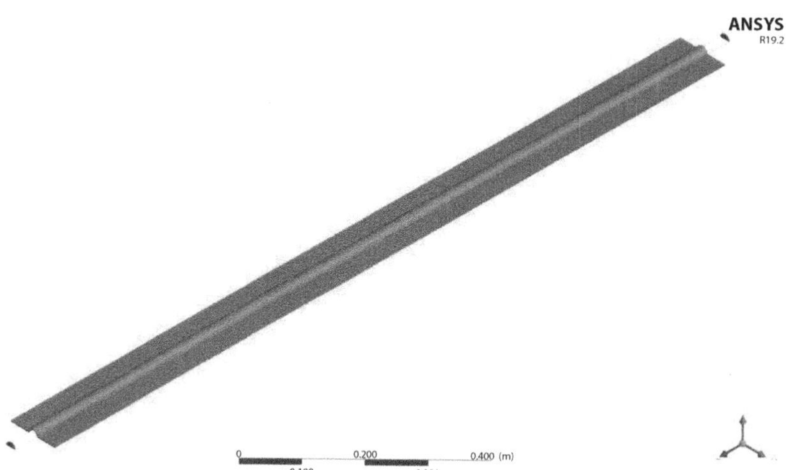

Figure 29.7 Modeling.

through the top glass cover, sider insulated wall and bottom insulation cover is considered in these numerical analyses. This is steady state heat transfer analysis for the fluid flow through the identical riser tube fitted under the absorber plate. The thermal resistance of the welded portion between the riser tube and absorber plate is neglected. The fluid flow is considered as incompressible within the computation domain. The thermophysical properties are given the material selection option, and no slip boundary condition is applied on the riser tubes.

The geometric model for the computational analysis is created in the ANSYS Workbench design modeller and the created model is discretized into fine mesh in the ANSYS meshing window [13]. Figure 29.8 shows the meshing of the rise tube model. The water flow along the tube axis is only applicable. The solar radiation intensity of 1000 W/m^2 is applied as the heat source. The convection heat transfer losses through the bottom insulation cover is applied. The velocity inlet condition is given for the fluid flow entrance and the pressure outlet with zero-gauge pressure boundary condition is given on the exit of the riser tube. The free stream temperature is set to the default value as 303 K [14, 15].

The SIMPLE algorithm condition is set for the pressure and velocity coupling equations. The solar radiation supplies thermal energy and the heat energy is reached by the working fluid by the conductive heat transfer through the absorber plate wall and riser tubes [16]. The absorptivity of the copper absorber plate is set as 0.95 and the total convective heat losses to the surroundings is applied on the back side of the bottom wall. The residuals for continuity, energy, and momentum equation convergence criteria are set as 10^{-6} [17].

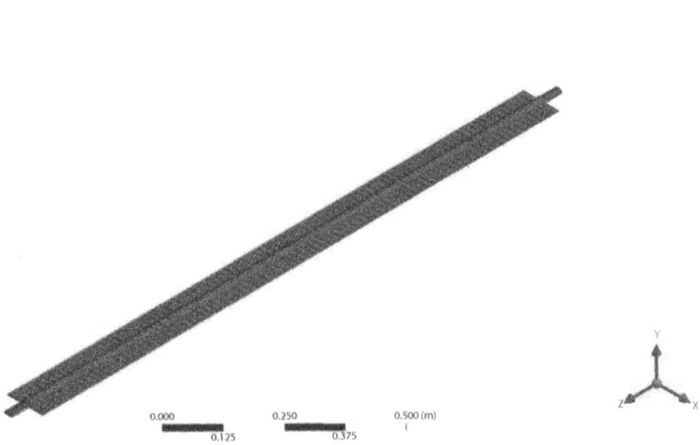

Figure 29.8 Meshing.

First of all, the numerical analysis is done for water as a working fluid. The fluid flow velocity is varied to find the thermal behavior of water in solar flat plate collectors. Then copper nanofluids are set as working fluid [18]. The properties of nanofluids are calculated from the corresponding

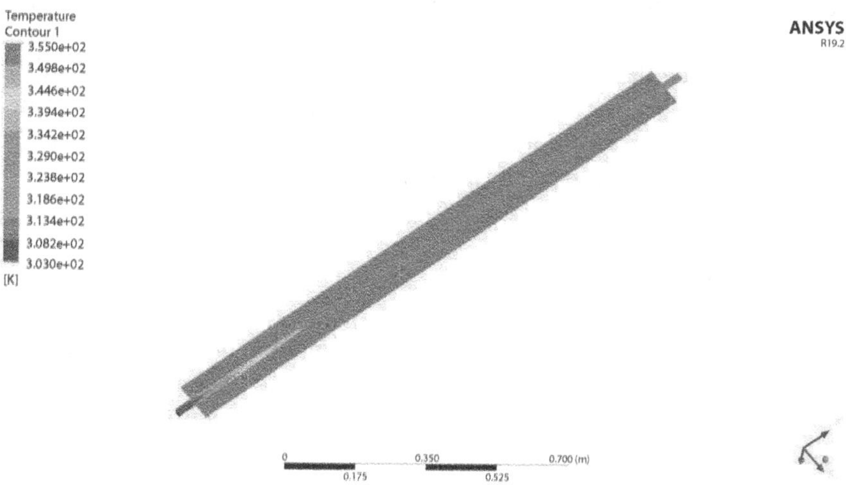

Figure 29.9 Temperature contour.

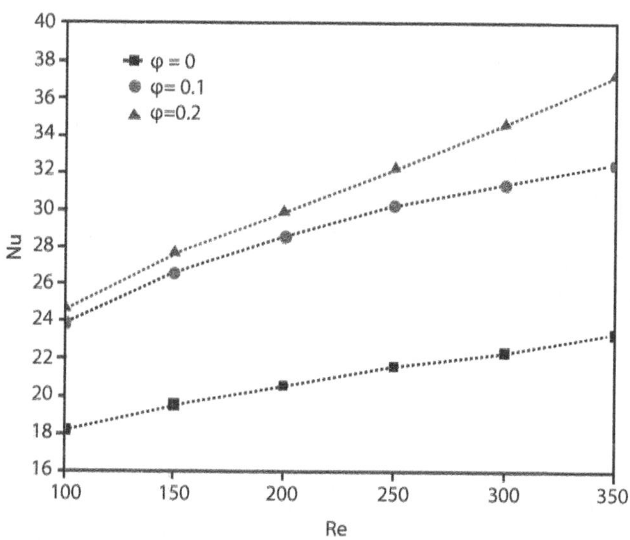

Figure 29.10 Variation of Nusselt number of the riser tube for various Reynolds number.

equations in the nanofluids theory. The breath, width and thickness of the solar flat plate collector is taken as 1.5, 0.1, and 0.002 m [19–21].

The exit temperature of the flat plate collector is taken as the output results. From the result, it is identified that the high temperature outlet is attained by using nanofluids as working fluid. The nanofluids performance is superior to conventional water due to the higher thermal conductivity and diffusivity thermal properties [21–27]. Figure 29.9 describes the temperature variation along the flat plate riser tube span on water is used as working fluid. The maximum outlet temperature of nanofluids is attained as 79.14°C for 5×10^{-6} kg/s and the maximum outlet temperature of 70.35°C is attained for water. Figure 29.10 shows the temperature change with respect to the corresponding Reynolds number.

29.6 Conclusion

The work presented in this chapter discussed in more detail the application of nanofluids in heat transfer and their performance evaluation using computational fluid dynamics (CFD). The review of literature demonstrates that nanofluids significantly improve heat transfer performance numerically, which is consistent with experimental results. Nonetheless, some modifications could be made to ensure that numerical processes produce more accurate results. Among researchers, some argue that a two-phase model is necessary to achieve the most accurate results. It is preferred to solve a two-phase model for a nanofluid because some slip mechanisms, such as the Dufour effect or thermophoresis, result in an increment in heat transfer. The results are also influenced by the flow regime, as turbulent flow yields a larger average. Furthermore, new computational approaches like the lattice Boltzmann method (LBM) can represent a nanofluid flow in a range of geometries. Even though this approach, which helps solve microscopic equations for the particles in the flow, yields findings that are similar to those acquired from experiments, future numerical studies employing LBM to evaluate its validity to that of standard finite volume methods may be conducted.

References

1. Sommerfeld, M. and Decker, S., State of the art and future trends in CFD simulation of stirred vessel hydrodynamics. *Chem. Eng. Technol.: Industrial*

Chemistry-Plant Equipment-Process Engineering-Biotechnology, 27, 3, 215–224, 2004.

2. Shourangiz-Haghighi, A., Haghnegahdar, M.A., Wang, L., Mussetta, M., Kolios, A., Lander, M., State of the art in the optimisation of wind turbine performance using CFD. *Arch. Comput. Methods Eng.*, 27, 2, 413–431, 2020.
3. Wong, K.V. and De Leon, O., Applications of nanofluids: Current and future, in: *Nanotechnology and Energy*, pp. 105–132, 2017.
4. Saidur, R., Leong, K.Y., Mohammed, H.A., A review on applications and challenges of nanofluids. *Renewable Sustainable Energy Rev.*, 15, 3, 1646–1668, 2011.
5. Vajjha, R.S., Das, D.K., Mahagaonkar, B.M., Density measurement of different nanofluids and their comparison with theory. *Pet. Sci. Technol.*, 27, 6, 612–624, 2009.
6. Bakthavatchalam, B., Habib, K., Saidur, R., Saha, B.B., Irshad, K., Comprehensive study on nanofluid and ionanofluid for heat transfer enhancement: A review on current and future perspective. *J. Mol. Liq.*, 305, 112787, 2020.
7. Wang, Z., Wu, Z., Han, F., Wadsö, L., Sundén, B., Experimental comparative evaluation of a graphene nanofluid coolant in miniature plate heat exchanger. *Int. J. Therm. Sci.*, 130, 148–156, 2018.
8. Pak, B. and Cho, Y.I., Hydrodynamic and heat transfer study of dispersed fluids with submicron metallic oxide particle. *Exp. Heat Transfer*, 1, 151–170, 1998.
9. Ali, M., El-Leathy, A.M., Al-Sofyany, Z., The effect of nanofluid concentration on the cooling system of vehicles radiator. *Adv. Mech. Eng.*, 1–13, 2014, Article ID 962510.
10. Sharma, V., Kumar, R.N., Thamilrarasan, K., Bhaskar, G.V., Devra, B., Heat reduction from IC engine by using Al2O3 nanofluid in engine cooling system. *Am. J. Eng. Res. (AJER)*, 03, 04, 173–177, 2014.
11. Ramezanizadeh, M., Nazari, M.A., Ahmadi, M.H., Açıkkalp, E., Application of nanofluids in thermosyphons: A review. *J. Mol. Liq.*, 272, 395–402, 2018.
12. Choi, S.U., Nanofluids: From vision to reality through research. *J. Heat Transfer*, 131, 3, 033106, 2009.
13. Appadurai, M. and Raj, E.F.I., Finite element analysis of composite wind turbine blades, in: *2021 7th International Conference on Electrical Energy Systems (ICEES)*, 2021, February, IEEE, pp. 585–589.
14. Appadurai, M., Raj, E.F.I., Venkadeshwaran, K., Finite element design and thermal analysis of an induction motor used for a hydraulic pumping system. *Mater. Today: Proc.*, 45, 7100–7106, 2021.
15. Appadurai, M. and Fantin Irudaya Raj, E., Finite element analysis of lightweight robot fingers actuated by pneumatic pressure, in: *Recent Advances in Manufacturing, Automation, Design and Energy Technologies*, pp. 379–385, Springer, Singapore, 2022.

16. Appadurai, M. and Raj, E., Epoxy/silicon carbide (sic) nanocomposites based small scale wind turbines for urban applications. *Int. J. Energy Environ. Eng.*, 13, 1, 191–206, 2022.
17. Appadurai, M., Raj, E., Jenish, I., Application of aluminium oxide–Water nanofluids to augment the performance of shallow pond: A numerical study. *Process Integr. Optim. Sustain.*, 6, 1, 211–222, 2022.
18. Deivakani, M., Kumar, S.S., Kumar, N.U., Raj, E.F.I., Ramakrishna, V., VLSI implementation of discrete cosine transform approximation recursive algorithm. *J. Phys.: Conf. Ser.*, 1817, 1, 012017, 2021, March, IOP Publishing.
19. Raj, E.F.I., and Balaji, M., Analysis and classification of faults in switched reluctance motors using deep learning neural networks. *Arab. J. Sci. Eng.*, 46, 2, 1313–1332, 2021.
20. Khatri, R., Jangid, R., Tomar, P.S., Sharma, S.S., Experimental investigations on direct absorption solar flat plate collector using Al2O3 nanofluid. *Int. J. Nanotechnol.*, 18, 11-12, 968–979, 2021.
21. Ashour, A.F., El-Awady, A., Tawfik, M.A., Numerical investigation on the thermal performance of a flat plate solar collector using ZnO & CuO water nanofluids under Egyptian weathering conditions. *Energy*, 240, 122743, 2022.
22. Jenish, I., Sahayaraj, A.F., Suresh, V., Appadurai, M., Irudaya Raj, E.F., Nasif, O., Kumaravel, A.K., Analysis of the hybrid of mudar/snake grass fiber-reinforced epoxy with nano-silica filler composite for structural application. *Adv. Mater. Sci. Eng.*, 2022, 2022.
23. Alghurabe, M.J., Al-Shamkhee, D.M., Alsahlani, A.A., Experimental and numerical study of thermal performance for flat plate solar water heater in Najaf. *IOP Conf. Ser.: Earth Environ. Sci.*, 877, 1, 012042, 2021, November.
24. Yilmaz, A. and Er, A., Thermal analysis of solar air collectors designed in different types with different flow rates using aluminum cans. *Energy Sources, Part A: Recovery, Util. Environ. Eff.*, 44, 2, 1–17, 2022.
25. Jenish, I., Felix Sahayaraj, A., Appadurai, M., Fantin Irudaya Raj, E., Suresh, P., Raja, T., Manikandan, V., Fabrication and experimental analysis of treated snake grass fiber reinforced with polyester composite. *Adv. Mater. Sci. Eng.*, 2021, 2021.
26. Tong, Y., Lee, H., Kang, W., Cho, H., Energy and exergy comparison of a flat-plate solar collector using water, Al2O3 nanofluid, and CuO nanofluid. *Appl. Therm. Eng.*, 159, 113959, 2019.
27. Shoeibi, S., Numerical analysis of optimizing a heat sink and nanofluid concentration used in a thermoelectric solar still: An economic and environmental study. *Environ. Res. Eng. Manage.*, 77, 2, 110–122, 2021.

About the Editors

Sharmistha Ghosh, PhD, a professor at the Institute of Engineering and Management, India. She received her doctorate in mathematics from the Indian Institute of Technology, India. Her major field of study includes fuzzy and vague databases as well as computational fluid dynamics, and she has published several papers in scientific journals. She is also the editor of several scientific journals and works as reviewer of journals as well as doctoral theses in India as well abroad.

M. Niranjanamurthy, PhD, is an assistant professor in the Department of Computer Applications, M S Ramaiah Institute of Technology, Bangalore, Karnataka. He earned his PhD in computer science at JJTU, Rajasthan, India. He has over 11 years of teaching experience and two years of industry experience as a software engineer. He has published several books, and he is working on numerous books for Scrivener Publishing. He has published over 60 papers for scholarly journals and conferences, and he is working as a reviewer in 22 scientific journals. He also has numerous awards to his credit.

Krishanu Deyasi is an associate professor in the Department of Basic Sciences and Humanities at the Institute of Engineering and Management, India. He earned his PhD from the Indian Institute of Science Education and Research, and he has postdoctoral experience from The Institute of Mathematical Sciences, India. He has written three books and has published papers in scientific journals. He is also an editor for several scientific journals.

Biswadip Basu Mallik is a senior assistant professor of mathematics in the Department of Basic Sciences and Humanities at the Institute of Engineering and Management, India. He has been involved in teaching and research for more than 21 years and has published several research papers in various scientific journals along with book chapters with various publishers. He has authored five and has five patents to his credit. He is a managing editor of the *Journal of Mathematical Sciences & Computational Mathematics* and is an editorial board member and reviewer for several scientific journals.

Santanu Das is as an assistant professor in the Department of Basic Sciences and Humanities at the Institute of Engineering and Management, India. He completed his undergraduate degree from and is pursuing his doctorate from Jadavpur University.

Index

Also of Interest

By the same editors

DATA WRANGLING: Concepts, Applications, and Tools, edited by M. Niranjanamurthy, Kavita Sheoran, Geetika Dhand, and Prabhjot Kaurk, ISBN: 9781119879688. Written and edited by some of the world's top experts in the field, this exciting new volume provides state-of-the-art research and latest technological breakthroughs in next-data wrangling, its theoretical concepts, practical applications, and tools for solving everyday problems.

ADVANCES IN DATA SCIENCE AND ANALYTICS, Edited by M. Niranjanamurthy, Hemant Kumar Gianey, and Amir H. Gandomi, ISBN: 9781119791881. Presenting the concepts and advances of data science and analytics, this volume, written and edited by a global team of experts, also goes into the practical applications that can be utilized across multiple disciplines and industries, for both the engineer and the student, focusing on machining learning, big data, business intelligence, and analytics.

WIRELESS COMMUNICATION SECURITY: Mobile and Network Security Protocols, Edited by Manju Khari, Manisha Bharti, and M. Niranjanamurthy, ISBN: 9781119777144. Presenting the concepts and advances of wireless communication security, this volume, written and edited by a global team of experts, also goes into the practical applications for the engineer, student, and other industry professionals.

MEDICAL IMAGING, Edited by H. S. Sanjay, and M. Niranjanamurthy ISBN: 9781119785392.Written and edited by a team of experts in the field, this is the most comprehensive and up-to-date study of and reference for the practical applications of medical imaging for engineers, scientists, students, and medical professionals.

SECURITY ISSUES AND PRIVACY CONCERNS IN INDUSTRY 4.0 APPLICATIONS, Edited by Shibin David, R. S. Anand, V. Jeyakrishnan, and M. Niranjanamurthy, ISBN: 9781119775621. Written and edited by a team of international experts, this is the most comprehensive and up-to-date coverage of the security and privacy issues surrounding Industry 4.0 applications, a must-have for any library.

Check out these other related titles from Scrivener Publishing

CONVERGENCE OF DEEP LEARNING IN CYBER-IOT SYSTEMS AND SECURITY, Edited by Rajdeep Chakraborty, Anupam Ghosh, Jyotsna Kumar Mandal and S. Balamurugan, ISBN: 9781119857211. In-depth analysis of Deep Learning-based cyber-IoT systems and security which will be the industry leader for the next ten years.

MACHINE INTELLIGENCE, BIG DATA ANALYTICS, AND IOT IN IMAGE PROCESSING: Practical Applications, Edited by Ashok Kumar, Megha Bhushan, José A. Galindo, Lalit Garg and Yu-Chen Hu, ISBN: 9781119865049. Discusses both theoretical and practical aspects of how to harness advanced technologies to develop practical applications such as drone-based surveillance, smart transportation, healthcare, farming solutions, and robotics used in automation.

MACHINE LEARNING TECHNIQUES AND ANALYTICS FOR CLOUD SECURITY, Edited by Rajdeep Chakraborty, Anupam Ghosh and Jyotsna Kumar Mandal, ISBN: 9781119762256. This book covers new methods, surveys, case studies, and policy with almost all machine learning techniques and analytics for cloud security solutions.

Printed and bound by CPI Group (UK) Ltd, Croydon, CR0 4YY

17/09/2026

14964274-0008